Halophiles and Hypersaline Environments

Antonio Ventosa • Aharon Oren • Yanhe Ma
Editors

Halophiles and Hypersaline Environments

Current Research and Future Trends

Springer

Editors
Professor Dr. Antonio Ventosa
University of Sevilla
Faculty of Pharmacy
Department of Microbiology
and Parasitology
c/ Profesor García González 2
41012 Sevilla, Spain
ventosa@us.es

Professor Dr. Aharon Oren
The Hebrew University of Jerusalem
The Institute of Life Sciences
Department of Plant and Environmental
Sciences
91904 Jerusalem, Israel
orena@cc.huji.ac.il

Professor Dr. Yanhe Ma
Chinese Academy of Sciences
Institute of Microbiology
1, Beichen West Road
Chaoyang District
Beijing 100101, China
mayanhe@im.ac.cn

ISBN 978-3-642-20197-4 e-ISBN 978-3-642-20198-1
DOI 10.1007/978-3-642-20198-1
Springer Heidelberg Dordrecht London New York

Library of Congress Control Number: 2011931727

Cover design: deblik Berlin, Germany

Printed on acid-free paper

Springer is part of Springer Science+Business Media (www.springer.com)

Preface

The chapters in this book deal with many aspects of the life of the diverse world of halophilic microorganisms and of the hypersaline environments in which they live, and present an overview of current research and future trends. Although we probably will never be able to obtain a full understanding of all the secrets of those diverse microorganisms that inhabit salt lakes and other hypersaline environments with salt concentrations up to saturation, considerable progress has been made.

When one looks back in the history of science, it often appears that our predecessors many decades ago knew and understood interesting phenomena that were later forgotten. Lourens Baas Becking (1895–1963) is an excellent example. He is known to most biologists for his statement "*everything is everywhere:* but, *the environment selects.*" However, it is seldom realized that many of Baas Becking's studies dealt with hypersaline environments. The longest chapter in his 1934 book "Geobiologie of inleiding tot de milieukunde" discusses halophilic microorganisms and their environments, and contains many surprising statements and observations mostly forgotten today. Baas Becking's career took him from the University of Utrecht to Stanford University, the University of Leiden, The Netherlands Indies, and Australia. The samples he collected from salt lakes and salterns during his travels formed the basis for his observations on *Dunaliella* and other salt flagellates, halophilic archaea, *Artemia*, and other life forms in high-salt environments (Chap. 2).

A group of bacteria encountered in most hypersaline environments is the family *Halomonadaceae*, which now encompasses 90 species grouped in 10 genera. In Chap. 3, de la Haba et al. review the phylogenetic relationships and taxonomic classification of the members of this family, as well as their biotechnological applications and other aspects of interest. Our understanding of the microbial diversity in hypersaline environments has greatly increased in recent years, not only based on culture-independent molecular techniques, but also on "old-fashioned" but increasingly relevant culturing methods. Thus, Grant and coworkers (Chap. 4) summarize a large project elucidating the microbial diversity in saline and saline/alkaline lakes in Inner Mongolia, China, using both culture-independent (molecular analysis techniques based on the direct isolation of DNA) and culture-dependent methods. A relatively recent addition to the list of halophilic prokaryotes is

Salinibacter ruber, an extremely halophilic representative of the *Bacteroidetes* that shares many properties with the haloarchaea (family *Halobacteriaceae*). Peña and coworkers have studied its microdiversity by comparing the genomic sequences of closely related *S. ruber* strains with a main focus on the species pangenome and the putative role of phages and lateral gene transfer in the shaping of its microdiversity (Chap. 5). Studies of the lipids present in hypersaline lakes can contribute much important information on the nature of the microbial communities inhabiting them. Mass spectrometry analysis of the lipid extracts of saltern biomass represents a powerful tool to obtain information on the presence of various archaeal and bacterial microorganisms in saltern ponds. As shown by Lopalco et al. (Chap. 6), ESI-MS lipid profiling by a shotgun lipidomic approach has allowed the discovery of new lipid molecules in the membranes of archaeal and bacterial halophiles. The recent introduction of the MALDI-TOF/MS technique in lipid analysis offers further possibilities of implementing the knowledge of lipid biology of halophilic prokaryotes. Fungi can also be an important component of hypersaline ecosystems. Plemenitaš and Nina Gunde-Cimerman describe the molecular mechanisms of adaptations of the extremely halotolerant black yeast *Hortaea werneckii* to life at high salt concentrations. *Hortaea* is an excellent model organism to study the mechanisms of salt tolerance in eukaryotes and a promising source of transgenes for osmotolerance improvement of industrially important yeasts, as well as agricultural crops. The organism maintains increased membrane fluidity and low intracellular sodium concentrations over a wide range of salinity, and accumulates glycerol and other small molecules as compatible solutes (Chap. 7).

Viruses are present in large numbers in hypersaline ecosystems. Roine and Oksanen present the diversity of halophilic viruses in salterns. Although most of the described haloarchaeal viruses are head–tail viruses, direct microscopic examination of environmental samples suggests a greater diversity. Following a review of the existing knowledge of the previously described head–tail viruses, more in-depth information is given on the structurally well characterized icosahedral virus SH1 and the group of enveloped viruses exemplified by HRPV-1 and HHPV-1 (Chap. 8). Baxter and coworkers have studied the haloviruses of Great Salt Lake, Utah, as a model for understanding viral diversity in hypersaline environments, using a culture-independent approach, coupling fractionation and negative staining transmission electron microscopy. Examination of the haloviruses pelleted from brine revealed the three major morphological types previously reported in other hypersaline waters: fusiform, spherical, and head–tail morphotypes. Partial success was accomplished in separating the different forms by cesium chloride density banding. In addition, abundant filamentous viruses were encountered, suggesting that "filamentous" should be considered a major new category of halovirus forms (Chap. 9).

In haloarchaea three different mechanisms for translation initiation work simultaneously, including a novel mechanism operating on leadered transcripts lacking a Shine Dalgarno motif. As discussed by Soppa, these three mechanisms function with differential efficiencies under various conditions, and thus haloarchaea apply translational regulons for the regulation of gene expression. Translatome analyses revealed that in two haloarchaeal species 20% and 10% of all transcripts, respectively, are subject to growth phase-dependent translational regulation. Reporter gene assays showed that 5′- and 3′-untranslated regions are sufficient and are both required for translational regulation, implying transcript circularization in vivo at least for translationally regulated transcripts (Chap. 10).

Several pathways have evolved in the three domains of life to facilitate membrane protein insertion and the transport of proteins across lipid membranes. Haloarchaea employ the universally conserved Sec pathway, which transports unfolded proteins, for the transport of biologically important substrates into and across the membrane. However, they also extensively employ the twin arginine translocation (Tat) system, which transports substrates across the lipid bilayer in a folded conformation. Most haloarchaeal Tat substrates appear to be anchored to cytoplasmic membranes via lipid modifications. In silico analyses suggest that the prominent use of the Tat pathway and the lipid tethering of Tat substrates are traits unique to halophilic archaea. The chapter by Dilks et al. discusses the selective pressures that may have led to these unique adaptations as well as possible explanations for why they are not observed in halobacteria (Chap. 11). Ever since the discovery of the first glycosylated archaeal protein, namely the *Halobacterium salinarum* surface-layer glycoprotein, some 35 years ago, research on haloarchaea has been at the forefront of efforts to decipher the archaeal version of N-glycosylation, a universal post-translational modification. Now, with the availability of sufficient numbers of genome sequences and the development of appropriate experimental tools, the possibility for detailed molecular analysis of archaeal N-glycosylation pathways is being realized, using haloarchaeal species as model systems. In the chapter by Eichler and coworkers the current understanding of N-glycosylation in archaea and the contribution of studies on *Haloferax volcanii* to such endeavors are described (Chap. 12).

Gas vesicle formation by *Halobacterium salinarum* requires the expression of 14 *gvp* genes that are arranged in the vac region. Two different vac regions, p-vac and c-vac, are found in wild-type strain PHH1 that forms gas vesicles by expressing p-vac. The c-vac region is only expressed in the p-vac deletion strain PHH4. As described by Pfeifer and her coworkers in Chap. 13, gas vesicle formation is influenced by environmental factors such as oxygen supply and temperature. Anoxic cultures grown in the presence of arginine show a reduced gas vesicle formation. Small groups of gas vesicles appear in PHH1, but PHH4 completely lacks gas vesicles. Heat shock inhibits gas vesicle formation in both strains, whereas a temperature of 15°C results in the overproduction of gas vesicles in PHH1 but not in PHH4. All these factors influence the amount of transcripts derived from p-vac or c-vac implying that this level is the major point of regulation.

The anaerobic halophilic alkalithermophiles are a novel group of "poly-extremophiles" that tolerate high salinity, alkaline pH and elevated temperatures. The recent isolation of novel anaerobic halophilic alkalithermophiles belonging to the genera *Natranaerobius* and *Natronovirga* (forming the order *Natranaerobiales*) has provided a platform for detailed biochemical and bioenergetic experiments, allowing a greater understanding of the novel adaptive mechanisms undoubtedly employed by the poly-extremophiles. The chapter by Mesbah and Wiegel highlights the various adaptive mechanisms used by the model anaerobic halophilic alkalithermophile, *Natranaerobius thermophilus*. *N. thermophilus* grows optimally at 3.3 M Na^+, pH 9.5, and 53°C. Biochemical and physiological analyses revealed that *N. thermophilus* combines two mechanisms for adaptation to high salinity. It also shows an unusual pattern of cytoplasm acidification, and utilizes a dual mechanism for cytoplasm acidification based on an unusually large number of cation/proton antiporters and cytoplasmic buffering, a feature not previously observed in other alkalithermophiles. Both *Natranaerobius thermophilus* and *Natronovirga wadinatrunensis* showed significant resistance to ultra-violet radiation. This combination of different strategies undoubtedly allows the halophilic alkalithermophiles to survive and proliferate when faced with multiple environmental extremes commonly encountered in their natural habitat (Chap. 14).

Also the study of less salt-tolerant microorganisms can contribute much to our understanding of the mechanisms of salt adaptation. The soil-dwelling bacterium *Bacillus subtilis* inhabits an ecological niche subjected to frequent changes in osmotic and saline conditions that are caused by rainfall and desiccation. Such changes elicit water fluxes across the cytoplasmic membrane and can drive up turgor under hypo-osmotic conditions to such an extent that the cell will rupture, or under hyper-osmotic conditions cause the dehydration of the cytoplasm, a reduction in turgor and eventually growth arrest and cell death. Proteome and genome-wide transcriptional profiling studies described by Pittelkow and Bremer (Chap. 15) have highlighted the complexity and multifaceted nature of the osmotic stress response systems of *B. subtilis*. However, it is beyond doubt that effective water management by the cell is the cornerstone of its acclimatization to either sudden or sustained rises in the environmental osmolarity and the osmotic downshift that inevitably will follow hyperosmotic growth conditions. The accumulation and expulsion of ions and compatible solutes play key roles in these cellular osmotic adjustment processes.

Carotenoids are widely distributed in extremophiles. Among these organisms especially C_{30} and C_{50} carotenoids are found. Köcher and Müller describe the structure and function of carotenoids in halophiles with focus on the moderately halophilic bacterium *Halobacillus halophilus*, which produces an unusual C_{30} carotenoid. The structure was solved by HR-MS and NMR analyses as methyl glucosyl-3,4-dehydro-apo-8′-lycopenoate. Six genes could be identified that are involved in the biosynthesis of carotenoids. Together with the structural analyses of intermediates of methyl glucosyl-3,4-dehydro-apo-8′-lycopenoate produced by a pigment mutant, a putative and unique biosynthesis pathway could be postulated. The isolated carotenoid and its biosynthesis intermediates showed a high antioxidative activity, and the protective function of these pigments could be demonstrated for *H. halophilus* (Chap. 16). Another carotenoid of great interest is xanthorhodopsin of *Salinibacter ruber*. It provides a simple system for collection of light by a carotenoid antenna, transfer of electronic excitation from the carotenoid to a retinal, and utilization of the energy gained for the active transport of protons across the membrane. As a model system, this protein, a member of the heptahelical transmembrane family of prokaryotic rhodopsins, poses numerous questions about carotenoid binding, energy transfer in donor-acceptor pairs and about bacterial rhodopsins in general, which we are beginning to answer. This review chapter by Lanyi and Balashov (Chap. 17) recounts recent advances in steady-state and ultrafast spectroscopy as well as X-ray crystallography of xanthorhodopsin, and what they reveal about this and other homologous retinal proteins.

The enzymatic degradation of lignocellulolytic biomass by halophilic microorganisms is described by Begemann and colleagues. Modern recombinant screening techniques and data mining of genomes have the potential to yield many halophilic lignocellulytic enzymes. There is great potential for developing biotechnologies with halophilic lignocellulolytic enzymes especially in regards to treating biomass, leading to innovations in fields such as bio-fuel production (Chap. 18). Enzymes from halophilic microorganisms have interesting potentials for applications in industry and biotechnology, as reviewed by Bonete and Martínez-Espinosa (Chap. 19). During the past few decades, the microbial communities inhabiting extreme environments aroused interest owing to the unique properties of the biocatalysts they produce (extremozymes). These extremozymes can cope with industrial process conditions (high temperatures, high salt concentrations, low water availability, etc.) due to their extreme stability under the mentioned parameters. The haloarchaea are therefore a valuable source of novel enzymes for biotechnology. Their peculiar physiology involving extreme adaptation to the salty environments has led to the development of interests in haloarchaeal enzymes, mainly in those processes carried out in non-aqueous media.

The chapters in this book are based on presentations given during the "Halophiles 2010" symposium, held in Beijing, China, June 29–July 3, 2010. The Beijing symposium was the most recent in a series of scientific meetings on halophilic microorganisms, the first of which was held in Rehovot, Israel, in 1978. The final chapter of this book presents a short history of the symposia on halophilic microorganisms from the personal point of view of one of the editors of this book.

We would like to express our gratitude to all chapter authors for their collaboration; they have been extremely helpful during the preparation of this book. Besides, we are grateful to Springer and the people involved in the production of this book.

We recently learned that Prof. Helge Larsen from Trondheim, Norway, one of the founders of halophilic microbiology, had passed away in March 2005 (Chap. 1). It is sad to note that nobody within the community of halophile scientists was at the time aware of his death. We dedicate this volume to Helge Larsen's memory.

Sevilla, Spain — Antonio Ventosa
Jerusalem, Israel — Aharon Oren
Beijing, China — Yanhe Ma
April 2011

Contents

Contributors

Rudolf Amann Max Planck Institute for Marine Microbiology, Celsiusstrasse 1, Bremen, Germany

Roberto Angelini Department of Medical Biochemistry, Biology and Physics, University of Bari, Bari, Italy

Josefa Antón Departamento de Fisiología, Genética y Microbiología, Universidad de Alicante, Apartado 99, Alicante, Spain, anton@ua.es

Sergei P. Balashov Department of Physiology & Biophysics, University of California, Irvine, CA 92697, USA

Maristella Baronio Department of Medical Biochemistry, Biology and Physics, University of Bari, Bari, Italy

Bonnie K. Baxter Great Salt Lake Institute, Westminster College, 1840 S 1300 E, Salt Lake City, UT, USA, bbaxter@westminstercollege.edu

Matthew B. Begemann Microbiology Doctoral Training Program, University of Wisconsin, Madison, WI 53706, USA

María José Bonete División de Bioquímica y Biología Molecular, Facultad de Ciencias, Universidad de Alicante, Ap. 99, E-03080 Alicante, Spain, mjbonete@ua.es

Erhard Bremer Department of Biology, Laboratory for Microbiology, Philipps-University Marburg, Karl-von-Frisch-Strasse 8, 35032 Marburg, Germany, bremer@biologie.uni-marburg.de

Doron Calo Department of Life Sciences, Ben Gurion University, Beersheva 84105, Israel

Angela Corcelli Department of Medical Biochemistry, Biology and Physics, University of Bari, Bari, Italy; Institute for Chemical-Physical Processes, National Research Council (IPCF-CNR), Bari, Italy, a.corcelli@biologia.uniba.it

Don A. Cowan Institute for Microbial Biotechnology and Metagenomics, University of the Western Cape, Cape Town, South Africa

Kieran Dilks Department of Biology, University of Pennsylvania, Philadelphia, PA 19104–6018, USA

Joaquín Dopazo Functional Genomics Node (INB), Centro de Investigación Príncipe Felipe (CIPF), Valencia, Spain

Jerry Eichler Department of Life Sciences, Ben Gurion University, Beersheva 84105, Israel, jeichler@bgu.ac.il

Karin Faist Mikrobiologie & Genetik, Technische Universität Darmstadt, Schnittspahnstrasse 10, 64287 Darmstadt, Germany

Regina Frommherz Mikrobiologie & Genetik, Technische Universität Darmstadt, Schnittspahnstrasse 10, 64287 Darmstadt, Germany

Toni Gabaldón Bioinformatics and Genomics Program, Centre for Genomics Regulation (CRG), UPF, Dr. Aiguader, 83, 08003 Barcelona, Spain

Maria Ines Gimenez Instituto de Investigaciones Biológicas, Facultad de Ciencias Exactas y Naturales, Universidad Nacional de Mar del Plata, Funes 3250 4 degrees Nivel, Mar del Plata, 7600, Argentina

Frank Oliver Glöckner Max Planck Institute for Marine Microbiology, Celsiusstrasse 1, Bremen, Germany; Jacobs University Bremen GmbH, Campusring 1, Bremen, Germany

William D. Grant Department of Infection, Immunity and Inflammation, University of Leicester, Leicester, LE1 9HN, UK, wdg1@leicester.ac.uk

Jack D. Griffith Department of Immunology and Microbiology, University of North Carolina at Chapel Hill, Chapel Hill, NC 27599, USA

Nina Gunde-Cimerman Department of Biology, Biotechnical Faculty, University of Ljubljana, Ljubljana, Slovenia; Centre of Excellence for Integrated Approaches in Chemistry and Biology of Proteins (CIPKeBiP), Jamova 39, SI-1000 Ljubljana, Slovenia, nina.gunde-cimerman@bf.uni-lj.si

M. Carmen Gutiérrez Department of Microbiology and Parasitology, Faculty of Pharmacy, University of Sevilla, 41012 Sevilla, Spain

Rafael R. de la Haba Department of Microbiology and Parasitology, Faculty of Pharmacy, University of Sevilla, 41012 Sevilla, Spain, rrh@us.es

Shaun Heaphy Department of Infection, Immunity and Inflammation, University of Leicester, Leicester LE1 9HN, UK

Torsten Hechler Mikrobiologie & Genetik, Technische Universität Darmstadt, Schnittspahnstrasse 10, 64287, Darmstadt, Germany

Jaime Huerta-Cepas Bioinformatics and Genomics Program, Centre for Genomics Regulation (CRG), UPF, Dr. Aiguader, 83, 08003 Barcelona, Spain

Brian E. Jones Genencor International BV, Archimedesweg 30, Leiden 2333CN, The Netherlands

Lina Kaminski Department of Life Sciences, Ben Gurion University, Beersheva 84105, Israel

Lina Kandiba Department of Life Sciences, Ben Gurion University, Beersheva 84105, Israel

Saskia Köcher Molecular Microbiology & Bioenergetics, Institute of Molecular Biosciences, Johann Wolfgang Goethe University Frankfurt, Max-von-Laue-Strasse 9, 60438 Frankfurt am Main, Germany

Zvia Konrad Department of Life Sciences, Ben Gurion University, Beersheva 84105, Israel

Janos K. Lanyi Department of Physiology & Biophysics, University of California, Irvine, CA 92697, USA, jklanyi@uci.edu

Simona Lobasso Department of Medical Biochemistry, Biology and Physics, University of Bari, Bari, Italy

Patrizia Lopalco Institute for Microelectronics and Microsystems National Research Council (IMM-CNR), Lecce, Italy

Marianna Lucio Helmholtz Zentrum München, German Research Center for Environmental Health, Institute of Ecological Chemistry, Neuherberg, Germany

M. Carmen Márquez Department of Microbiology and Parasitology, Faculty of Pharmacy, University of Sevilla, 41012 Sevilla, Spain

Volker Müller Molecular Microbiology & Bioenergetics, Institute of Molecular Biosciences, Johann Wolfgang Goethe University Frankfurt, Max-von-Laue-Strasse 9, 60438 Frankfurt am Main, Germany, vmueller@bio.uni-frankfurt.de

Yanhe Ma Chinese Academy of Sciences, Tianjin Institute of Industrial Biotechnology, Institute of Microbiology, Beijing 100101, China

Hilla Magidovich Department of Life Sciences, Ben Gurion University, Beersheva 84105, Israel

Mihnea R. Mangalea Department of Immunology and Microbiology, University of North Carolina at Chapel Hill, Chapel Hill, NC 27599, USA

Rosa María Martínez-Espinosa División de Bioquímica y Biología Molecular, Facultad de Ciencias, Universidad de Alicante, Ap. 99, E-03080 Alicante, Spain

Larissa Marschaus Mikrobiologie & Genetik, Technische Universität Darmstadt, Schnittspahnstrasse 10, 64287 Darmstadt, Germany

Noha M. Mesbah Department of Biochemistry, Suez Canal University, Ismailia 41522, Egypt, noha_mesbah@pharm.suez.edu.eg

Inmaculada Meseguer Departamento de Producción Vegetal y Microbiología, Universidad Miguel Hernández. Avda. de la Universidad s/n., Elche (Alicante), Spain

Melanie R. Mormile Department of Biological Sciences, Missouri University of Science and Technology, Rolla, MO 65409, USA, mmormile@mst.edu

Miriam-Nirvana Nagoulat Department of Biology, La Sierra University, 4500 Riverwalk Pkwy, Riverside, CA 92515, USA

Shai Naparstek Department of Life Sciences, Ben Gurion University, Beersheva 84105, Israel

Hanna M. Oksanen Department of Biosciences and Institute of Biotechnology, University of Helsinki, P.O. Box 56, (Viikinkaari 5), 00014 Helsinki, Finland

Aharon Oren Department of Plant and Environmental Sciences, The Institute of Life Sciences, and the Moshe Shilo Minerva Center for Marine Biogeochemistry, The Hebrew University of Jerusalem, Jerusalem 91904, Israel, orena@cc.huji.ac.il

Eulyn Pagaling Department of Infection, Immunity and Inflammation, University of Leicester, Leicester LE1 9HN, UK

Varun G. Paul Geology and Geophysics Program, Missouri University of Science and Technology, Rolla, MO 65409, USA

Arantxa Peña Departamento de Fisiología, Genética y Microbiología, Universidad de Alicante, Apartado 99, Alicante, Spain

Felicitas Pfeifer Mikrobiologie & Genetik, Technische Universität Darmstadt, Schnittspahnstrasse 10, 64287 Darmstadt, Germany, pfeifer@bio.tu-darmstadt.de

Marco Pittelkow Department of Biology, Laboratory for Microbiology, Philipps-University Marburg, Karl-von-Frisch-Strasse 8, 35032 Marburg, Germany

Ana Plemenitaš Institute of Biochemistry, Faculty of Medicine, University of Ljubljana, Ljubljana, Slovenia

Mechthild Pohlschröder Department of Biology, University of Pennsylvania, Philadelphia, PA 19104–6018, USA, pohlschr@sas.upenn.edu

Elina Roine Department of Biosciences and Institute of Biotechnology, University of Helsinki, P.O. Box 56, (Viikinkaari 5), 00014 Helsinki, Finland, elina.roine@helsinki.fi

Ramon Rosselló-Mora Marine Microbiology Group, Institut Mediterrani d'Estudis Avançats (CSIC-UIB), C/ Miquel Marqués 21, 07190 Esporles, Illes Balears, Spain

Shereen Sabet Department of Biology, La Sierra University, 4500 Riverwalk Pkwy, Riverside, CA 92515, USA

Cristina Sánchez-Porro Department of Microbiology and Parasitology, Faculty of Pharmacy, University of Sevilla, 41012 Sevilla, Spain, sanpor@us.es

Fernando Santos Departamento de Fisiología, Genética y Microbiología, Universidad de Alicante, Apartado 99, Alicante, Spain

Margarete Schüler Max Planck Institute for Marine Microbiology, Celsiusstrasse 1, Bremen, Germany

Philippe Schmitt-Kopplin Helmholtz Zentrum München, German Research Center for Environmental Health, Institute of Ecological Chemistry, Neuherberg, Germany

Jörg Soppa Institute for Molecular Biosciences, Goethe-University, Biocentre, Max-von-Laue-Strasse 9, D-60438 Frankfurt, Germany, soppa@bio.uni-frankfurt.de

Hanno Teeling Max Planck Institute for Marine Microbiology, Celsiusstrasse 1, Bremen, Germany

Katharina Teufel Mikrobiologie & Genetik, Technische Universität Darmstadt, Schnittspahnstrasse 10, 64287, Darmstadt, Germany

Manuela Tripepi Department of Biology, University of Pennsylvania, Philadelphia, PA 19104–6018, USA

Antonio Ventosa Department of Microbiology and Parasitology, University of Sevilla, 41012 Sevilla, Spain, ventosa@us.es

Daniel J. Vidt Mining and Nuclear Engineering, Missouri University of Science and Technology, Rolla, MO 65409, USA

Juergen Wiegel Department of Microbiology, University of Georgia, Athens, GA 30602, USA, jwiegel@uga.edu

Smaranda Willcox Department of Immunology and Microbiology, University of North Carolina at Chapel Hill, Chapel Hill, NC 27599, USA

Sophie Yurist-Doutsch Department of Life Sciences, Ben Gurion University, Beersheva 84105, Israel

Chapter 1
Helge Larsen (1922–2005) and His Contributions to the Study of Halophilic Microorganisms

Aharon Oren, Antonio Ventosa, and Yanhe Ma

1.1 Introduction

In 1972, the late Helge Larsen was honored by The Netherlands Society for Microbiology to deliver the fourth A.J. Kluyver memorial lecture. The somewhat enigmatic title of his presentation – "The halobacteria's confusion to biology" (Larsen 1973) was based on "The microbe's contribution to biology," the title of the famous series of essays by Kluyver and van Niel (1956). It was Albert Jan Kluyver who had encouraged Larsen to start a thorough study of the halophiles: "In the early fall of 1954 I had the good fortune that Professor Kluyver came to Trondheim upon an invitation from Norway Institute of Technology, to take part in my promotion to the status of *doctor technicae*. At that time, I had just been approached by representatives of the salt fish industry in Norway, who had inquired of me whether I would look into the problem of microbial deterioration of salt fish. I happened to discuss this matter with Kluyver and I clearly remember the enthusiasm with which he recommended me to take up work on the microbes of salt and salted products" (Larsen 1973).

Helge Larsen's diverse studies of the world of microorganism inhabiting hypersaline habitats made him the greatest expert of his time on the halophiles. His chapter on "Halophilism" in "The Bacteria" Vol. IV (Larsen 1962) was the first comprehensive review written on halophilic microorganisms and their modes of adaptation to life at high salinity, and this visionary article introduced many to the fascinating world of the halophiles. Later general articles (Larsen 1981, 1986) and

A. Oren (✉)
Department of Plant and Environmental Sciences, The Institute of Life Sciences, The Hebrew University of Jerusalem, 91904 Jerusalem, Israel
e-mail: orena@cc.huji.ac.il

A. Ventosa
Department of Microbiology and Parasitology, University of Sevilla, 41012 Sevilla, Spain

Y. Ma
Institute of Microbiology, Chinese Academy of Sciences, Tianjin Institute of Industrial Biotechnology, 1 Beichen West Road, Beijing 100101, China

A. Ventosa et al. (eds.), *Halophiles and Hypersaline Environments*,
DOI 10.1007/978-3-642-20198-1_1, © Springer-Verlag Berlin Heidelberg 2011

specialized reviews on the biochemistry (Larsen 1967) and ecology (Larsen 1980) provided timely updates on the field of halophile microbiology.

1.2 Helge Larsen's Life and Studies

Helge Larsen (Fig. 1.1) was born in Ålesund, Norway on 25 April 1922. He graduated as a chemical engineer from the Norwegian Institute of Technology in Trondheim in 1944. From 1947 until 1950 he worked with Cornelis van Niel at the Hopkins Marine Station, Pacific Grove, CA, where he studied the biology of the green phototrophic sulfur bacteria. He obtained his Ph.D. degree in 1954 based on a thesis "On the microbiology and biochemistry of photosynthetic green sulfur bacteria." In 1955, he was appointed professor of biochemistry at the Norwegian Institute of Technology (now the Norwegian University of Science and Technology) in Trondheim, where he remained until his retirement in 1992.

In Trondheim, he worked on a number of projects of applied nature, developing modern fermentation technology in Norway. Early in his career he devised an efficient process for the industrial production of itaconic acid using *Aspergillus terreus*, and in the last years of his work he devoted part of his time to the development of an industrial process for the production of lysine.

His interest in the halophiles was also aroused because of applied aspects. From his early youth he was acquainted with the discoloration and spoilage of salted fish. By 1955, he had recognized that the color phenomena on dried cod known as brunmidd ("brown mite") and rødmidd ("red mite") were, respectively, caused by a halotolerant fungus and by an extremely halophilic prokaryote (Larsen 1986). It was realized that spoilage of the fish by the red microorganisms could be prevented by using rock salt instead of traditionally used sea salt which turned out to be the source of the red halophiles. Soon Larsen started to ask fundamental

Fig. 1.1 Portrait of Prof. Helge Larsen (photograph: Det Norske Videnskaps Akademi; http://www.dnva.no/c27004/artikkel/vis.html?tid=27016)

questions about the nature of the halophilic microorganisms and their modes of adaptation to life in hypersaline environments. His extensive studies, described below, have contributed much to our understanding of the physiology, biochemistry, taxonomy, and ecology of the halophiles.

1.2.1 Studies on the Cell Envelope of Halophilic Microorganisms

Larsen's studies on the lysis of *Halobacterium salinarum* upon dilution with distilled water showed that Na^+ is specifically needed at high concentrations to keep the cell envelope intact. Attempts to demonstrate the presence of peptidoglycan in the cell wall failed, and Larsen and his coworkers showed that instead the cell is surrounded by an envelope composed of (glyco)protein subunits. The lysis phenomenon upon dilution was found to be due not to osmotic swelling and bursting of the cell, but to the disintegration of the envelope (Mohr and Larsen 1963a, b; Steensland and Larsen 1969). Different enzyme inhibitors tested did not affect lysis, excluding the possibility that enzymes are involved in the lysis phenomenon. In summary, "The observations support the conclusion that the globular lipoprotein particles, which constitute the bulk of the material of the cell wall of these bacteria, are bound together mainly by electrostatic forces and secondary bonds. When the cells are exposed to hypotonic solutions, or to ions which bind strongly to proteins, or to chemicals which are believed to break secondary bonds between protein molecules, the linkages binding the lipoprotein particles together are weakened so that the wall structure disintegrates. Only in the presence of high concentrations of sodium and chloride ions, or other ions which bind loosely to proteins, is it possible for the proteinaceous particles of the cell wall to associate in an orderly array" (Mohr and Larsen 1963b).

In contrast, the cells of the red extremely halophilic *Halococcus morrhuae* did not lyse in distilled water. Also, here the attempts to show presence of peptidoglycan failed, and instead the cells were shown to be surrounded by a thick wall composed of hexoses and hexosamines (Steensland and Larsen 1971).

At the time these studies were performed the concept of the Archaea as the third domain of life did not yet exist. Only in the late 1970s it became clear that the special nature of the envelopes of these organisms, and notably the lack of peptidoglycan, is also linked to their phylogenetic affiliation.

1.2.2 The Function of the Carotenoid Pigments of the Extreme Halophiles

To elucidate the function of the red carotenoid pigment of *Halobacterium salinarum*, Larsen compared the properties of the red wild type and a carotenoid-less white mutant. Both the red and the white cells survived exposure to the light intensity of full sunlight. However, when they were grown together at high light intensities, the pigmented wild type had a selective advantage and outcompeted the white mutant (Dundas and Larsen 1962). In the presence of the photosentitizer phenosafranine, the colorless mutant was rapidly killed at high light, while the red cells survived (Dundas and Larsen 1963). These experiments proved that the carotenoid pigments of many extreme halophiles provide protection against the damaging action of sunlight.

1.2.3 The Properties of the Gas Vesicles of Halobacterium

In a study of the gas vesicles of *Halobacterium salinarum*, Larsen characterized the conditions for production of new gas vesicles after collapse of the existing vesicles following application or pressure. He also devised a method for the isolation of gas vesicles from extreme halophiles, based on the lysis of the cell envelope in alkaline solution, a treatment that leaves the vesicles intact. These can then be collected from the surface of the solution (Larsen et al. 1967).

1.2.4 Taxonomy of Halophilic Microorganisms

Larsen had a profound interest in the taxonomy of the red extreme halophiles, and he was an active member of the International Committee on the Systematics of Bacteria subcommittee on the taxonomy of *Halobacteriaceae* in the years 1986–1994. He named the order *Halobacteriales* (Grant and Larsen 1989a) and described *Halobacterium volcanii*, later renamed *Haloferax volcanii*, a species of pleomorphic flat cells, isolated from the Dead Sea. This organism requires less NaCl than do *Halobacterium salinarum* and *Halococcus morrhuae*, but shows a high tolerance to magnesium, the dominant cation in Dead Sea water (Mullakhanbhai and Larsen 1975).

For Bergey's Manual of Systematic Bacteriology Larsen wrote descriptions of the family *Halobacteriaceae* (Larsen 1984a; Larsen and Grant 1989), the genus *Halobacterium* (Larsen 1984b; Larsen and Grant 1989), the genus *Halococcus* (Larsen 1984c, 1989), the genus *Haloarcula* (Grant and Larsen 1989b), and the genus *Haloferax* (Grant and Larsen 1989c).

As thanks for Helge Larsen's contributions to the field, two recently characterized new species of extremely halophilic Archaea were named in his honor: *Halostagnicola larsenii*, an extremely halophilic archaeon from a saline lake in Inner Mongolia (Castillo et al. 2006) and *Haloferax larsenii*, an extremely halophilic archaeon from a solar saltern in China (Xu et al. 2007).

1.2.5 *Studies on* Dunaliella

During his stay in the laboratory of A. Duncan Brown as a guest professor at the University of Wollongong, NSW, Australia, Larsen studied a salt-sensitive mutant of *Dunaliella tertiolecta*. This mutant was shown to require high concentrations of carbon dioxide for growth and has low carbonic anhydrase activity (Brown et al. 1987).

1.2.6 *Microbiological Studies of the Deterioration of Fish*

Larsen's work on microorganisms associated with the spoilage of fish was not restricted to the extreme halophiles. Studies of post-mortem changes included the characterization of the release of ammonia and other malodorous compounds (Strøm and Larsen 1979). The formation of bad-smelling trimethylamine from trimethylamine oxide during the deterioration of fish was already known for a long time. Larsen was the first to recognize the true nature of the process as a novel mechanism of anaerobic respiration. Trimethylamine oxide, a compound that serves as an osmotic stabilizer in fish tissues, can serve as electron acceptor for respiration by *Proteus* and some other prokaryotes, enabling anaerobic growth on non-fermentable substrates (Strøm et al. 1979).

1.3 Final Comments

Helge Larsen passed away on 11 March 2005. It is sad to note that nobody within the community of halophile scientists was at the time aware of Helge's death, and the obituary in Norwegian published on the website of the Norwegian Academy of Sciences (Eimhjellen 2005) remained unnoticed until recently (Oren 2010). The portrait reproduced in Fig. 1.1 was derived from that website.

The first international symposium that dealt with all aspects of the life of halophilic microorganisms was held in Obermarchtal, Germany, in 1985. There, two of the editors of this book (A.V. and A.O.) had the pleasure meeting "Mister Halophile," as he was known at the time, discussing their work with him, and being stimulated by his enthusiasm. In view of Helge's great contributions to the field we dedicate this volume to his memory.

References

Brown AD, Goyal A, Larsen H, Lilley RMcC (1987) A salt-sensitive mutant of *Dunaliella tertiolecta*. Arch Microbiol 147:309–314

Castillo AM, Gutiérrez MC, Kamekura M, Xue Y, Ma Y, Cowan DA, Jones BE, Grant WD, Ventosa A (2006) *Halostagnicola larsenii* gen. nov., sp. nov., an extremely halophilic archaeon from a saline lake in Inner Mongolia, China. Int J Syst Evol Microbiol 56:1519–1524

Dundas ID, Larsen H (1962) The physiological role of the carotenoid pigments of *Halobacterium salinarium*. Arch Mikrobiol 44:233–239

Dundas ID, Larsen H (1963) A study on the killing by light of photosensitized cells of *Halobacterium salinarium*. Arch Mikrobiol 46:19–28

Eimhjellen KE (2005) Minnetale over professor emeritus Helge Larsen holdt i fellesmøteden 17 November 2005. Det Norske Videnskaps-Akademi. http://www.dnva.no/c27004/artikkel/vis.html?tid=27016. Cited 21 Sept 2010

Grant WD, Larsen H (1989a) Group III. Extremely halophilic archaeobacteria, order *Halobacteriales* ord. nov. In: Staley JT, Bryant MP, Pfennig N, Holt JG (eds) Bergey's manual of systematic bacteriology, vol 3. Williams & Wilkins, Baltimore, pp 2216–2233

Grant WD, Larsen H (1989b) Genus II. *Haloarcula* Torreblanca, Rodriguez-Valera, Juez, Ventosa, Kamekura and Kates 1986b, 537[VP] (Effective publication: Torreblanca, Rodriguez-Valera, Juez, Ventosa, Kamekura and Kates 1986[a], 98). In: Staley JT, Bryant MP, Pfennig N, Holt JG (eds) Bergey's manual of systematic bacteriology, vol 3. Williams & Wilkins, Baltimore, pp 2224–2226

Grant WD, Larsen H (1989c) Genus III. *Haloferax* Torreblanca, Rodriguez-Valera, Juez, Ventosa, Kamekura and Kates 1986b, 537[VP] (Effective publication: Torreblanca, Rodriguez-Valera, Juez, Ventosa, Kamekura and Kates 1986[a], 98). In: Staley JT, Bryant MP, Pfennig N, Holt JG (eds) Bergey's manual of systematic bacteriology, vol 3. Williams & Wilkins, Baltimore, pp 2226–2268

Kluyver AJ, van Niel CB (1956) The microbe's contribution to biology. Harvard University Press, Cambridge, MA

Larsen H (1962) Halophilism. In: Gunsalus IC, Stanier RY (eds) The bacteria, vol IV. Academic Press, New York, pp 297–342

Larsen H (1967) Biochemical aspects of extreme halophilism. In: Rose AH, Wilkinson JF (eds) Advances in microbial physiology, vol 1. Academic Press, London, pp 97–132

Larsen H (1973) The halobacteria's confusion to biology. Antonie van Leeuwenhoek 39:383–396

Larsen H (1980) Ecology of hypersaline environments. In: Nissenbaum A (ed) Hypersaline brines and evaporitic environments. Elsevier, Amsterdam, pp 23–39

Larsen H (1981) The family *Halobacteriaceae*. In: Starr MP, Stolp H, Trüper HG, Balows A, Schlegel HG (eds) The prokaryotes – a handbook on habitats, isolation, and identification of bacteria. Springer, Berlin, pp 985–994

Larsen H (1984a) Family V. *Halobacteriaceae* Gibbons 1974. In: Krieg NR, Holt JG (eds) Bergey's manual of systematic bacteriology, vol 1. Williams & Wilkins, Baltimore, pp 261–262

Larsen H (1984b) Genus I. *Halobacterium* Elazari-Volcani 1957, 207[AL]. In: Krieg NR, Holt JG (eds) Bergey's manual of systematic bacteriology, vol 1. Williams & Wilkins, Baltimore, pp 262–266

Larsen H (1984c) Genus II. *Halococcus* Schoop 1935, 817[AL]. In: Krieg NR, Holt JG (eds) Bergey's manual of systematic bacteriology, vol 1. Williams & Wilkins, Baltimore, pp 266–267

Larsen H (1986) Halophilic and halotolerant microorganisms – an overview and historical perspective. FEMS Microbiol Rev 39:3–7

Larsen H (1989) Genus IV. *Halococcus* Schoop 1935, 817[AL]. In: Staley JT, Bryant MP, Pfennig N, Holt JG (eds) Bergey's manual of systematic bacteriology, vol 3. Williams & Wilkins, Baltimore, pp 2226–2228

Larsen H, Grant WD (1989) Genus I. *Halobacterium* Elazari-Volcani 1957, 207[AL]. In: Staley JT, Bryant MP, Pfennig N, Holt JG (eds) Bergey's manual of systematic bacteriology, vol 3. Williams & Wilkins, Baltimore, pp 2228–2230

Larsen H, Omang S, Steensland H (1967) On the gas vacuoles of the halobacteria. Arch Mikrobiol 59:197–203

Mohr V, Larsen H (1963a) On the structural transformations and lysis of *Halobacterium salinarium* in hypotonic and isotonic solutions. J Gen Microbiol 31:267–280

Mohr V, Larsen H (1963b) The mechanism of lysis of *Halobacterium salinarium* in hypotonic solutions. Acta Chem Scand 17:888

Mullakhanbhai MF, Larsen H (1975) *Halobacterium volcanii* spec. nov., a Dead Sea halobacterium with a moderate salt requirement. Arch Microbiol 104:207–214

Oren A (2010) Helge Larsen (1922–2005). Extremophiles 14:483–484

Steensland H, Larsen H (1969) A study of the cell envelope of the halobacteria. J Gen Microbiol 55:325–336

Steensland H, Larsen H (1971) The fine structure of the extremely halophilic cocci. Kong Norske Vidensk Selsk Skr 8:1–5

Strøm AR, Larsen H (1979) Anaerobic fish spoilage by bacteria. Biochemical changes in herring extracts. J Appl Bacteriol 46:269–277

Strøm AR, Olafsen JA, Larsen H (1979) Trimethylamine oxide: a terminal electron acceptor in anaerobic respiration of bacteria. J Gen Microbiol 112:315–320

Xu X-W, Wu Y-H, Wang C-S, Oren A, Zhou P-J, Wu M (2007) *Haloferax larsenii* sp. nov., an extremely halophilic archaeon from a solar saltern. Int J Syst Evol Microbiol 57:717–720

Chapter 2
The Halophilic World of Lourens Baas Becking

Aharon Oren

2.1 Introduction

Most biologists know the name Lourens Baas Becking only thanks to his famous statement "*everything is everywhere:* but, *the environment selects.*" This often-quoted sentence ("*alles is overal:* maar *het milieu selecteert*") appeared in a little book of collected lectures, entitled "Geobiologie of inleiding tot de milieukunde" (Geobiology or introduction to environmental science) (Baas Becking 1934); it is not generally associated with halophilic microorganisms. The book contains chapters on the influence of environmental parameters such as radiation, temperature, and chemical characteristics on microorganisms and other forms of life; it discusses the biogeochemical cycles of the elements, and special chapters are devoted to oligotrophic and eutrophic fresh waters, the role of microorganisms in the marine environment, and last but not least, on the biology of hypersaline environments. The book was based on a series of lectures that Baas Becking had given for "Diligentia," a society of scientifically interested laymen in The Hague.

It is seldom realized that many of Baas Becking's studies dealt with salt lakes and salterns and their biota. His first publication on the topic, entitled "On organisms living in concentrated brine," dates from 1928. Many will know his interesting essay on "Historical notes on salt and salt-manufacture" (Baas Becking 1931a), in which he stressed the importance of the biota in saltern evaporation ponds (*Dunaliella* and other algae, red halophilic prokaryotes, the brine shrimp *Artemia*) in the salt production process. But nowadays hardly anybody is aware of the fact that the longest chapter in this 1934 book ("De Pekel" – The Brine) deals with halophilic microorganisms and their environments. It contains many surprising statements and observations that appear to be mostly forgotten today

A. Oren (✉)
Department of Plant and Environmental Sciences, The Institute of Life Sciences, and the Moshe Shilo Minerva Center for Marine Biogeochemistry, The Hebrew University of Jerusalem, 91904 Jerusalem, Israel
e-mail: orena@cc.huji.ac.il

A. Ventosa et al. (eds.), *Halophiles and Hypersaline Environments*,
DOI 10.1007/978-3-642-20198-1_2, © Springer-Verlag Berlin Heidelberg 2011

but are worth examining; some ideas even turned out to be prophetic, and the basis of some of the observations became clear only much later.

This essay provides an overview of Baas Becking's contributions to our understanding of halophilic microorganisms and their interactions with their environments, based both on his 1934 book and on his many publications in scientific journals between 1928 and 1960.

2.2 Scientific Work on Four Continents: The Life of Lourens Baas Becking

Lourens Gerhard Martinus Baas Becking (Fig. 2.1) was born in 1895 in Deventer, the Netherlands. His scientific career took him to places all over the globe (Went 1963; Quispel 1998). After his basic studies at the University of Utrecht, he moved to Columbia University, New York, where he worked in the laboratory of the famous geneticist Thomas Hunt Morgan. His Ph.D. thesis (University of Utrecht) entitled "Radiation and Vital Phenomena" was mainly based on his work in the USA. After a short stint as assistant professor of plant physiology in Utrecht in 1923 he was appointed professor at Stanford University, Palo Alto, where he taught economic botany and plant physiology. He then became director of the Jacques Loeb Marine Laboratory, Pacific Grove, CA, where Cornelis van Niel (1897–1985) became his first assistant. His work with van Niel strengthened his association with the "Delft School of Microbiology" initiated by Martinus Beijerinck (1851–1931) and Albert Jan Kluyver (1888–1956). The application of Beijerinck's approach of enrichment cultures using selective growth media (Beijerinck 1913; van Niel 1949; de Wit and Bouvier 2006; O'Malley 2007) led to Baas Becking's famous "Everything is everywhere..." statement. During his years in California, Baas Becking

Fig. 2.1 Portrait of Lourens Baas Becking. From Quispel 1998. Reproduced by permission of International Microbiology, published in Vol. 1, p. 69, 1998

visited many salt lakes, and his observations there formed the basis for most of his studies on halophilic microorganisms discussed below.

In 1930, Baas Becking became professor of general botany at the University of Leiden, the Netherlands. It is interesting to note the title of his inaugural address: "Gaia of Leven en Aarde" – Gaia or Life and Earth (Baas Becking 1931b), long before Lovelock and Margulis (1974) presented their Gaia theory (see also de Wit and Bouvier 2006). It showed Baas Becking's profound recognition that organisms modify their environment and that such phenomena are important on a global scale.

In 1940, he was appointed director of the botanical gardens at Buitenzorg (Bogor), Java, the Netherlands Indies (now Indonesia), but because of World War II his journey to the Far East was postponed. Then followed a period starting in 1948 in which Baas Becking served as vice-president of the scientific council of the "South Pacific Commission" and moved to New Caledonia. His last position was in Canberra, Australia, where the Baas Becking Geobiological Laboratory (Parkes, Australian Capital Territory) was established in the late 1960s, jointly operated by the Commonwealth Scientific and Industrial Research Organization's Division of Mineralogy, the Bureau of Mineral Resources, and the Australian Mineral Industries Research Association. This laboratory existed until 1985.

Lourens Baas Becking passed away in Canberra, Australia, in 1963.

2.3 Baas Becking's Studies of Salt Lakes and Their Biota

2.3.1 Dispersal and Biogeography of Halophiles

In his 1934 book Baas Becking listed no less than 30 salt lakes – 15 coastal hypersaline environments on five continents and 15 inland saline lakes on three continents from which he had studied samples. He himself had collected samples from Mono Lake, Owens Lake, Searles Lake, and solar salterns in California, from Pyramid Lake and Soda Lake, Nevada, and Great Salt Lake, Utah. He further examined brines sent to him from hypersaline lakes in Brazil, Italy, France, North Africa, India, Java, Australia, Romania, and other locations.

The availability of so many samples from salt lakes all over the world provided Baas Becking plenty of opportunity to test his "Everything is everywhere; *but*, the environment selects" hypothesis, studies inspired by Beijerinck's enrichment culture and selective culturing techniques. Observations made in the Californian salt lakes and additional samples from lakes all over the world made it clear that certain organisms were invariably present in salt lakes of similar chemical properties, but geographically widely dispersed. This is particularly true for the flagellate alga *Dunaliella viridis* and the brine shrimp *Artemia*. Baas Becking used these cases as striking illustrations of his "Everything is everywhere..." statement. He wrote: "De zoutorganismen zijn dus een fraaie illustratie van het beginsel 'alles is overal', hetgeen nog duidelijker blijkt na verdere beschouwing van eenige vertegenwoordigers" (The salt organisms are thus a beautiful illustration of the principle that "everything is everywhere," which becomes even more obvious upon further examination of some representatives) (these and following translations from Baas Becking 1934: AO). However, Baas Becking was also aware of the limitations to the dispersal of microorganisms, explaining why still some potentially suitable environments were not (yet) colonized: "There thus are rare and less rare microbes. Perhaps there are very rare microbes, i.e., microbes whose possibility of dispersion is limited for whatever reason."

His studies of the brines and the associated salt produced at the factory in Boekelo, the Netherlands, provide an interesting example. Here salt was gained by injecting freshwater into an underground Permian salt deposit, pumping up the brine and evaporating the water under reduced pressure. Baas Becking commented: "Although the salt factory in Boekelo is operative already for over 12 years, the author could not yet find any specific salt organisms there (summer 1932). Doesn't this mean that the microbes, and the smaller organisms in general, do also have their areas of distribution?" He then provided the following answer to the paradox: "No: the environment selects. When our description of the environment is not yet sufficiently advanced and is not sufficiently specific, and the organism occurs less frequently, it is possible that the organism does not appear in enrichment cultures; it is not necessary to assume that it may be restricted to a certain geographic area." In short, the explanation may be that we do not know enough about the properties of the organisms and about the physical and chemical parameters that determine their limits of existence, to draw far-reaching conclusions.

The "everything is everywhere..." statement was based on the assumption that the enrichment culturing technique can disclose the "hidden reality" of microbial diversity, implying that in a given environmental setting most of the microbial species are only latently present (De Wit and Bouvier 2006). Our understanding of the mechanisms through which halophilic microorganisms, which need the constant presence of high salt for structural stability and viability, can colonize new environments is still limited. The recent study by Brito-Echeverría et al. (2009) that showed the presence of *Halococcus* spp. in the nostrils salt glands of the seabird *Calonextris diomedea* provides at least a partial answer to the question how extremely halophilic microorganisms may travel around the world (see also Kushner 1985); this finding also explains why *Halococcus* spp. have repeatedly been isolated from seawater (Rodriguez-Valera et al. 1979; Ventosa et al. 1984). One of Baas Becking's last published papers (Baas Becking and Kaplan 1956) contains the following intriguing statement: "While 'everything is everywhere' at least as far as soil and water bacteria are concerned, bacteria cannot wait forever." This statement was followed by examples of cases in which enrichments (using different materials as inoculum) failed to yield the desired sulfate reducers or sulfur-oxidizing autotrophs.

Today, following the development of cultivation-independent molecular techniques, it is becoming increasingly clear that it is physically impossible that "everything is everywhere," as the number of possible species (or phylotypes) by far exceeds the number of microorganisms that occur in a milliliter of water or in a gram of soil. "Molecular biogeography" has become an interesting branch of microbial ecology, which also has its impact on the study of the dispersion, the geographical variation of the extremely halophilic Archaea, and even on the species concept for such forms of life (Papke et al. 2007; see also O'Malley 2007).

2.3.2 *The Influence of Ionic Composition of Hypersaline Environments*

A recurring theme in Baas Becking's publications on life at high salt concentrations is the effect of specific ions and the antagonistic actions between the ions that define the possibilities for the different organisms to develop.

The lyotropic series or Hofmeister series (Hofmeister 1888) that defines the salting-in/salting-out effect of different ions ($F^- > SO_4^{2-} > HPO_4^{2-} > Cl^- > NO_3^- > Br^- > ClO_3^- > I^- > ClO_4^-$; $NH_4^+ > K^+ > Na^+ > Mg^{2+} > Ca^{2+}$) has gained a renewed interest in recent years (Leberman and Soper 1995; Parsegian 1995). Baas Becking spent much effort testing the effects of different cations and anions on the growth and behavior of halophilic microorganisms (and macroorganisms such as *Artemia* as well). Especially calcium ions were found very toxic, but their inhibitory effect is relieved when sodium ions are present in sufficiently high concentrations. Magnesium and potassium were generally found to have little toxicity.

Not always were Baas Becking's theories based on specific ion effects and his antagonisms correct. He did not visit the Dead Sea himself, but based on the examination of Dead Sea water samples he concluded that the sterile nature of the Dead Sea is caused by the high calcium concentration of its waters. However, just 2 years after his book was published, Wilkansky (Elazari Volcani) showed the Dead Sea to be inhabited by a diverse community of microorganisms, well adapted to life in this unusual habitat (Wilkansky 1936; Elazari-Volcani 1940).

2.3.3 *Studies on* Dunaliella *and Other Halophilic Flagellate Algae*

Baas Becking devoted extensive studies to the physiology of *Dunaliella viridis* and *D. salina*, which he considered to be a variant of *D. viridis* (Fig. 2.2a). He also studied other algae found in hypersaline brines such as *Asteromonas gracilis*, another naked green flagellate that occurs in marine brine in temperate climates, earlier described from salt lakes near the Black Sea and found by Baas Becking in very concentrated brine, including the magnesium "bittern" ponds, of the Leslie Salt Works in San Francisco Bay and in sea salt of Setubal, Portugal (Fig. 2.2b).

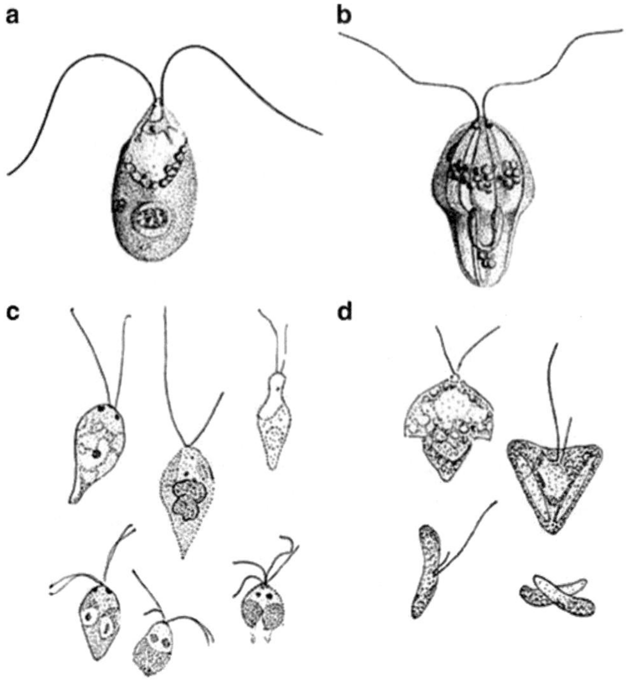

Fig. 2.2 Baas Becking's drawings of phototrophic flagellate algae growing in brines: (**a**) *Dunaliella viridis* Teodoresco. Material from Brazil (Baas Becking 1934); (**b**) *Asteromonas gracilis* Artari from Leslie Salt Works, San Mateo, California (Baas Becking 1934); (**c**) *Dunaliella piercei* from a salt lake near Marina, California (Nicolai and Baas Becking 1935); (**d**) *Boekelovia hooglandii* from a saline swimming pool in Boekelo, the Netherlands (Nicolai and Baas Becking 1935)

In his essay on the history of salt making he noted that already Pliny may have described salt that contains *Dunaliella* (Baas Becking 1931a).

Physiological studies on *Dunaliella* centered on the following aspects (1) the influence of the concentration; (2) the influence of change in concentration; (3) the influence of different salts; (4) antagonism of the elements; (5) the influence of temperature (Baas Becking 1930, 1931c, 1934). Optimal growth was found at pH 9, but survival was possible at extremes of pH: viability was retained after 24 h at pH 3.5–4 and at pH 11 and possibly even higher. In a later study, Baas Becking reported that *Dunaliella* requires at least 5 μg/l boron for growth (Baas Becking and Kaplan 1956). The observed versatility of *D. viridis* and its ability to quickly adjust to changes in the environmental conditions made Baas Becking comment: "there appears to be no reason why such organisms should exist; that they do exist proves either that the organism is utterly unusual or that the rules derived by and for organisms living in fresh water or sea water are inapplicable to these beings. The latter assumption appears to be worthy of further consideration..." (Baas Becking 1930).

Table 2.1 Survival of brine organisms as affected by various anions

Substance	Molarity	Concentration in weight percentage	Survival time	Remarks
KI	2.5	41.5	∞	Phototaxis normal
KBr	2.5	30.0	∞	Phototaxis normal
KCN	2.5	16.23	∞	*D. viridis* only[a]
NaCNS [Na-thiocyanate]	2.5	20.27	∞	Phototaxis normal
$NaNO_3$	2.5	21.25	∞	Phototaxis normal
Na(HCOO) [Na-formate]	2.5	37.00	∞	Phototaxis normal
$K_2C_2O_7$ [possibly the intention was $K_2Cr_2O_7$; K-dichromate]	(in 2.5 M NaCl)	5.00	∞	Phototaxis normal
$K_2Cr_2O_4$ [possibly the intention was K_2CrO_4; K-chromate]	2.5	48.55	∞	Phototaxis normal
$NaBO_2$ [Na-metaborate]	(in 2.5 M NaCl)	1.39	∞	Phototaxis normal
$Na_2B_2O_7$ [??, possibly $Na_2B_4O_7$; Na-tetraborate]	(in 2.5 M NaCl)	.43	∞	Phototaxis normal
$KMnO_4$	(in 2.5 M NaCl)	2.5	0	Instants death, stains brown in toto, shrinks

Adapted from Baas Becking 1930
∞ This symbol means that the organisms were still alive after 6 h
[a]Caudate *Dunaliella* and *Trichomastix* die in 35 min

Baas Becking observed that *Dunaliella* is resistant to very high concentrations of toxic anions such as chromate, cyanide, and thiocyanate: "Anions have little or no influence on *Dunaliella*. It is intriguing to watch this alga swimming for hours in concentrations of cyanide, thiocyanate, or chromate" (Baas Becking 1934). Table 2.1 summarizes Baas Becking's observations on the viability of *Dunaliella* when suspended in concentrated solutions of unusual anions. These include chaotropic anions as defined by the Hofmeister series that are toxic to most other forms of life already at low concentrations. This observation was largely forgotten, but the resistance of *Dunaliella* cells to high concentrations of otherwise toxic anions can easily be confirmed in simple experiments, as shown in Table 2.2.

The true physiological basis for the survival of *Dunaliella* in molar concentrations of chromate, cyanide, and other unusual anions is yet to be elucidated, but Baas Becking's explanation for his findings is interesting. He attributed the phenomenon to the presence of a strong negative charge of the cell membrane and of the flagella, which can be neutralized by cations in a lyotropic series, rendering the membrane impermeable to anions and selectively permeable to cations (Baas Becking 1930). This is an interesting statement as it was not based on direct measurements of the surface charge of the membrane. Only much later it was proven that indeed the surface layers of *Dunaliella* carry a large excess of negative charges as compared

Table 2.2 The effect of thiocyanate, chromate and cyanide exposure on motility of *Dunaliella*

		Motility score after	
Compound added	Final concentration	10 min	60 min
KSCN	0.17 M	+++	+++
	0.34 M	+++	+++
	0.59 M	+++	+++
Na_2CrO_4	0.10 M	+++	+++
	0.21 M	+++	+++
	0.35 M	+++	+++
KCN	0.08 M	++	++
	0.18 M	+	+
	0.33 M	−	−

A culture of a Dead Sea isolate of *Dunaliella* sp., grown in medium containing 72 g/l total salts (56.3 g/l NaCl), pH 7.5, was supplemented with different concentrations of potassium thiocyanate, sodium chromate, or potassium cyanide. After 10 and 60 min incubation in the light at room temperature the motility of the cells was checked microscopically and compared to a control without additives. +++: no decrease in motility; ++, few cells lost motility; +, most cells non-motile; −: no more motile cells observed (Oren, unpublished results)

to non-halophilic algae. For example, membrane-bound proteins such as carbonic anhydrase and a transferrin-like protein have a great excess of acidic amino acids, conferring a large negative charge to the proteins (Gokhman et al. 1999). The first report showing that halophilic proteins have a large excess of acidic amino acids and are therefore highly negatively charged was published by Reistad in 1970, making Baas Becking's statement even more surprising.

Much of the behavior of *Dunaliella* and other halophilic microorganisms in their natural habitats was explained by Baas Becking on the basis of antagonism between the different ions in the brines. Calcium was found to be particularly toxic, especially in slightly acidic medium, but magnesium can relieve the calcium toxicity. The higher the NaCl concentrations, the more magnesium is required to detoxify calcium to enable growth of *Dunaliella*. In 1 M NaCl, the antagonistic relation Mg:Ca is 4–5, while in 4 M NaCl the proportion becomes many times as great (20:1). The red Archaea were found to require high salinity, pH and magnesium, and they did not develop in cultures containing calcium. Cyanobacteria grew best in lower NaCl and low Mg/Ca ratios. "Thus, during the evaporation of sea water, we may predict the environment will become suitable, in succession for: *blue-green algae and fungus* → *Dunaliella viridis*, first swarmers and zygotes, then palmella → *D. salina* → *red bacteria*" (Baas Becking 1931c).

Baas Becking also described a few novel types of halophilic flagellate algae. The best known is a new species of *Dunaliella*, *D. piercei* (Nicolai and Baas Becking 1935). It was first isolated from a piece of algal mat ("meteor paper," see also Oren 2010) that had formed in Marina Lake, a small salt lake in California. Inoculation of this material into growth media yielded not only *Dunaliella viridis* but also another green flagellate with flattened cells, 12–21 × 5.4–10.8 μm in size and 2–4 μm thick (Fig. 2.2c). *D. piercei* is still recognized as representative of one of the major subgroups within the diverse genus *Dunaliella* (Borowitzka and Siva 2007; González et al. 2009).

To honor Baas Becking's contribution to *Dunaliella* research, a new species of the genus was named in his honor as *D. baasbeckingii* (Masyuk 1973).

Another interesting (slightly) halophilic phototrophic flagellate described by Baas Becking is *Boekelovia hooglandii.* Soon after the above-mentioned salt industry was established in Boekelo, the Netherlands, a salt-water swimming pool was constructed nearby, containing water of 3% NaCl with some sulfate present as well, but almost devoid of potassium and magnesium salts. The water had a pH of 9. In the summer of 1934, a brownish-orange bloom developed in the swimming pool, consisting of a triangular *Ochromonas*-like chrysomonad, 6–7 μm in diameter, and carrying two flagella of different length (Fig. 2.2d). The organism was named in honor of engineer Jelle Hoogland, the director of the Boekelo salt industry. It could live at 0.4–10% NaCl, but was killed at 14% salt. The alga could not be kept in culture more than a few days (Nicolai and Baas Becking 1935).

2.3.4 Cyanobacteria

In his essay on the history of salt making, Baas Becking (1931a) mentioned the importance of benthic cyanobacterial mats in coastal solar saltern ponds: "In some places in France, Italy, Spain, and Portugal, people exploit the natural growth of a cyanobacterium, *Microcoleus chthonoplastes*, which covers the bottom of the salt pan like a layer of linoleum (meteor paper). It is easy to harvest pure salt on top of this mat. When a new saltern is prepared, people do not forget this mat. Like the green alga *Lochmiopsis*, *Microcoleus* is halotolerant, and can only grow below 2 M NaCl." The halophilic *M. chthonoplastes*, recently renamed *Coleofasciculus chthonoplastes* (Siegesmund et al. 2008), is not restricted to salt concentrations below 2 M, and is sometimes found even on the bottom of salt-saturated saltern ponds. This is, for example, the case in the saltern ponds of Sečovlje on the Adriatic coast on the border between Slovenia and Kroatia. These ponds, which have been in operation since the Middle Ages using the same traditional methods of salt production and salt harvesting, depend on the presence of the benthic "petola" mat, consisting mainly of *Microcoleus/Colefasciculus* (Gunde-Cimerman et al. 2005; Schneider 1979; Schneider and Herrmann 1979). The petola prevents mixing of crystallized halite with the mud on the bottom of the ponds and prevents incorporation of undesirable ions such as iron and manganese into the halite crystals. The properties of this mat and the function of the filamentous cyanobacteria found in it were the subject of a recent study (Tkavc et al. 2010).

Other types of cyanobacteria recorded in Baas Becking's publications are the unicellular *Aphanothece*-type cells found in many natural salt lakes and salterns (Baas Becking 1934) and the filamentous halotolerant cyanobacterium *Nodularia spumigena* he found in Lake Eyre, South Australia (Baas Becking and Kaplan 1956), an organism found today in large masses in the South Arm of Great Salt Lake, Utah (Roney et al. 2009).

2.3.5 Red Halophilic Archaea

Red brines have always been associated with salt making, and in his essay on the history of salt production, Baas Becking (1931a) cites the ancient Chinese treatise on pharmacology – the "Peng-Tzao-Kan-Mu" ("man-plant-classification"), compiled by Li Shih-Chen in the sixteenth century, and dating back to around 2700 B.C. according to some authorities: "An embankment is made and ditches to draw clear sea water. It is left for a long time until the color becomes red. If the south wind blows with force during the summer and autumn the salt may grain over night. If the south wind does not come all the profits are lost."

In his 1931 essay and in his 1934 book, Baas Becking discussed the cause of red brines. Considerable confusion existed on the nature of the organisms responsible for the red coloration of saltern brines, and to some extent conflicting opinions exist even today (Oren 2009). Baas Becking was well aware of the existing confusion on the topic when he wrote: "Pink, red and purple bacteria seem to be the chief cause of red brines. It is well to define the groups more clearly, as much confusion may result from the term 'red bacteria'." Carotenoid-rich *Dunaliella* cells may be in part responsible for the color of the brines, but Baas Becking recognized the far greater importance of the red Archaea ("stokvischbacteriën" – "dried cod bacteria" that cause the reddening of salted fish) such as "*Bacterium halobium* Petter" (now *Halobacterium salinarum*) and *Sarcina morrhuae* (now *Halococcus morrhuae*) Klebahn. Baas Becking was well acquainted with the work of Helena Petter (1931, 1932) on these extreme halophiles, and he adopted in his 1934 book a figure drawn by Petter showing *Halobacterium* cells with gas vacuoles, supposed to help the aerobic cells to keep afloat in the oxygen-poor brines (Fig. 2.3). He also stated

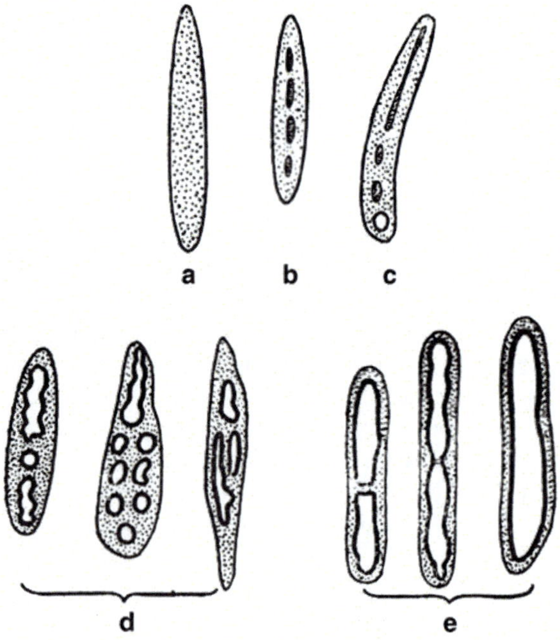

Fig. 2.3 *Bacterium halobium* Petter (*Halobacterium salinarum*). Bacteria of different age. Forms with and without gas vacuoles (based on Petter) (Baas Becking 1934)

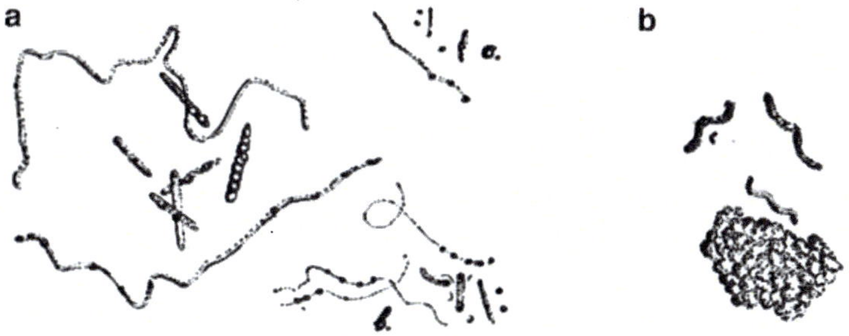

A. Kleurlooze zwavelbacteriën uitzeewater pekel van Venezuela Brazilië en Portuga. × 500
B. Purperen spirillen (Rhodospirillum halophilum B-B) uit Searles' meer, Californië. × 500 (orig.)

Fig. 2.4 (**a**) Colorless sulfur bacteria from seawater brine from Venezuela, Brazil and Portugal; (**b**) Purple spirilla [designated *Rhodospirillum halophilum*, a name for which there is no record in the Index Bergeyana or other handbooks on bacterial nomenclature] from Searles Lake, California (Baas Becking 1934)

that in some hypersaline lakes, the dark red colors of the brines can be attributed to the presence of purple bacteria, facultative saprophytes, unable to live without light under anaerobic conditions (Fig. 2.4).

When halite crystallizes, red halophilic Archaea are often trapped within fluid inclusions within the salt crystals. Baas Becking (1931a) mentioned crystals of NaCl (with adhering carbonates) from Searles' Lake, California, which were inside still a dark purple. Interestingly, he attributed the color to the presence of a large *Thiospirillum*, a genus currently consisting of only of *T. jenense*, of yellow, not purple cells; in the past red species were recognized as well, such as "*T. sanguineum*."

2.3.6 The Sulfur Cycle in Hypersaline Environments

Bacteria involved in the sulfur cycle in hypersaline environments, including purple sulfur bacteria, green sulfur bacteria, aerobic chemoautotrophic sulfur oxidizers, and sulfate reducers (Fig. 2.4), were mentioned in several of Baas Becking's publications. He noted that sulfate-reducing bacteria are highly adaptable, and that neither temperature nor high salt concentrations appear to inhibit the activity of these bacteria, so that a thick layer of black mud can be found below nearly all saltern ponds worldwide (Baas Becking 1934). Sulfate reducers can indeed be found up to quite high salinities, but the statement that salt at high concentrations does not affect sulfate reduction rates cannot be maintained without certain restrictions (Oren 1999, 2011).

One of Baas Becking's last publications dealt with the origin of the sulfur nodules of Lake Eyre, South Australia (Baas Becking and Kaplan 1956). Nodules of elemental sulfur are found in the sediments of Lake Eyre, encased in a mass of gypsum crystals. It was shown that the sulfur had originated from microbial transformation of gypsum. Sulfate-reducing and sulfur-oxidizing bacteria could be isolated from the system, growing well in 20–25% salt. The brine above the salt crust was salt-saturated in 1954 when the study was performed. Enrichment cultures were set up for halophilic sulfate reducers growing on hydrogen and on lactate, and formation of hydrogen from glucose by fermentative bacteria was also demonstrated up to 25% salt. The authors summarized the outcome of these experiments, stating: "Most of the biological evidence was obtained from enrichment cultures and, in some cases, pure cultures of various bacterial groups. As most of the results will be published elsewhere, a brief statement will suffice" (Baas Becking and Kaplan 1956). Unfortunately, no such additional information was ever published as far as I could ascertain.

2.3.7 Other Halophilic and Salt-Tolerant Microorganisms

In his 1934 book and in his study on the sulfur cycle of Lake Eyre (Baas Becking and Kaplan 1956), Baas Becking mentioned a variety of additional microorganisms and microbially mediated processes he had encountered in salt lakes worldwide and/or were studied in enrichment cultures, for which media with 5, 10, 15, 20, and 25% salt were used:

- Denitrifying bacteria. Surprisingly, Baas Becking found only little denitrification potential in Lake Eyre at salt concentrations above 10% (Baas Becking and Kaplan 1956), in spite of the fact that even some extreme halophiles are able to produce gas from nitrate at salt concentrations near saturation (Oren 1999, 2011).
- Methane production from acetate, a process he found to be restricted to the low salinity range (Baas Becking and Kaplan 1956), an observation that agrees with our current understanding of the process (Oren 1999).
- Aerobic and anaerobic degradation of cellulose, reported to proceed up to the highest salinities (Baas Becking and Kaplan 1956). Even today very little is known about cellulose degradation in hypersaline environments, and no cultures of true halophiles appear to be available that grow on cellulose.
- Halophilic fungi. The filamentous fungus *Oospora halophila* (also known as *Scopulariopsis halophilica* or *Basipetospora halophila*) was reported by Baas Becking (1934) from Great Salt Lake, Utah; he considered it to have a cosmopolitan distribution in high-salt environments.
- Different amoeboid, ciliate, and flagellate protozoa (Fig. 2.5).
- Diatoms such as *Pleurosigma* and *Amphora* spp.

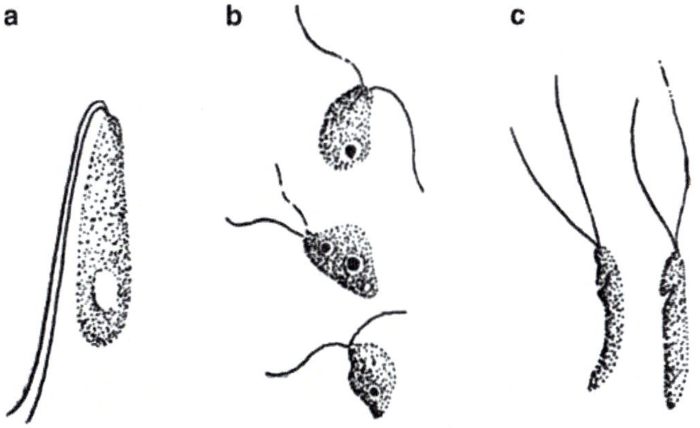

Fig. 2.5 Colorless brine flagellates. (**a**) *Amphimonas cuneatus* Namys. (Marina, California); (**b**) *Amphimonas polymorphus* Namys. (Venezuela); (**c**) *Pleurostoma parvulum* Namys. (Searles, California) (Baas Becking 1934)

2.3.8 Artemina

Baas Becking was fascinated not only by the microorganisms that inhabit the brines of salterns and natural salt lakes, but also by the brine shrimp *Artemia* often found in such environments. In his 1931 essay on historical notes on salt and salt manufacture, he explained the importance that this little crustacean may have in the salt-making process: "A few years ago the foreman of a near-by salt works came to our laboratory to ask for a few *Artemiae*. When asked for what purpose he wanted them, he declared that he could not make salt without the 'brine-worms'... At that time we thought the man foolishly superstitious, little realizing that these organisms were used by the old English briner and called by them 'clearer-worms.'" Baas Becking performed some experiments on the ability of *Artemia* to clarify brines from particles that may interfere with the salt-making process: "I have repeated this experiment with fine and rather stable suspensions of barium sulfate, calcium carbonate and calcium sulfate with the same result: the liquid which passes through their digestive tract is freed from its particles, which coagulate in small pellets. Five artemiae cleared a milky white suspension of barium sulfate (100 cc) in 24 h while the controls remained unchanged. The pellets on the bottom contained the precipitated matter."

Baas Becking noted that the internal salt concentration of *Artemia* tissues was never higher than 0.9%, even in animals living in brines of 10–25% salt (Warren et al. 1938). He also tested the limits of salt concentrations and other environmental conditions that support development of eggs and nauplii of *Artemia salina*, with special emphasis on the antagonistic effects of different ions. Eggs hatched over a very broad pH range, from acid solutions at pH 2–0.1 M NaOH solutions. $CaCl_2$ and $MgCl_2$ were found toxic. Potassium ions were toxic as well, but inhibition by K^+ could be relieved by presence of excess Na^+ to a smaller extent even by Mg^{2+}

and Ca^{2+}. The results of these studies were summarized in complex diagrams showing which combinations of Na^{+}, K^{+}, Mg^{2+} and Ca^{2+}concentrations did and which did not support *Artemia* development (Boone and Baas Becking 1931).

2.4 Final Comments

Much of Baas Becking's work on salt lakes and halophilic microorganisms centered on the adaptation of halophiles to life at the limits: high salinity, extremes of pH, presence of toxic ions, etc. He was probably the first to realize that nature has devised more than one strategy to enable (micro)organisms to live at high salt concentrations, when he wrote: "These simple experiments show that the mechanism on which halophily is based can be completely different for each organism, and that nature has achieved salt resistance in very different ways" (Baas Becking 1934). The full impact of this statement became clear only several decades later, when it became clear that some halophiles use inorganic salt to provide osmotic balance with their environments, while others use one or more of a variety of organic "compatible" solutes.

Another interesting statement found in Baas Becking's 1934 book deals with the interrelationships between salinity and temperature defining the limits of growth of halophiles: "Still, an investigation of the temperature tolerance of salt organisms is not without interest, especially also because our great compatriot Hugo de Vries already mentioned in his Ph.D. thesis that even in higher plants the thermotolerance increases when the cells are investigated in salt solutions." I have not been able to find such a statement in the Ph.D. thesis mentioned (de Vries 1870), but Baas Becking's statement, although not documented at the time by experimental evidence, was proven completely correct based on publications from the 1970s onwards. Both in the archaeon *Haloferax volcanii* and in the bacterium *Planococcus halophilus* the temperature optimum is shifted to higher values at increased salt concentrations, and the higher the temperature, the more salt is needed for growth (Mullakhanbhai and Larsen 1975; Novitsky and Kushner 1975, 1976). A similar phenomenon was reported for green algae of the genus *Dunaliella* (Borowitzka and Borowitzka 1988). Also among marine fungi an increase in salinity tolerance with an increase of incubation temperature is a widespread phenomenon, known as the "*Phoma*-pattern" (Lorenz and Molitoris 1992). Until the end of his life Baas Becking remained fascinated by the exploration of the environmental limits to life, as shown by the publication a few years before his death of a long paper entitled "Limits of the natural environment in terms of pH and oxidation-reduction potentials" (Baas Becking et al. 1960).

The 1934 collection of lectures on "Geobiologie of inleiding tot de milieukunde" published only in the Dutch language, has remained Baas Becking's only book. In the 1950s he started work on a manuscript for a newly updated "Geobiology" book in English, which unfortunately was never completed. About 750 manuscript pages were discovered in 1976, and these are kept in the library of the Royal

Netherlands Academy of Sciences in Amsterdam (Quispel 1998). To my knowledge nobody has ever seriously studied this document. It may be interesting to subject this unpublished manuscript to a thorough examination: it may well contain more unexpected views and ideas about the nature of the halophilic microorganisms and their mode of adaptation to life at high salt concentrations.

Acknowledgments My copy of Baas Becking's "Geobiologie of inleiding tot de milieukunde" was presented to me by my former student Danny Ionescu. Thanks to this wonderful gift I realized the largely underrated contribution of Baas Becking to our understanding of the halophiles, which resulted in the writing of this essay. I thank Dr. Brian E. Jones (Genencor International BV, Leiden, The Netherlands) for helpful comments.

References

Baas Becking LGM (1928) On organisms living in concentrated brine. Tijdschr Ned Dierkund Ver Ser III 1:6–9

Baas-Becking LGM (1930) Observations on *Dunaliella viridis* Teodoresco. In: Contributions in marine science. Stanford University, Palo Alto, CA pp 102–114

Baas-Becking LGM (1931a) Historical notes on salt and salt-manufacture. Sci Month 32:434–446

Baas Becking LGM (1931b) Gaia of leven en aarde. Ambtsrede R.U. Leiden. Nijhoff, The Hague

Baas-Becking LGM (1931c) Salt effects on swarmers of *Dunaliella viridis* Teod. J Gen Physiol 14:765–779

Baas Becking LGM (1934) Geobiologie of inleiding tot de milieukunde. W.P. van Stockum & Zoon, Den Haag

Baas-Becking LGM, Kaplan IR (1956) The microbiological origin of the sulphur nodules of Lake Eyre. Trans R Soc S Aust 79:52–65

Baas Becking LGM, Kaplan IR, Moore D (1960) Limits of the natural environment in terms of pH and oxidation-reduction potential. J Geol 68:243–284

Beijerinck MW (1913) De infusies en de ontdekking der bakteriën. Jaarboek voor 1913 van de Koninklijke Akademie van Wetenschappen gevestigd te Amsterdam. Johannes Müller, Amsterdam. Reprinted in: Verzamelde geschriften van M.W. Beijerinck ter gelegenheid van zijn 70sten verjaardag, vijfde deel, Delft, 1921, pp 119–140

Boone E, Baas-Becking LGM (1931) Salt effects on eggs and nauplii of *Artemia salina* L. J Gen Physiol 14:753–763

Borowitzka MA, Borowitzka LJ (1988) *Dunaliella*. In: Borowitzka LJ, Borowitzka MA (eds) Microbial biotechnology. Cambridge University Press, Cambridge, UK, pp 27–88

Borowitzka MA, Siva CJ (2007) The taxonomy of *Dunaliella* (Chlorophyta, Dunaliellales) with emphasis on the marine and halophilic species. J Appl Phycol 19:567–590

Brito-Echeverría J, López-López A, Yarza P, Antón J, Rosselló-Móra R (2009) Occurrence of *Halococcus* spp. in the nostrils salt glands of the seabird *Calonextris diomedea*. Extremophiles 13:557–565

de Vries H (1870) De invloed der temperatuur op de levensverschijnselen der planten. Ph.D. Thesis, Leiden. Martinus Nijhoff, 's Gravenhage

de Wit R, Bouvier T (2006) '*Everything is everywhere*, but, *the environment selects*': what did Baas Becking and Beijerinck really say? Environ Microbiol 8:755–758

Elazari-Volcani B (1940) Studies on the microflora of the Dead Sea. Ph.D. thesis, The Hebrew University of Jerusalem (in Hebrew)

Gokhman I, Fisher M, Pick U, Zamir A (1999) New insights into the extreme salt tolerance of the unicellular green alga *Dunaliella*. In: Oren A (ed) Microbiology and biogeochemistry of hypersaline environments. CRC, Boca Raton, FL, pp 203–213

González MA, Gómez PI, Polle JEW (2009) Taxonomy and phylogeny of the genus *Dunaliella*. In: Ben-Amotz A, Polle JEW, Subba Rao DV (eds) The alga *Dunaliella*. Biodiversity, physiology, genomics and biotechnology. Science Publishers, Enfield, NH, pp 15–44

Gunde-Cimerman N, Oren A, Plemenitaš A (2005) Mikrosafari – Čudoviti svet mikroorganizmov solin – The beautiful world of microorganisms in the salterns. Bilingual Slovenian/English publication. Državna Založba Slovenije, Ljubljana

Hofmeister F (1888) Zur Lehre von der Wirkung der Salze. Arch Exp Pathol Pharmakol 24: 247–260

Kushner DJ (1985) The *Halobacteriaceae*. In: Woese CR, Wolfe RS (eds) The bacteria, vol VIII, Archaebacteria. Academic Press, Orlando, pp 171–214

Leberman R, Soper AK (1995) Effect of high salt concentrations on water structure. Nature 378:364–366

Lorenz R, Molitoris HP (1992) Combined influence of salinity and temperature (*Phoma*-pattern) on growth of marine fungi. Can J Bot 70:2111–2115

Lovelock JA, Margulis L (1974) Atmospheric homeostasis by and for the biosphere: the Gaia hypothesis. Tellus 26:2–9

Masyuk NP (1973) New taxons for the genus *Dunaliella* Teod., part 2. Ukr Bot Zhur 30:345–354

Mullakhanbhai M, Larsen H (1975) *Halobacterium volcanii*, spec. nov., a Dead Sea halobacterium with a moderate salt requirement. Arch Microbiol 104:207–214

Nicolai E, Baas Becking LGM (1935) Einige Notizen über Salzflagellaten. Arch Protistenkd 85:319–328

Novitsky TJ, Kushner DJ (1975) Influence of temperature and salt concentration on the growth of a facultatively halophilic "*Micrococcus*" sp. Can J Microbiol 21:107–110

Novitsky TJ, Kushner DJ (1976) *Planococcus halophilus* sp. nov., a facultatively halophilic coccus. Int J Syst Bacteriol 26:53–57

O'Malley MA (2007) The nineteenth century roots of 'everything is everywhere'. Nat Rev Microbiol 5:647–651

Oren A (1999) Bioenergetic aspects of halophilism. Microbiol Mol Biol Rev 63:334–348

Oren A (2009) Microbial diversity and microbial abundance in salt-saturated brines: why are the waters of hypersaline lakes red? In: Oren A, Naftz DL, Palacios P, Wurtsbaugh WA (eds) Saline lakes around the world: unique systems with unique values. The SJ and Jessie E Quinney Natural Resources Research Library, College of Natural Resources, Utah State University, Utah, pp 247–255

Oren A (2010) Paper from outer space – on 'meteorpapier' and microbial mats. In: Seckbach J, Oren A (eds) Microbial mats. Modern and ancient microorganisms in stratified systems. Springer, Dordrecht, pp 29–38

Oren A (2011) Thermodynamic limits to microbial life at high salt concentrations. Environ Microbiol/Microb Biotechnol, in press (DOI: 10.111/j.1462-2920-2010.02365.x)

Papke RT, Zhaxybayeva O, Feil EJ, Sommerfeld K, Muise D, Doolittle WF (2007) Searching for species in haloarchaea. Proc Natl Acad Sci USA 104:14092–14097

Parsegian VA (1995) Hopes for Hofmeister. Nature 378:335–336

Petter HFM (1931) On bacteria of salted fish. Proc Kon Akad Wetensch B 34:1417–1423

Petter HFM (1932) Over roode en andere bacterieën van gezouten visch. Ph.D. thesis, University of Utrecht

Quispel A (1998) Lourens G. M. Baas Becking (1895–1963), inspirator for many (micro) biologists. Int Microbiol 1:69–72

Reistad R (1970) On the composition and nature of the bulk protein of extremely halophilic bacteria. Arch Mikrobiol 71:353–360

Rodriguez-Valera F, Ruiz-Berraquero F, Ramos-Cormenzana A (1979) Isolation of extreme halophiles from seawater. Appl Environ Microbiol 38:164–165

Roney HC, Booth GM, Cox PA (2009) Competitive exclusion of cyanobacterial species in the Great Salt Lake. Extremophiles 13:355–361

Schneider J (1979) Stromatolithische Milieus in Salinen der Nord-Adria (Sečovlje, Portorož, Jugoslawien) 1. Oldenburger Symposium über Cyanobakterien, 1977, - Taxonomische Stellung und Ökologie. Universität Oldenburg, Oldenburg

Schneider J, Herrmann GA (1979) Saltworks – natural laboratories for microbiological and geochemical investigations during the evaporation of seawater. In: Coogan AH, Hauder L (eds) Fifth international symposium on salt. Northern Ohio Geological Society, Cleveland, Ohio, pp 371–381

Siegesmund MA, Johansen JR, Karsten U, Friedl T (2008) *Coleofasciculus* gen. nov. (Cyanobacteria): morphological and molecular criteria for revision of the genus *Microcoleus* Gomont. J Phycol 44:1572–1585

Tkavc R, Turk M, Visscher P, Gostinčar C, Oren A, Gunde-Cimerman N (2010) Prokaryotic communities in microbial mat "petola" from Sečovlje salterns (Slovenia). FEMS Microbiol Ecol 75:48–62

van Niel CB (1949) The 'Delft School' and the rise of general microbiology. Bacteriol Rev 13: 161–174

Ventosa A, Rodriguez-Valera F, Poindexter JS, Reznikoff WS (1984) Selection for moderately halophilic bacteria by gradual salinity increases. Can J Microbiol 30:1279–1282

Warren H, Kuenen D, Baas Becking LGM (1938) On the relation between internal and external medium in *Artemia salina* (L.) var. *principalis* Simon. Proc Kon Ned Akad Wetensch Amsterdam afd Natuurkunde 41:873–878

Went FW (1963) Dr. L.G.M. Baas Becking. Nature 198:134

Wilkansky B (1936) Life in the Dead Sea. Nature 138:467

Chapter 3
Taxonomy, Phylogeny, and Biotechnological Interest of the Family *Halomonadaceae*

Rafael R. de la Haba, Cristina Sánchez-Porro, and Antonio Ventosa

3.1 Introduction

Traditionally, halophilic microorganisms have been isolated from salted foods, although studies carried out during the last 3 decades have permitted their isolation and characterization from hypersaline environments, mainly saline lakes, salterns, and saline soils. The responses of microorganisms to increasing NaCl concentrations are quite different, but some patterns are observed, and thus, on the basis of their optimal NaCl requirements, microorganisms are classified in different categories: non halophilic, which require ~0.5–1% NaCl for growth (some may tolerate high NaCl concentrations and are defined as halotolerant); slightly halophilic, with optimal NaCl requirements between 1 and 3%; moderately halophilic, growing optimally in media with 3–15% NaCl; and extremely halophilic, requiring more than 15% NaCl for optimal growth (Kushner 1978).

Moderately and extremely halophilic microorganisms are represented by bacteria and archaea (Ventosa et al. 1998; de la Haba et al. 2011). The family *Halomonadaceae* includes currently 90 species grouped in ten genera and, with some exceptions, they are halophilic and in some cases also alkaliphilic, *Halomonas* being the most important genus with more than 60 species (de la Haba et al. 2011). Some reviews about this family have been published (Garrity et al. 2005b; Arahal and Ventosa 2006). In this chapter, we review the current phylogeny and taxonomy of the family *Halomonadaceae*. Besides, some current and future biotechnological applications of species of this microbial group are discussed.

R.R. de la Haba • C. Sánchez-Porro • A. Ventosa (✉)
Department of Microbiology and Parasitology, Faculty of Pharmacy, University of Sevilla, 41012 Sevilla, Spain
e-mail: ventosa@us.es

A. Ventosa et al. (eds.), *Halophiles and Hypersaline Environments*,
DOI 10.1007/978-3-642-20198-1_3,

3.2 Phylogeny of the Family *Halomonadaceae*

3.2.1 Studies Based on 16S/23S rRNA Genes

The first phylogenetic study conducted within this family was carried out by Franzmann et al. (1988) based on the 16S rRNA oligonucleotide cataloging technique. As a result, a new family, *Halomonadaceae*, was proposed to accommodate the moderately halophilic and marine bacteria of the genera *Deleya* and *Halomonas*. Later, five more important phylogenetic studies have been performed on the *Halomonadaceae* (Dobson et al. 1993; Mellado et al. 1995b; Dobson and Franzmann 1996; Arahal et al. 2002b; de la Haba et al. 2010a).

Since the 1990s, 16S rRNA gene sequence comparison became the most widely used method to establish phylogenetic relationships between members of the family *Halomonadaceae*. Dobson et al. (1993) obtained the 16S rRNA sequences of *Halomonas* (*Deleya*) *aquamarina*, *Halomonas* (*Deleya*) *halophila*, *Halomonas* (*Deleya*) *marina*, *Halomonas elongata*, *Halomonas meridiana*, *Halomonas subglaciescola*, and *Halomonas* (*Halovibrio*) *variabilis* and analyzed them together with the sequences of *Halomonas halmophila* and other species of the *Gammaproteobacteria*. This study indicated that the members of the genera *Halomonas* and *Deleya*, as well as the species *Halovibrio variabilis*, shared levels of similarity between their 16S rRNA sequences ranging from 92.6 to 100% and they did not form separate monophyletic subgroups, confirming the lack of any phylogenetic support for retention of these taxa as separate genera and suggesting their unification into a single genus. Furthermore, they also identified several characteristic sequence signatures (among them an extremely rare cytosine residue at position 486 [*Escherichia coli* numbering]), which defined this family in the class *Gammaproteobacteria*. These 16S rRNA signatures for the *Halomonadaceae* have been readapted in other more recent studies (Dobson and Franzmann 1996; Arahal et al. 2002b; Ntougias et al. 2007; Ben Ali Gam et al. 2007), as new members have been described and, currently, they can be redefined as follows: position 484 (A or G), position 486 (C or U), position 640 (A or G), position 660 (A), position 668 (A), position 669 (A), position 737 (U), position 738 (U), position 745 (U), position 776 (U), position 1124 (U or G), position 1297 (U), position 1298 (C), position 1423 (A), position 1424 (C or U), position 1439 (U or C), position 1462 (A or G), position 1464 (C or U). This feature has been used in some studies as a characteristic key for differentiation of genera within *Halomonadaceae* (Dobson and Franzmann 1996; Arahal et al. 2002b; Ntougias et al. 2007; Ben Ali Gam et al. 2007). However, it must be taken into consideration that this is an arbitrary feature, which often must be changed when new isolates are characterized and requires the emendation of the family. If 16S rRNA is to be used as a delineating trait at the rank of family or any other, it makes much more sense to use (almost) complete sequences and not only a few nucleotides of it.

Mellado et al. (1995b) determined the phylogenetic positions of *Chromohalobacter marismortui* (four strains), *Halomonas* (*Volcaniella*) *eurihalina*, and

Halomonas (*Deleya*) *salina* by sequencing their 16S rRNA gene and comparing the resulting data with data for other close relatives obtained from 16S rRNA sequence databases. They demonstrated, on the one hand, that *Chromohalobacter marismortui* belonged to the family *Halomonadaceae* and had the characteristic 16S rRNA signatures defined for this family and, on the other hand, that *Deleya salina* was more closely related to *Halomonas elongata* than to *Deleya aquamarina*, the type species of the genus *Deleya*. They also proposed the reclassification of *Volcaniella eurihalina* as *Halomonas eurihalina* but highlighted the need for a polyphasic approach to determine the natural taxonomic position of members of the *Halomonadaceae*.

Further studies conducted by Dobson and Franzmann (1996) resulted in seven new 16S rRNA sequences, corresponding to the type strains of *Halomonas* (*Deleya*) *cupida*, *Halomonas* (*Deleya*) *pacifica*, *Halomonas* (*Deleya*) *salina*, *Halomonas* (*Deleya*) *venusta*, *Halomonas halodurans*, and *Halomonas eurihalina* and to the strain ACAM 21, a representative biovar of *Halomonas subglaciescola*. They proved that the members of the genera *Deleya*, *Halomonas*, and *Halovibrio* and the species *Paracoccus halodenitrificans* formed a monophyletic group within the class *Gammaproteobacteria* with levels of 16S rRNA sequence similarity ranging from 91.5 to 100% and they, therefore, proposed the unification of all these organisms into a single genus, *Halomonas*. In addition, the genus *Zymobacter* was placed in the family *Halomonadaceae*.

The 23S rRNA gene sequence was first used as a phylogenetic marker within the family *Halomonadaceae* by Arahal et al. (2002b). These authors evaluated the phylogenetic status of this family by comparative 23S and 16S rRNA analyses. The complete sequences of the 23S rRNA of 18 species of the family were determined and the complete 16S rRNA sequences of seven species of *Halomonas* were resequenced due to the several ambiguous positions contained in the previously available sequences of these strains. A combination of different treeing methods and filters was used to elucidate the most stable branchings. There was good agreement between the 23S rRNA- and 16S rRNA-derived trees. According to this study, the genus *Halomonas* was not monophyletic and two phylogenetic groups (respectively, containing five and seven species) were distinguishable. Group 1 comprised *Halomonas elongata* (the type species of the genus), *H. eurihalina*, *H. halmophila*, *H. halophila*, and *H. salina* and exhibited a 98.2% mean 23S rRNA or 16S rRNA sequence similarity. Group 2 was formed by the species *H. aquamarina*, *H. meridiana*, *H. magadiensis*, *H. variabilis*, *H. venusta*, *H. halodurans*, and *H. subglaciescola*, being 97.6% and 97.4% the 23S rRNA and 16S rRNA average sequence, respectively. Six other species in this genus (*Halomonas pacifica*, *H. halodenitrificans*, *H. cupida*, *H. desiderata*, *H. campisalis*, and *H. pantelleriensis*) could not be assigned to either of the above-mentioned groups and did not form a group themselves, sharing relatively low values of sequence similarity with the strains included in groups 1 and 2 or even between themselves (92.9–96.4% for 23S rRNA and 91.7–96.7% for 16S rRNA). These results were in agreement with the phenotypic heterogeneity reported for the species of the genus *Halomonas* and with the wide range of DNA G + C content described for this genus (52–68 mol%). Furthermore, the species *Halomonas marina* formed a separate branch at a deeper level than the other *Halomonas* species, bearing lower than 95% sequence similarity (23S rRNA or 16S rRNA) in all cases, which is generally accepted as a reference value for genus delineation (Ludwig et al. 1998). Consequently, it was suggested that *H. marina* should be ascribed to a separate genus and, indeed, according to this and other data, this organism was later proposed as the type species of a new genus, *Cobetia* (Arahal et al. 2002a). Concerning the genus *Chromohalobacter*, the four species comprising this genus clustered together in a node clearly separated from all *Halomonas* species and with similarity values below 95% in all cases. The mean 23S/16S rRNA similarity of the genus *Chromohalobacter* was 98.6/98.5%. Finally, the sequences of *Zymobacter palmae* and *Carnimonas nigrificans* formed a very stable cluster in a deeper branch of the trees, and therefore, the inclusion of the genus *Carnimonas* in the family *Halomonadaceae* was proposed. Their 16S rRNA sequence similarity was 93.5% and even lower values were obtained when comparing any of the two with the other members of the family.

Very recently, de la Haba et al. (2010a) reevaluated and updated the phylogenetic relationships within the family *Halomonadaceae* based on 23S and 16S rRNA sequences, including the 49 new species with names validly published since the studies by Arahal et al. (2002b) until the date of writing. In addition to the 23S rRNA sequences obtained in previous studies (Arahal et al. 2002b; Lee et al. 2005; Sánchez-Porro et al. 2009), 28 new 23S rRNA complete sequences were used. Moreover, following the recommended minimal standards of the family *Halomonadaceae* (Arahal et al. 2007), seven already available 16S rRNA sequences of the type strains were sequenced again to resolve undetermined positions and reach the established quality standards. Besides, some suggestions about the most recommended sequences to be used for future comparative phylogenetic analysis were included. The sequences were compared to more than 330,000 complete or almost complete rRNA sequences. Furthermore, three different tree construction methods, branch support tests, and several outgroups were used in order to assess the robustness of the topologies. According to this study, there was an excellent agreement between the data based on both ribosomal RNA sequences, but 23S rRNA showed a higher resolution in the differentiation of the species of the family *Halomonadaceae* as the average similarity scored was one unit higher for the 16S rRNA than for the 23S rRNA sequences, suggesting a slower evolutionary rate for the 16S fraction of the rRNA gene. Resulting data confirmed the earlier studies showing that the genus *Halomonas* was not monophyletic and comprised two clearly separated phylogenetic groups containing a larger number of species (12 and 16, respectively) than the ones defined by Arahal et al. (2002b). Group 1 or *Halomonas* sensu stricto formed by the five species mentioned above and *H. caseinilytica*, *H. sabkhae*, *H. almeriensis*, *H. organivorans*, *H. koreensis*, *H. maura*, and *H. nitroreducens*, all having a 97.0% (for 23S rRNA) and a 97.8% (for 16S rRNA) average sequence similarity. Group 2 included the species *Halomonas aquamarina*, *H. meridiana*, *H. axialensis*, *H. magadiensis*, *H. hydrothermalis*, *H. alkaliphila*, *H. venusta*, *H. boliviensis*, *H. neptunia*,

H. variabilis, *H. sulfidaeris*, *H. subterranea*, *H. janggokensis*, *H. gomseomensis*, *H. arcis*, and *H. subglaciescola*, displaying mean 23S/16S rRNA values of similarities of 97.5 and 97.4%, respectively. Phylogenetic distances between group 1 and 2 were large enough as to suggest that both groups could constitute two different genera; however, neither chemotaxonomic nor more general phenotypic studies permitted their separation.

The other 27 species at that time assigned to the genus *Halomonas* (*H. alimentaria*, *H. anticariensis*, *H. campaniensis*, *H. campisalis*, *H. cerina*, *H. cupida*, *H. daqingensis*, *H. denitrificans*, *H. desiderata*, *H. gudaonensis*, *H. halocynthiae*, *H. halodenitrificans*, *H. halodurans*, *H. kenyensis*, *H. korlensis*, *H. kribbensis*, *H. lutea*, *H. mongoliensis*, *H. muralis*, *H. pacifica*, *H. pantelleriensis*, *H. saccharevitans*, *H. salaria*, *H. salifodinae*, *H. shengliensis*, *H. taeanensis*, and *H. ventosae*) did not appear clearly included into either of these two phylogenetic groups. One of these species, *Halomonas salaria*, always formed a separate group with the species *Chromohalobacter salarius* and *Salinicola socius*, sharing 16S rRNA sequence similarities above 97.7% among them. On the contrary, similarity values between this group and any of the *Halomonas-Chromohalobacter* species were low enough (below 96.5%) as to justify their placement in a single genus. In fact, data from these and further studies led to propose the transfer of *Halomonas salaria* and *Chromohalobacter salarius* to the genus *Salinicola*, as *Salinicola salarius* and *Salinicola halophilus*, respectively (de la Haba et al. 2010b). With respect to the species *Halomonas halodurans*, it was included within the group 2 defined by Arahal et al. (2002b) due to the low quality of the 16S rRNA gene sequence used. Resequencing of the 16S rRNA as well as the analysis of the 23S rRNA of this species by de la Haba et al. (2010a) demonstrated that it constituted a very robust and stable cluster with *Cobetia marina*, having exactly the same 16S/23S rRNA sequences. Nevertheless, there was not sufficient evidence to differentiate if *Halomonas halodurans* constituted the same or a different species than *Cobetia marina* and thus, more extensive studies are required to define their taxonomic delineation. The authors also highlighted a possible recombination event affecting the species *Halomonas anticariensis*, which might have happened in one of the 23S or 16S rRNA genes.

Concerning the genus *Chromohalobacter*, all the species described within this genus clustered together (the mean 16S/23S rRNA similarity of this group was 98.0/97.8%) with the only exception of *Chromohalobacter salarius*, as previously commented. Therefore, the taxonomic position of *Chromohalobacter* as an independent genus within the family *Halomonadaceae* was confirmed. Regarding the genus *Modicisalibacter*, unlike the 16S rRNA, 23S rRNA gene sequence analysis placed the single species of this genus, *M. tunisiensis*, into the group 1 of *Halomonas*, suggesting a horizontal gene transfer occurred with another species of the genus *Halomonas*. Finally, the phylogenetic distinctness of the rest of the genera included in the family *Halomonadaceae* (*Carnimonas*, *Cobetia*, *Halotalea*, *Kushneria*, *Salinicola* or *Zymobacter*) was confirmed in this study, being stable in all the trees produced from all methods of analysis (Fig. 3.1).

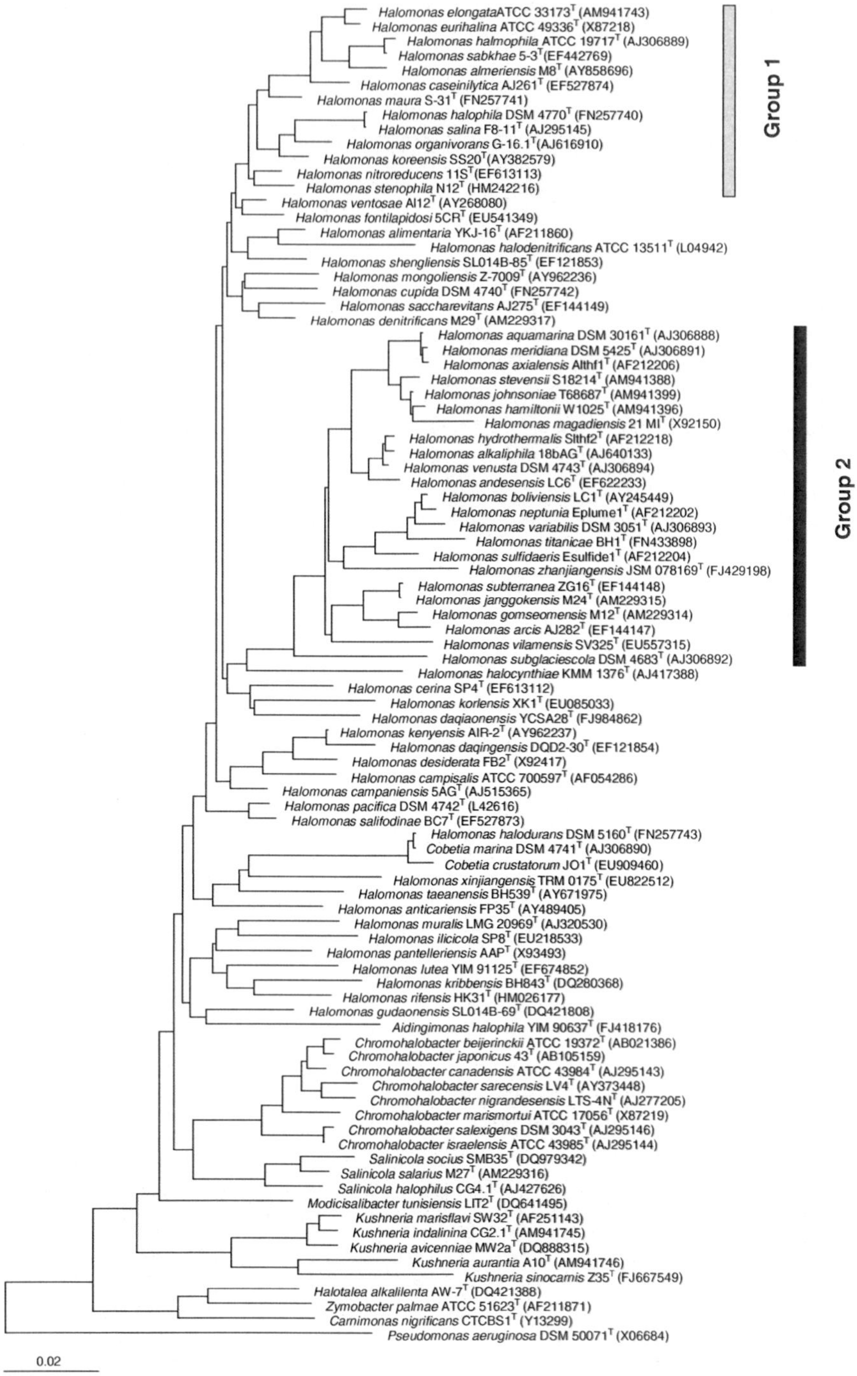

Fig. 3.1 Neighbour-joining phylogenetic tree, based on the 16S rRNA gene sequence comparison, showing the phylogenetic relationship among members of the family *Halomonadaceae*.

In the meanwhile, a new genus and 11 new species have been proposed within the family *Halomonadaceae*: *Aidingimonas halophila* (bearing a 95% or lower 16S rRNA sequence similarity with respect to the other members of the family; Wang et al. 2009), *Cobetia crustatorum* (98.9% 16S rRNA sequence similarity with *Cobetia marina*; Kim et al. 2010b), *Halomonas andesensis* (which falls into group 2 of *Halomonas*; Guzmán et al. 2010), *Halomonas fontilapidosi* (its closest relative is *H. ventosae*; González-Domenech et al. 2009), *Halomonas hamiltonii*, *H. johnsoniae*, and *H. stevensii* (which fall into group 2 of *Halomonas*; Kim et al. 2010a), *Halomonas ilicicola* (which cannot be ascribed to any of the groups of *Halomonas*; Arenas et al. 2009), *Halomonas titanicae* (which also fall into group 2 of *Halomonas*; Sánchez-Porro et al. 2010), *Halomonas xinjiangensis* (its closest relative is *H. anticariensis*; Guan et al. 2010), *Halomonas zhanjiangensis* (again included into group 2 of *Halomonas*; Chen et al. 2009), and *Kushneria sinocarnis* (its closest relative is *K. aurantia*, exhibiting 93.4% and 95.3% sequence similarity for 23S and 16S rRNA, respectively; Zou and Wang 2010). Besides, up to 31 January 2011, four other new species of *Halomonas* are listed as paper in press in the International Journal of Systematic and Evolutionary Microbiology (http://ijs.sgmjournals.org/papbyrecent.dtl): *H. daqiaonensis* (the most closely related species is *H. ventosae*; Qu et al. 2011), *H. rifensis* (96.5% 16S rRNA sequence similarity with *H. anticariensis*; Amjres et al. 2011), *H. stenophila* (which can be ascribed to the group 1 of *Halomonas*; Llamas Company et al. 2011), and *H. vilamensis* (clustering within the group 2 of *Halomonas*; Menes et al. 2011) (Fig. 3.1).

From the phylogenetic point of view, although 23S and 16S rRNA genes allow to observe some well-defined relationships within members of the *Halomonadaceae*, they still do not permit to resolve the phylogenetic relationships of very closely related species. Besides, the continuous and rapidly increasing in the numbers of genera and species within this family make it even more complicated.

3.2.2 *Multilocus Sequence Analysis*

The inclusion of protein-encoding genes, called "housekeeping genes," as alternative phylogenetic markers to rRNA genes has been suggested to carry out a multilocus sequence analysis (MLSA), which has been developed for species circumscription in many taxa and may help to clarify the phylogenetic relationships between the members of the family *Halomonadaceae* (Stackebrandt et al. 2002; Zeigler 2003; Arahal et al. 2007; Tindall et al. 2010). The recommended features

Fig. 3.1 (Continued) An explanation about the methodology followed for sequence analysis can be found in de la Haba et al. (2010a). The accession numbers of the sequences used are shown in parentheses after the strain designation. *Pseudomonas aeruginosa* DSM 50071^T was used as outgroup. *Halomonas* species that can be ascribed to the 16S rRNA groups 1 and 2 defined by de la Haba et al. (2010a) have been marked accordingly. The scale bar represents 0.02 substitutions per nucleotide position that should possess the housekeeping genes to be suitable for MLSA are: wide distribution among the prokaryotes, single copy, adequate length to be phylogenetically informative, located separately on the main chromosome, and relatively high degree of conservation, but faster evolutionary rate than ribosomal genes (Zeigler 2003). The phylogenies inferred based on housekeeping genes are not always in agreement with 16S rRNA gene sequence phylogeny (Martens et al. 2007).

The first attempt of MLSA within the *Halomonadaceae* was performed by Okamoto et al. (2004), but their studies were focused on a single species. They determined the sequences of *gyrB*, *ectB*, and *ectC* genes of ten strains of *Halomonas variabilis* finding that phylogenetic trees based on *gyrB* and *ectB* genes were very similar to those based on 16S rRNA gene, but the *ectC*-based trees were inconsistent with the other topologies, so *ectC* was inferred to derive from horizontal transfer.

Very recently, de la Haba et al. (2011) carried out a MLSA study of the family *Halomonadaceae* based on 16S rRNA, 23S rRNA, *atpA*, *gyrB*, *rpoD*, and *secA* gene sequences from 52 representative species. They retrieved from public databases the 16S and 23S rRNA gene sequences recommended by de la Haba et al. (2010a) and developed specific PCR primers for the other four genes. Multiple sequence alignments were made taking into account the corresponding amino acid alignments for protein-coding genes. Not only were several treeing methods used to elucidate the most stable topology, but also branch support and statistical tests. Data showed that *secA* gene had the highest evolutionary rate, followed by *rpoD*, *gyrB*, and *atpA*; finally, 23S and 16S rRNA genes were the most evolutionary conserved. Sequence similarity values demonstrated the enormous heterogeneity within the genus *Halomonas*, presenting a large overlap between intrageneric and intergeneric similarities, which suggested that it might be divided into two or more genera; however, no clearly differential phenotypic, chemotaxonomic and genotypic features could be found to support this division.

With respect to the phylogenetic trees, statistically significant differences between the topologies of the six studies genes were found. This was probably due to lateral gene transfer events, indicating the important role of recombination in the evolution of members of the family *Halomonadaceae*. However, most genera and intrageneric groups were conserved regardless of the gene or algorithm used, confirming the strong stability of these groups. Except *Halomonas* and *Modicisalibacter*, all the other analyzed genera within the *Halomonadaceae* (*Carnimonas*, *Chromohalobacter*, *Cobetia*, *Halotalea*, *Kushneria*, *Salinicola*, and *Zymobacter*) were phylogenetically coherent.

Analysis of the six concatenated gene sequences minimized the impact of recombination events and phylogenetically was in agreement with the taxonomic scheme of this family. Therefore, concatenation of all six genes was shown to be a very good tool for delineation at a broad scale of taxonomic relationships at the family *Halomonadaceae* level.

Due to the large set of *Halomonadaceae* species included in this study, only the type strain of each species was analyzed, being impossible to determinate the intraspecies topology and the intraspecific resolution capability of the genes. Therefore, further studies focused on a specific group of this family and including both type and reference strains seem necessary. In addition, future MLSA approaches should determine interspecies and intraspecies DNA–DNA hybridization values under the same methodology and laboratory conditions and correlate these data against the MLSA results in order to validate the usefulness of MLSA as an alternative to DNA–DNA hybridization for species circumscription within the family *Halomonadaceae*.

3.3 Taxonomy of the Family *Halomonadaceae*

Species of the family *Halomonadaceae*, together with the families *Oceanospirillaceae*, *Alcanivoraceae*, *Hahellaceae*, *Litoricolaceae*, *Oleiphilaceae*, and "*Saccharospirillaceae*" are members of the order *Oceanospirillales* within the class *Gammaproteobacteria* (Garrity et al. 2005a; Kim et al. 2007a), and they are related to genera such as *Marinospirillum*, *Marinomonas*, *Neptunomonas* or *Oceanospirillum*.

The family *Halomonadaceae* was firstly proposed by Franzmann et al. (1988) and later emended by Dobson and Franzmann (1996), Ntougias et al. (2007), and Ben Ali Gam et al. (2007). The oldest species described and later included within this family were *Pseudomonas beijerinckii* Hof 1935 (currently *Chromohalobacter beijerinckii*), "*Chromobacterium marismortui*" Elazari-Volcani 1940 (*Chromohalobacter marismortui*), "*Arthrobacter marinus*" Cobet et al. 1970 (earlier synonym of *Cobetia marina*), "*Achromobacter aquamarinus*" ZoBell and Upham 1944 (now *Halomonas aquamarina*), *Flavobacterium halmophilum* Elazari-Volcani 1940 (basonym of *Halomonas halmophila*), and "*Micrococcus halodenitrificans*" Robinson and Gibbons 1952 (*Halomonas halodenitrificans*).

In 1972, Baumann et al. carried out an extensive characterization of Gram-negative, motile, non-fermentative marine bacteria which lead to the description of the species *Alcaligenes aestus*, *A. cupidus*, *A. pacificus*, and *A. venustus* as well as to the reclassification of *Arthrobacter marinus* as *Pseudomonas marina*. Some years later, Baumann et al. (1983) proposed the description of the genus *Deleya* to place these five species within this new genus. Only 3 years earlier Vreeland et al. (1980) had described the genus *Halomonas*, with the moderately halophilic bacterium, *H. elongata*. Further studies by Franzmann et al. (1988) concluded with the creation of the family *Halomonadaceae* to accommodate the genera *Halomonas* and *Deleya*, which, respectively, contained at that time four and six species.

After that, several new genera closely related to *Halomonas* and *Deleya* were proposed, all of them containing a single species: *Halovibrio*, with the species *H. variabilis* (Fendrich 1988); *Chromohalobacter*, including the species *C. marismortui* previously named "*Chromobacterium marismortui*" (Ventosa et al. 1989); *Volcaniella*, comprising *V. eurihalina* (Quesada et al. 1990); and

Zymobacter, with *Z. palmae* as the single species (Okamoto et al. 1993). Subsequently, 16S rRNA gene sequence phylogenetic analysis permitted the placement of the species of the genus *Chromohalobacter* into the family *Halomonadaceae* and the reclassification of *Volcaniella eurihalina* into the genus *Halomonas* (Mellado et al. 1995b). Later, Dobson and Franzmann (1996) transferred the species of the genus *Deleya*, as well as *Halovibrio variabilis* and *Paracoccus halodenitrificans* (originally described as "*Micrococcus halodenitrificans*"), to the genus *Halomonas*, resulting in an increase in the number of *Halomonas* species to 15; besides, in this study they proposed the placement of the genus *Zymobacter* into the family *Halomonadaceae*, that together with the genera *Chromohalobacter* and *Halomonas* conformed this family.

The genus *Carnimonas*, containing one single species isolated from cured meat products, *C. nigrificans* (Garriga et al. 1998), was initially excluded from the family *Halomonadaceae*, but Arahal et al. (2002b) proposed that it should be classified within this family. The same year, the species *Halomonas marina* (initially described as "*Arthrobacter marinus*" and subsequently reclassified as *Pseudomonas marina* and *Deleya marina*) was transferred into the genus *Cobetia*, as *C. marina* (Arahal et al. 2002a) which was also included as a member of this family.

In 2007, three new genera belonging to the family *Halomonadaceae* were created. Ntougias et al. (2007) isolated a strain from alkaline sludge-like wastes ("alpeorujo" or "alperujo") of two-phase olive oil extraction, characterization of which concluded with the creation of a new genus and species, *Halotalea alkalilenta*. These authors also emended the description of the family *Halomonadaceae* to include this organism in it. Just 1 month later, Ben Ali Gam et al. (2007) carried out a screening from an oilfield-water injection sample, result of which a new genus of this family was proposed, *Modicisalibacter*, with a single species, *M. tunisiensis*. They also redefined the current 16S rRNA signature nucleotide characteristics of the family as follows: position 484 (A or G), position 486 (C or U), position 640 (A or G), position 660 (A), position 668 (A), position 669 (A), position 737 (U), position 738 (U), position 745 (U), position 776 (U), position 1124 (U or G), position 1297 (U), position 1298 (C), position 1423 (A), position 1424 (C or U), position 1439 (U or C), position 1462 (A or G), and position 1464 (C or U). Last of all, the genus *Salinicola*, with the only species *S. socius*, was described and placed into the *Halomonadaceae* (Anan'ina et al. 2007). Recently, de la Haba et al. (2010b) transferred the species *Halomonas salaria* and *Chromohalobacter salarius* to the genus *Salinicola*, as *S. salarius* and *S. halophilus*, respectively, on the basis on 23S and 16S rRNA gene sequence analysis in addition to other phenotypic, chemotaxonomic and genotypic features.

Sánchez-Porro et al. (2009) investigated the microbiota from the surface of leaves of the black mangrove *Avicennia germinans*. One of the strains was subjected to a polyphasic taxonomic study, which concluded with the description of *Kushneria aurantia*, a novel genus and species of the family *Halomonadaceae*, as well as a proposal for reclassification of *Halomonas marisflavi* as *Kushneria marisflavi*, of *Halomonas indalinina* as *Kushneria indalinina* and of *Halomonas avicenniae* as *Kushneria avicenniae*. Finally, the last genus included within the

Halomonadaceae was *Aidingimonas*, comprising one species, *A. halophila*, described based on two isolates (Wang et al. 2009).

As of January 2011, there were ten validly published genera names within the family *Halomonadaceae*: *Halomonas* (the type genus of the family, containing 63 species), *Chromohalobacter* (which comprises eight species, including the reclassified *Halomonas canadensis*, *Halomonas israelensis* [Arahal et al. 2001a], and *Pseudomonas beijerinckii* [Peçonek et al. 2006]), *Kushneria* (five species), *Salinicola* (grouping three species), *Cobetia* (two species), *Aidingimonas*, *Carnimonas*, *Halotalea*, *Modicisalibacter* and *Zymobacter* (each containing only one species). Moreover, the list of articles in press of the International Journal of Systematic and Evolutionary Microbiology (http://ijs.sgmjournals.org/papbyrecent.dtl) includes the proposals of four new *Halomonas* species for which the following names have been proposed: *H. vilamensis*, *H. daqiaonensis*, *H. stenophila*, and *H. rifensis*. A list of all these species, including additional data, is shown in Table 3.1. At the time of writing, and to the best of our knowledge, there are descriptions of six novel species of *Halomonas* whose names have not yet been validated: "*H. glaciei*" (Reddy et al. 2003), "*H. alkaliantarctica*" (Poli et al. 2007), "*H. sinaiensis*" (Romano et al. 2007), "*H. profundus*" (Simon-Colin et al. 2008), "*H. sediminis*" (Huang et al. 2008), and "*H. chromatireducens*" (Shapovalova et al. 2009). Sorokin and Tindall (2006) mentioned an existent problem between two strains of halophilic bacteria, *Pseudomonas halophila* DSM 3050^{T} and *Halomonas variabilis* DSM 3051^{T}. They demonstrated that the characteristics of strain DSM 3050^{T} corresponded to the original description of *Halomonas variabilis* and those of DSM 3051^{T} to *Pseudomonas halophila*. If the Judicial Commission accepts this proposal, (a) the type strain of the species *Halomonas variabilis* should be DSM 3050 (instead of DSM 3051) and the name *Halomonas variabilis* should be rejected and (b) the type strain of *Pseudomonas halophila* should be DSM 3051 and it should be transferred to the genus *Halomonas* as *H. utahensis* nom. nov.

In addition, other two taxonomically important studies concerning the family *Halomonadaceae* should be taken into consideration. On the one hand, a chemotaxonomic characterization (Franzmann and Tindall 1990), which concluded that on the basis of respiratory quinone, polar lipid, and fatty acid compositions, no clear distinction existed at the genus level. On the other hand, the article of Mata et al. (2002), presenting a detailed phenotypic characterization of the type strains of all *Halomonas* species recognized at that time and the intraspecific variation of four of those species by studying 87 additional strains. They compared a total of 234 morphological, physiological, biochemical, nutritional and antimicrobial susceptibility tests and demonstrated the phenotypic heterogeneity of the *Halomonas* species. Since it is the largest and more complete phenotypic study ever conducted in the *Halomonadaceae*, it should be taken as a model for future taxonomic description within this bacterial group.

An invaluable resource to guide authors in preparing proposal of novel taxa are the recommended minimal standards for describing new taxa within the family *Halomonadaceae* (Arahal et al. 2007), which have been endorsed by the

Table 3.1 Validly published genera and species names of the family *Halomonadaceae* (as to 31 January 2011)

Genus and species name	References	Type strain designation(s)
Aidingimonas		
A. halophila[a]	Wang et al. (2009)	YIM 90637 = CCTCC AB 207002 = KCTC 12885
Carnimonas		
Car. nigrificans[a]	Garriga et al. (1998)	CTCBS1 = ATCC BAA-78 = CECT 4437 = CIP 105703
Chromohalobacter		
Chr. beijerinckii	Peçonek et al. (2006)	ATCC 19372 = CCUG 49679 = CIP 106957 = DSM 7218 = JCM 13305 = JCM 21422 = LMG 2148 = NBRC 103041 = NCCB 35008 = NCIMB 9041 = NRRL B-3153
Chr. canadensis	Arahal et al. (2001a)	ATCC 43984 = CECT 5385 = CCM 4919 = CIP 105571 = DSM 6769 = LMG 19547 = NCIMB 13767 = NRCC 41227
Chr. israelensis	Arahal et al. (2001a)	Ba1 = ATCC 43985 = CECT 5287 = CCM 4920 = CIP 106853 = DSM 6768 = LMG 19546 = NCIMB 13766
Chr. japonicus	Sánchez-Porro et al. (2007)	43 = CCM 7416 = CECT 7219
Chr. marismortui[a]	Ventosa et al. (1989)	CCM 3518 = ATCC 17056 = DSM 6770 = JCM 21220 = LMG 3935 = NBRC 103155
Chr. nigrandesensis	Prado et al. (2006)	LTS-4N = CECT 5315 = DSM 14323
Chr. salexigens	Arahal et al. (2001b)	ATCC BAA-138 = CECT 5384 = CCM 4921 = CIP 106854 = DSM 3043 = NCIMB 13768 = 1H11
Chr. sarecensis	Quillaguamán et al. (2004a)	LV4 = ATCC BAA-761 = CCUG 47987 = DSM 15547
Cobetia		
Cob. crustatorum	Kim et al. (2010b)	JO1 = JCM 15644 = KCTC 22486
Cob. marina[a]	Arahal et al. (2002a)	219 = ATCC 25374 = CCUG 49558 = CCUG 49558 = CECT 4278 = CIP 104765 = DSM 4741 = LMG 2217 = NBRC 102605 = NCIMB 1877
Halomonas[b]		
H. alimentaria	Yoon et al. (2002)	YKJ-16 = DSM 15356 = KCCM 41042 = JCM 10888
H. alkaliphila	Romano et al. (2006)	18bAG = ATCC BAA-953 = DSM 16354
H. almeriensis	Martínez-Checa et al. (2005)	M8 = CECT 7050 = LMG 22904
H. andesensis	Guzmán et al. (2010)	LC6 = CCUG 54844 = DSM 19434 = LMG 24243
H. anticariensis	Martínez-Cánovas et al. (2004a)	FP35 = CECT 5854 = LMG 22089
H. aquamarina	Dobson and Franzmann (1996)	ZoBell and Upham 558 = ATCC 14400 = CCUG 16157 = CIP 105454 = DSM 30161 = IAM 12550 = LMG 2853 = NCIMB 557

(continued)

Table 3.1 (continued)

Genus and species name	References	Type strain designation(s)
H. arcis	Xu et al. (2007)	AJ282 = CGMCC 1.6494 = JCM 14607 = LMG 23978
H. axialensis	Kaye et al. (2004)	Althf1 = ATCC BAA-802 = CECT 5812 = DSM 15723
H. boliviensis	Quillaguamán et al. (2004b)	LC1 = ATCC BAA-759 = DSM 15516
H. campaniensis	Romano et al. (2005)	5AG = ATCC BAA-966 = DSM 15293
H. campisalis	Mormile et al. (1999)	4A = ATCC 700597 = CIP 106639
H. caseinilytica	Wu et al. (2008b)	AJ261 = CGMCC 1.6773 = JCM 14802
H. cerina	González-Domenech et al. (2008b)	SP4 = CECT 7282 = LMG 24145
H. cupida	Dobson and Franzmann (1996)	79 = ATCC 27124 = CCUG 16075 = CIP 103199 = DSM 4740 = JCM 20632 = LMG 3448 = NBRC 102219
H. daqiaonensis	Qu et al. (2011)	YCSA28 = CGMCC 1.9150 = NCCB 100305 = MCCC 1B00920
H. daqingensis	Wu et al. (2008a)	DQD2-30 = CGMCC 1.6443 = LMG 23896
H. denitrificans	Kim et al. (2007b)	M29 = DSM 18045 = KCTC 12665
H. desiderata	Berendes et al. (1996)	FB2 = CIP 105505 = DSM 9502 = LMG 19548
H. elongata[a]	Vreeland et al. (1980)	1H9 = ATCC 33173 = CIP 104264 = DSM 2581 = NBRC 15536 = JCM 21044 = LMG 9076
H. eurihalina	Mellado et al. (1995b)	F9-6 = ATCC 49336 = CIP 106091 = DSM 5720
H. fontilapidosi	González-Domenech et al. (2009)	5CR = CECT 7341 = LMG 24455
H. gomseomensis	Kim et al. (2007b)	M12 = DSM 18042 = KCTC 12662
H. gudaonensis	Wang et al. (2007b)	SL014B-69 = CGMCC 1.6133 = LMG 23610
H. halmophila	Dobson et al. (1990)	ACAM 71 = ATCC 19717 = CIP 105455 = DSM 5349 = NBRC 15537 = JCM 21222 = LMG 4023 = NCIMB 1971
H. halocynthiae	Romanenko et al. (2002)	KMM 1376 = DSM 14573 = CIP 107736
H. halodenitrificans	Dobson and Franzmann (1996)	ATCC 13511 = CIP 105456 = DSM 735 = CCM 286 = CECT 5012 = IAM 13950 = KCTC 5069 = NBRC 14912
H. halodurans	Hebert and Vreeland (1987)	ATCC 29686 = DSM 5160 = NBRC 15607 = LMG 10144
H. halophila	Dobson and Franzmann (1996)	F5-7 = ATCC 49969 = CCM 3662 = CIP 103512 = DSM 4770 = JCM 20791 = LMG 6456 = NBRC 102604
H. hamiltonii	Kim et al. (2010a)	W1025 = DSM 21196 = KCTC 22154
H. hydrothermalis	Kaye et al. (2004)	Slthf2 = ATCC BAA-800 = CECT 5814 = DSM 15725
H. ilicicola	Arenas et al. (2009)	SP8 = CCM 7522 = CECT 7331 = DSM 19980
H. janggokensis	Kim et al. (2007b)	M24 = KCTC 12663 = DSM 18043
H. johnsoniae	Kim et al. (2010a)	T68687 = DSM 21197 = KCTC 22157
H. kenyensis	Boltyanskaya et al. (2007)	AIR-2 = DSM 17331 = VKM B-2354

(continued)

Table 3.1 (continued)

Genus and species name	References	Type strain designation(s)
H. koreensis	Lim et al. (2004)	SS20 = JCM 12237 = KCTC 12127
H. korlensis	Li et al. (2008)	XK1 = CGMCC 1.6981 = DSM 19633
H. kribbensis	Jeon et al. (2007)	BH843 = DSM 17892 = KCTC 12584
H. lutea	Wang et al. (2008a)	YIM 91125 = CCTCC AB 206093 = KCTC 12847
H. magadiensis	Duckworth et al. (2000)	21 MI = CIP 106823 = CIP 106874 = DSM 15367 = NCIMB 13595
H. maura	Bouchotroch et al. (2001)	S-31 = ATCC 700995 = CECT 5298 = DSM 13445
H. meridiana	James et al. (1990)	ACAM 246 = ATCC 49692 = CIP 104043 = DSM 5425 = NBRC 15608 = UQM 3352
H. mongoliensis	Boltyanskaya et al. (2007)	Z-7009 = DSM 17332 = VKM B-2353
H. muralis	Heyrman et al. (2002)	LMG 20969 = CIP 108825 = DSM 14789
H. neptunia	Kaye et al. (2004)	Eplume1 = ATCC BAA-805 = CECT 5815 = DSM 15720
H. nitroreducens	González-Domenech et al. (2008a)	11S = CECT 7281 = LMG 24185
H. organivorans	García et al. (2004)	G-16.1 = CCM 7142 = CECT 5995
H. pacifica	Dobson and Franzmann (1996)	62 = ATCC 27122 = CIP 103200 = DSM 4742 = JCM 20633 = LMG 3446 = NBRC 102220
H. pantelleriensis	Romano et al. (1996)	AAP = ATCC 700273 = CIP 105506 = DSM 9661 = LMG 19550
H. rifensis	Amjres et al. (2011)	HK31 = CECT 7698 = LMG 25695
H. sabkhae	Kharroub et al. (2008)	5–3 = CECT 7246 = DSM 19122 = LMG 24084
H. saccharevitans	Xu et al. (2007)	AJ275 = CGMCC 1.6493 = JCM 14606 = LMG 23976
H. salifodinae	Wang et al. (2008b)	BC7 = CGMCC 1.6774 = JCM 14803
H. salina	Dobson and Franzmann (1996)	F8-11 = ATCC 49509 = CIP 106092 = DSM 5928 = JCM 21221
H. shengliensis	Wang et al. (2007a)	SL014B-85 = CGMCC 1.6444 = LMG 23897
H. stenophila	Llamas Company et al. (2011)	N12 = CECT 7744 = LMG 25812
H. stevensii	Kim et al. (2010a)	S18214 = DSM 21198 = KCTC 22148
H. subglaciescola	Franzmann et al. (1987)	ACAM 12 = ATCC 43668 = CIP 104042 = DSM 4683 = NBRC 14766 = JCM 21045 = LMG 8824 = UQM 2926
H. subterranea	Xu et al. (2007)	ZG16 = CGMCC 1.6495 = JCM 14608 = LMG 23977
H. sulfidaeris	Kaye et al. (2004)	Esulfide1 = ATCC BAA-803 = CECT 5817 = DSM 15722
H. taeanensis	Lee et al. (2005)	BH539 = DSM 16463 = KCTC 12284
H. titanicae	Sánchez-Porro et al. (2010)	BH1 = ATCC BAA-1257 = CECT 7585 = JCM 16411 = LMG 25388
H. variabilis	Dobson and Franzmann (1996)	Isolate III = ATCC 49240 = CIP 105504 = DSM 3051 = IAM 14440 = JCM 21223 = NBRC 102410

(continued)

Table 3.1 (continued)

Genus and species name	References	Type strain designation(s)
H. ventosae	Martínez-Cánovas et al. (2004b)	Al12 = CECT 5797 = DSM 15911
H. venusta	Dobson and Franzmann (1996)	86 = ATCC 27125 = CCUG 16063 = CIP 103201 = DSM 4743 = JCM 20634 = LMG 3445 = NBRC 102221
H. vilamensis	Menes et al. (2011)	SV325 = DSM 21020 = LMG 24332
H. xinjiangensis	Guan et al. (2010)	TRM 0175 = CCTCC AB 208329 = KCTC 22608
H. zhanjiangensis	Chen et al. (2009)	JSM 078169 = CCTCC AB 208031 = DSM 21076 = KCTC 22279
Halotalea		
Halot. alkalilenta[a]	Ntougias et al. (2007)	AW-7 = CECT 7134 = DSM 17697
Kushneria		
K. aurantia[a]	Sánchez-Porro et al. (2009)	A10 = CCM 7415 = CECT 7220
K. avicenniae	Sánchez-Porro et al. (2009)	MW2a = CCM 7396 = CECT 7193 = CIP 109711
K. indalinina	Sánchez-Porro et al. (2009)	CG2.1 = CECT 5902 = CIP 109528 = DSM 14324 = LMG 23625
K. marisflavi	Sánchez-Porro et al. (2009)	SW32 = CIP 107103 = DSM 15357 = JCM 10873 = KCCM 80003
K. sinocarnis	Zou and Wang (2010)	Z35 = CCTCC AB 209027 = DSM 23229 = NRRL B-59197
Modicisalibacter		
M. tunisiensis[a]	Ben Ali Gam et al. (2007)	LIT2 = CCUG 52917 = CIP 109206
Salinicola		
S. halophilus	de la Haba et al. (2010b)	CG4.1 = CECT 5903 = LMG 23626
S. salarius	de la Haba et al. (2010b)	M27 = DSM 18044 = KCTC 12664
S. socius[a]	Anan'ina et al. (2007)	SMB35 = DSM 19940 = VKM B-2397
Zymobacter		
Z. palmae[a]	Okamoto et al. (1993)	T109 = ATCC 51623 = DSM 10491 = IAM 14233 = JCM 21091 = NBRC 102412

Basonyms/synonyms of microorganisms that have been transferred to other genera are not included. For genera/species whose descriptions have been emended only the most recent reference is included

Abbreviations of culture collections are: *ACAM* Australian Collection of Antarctic Microorganisms; *ATCC* American Type Culture Collection; *CCM* Czech Collection of Microorganisms; *CCTCC* China Center for Type Culture Collection; *CCUG* Culture Collection University of Göteborg; *CECT* Colección Española de Cultivos Tipo; *CIP* Collection de L'Institut Pasteur; *CGMCC* China General Microbiological Culture Collection Center; *DSM* Deutsche Sammlung von Mikroorganismen und Zellkulturen; *IAM* Institute of Applied Microbiology; *JCM* Japan Collection of Microorganisms; *KCCM* Korean Culture Center of Microorganisms; *KCTC* Korean Collection for Type Cultures; *KMM* Collection of Marine Microorganisms; *LMG* Belgian Co-ordinated Collections of Microorganisms; *MCCC* Marine Culture Collection of China; *NBRC* NITE Biological Resource Center; *NCCB* The Netherlands Culture Collection of Bacteria; *NCIMB* The National Collection of Industrial, Marine and Food Bacteria; *NRCC* National Research Council of Canada; *NRRL* Agricultural Research Service Culture Collection; *UQM* Australian Collection of Microorganisms; *VKM* All-Russian Collection of Microorganisms

[a]Type species of the genus

[b]Type genus of the family

International Committee on Systematics of Prokaryotes-Subcommittee on the Taxonomy of *Halomonadaceae*. The recommendations given were aimed at preventing the establishment of insufficiently characterized "new" taxa, which are later difficult to handle by other microbiologists and quite often represent a source of confusion in taxonomic classification. Obviously, there is no intent to limit the characterization of new isolates to the characteristics that are indicated. Moreover, these standards are not compulsory, but adhering to them will benefit all. This article does not only present a set of suggested traits, but also the techniques and methodologies that should be employed. The adoption of polyphasic approaches that integrate phenotypic (including chemotaxonomic markers) with genotypic and phylogenetic methods is proposed and researchers are encouraged to describe species based on more than a single strain.

3.4 Interest of the Family *Halomonadaceae*

Member of the family *Halomonadaceae* have promising applications in biotechnology as a source of enzymes, compatible solutes, biosurfactants or exopolysaccharides among other products. Moreover, they offer a number of advantages to be used as cell factories for the production of recombinant proteins. They present some characteristics, such as high salt tolerance, decreasing to a minimum the necessity for aseptic conditions, resulting in cost-reducing conditions; most of them are very easy to grow and maintain in the laboratory and their nutritional requirements are simple being the majority able to use a large range of compounds as the sole carbon and energy sources (Ventosa and Nieto 1995; Ventosa et al. 1998). Efforts of different researchers have made possible the genetic manipulation of this bacterial group. The presence of plasmids (Fernandez-Castillo et al. 1992; Mellado et al. 1995a; Vargas et al. 1995, 1999a, b; Argandoña et al. 2003) and the isolation of stable mutants (Kunte and Galinski 1995; Cánovas et al. 1997a, b) have been described. The physical sizes of the chromosome of different species (Mellado et al. 1998; Llamas et al. 2002) have also been reported. Shuttle vectors have been constructed (Mellado et al. 1995c; Vargas et al. 1995) and transposon mutagenesis procedures have been established (Llamas et al. 2000). The influence of salt concentration on the susceptibility of moderately halophilic bacteria to antimicrobial compounds and their potential use for genetic transfer studies has been reported (Coronado et al. 1995) and also a few reporter genes have been described to be used in this family such as the ice nucleation gene *inaZ* or the *gfp* gene (Arvanitis et al. 1995; Tegos et al. 1997, 2000; Douka et al. 2001). Few complete genomes are currently available, including *Halomonas elongata* DSM 2581^{T} (Schwibbert et al. 2010) and *Chromohalobacter salexigens* DSM 3043^{T} (Ates et al. 2011), although several other are in progress. Members of *Halomonadaceae* are considered among the most interesting model organisms on moderately halophilic bacteria for the study of halophilism due to the wide range of NaCl requirements and the halophilic nature of some members of this family.

Besides their interest as model microorganisms for genomics and system biology studies, the members of the family *Halomonadaceae* have important applications and biotechnological potential. The species of the genera *Aidingimonas*, *Modicisalibacter*, *Halotalea* and *Salinicola* have not been described in any biotechnological application (Ntougias et al. 2007; Ben Ali Gam et al. 2007; Wang et al. 2009; Anan'ina et al. 2007) probably due to the fact that all these taxa have been described in recent years and no further studies have been carried out with them. We will review some interesting biotechnological features and potentialities of the others members of *Halomonadaceae*.

Members of the family *Halomonadaceae* have been isolated from a wide variety of natural habitats, such as marine and hypersaline waters, saline soils, salterns, foods, plants, animals, and oilfields. However, we should consider that very recently some *Halomonas* strains have been recognized as human pathogens, causing infections and contamination in a dialysis center, and later described as three new species of this genus: *H. stevensii*, *H. hamiltonii*, and *H. johnsoniae* (Stevens et al. 2009; Kim et al. 2010a).

3.4.1 Food Industry

The majority of members of the family *Halomonadaceae* have been isolated from marine or hypersaline environments (Ventosa et al. 1998) and rarely from cured-meat products. Six species of the family *Halomonadaceae* have been isolated from salted food products: *Halomonas halodenitrificans* was isolated from meat curing brines (Robinson and Gibbons 1952), *Halomonas alimentaria*, *Cobetia crustatorum*, and *Halomonas jeotgali* were isolated from jeotgal, a traditional Korean fermented seafood (Yoon et al. 2002; Kim et al. 2010b, c), *Chromohalobacter japonicus* was isolated for a Japanese salted food (Sánchez-Porro et al. 2007) and *Kushneria sinocarnis* was isolated from a piece of Chinese traditional salty pork that was cured with salt (Zou and Wang 2010). Although there are no specific studies about them, these and other related species could potentially be used to enhance flavor of salted foods.

Carnimonas nigrificans is the sole representative of the genus *Carnimonas* and it is responsible for the presence of black spot on the surface of food products in connection with organisms of different origin (Garriga et al. 1998). *Cobetia* sp. and *Halomonas* sp. have been identified as predominant spoilage bacteria in water-boiled salted duck during storage (Liu et al. 2010). Furthermore, a recent study carried out by Essghaier et al. (2009) used moderately halophilic bacteria, including *Halomona elongata*, in the biological control against the gray mold caused by *Botrytis cinerea*, an economically important disease of strawberries in Tunisia and worldwide. The use of such bacteria may constitute an important alternative to synthetic fungicides and can be exploited in commercial production and application under storage and greenhouse conditions.

The only representative of the genus *Zymobacter*, *Zymobacter palmae* is an ethanol-fermenting bacterium (Okamoto et al. 1993). It is a facultative anaerobe that ferments hexoses, α-linked di- and trisaccharides, and sugar alcohols (glucose, fructose, galactose, maltose, melibiose, sucrose, raffinose, mannitol and sorbitol). This organism produces approximately 2 mol of ethanol per mol of glucose (Okamoto et al. 1993). Pyruvate decarboxylase (PDC) is the key enzyme for all homo-fermentative ethanol pathways. This enzyme catalyzes the nonoxidative decarboxylation of pyruvate to acetaldehyde using Mg^{2+} and thiamine pyrophosphate (TPP) as cofactors. Acetaldehyde is reduced to ethanol by alcohol dehydrogenase (ADH) during NADH oxidation. ADH enzymes are widely distributed throughout nature (Reid and Fewson 1994). In contrast, PDC is common only to plants, yeasts and fungi; it is absent in animals and rare in prokaryotes (König 1998). *Z. palmae* has been reported to contain a PDC enzyme. It was originally isolated from palm sap and produces ethanol as a primary fermentation product from a variety of hexose sugars and saccharides (Okamoto et al. 1993). Its metabolic characteristics suggest it could serve as a useful new ethanol-fermenting bacterium, but its biotechnological exploitation would require certain genetic improvements. In that way, the PDC from *Z. palmae* has been cloned and characterized (Raj et al. 2002) and has been functionally expressed in *Lactococcus lactis* (Liu et al. 2005). Transformants of *Z. palmae* able to produce ethanol from cellobiose and xylose have also been obtained (Yanase et al. 2005, 2007). In recent studies, the putative three-dimensional structures of PDC dimer of *Z. palmae* PDC were generated based on the X-ray crystal structures of *Zymomonas mobilis* PDC, *Saccharomyces cerevisiae* PDC form-A and *Enterobacter cloacae* indolpyruvate decarboxylase in order to compare the quaternary structures of these microbial PCDs with respect to enzyme–substrate interactions, and subunit–subunit interfaces that might be related to the different biochemical characteristics. The models shed valuable information necessary for further improvement of PDC enzymes for industrial production of ethanol and other products (Shrestha et al. 2010).

3.4.2 Enzymes

The interest in salt-adapted enzymes has increased considerably in the last few years, and a variety of enzymes from halophilic microorganisms have been isolated and characterized, aiming at expansion of their commercial applications. Enzymes from halophilic microorganisms present the advantage to be stable over a range of salinities, constituting interesting catalysts from a biotechnological point of view. Several authors have previously reviewed the production of extracellular enzymes by moderately halophilic bacteria (Kamekura 1986; Ventosa et al. 1998; Mellado et al. 2003). Different screenings of bacteria from diverse areas of South Spain, Iran, and South China led to the isolation of moderately halophilic bacteria, including member of the family *Halomonadaceae* able to produce different extracellular hydrolytic enzymes (Sánchez-Porro et al. 2003; Rohban et al. 2009; Zhou et al. 2009). The only representative of the family *Halomonadaceae* able to produce an extracellular enzyme that has been studied in detail is *Halomonas meridiana*. This bacterium is able to produce an amylolytic enzyme that has optimal activity at 37°C, pH 7.0, being relatively stable under alkaline conditions, and 10% NaCl, although activity at salinity as high as 30% salts was reported. The main products resulting from the hydrolysis of starch were maltose and maltotriose (Coronado et al. 2000a). The *amyH* gene, encoding this α-amylase was cloned by functional complementation of a Tn1732-induced mutant deficient in extracellular amylase activity. *AmyH* encodes a 457-residue protein with a molecular mass of 50 kDa, assigned to family 13 of Henrissat's classification of glycosyl hydrolases (Henrissat and Bairoch 1993) and contains the four highly conserved regions in amylase enzymes. This enzyme has been cloned and expressed in the heterologous hosts *E. coli* and *Halomonas elongata* (Coronado et al. 2000b). Other genes of species of the family *Halomonadaceae* that have been cloned, expressed and characterized in *E. coli* are *cel8H*, which encodes a novel endoglucanase from *Halomonas* sp. S66-4 (Huang et al. 2010) or an alkaline phosphatase from *Halomonas* sp. 593 (Ishibashi et al. 2005, 2011).

Member of the family *Halomonadaceae* could serve as cell hosts for the production of heterologous proteins of biotechnological importance. In this sense, the heterologous expression of the *Bacillus licheniformis* α-amylase gene in *H. meridiana* and *H. elongata* has been reported (Coronado et al. 2000b). Moreover, an extracellular α-amylase gene from the hyperthermophilic archaeon *Pyrococcus woesei* has been expressed in *H. elongata*, under the control of a native *H. elongata* promoter (Frillingos et al. 2000). More recently, Rodríguez-Sáiz et al. (2007) expressed the β-carotene biosynthetic genes *crtE*, *crtY*, *cryI*, and *crtB* from *Pantoea agglomerans* and the cDNA encoding isopentenyl pyrophosphatase isomerase from *Haematococcus pluvialis* in *H. elongata*.

Studies carried out on the decolorization of textile azo dyes under high salt concentrations have permitted the isolation of *Halomonas* from effluents of textile industries (Asad et al. 2007) and coastal sediments contaminated by chemical wastewater (Guo et al. 2008).

An alkaline phosphatase (AP) with unusually high specific activity has been found to be produced by *Cobetia marina*. APs are widely found in various organisms, indicating their important role in metabolism of different phosphorus-containing organic compounds (Bjorkman and Karl 1994). The properties of this enzyme, such as a very high specific activity, no activation with divalent cations, resistance to high concentrations of inorganic phosphorus, as well as substrate specificity toward 5′ nucleotides make it advantageous as a tool for practical applications in comparison with the commercially available APs (Plisova et al. 2005).

3.4.3 *Exopolysaccharides*

Microbial exopolysaccharides (EPSs) have aroused great interest among biotechnologists because of their wide potential range of applications in fields such as medicine, pharmacy, foodstuff, cosmetics and the petroleum industry, where emulsifying, viscosifying, suspending, and chelating agents are required (Sutherland 1998). To date many species of the family *Halomonadaceae* have been described as producers of EPSs. Due to their great biotechnological interest, several studies related to the isolation of exopolysaccharide-producing bacteria from different areas such as Spain, India, Morocco and Chile have been carried out (Béjar et al. 1998; Bouchotroch et al. 1999; Martínez-Cánovas et al. 2004c; Joshi et al. 2008). The first studies on EPS´s in halophilic microorganisms were performed in *Halomonas maura* (Bouchotroch et al. 2001) and *Halomonas eurihalina* (Calvo et al. 2002). Other examples of *Halomonas* species able to produce EPSs are: *H. nitroreducens* (González-Domenech et al. 2008a), *H. cerina* (González-Domenech et al. 2008b), *H. sabkhae* (Kharroub et al. 2008), *H. alkaliphila* (Romano et al. 2006), *H. ventosae* (Martínez-Cánovas et al. 2004b; Mata et al. 2006), *H. anticariensis* (Martínez-Cánovas et al. 2004a; Mata et al. 2006), "*H. alkaliantarctica*" (Poli et al. 2007), *H. almeriensis* (Martínez-Checa et al. 2005) and the two recently described species, *H. rifensis* (Amjres et al. 2011), and *H. stenophila* (Ruiz-Ruiz et al. 2011; Llamas Company et al. 2011). Besides, other *Halomonas* strains able to produce EPSs are *Halomonas* strain CTSS (Poli et al. 2004), *Halomonas* sp. V3a′ (He et al. 2009), and *Halomonas* sp. strain TG39 (Gutierrez et al. 2009).

With respect to *Halomonas eurihalina,* the polymers best characterized so far are V2-7, H-96 and H-28 produced by strains F2-7, H-96 and H-28, respectively (Calvo et al. 1995, 1998; Béjar et al. 1996, 1998; Pérez-Fernández et al. 2000; Martínez-Checa et al. 2002, 2007; Llamas et al. 2003). These EPSs give moderately viscous solutions if they are suspended at neutral pH but the viscosity increases enormously in the acidic pH range (Calvo et al. 1995, 1998; Béjar et al. 1996). Another important property of the biopolymers synthesized by *H. eurihalina* is their ability to emulsify crude oil and other hydrocarbons much more effectively than do Tween 20, Tween 80 or Triton X-100 (Martínez-Checa et al. 2002, 2007). Pérez-Fernández et al. (2000) have described the effect of EPS V2-7 on the proliferation in vitro of human peripheral blood lymphocytes. The *carAB* genes, which encode carbamoyl-phosphate synthetase in *H. eurihalina* have been cloned and characterized; this enzyme is involved in the pathway for the synthesis of EPSs in strain F2-7 (Llamas et al. 2003). With respect to EPSs produced by *Halomonas maura* they have both viscosifying and emulsifying properties, particularly from strain S-30. EPS S-30, which has been designated mauran, gives highly viscous solutions and in addition is resistant to osmotic stress, changes in pH and to freezing–thawing processes (Arias et al. 2003). The *epsABCJ* genes that are involved in the biosynthesis of mauran have been cloned and characterized (Arco et al. 2005).

Recently, a highly exopolysaccharide-producing *Halomonas* strain has been reported as causing epizootics in larval cultures of the Chilean scallop *Argopecten purpuratus* (Rojas et al. 2009).

3.4.4 Compatible Solutes

Bacterial species within the family *Halomonadaceae* achieve osmotic equilibrium in hypersaline habitats by the synthesis and/or import of so-called "compatible solutes" since they provide osmotic balance without interfering with essential cellular processes and the normal metabolism. Although there is a relative diversity of compatible solutes across the kingdoms, moderate halophiles accumulate mainly some polyols (glucosylglycerol), amino acids (proline) and diamino acids, quaternary amines (betaines) and ectoines, which have been demonstrated to be the predominant compatible solutes accumulated by members of this family (Ventosa et al. 1998). Compatible solutes providing osmotic activities have the ability to protect proteins, membranes, and even whole cells in vivo and in vitro against denaturation, inactivation, and inhibition by heat or hyperosmotic stress. For these reasons, they have received much interest as general protecting agents with a wide range of biotechnological applications (Ventosa et al. 1998). The protective action of ectoines on biological compounds highlights their potential in fields related to molecular biology, agriculture, food processing, biotechnology, pharmacy, and medicine (Pastor et al. 2010). For example, it has been demonstrated that ectoine confers hyperosmotic tolerance in cultured tobacco cells and in ectoine-transformed tobacco plants (*Nicotiana tabacum*) (Nakayama et al. 2000; Moghaieb et al. 2006). In a recent review, the uses and biotechnological production of ectoines have been described in detail (Pastor et al. 2010).

The best-investigated compatible solute, ectoine, is biotechnologically produced by *Halomonas elongata* using a process named "bacterial milking", involving repetitive cycles of a fed-batch fermentation of *H. elongata* at 15% (w/v) NaCl to allow ectoine accumulation, followed by osmotic down shock at 3% (w/v) NaCl to release the osmolyte from the cells (Sauer and Galinski 1998). The physiology and genetics of ectoine biosynthesis in this bacterium have been studied in detail (Maskow and Babel 2001; Ono et al. 1999; Peters et al. 1990; Göller and Ofer 1998) as well as the glycine-betaine-synthesizing enzymes (Gadda and McAllister-Wilkins 2003). Also the compatible solute uptake system for ectoine accumulation (*teaABC*) from *H. elongata* has been characterized at the physiological and molecular level (Grammann et al. 2002; Tetsch and Kunte 2002). Furthermore, the structure of TeaA in complex with ectoine at a resolution of 1.55 Å has been described by Kuhlmann et al. (2008). Recently, a fourth *orf* (*teaD*) has been located adjacent to the *teaABC* locus encoding a putative universal stress protein (Schweikhard et al. 2010). Alternative processes have been reported for the production of these compatible solutes such as continuous synthesis and excretion of osmolytes by recombinant *E. coli* strains (Schubert et al. 2007; Bestvater et al. 2008), the production of ectoines by *Halomonas boliviensis* using a combined two-step fed-batch and milking process (Van-Thuoc et al. 2010a, b; Guzmán et al. 2009), *Halomonas salina* able to secrete ectoine at relatively low NaCl concentrations (Zhang et al. 2009), *Halomonas campisalis* observed to produce high concentration of compatible solutes (Aston and Peyton 2007), *Halomonas pantelleriensis* able to accumulated glycine betaine, ectoine, hydroxyectoine and glutamate (Romano et al. 2001) or *Cobetia marina* able to synthesize hydroxyectoine (Lang et al. 2009).

On the other hand, genetic engineering studies permitted the expression of the ectoine biosynthesis genes *ectABC* from *Halomonas* sp. BYS-1, or the ectoine synthase from *Halomonas* sp. Nj223 in *E. coli* (He et al. 2006; Chen et al. 2007). Also, the formation of hydroxyectoine in the industrial ectoine producer *H. elongata* has been improved by the heterologous expression of the ectoine hydroxylase gene *thpD* from *Streptomyces chrysomallus* (Prabhu et al. 2004). Nγ-acetyldiamino-butyrate, the precursor of the compatible solute ectoine, confers osmoprotection to *Salmonella enterica* serovar *typhimurium* (García-Estepa et al. 2006).

The moderate halophile *Chromohalobacter salexigens* is another interesting producer strain for ectoine and hydroxyectoine. The biochemical pathways of ectoine and hydroxyectoine synthesis have been described, but the details of the correspondent gene expression and enzyme regulation networks still remain to be fully elucidated (Cánovas et al. 1996, 1997a, 1998a, b, 1999, 2000; Calderón et al. 2004; García-Estepa et al. 2006; Vargas et al. 2008; Rodríguez-Moya et al. 2010). Recently, the synthesis and secretion of ectoine and hydroxyectoine using *Chromohalobacter salexigens* has been studied and optimized (Fallet et al. 2010). *Chromohalobacter salexigens* DSM 3043^{T} metabolism has been reconstructed based on genomic, biochemical and physiological information via a non-automated but interactive process. This work is the first comprehensive genome-scale metabolic model of a halophilic bacterium being a useful guide for identification and filling of knowledge gaps; the reconstructed metabolic network will accelerate the research on halophilic bacteria towards application of system biology approaches and design of metabolic engineering strategies (Ates et al. 2011). Recently, Schwibbert and coworkers (2011) reported the complete genome sequence of *Halomonas elongata* (~4 Mbp) and included experiments and analysis identifying and characterizing the entire ectoine metabolism. Based on the resulting genetic and biochemical data, a metabolic flux model of ectoine metabolism was derived that can be used to understand the way *H. elongata* survives under varying salt stresses and that provides a basis for a model-driven improvement of industrial ectoine production.

3.4.5 Polyhydroxyalkanoates

Polyhydroxyalkanoates (PHAs) are among the bioplastic biodegradable polymers which are attracting enormous attention due to increasing demand for environmentally compatible material from renewable resources. Thirty-two species of *Halomonas* are known to synthesize PHAs, although the capability of storing PHAs has not been determined in several species (Simon-Colin et al. 2008; Biswas et al. 2009; Quillaguamán et al. 2010). *Halomonas boliviensis* reaches PHA yields and volumetric productivities close to the highest reported so far. Furthermore,

H. boliviensis and other *Halomonas* species are able to co-produce PHA and osmolytes, i.e., ectoine and hydroxyectoine, in one single process (Quillaguamán et al. 2005, 2008). Production and characterization of a biodegradable polyhydroxybutyrate (PHB) by *Halomonas campisalis* has been described (Kulkarni et al. 2010). Recent studies reported the isolation of bacterial strains (some of them belonging to *Halomonadaceae*) from soil and marine environments for PHA production utilizing Jatropha biodiesel by-product as carbon source (Shrivastav et al. 2010) and a strain, *Halomonas* sp. KM-1, able to produce PHB by using glycerol (Kawata and Aiba 2010).

3.4.6 Bioremediation and Other Related Applications

Hypersaline waters and soils are often contaminated with heavy metals or other toxic compounds from anthropogenic sources. Hypersaline wastewaters are generated during the manufacture of chemicals such as pesticides, pharmaceuticals, and herbicides and during oil and gas recovery processes. Conventional microbiological treatment processes do not function at high salt concentrations, and therefore the use of moderately halophilic bacteria should be considered (Oren et al. 1992). Information on the biodegradation and/or biotransformation of contaminants in hypersaline ecosystems is scarce, and, as a result, the potential for accumulation and toxicity of contaminants in these atypical aqueous systems is unknown. The contaminant degradative capabilities and the toxicity response of member of the family *Halomonadaceae* is still very limited although a few studies have been carried out in the last years.

Hydrocarbon-degrading moderate halophiles have been isolated from a variety of environments. Phenol utilization by moderately halophilic bacteria under aerobic conditions has also been described. Some examples are *Halomonas* sp. isolated from a hypersaline soil in the Great Salt Lake, Utah (Hinteregger and Streichsbier 1997), *Halomonas venusta* (Muñoz et al. 2001) and the haloalkaliphile *Halomonas* sp. EF11 (Maskow and Kleinsteuber 2004). *Halomonas organivorans* is characterized by its ability to use a wide range of organic compounds such as benzoic acid, *p*-hydroxybenzoic acid, cinnamic acid, salicylic acid, phenylacetic acid, phenylpropionic acid, phenol, *p*-coumaric acid, ferulic acid, and *p*-aminosalicylic acid, and it could be useful for decontamination of polluted saline habitats (García et al. 2004). Benzoate and other aromatic compounds can be degraded by *Halomonas halodurans* by cleavage of aromatic rings (Rosenberg 1983). Benzoate, salicylate, phenol and cathechol biodegradation by the haloalkaliphile *Halomonas campisalis* has been reported. Improved understanding and characterization of these degradation processes at high pH and salt concentrations may lead to improved applications of haloalkaliphiles for industrial wastewater treatment (Alva and Peyton 2003; Oie et al. 2007). The utilization of arylaliphatic nitriles by *Halomonas nitrilicus* has also been demonstrated (Chmura et al. 2008) as well as the bioconversion of ferulic acid to vanillic acid by *Halomonas elongata* (Abdelkafi et al. 2006).

Chromohalobacter sp. strain HS-2 was isolated from a salted fermented clam and analyzed for the ability to grow on benzoate and *p*-hydroxybenzoate as a sole carbon and energy source. The genes encoding the benzoate metabolism were cloned, sequenced and analyzed (Kim et al. 2008). Recently, a phenol removal process from hypersaline wastewater in a membrane biological reactor has been reported. Microbial community analysis of the reactors revealed that members of the genera *Halomonas* and *Marinobacter* were the major components (Dosta et al. 2011). Chloroaromatic compounds can be also degraded by member of the family *Halomonadaceae*. An haloalkaliphilic *Halomonas* strain I-18 was shown to possess high activities of catechol 1,2-dioxygenase, muconate cycloisomerase, and dienelactone hydrolase and could also use other aromatic compounds including benzoic acid, 3-chlorobenzoic acid, and 4-chlorophenol (Maltseva et al. 1996).

Osman et al. (2010) have recently isolated from the polluted Maruiy Lake, Egypt, a copper-resistant *Halomonas* strain that could be a fast and efficient tool for copper bioremediation. This bacterium was also resistant to other metals such as nickel, cobalt and zinc and to a group of antibiotics.

Aluminum is the most abundant metallic element in the Earth's crust and is extensively used in various industries due to its versatility. Bauxite is the primary ore for aluminum extraction and contains a number of aluminum containing minerals; bauxite is refined by the Bayer process. Several strains of the genus *Halomonas* have been isolated from a pilot-scale bioreactor treating Bayer liquor organic wastes (McSweeney et al. 2011).

Paper pulp wastewater resulting from alkaline extraction of wheat straw, known as black liquor, is very difficult to be treated and causes serious environmental problems due to its high pH value and chemical oxygen demand. Lignin, hemicellulose, and cellulose are the main contributors to the high chemical oxygen demand values in black liquor. A microbial community which develops naturally in black liquor was investigated and the isolates were assigned to the genera *Halomonas* and *Bacillus* (Yang et al. 2008, 2010a). Recently, two halotolerant and alkaliphilic bacteria, designated as *Halomonas* sp. 19-A and Y2, were isolated from wheat straw black liquor and shown to be able to use guaiacol, dibenzo-*p*-dioxin, biphenyl, and fluorine (Yang et al. 2010b).

Some members of *Halomonadaceae* can use oxyanions as terminal electron acceptors, and several nitrate-reducing *Halomonas* have been isolated and studied. *Halomonas desiderata*, isolated from municipal sewage in Germany has been suggested for a combined process of chemical and biological degradation of nitrocellulose (Berendes et al. 1996). *Halomonas campisalis* reduces nitrate in the presence of lactate, acetate and glycerol at 12% NaCl and pH 10, being suggested as a useful organism to remove nitrate in processes at high pH and in the presence of high salt concentrations (Peyton et al. 2001). Up to date, several denitrifying *Halomonas* species have been described and González-Domenech et al. (2010) have proposed denitrification as an important taxonomic marker within the genus *Halomonas*.

Very recently, a bacterium that can grow using arsenic instead of phosphorus belonging to the genus *Halomonas*, isolated from Mono Lake, USA has been described. However, how arsenic insinuates itself into the structure of biomolecules is unclear, and the mechanisms by with such molecules operate are unknown (Wolfe-Simon et al. 2010).

Acknowledgments Research by authors was supported by grants from the Spanish Ministerio de Educación y Ciencia (CGL2010-19303), National Science Foundation (Grant DEB-0919290) and Junta de Andalucía (P06-CVI-01829). FEDER funds also supported their studies.

References

Abdelkafi S, Sayadi S, Ben Ali Gam Z, Casalot L, Labat M (2006) Bioconversion of ferulic acid to vanillic acid by *Halomonas elongata* isolated from table-olive fermentation. FEMS Microbiol Lett 262:115–120

Alva VA, Peyton BM (2003) Phenol and catechol biodegradation by the haloalkaliphile *Halomonas campisalis*: influence of pH and salinity. Environ Sci Technol 37:4397–4402

Amjres H, Béjar V, Quesada E, Abrini J, Llamas I (2011) *Halomonas rifensis* sp. nov., an exopolysaccharide-producing, halophilic bacterium isolated from Brikcha, a solar saltern in Chefchaouen, Morocco. Int J Syst Evol Microbiol. doi: 10.1099/ijs.0.027268-0

Anan'ina LN, Plotnikova EG, Gavrish EY, Demako VA, Evtushenko LI (2007) *Salinicola socius* gen. nov., sp. nov., a moderately halophilic bacterium from a naphthalene-utilizing microbial association. Microbiology (English translation of Mikrobiologiya) 76:324–330

Arahal DR, Ventosa A (2006) The family *Halomonadaceae*. In: Dworkin M, Falkow S, Rosenberg E, Schleifer K-H, Stackebrandt E (eds) The prokaryotes, vol 6, 3rd edn, *Proteobacteria*: gamma subclass. Springer, New York, pp 811–835

Arahal DR, García MT, Ludwig W, Schleifer KH, Ventosa A (2001a) Transfer of *Halomonas canadensis* and *Halomonas israelensis* to the genus *Chromohalobacter* as *Chromohalobacter canadensis* comb. nov. and *Chromohalobacter israelensis* comb. nov. Int J Syst Evol Microbiol 51:1443–1448

Arahal DR, García MT, Vargas C, Cánovas D, Nieto JJ, Ventosa A (2001b) *Chromohalobacter salexigens* sp. nov., a moderately halophilic species that includes *Halomonas elongata* DSM 3043 and ATCC 33174. Int J Syst Evol Microbiol 51:1457–1462

Arahal DR, Castillo AM, Ludwig W, Schleifer KH, Ventosa A (2002a) Proposal of *Cobetia marina* gen. nov., comb. nov., within the family *Halomonadaceae*, to include the species *Halomonas marina*. Syst Appl Microbiol 25:207–211

Arahal DR, Ludwig W, Schleifer KH, Ventosa A (2002b) Phylogeny of the family *Halomonadaceae* based on 23S and 16S rDNA sequence analyses. Int J Syst Evol Microbiol 52:241–249

Arahal DR, Vreeland RH, Litchfield CD, Mormile MR, Tindall BJ, Oren A, Bejar V, Quesada E, Ventosa A (2007) Recommended minimal standards for describing new taxa of the family *Halomonadaceae*. Int J Syst Evol Microbiol 57:2436–2446

Arco Y, Llamas I, Martínez-Checa F, Argandoña M, Quesada E, del Moral A (2005) *epsABCJ* genes are involved in the biosynthesis of the exopolysaccharide mauran produced by *Halomonas maura*. Microbiology 151:2841–2851

Arenas M, Bañón PI, Copa-Patiño JL, Sánchez-Porro C, Ventosa A, Soliveri J (2009) *Halomonas ilicicola* sp. nov., a moderately halophilic bacterium isolated from a saltern. Int J Syst Evol Microbiol 59:578–582

Argandoña M, Martínez-Checa F, Llamas I, Quesada E, del Moral A (2003) Megaplasmids in Gram-negative, moderately halophilic bacteria. FEMS Microbiol Lett 227:81–86

Arias S, del Moral A, Ferrer MR, Tallon R, Quesada E, Béjar V (2003) Mauran, an exopolysaccharide produced by the halophilic bacterium *Halomonas maura*, with a novel composition and interesting properties for biotechnology. Extremophiles 7:319–326

Arvanitis N, Vargas C, Tegos G, Perysinakis A, Nieto JJ, Ventosa A, Drainas C (1995) Development of a gene reporter system in moderately halophilic bacteria by employing the ice nucleation gene of *Pseudomonas syringae*. Appl Environ Microbiol 61:3821–3825

Asad S, Amoozegar MA, Pourbabaee AA, Sarbolouki MN, Dastgheib SM (2007) Decolorization of textile azo dyes by newly isolated halophilic and halotolerant bacteria. Bioresour Technol 98:2082–2088

Aston JE, Peyton BM (2007) Response of *Halomonas campisalis* to saline stress: changes in growth kinetics, compatible solute production and membrane phospholipid fatty acid composition. FEMS Microbiol Lett 274:196–203

Ates O, Toksoy Oner E, Arga KY (2011) Genome-scale reconstruction of metabolic network for a halophilic extremophile, *Chromohalobacter salexigens* DSM 3043. BMC Syst Biol 5:12

Baumann L, Baumann P, Mandel M, Allen RD (1972) Taxonomy of aerobic marine eubacteria. J Bacteriol 110:402–429

Baumann L, Bowditch RD, Baumann P (1983) Description of *Deleya* gen. nov. created to accommodate the marine species *Alcaligenes aestus*, *A. pacificus*, *A. cupidus*, *A. venustus*, and *Pseudomonas marina*. Int J Syst Bacteriol 33:793–802

Béjar V, Calvo C, Moliz J, Díaz-Martínez F, Quesada E (1996) Effect of growth conditions on the rheological properties and chemical composition of *Volcaniella eurihalina* exopolysaccharide. Appl Biochem Biotechnol 59:77–86

Béjar V, Llamas I, Calvo C, Quesada E (1998) Characterization of exopolysaccharides produced by 19 halophilic strains of the species *Halomonas eurihalina*. J Biotechnol 61:135–141

Ben Ali Gam Z, Abdelkafi S, Casalot L, Tholozan JL, Oueslati R, Labat M (2007) *Modicisalibacter tunisiensis* gen. nov., sp. nov., an aerobic, moderately halophilic bacterium isolated from an oilfield-water injection sample, and emended description of the family *Halomonadaceae* Franzmann et al. 1989 emend. Dobson and Franzmann 1996 emend. Ntougias et al. 2007. Int J Syst Evol Microbiol 57:2307–2313

Berendes F, Gottschalk G, Heine-Dobbernack E, Moore ERB, Tindall BJ (1996) *Halomonas desiderata* sp. nov., a new alkaliphilic, halotolerant and denitrifying bacterium isolated from a municipal sewage works. Syst Appl Microbiol 19:158–167

Bestvater T, Louis P, Galinski EA (2008) Heterologous ectoine production in *Escherichia coli*: by-passing the metabolic bottle-neck. Saline Systems 4:12

Biswas A, Patra A, Paul AK (2009) Production of poly-3-hydroxyalkanoic acids by a moderately halophilic bacterium, *Halomonas marina* HMA 103 isolated from solar saltern of Orissa, India. Acta Microbiol Immunol Hung 56:125–143

Bjorkman K, Karl D (1994) Bioavailability of inorganic and organic phosphorus compounds to natural assemblages of microorganisms in Hawaiian coastal water. Mar Ecol Prog Ser 111: 265–273

Boltyanskaya YV, Kevbrin VV, Lysenko AM, Kolganova TV, Tourova TP, Osipov GA, Zhilina TN (2007) *Halomonas mongoliensis* sp. nov. and *Halomonas kenyensis* sp. nov., new haloalkaliphilic denitrifiers capable of N_2O reduction, isolated from soda lakes. Microbiology 76: 739–747

Bouchotroch S, Quesada E, del Moral A, Béjar V (1999) Taxonomic study of exopolysaccharide-producing moderately halophilic bacteria isolated from hypersaline environments in Morocco. Syst Appl Microbiol 22:412–419

Bouchotroch S, Quesada E, Del Moral A, Llamas I, Béjar V (2001) *Halomonas maura* sp. nov., a novel moderately halophilic, exopolysaccharide-producing bacterium. Int J Syst Evol Microbiol 51:1625–1632

Calderón MI, Vargas C, Rojo F, Iglesias-Guerra F, Csonka LN, Ventosa A, Nieto JJ (2004) Complex regulation of the synthesis of the compatible solute ectoine in the halophilic bacterium *Chromohalobacter salexigens* DSM 3043^T. Microbiology 150:3051–3063

Calvo C, Ferrer MR, Martínez-Checa F, Béjar V, Quesada E (1995) Some rheological properties of the extracellular polysaccharide produced by *Volcaniella eurihalina* F2-7. Appl Biochem Biotechnol 55:45–54

Calvo C, Martínez-Checa F, Mota A, Béjar V, Quesada E (1998) Effect of cations, pH and sulfate content on the viscosity and emulsifying activity of the *Halomonas eurihalina* exopolysaccharide. J Ind Microbiol Biotechnol 20:205–209

Calvo C, Martínez-Checa F, Toledo FL, Porcel J, Quesada E (2002) Characteristics of bioemulsifiers synthesised in crude oil media by *Halomonas eurihalina* and their effectiveness in the isolation of bacteria able to grow in the presence of hydrocarbons. Appl Microbiol Biotechnol 60:347–351

Cánovas D, Vargas C, Csonka LN, Ventosa A, Nieto JJ (1996) Osmoprotectants in *Halomonas elongata*: high-affinity betaine transport system and choline-betaine pathway. J Bacteriol 178:7221–7226

Cánovas D, Vargas C, Iglesias-Guerra F, Csonka LN, Rhodes D, Ventosa A, Nieto JJ (1997a) Isolation and characterization of salt-sensitive mutants of the moderate halophile *Halomonas elongata* and cloning of the ectoine synthesis genes. J Biol Chem 272:25794–25801

Cánovas D, Vargas C, Ventosa A, Nieto JJ (1997b) Salt-sensitive and auxotrophic mutants of *Halomonas elongata* and *H. meridiana* by use of hydroxylamine mutagenesis. Curr Microbiol 34:85–90

Cánovas D, Vargas C, Csonka LN, Ventosa A, Nieto JJ (1998a) Synthesis of glycine betaine from exogenous choline in the moderately halophilic bacterium *Halomonas elongata*. Appl Environ Microbiol 64:4095–4097

Cánovas D, Vargas C, Calderón MI, Ventosa A, Nieto JJ (1998b) Characterization of the genes for the biosynthesis of the compatible solute ectoine in the moderately halophilic bacterium *Halomonas elongata* DSM 3043. Syst Appl Microbiol 21:487–497

Cánovas D, Borges N, Vargas C, Ventosa A, Nieto JJ, Santos H (1999) Role of N_{γ}-acetyldiaminobutyrate as an enzyme stabilizer and an intermediate in the biosynthesis of hydroxyectoine. Appl Environ Microbiol 65:3774–3779

Cánovas D, Vargas C, Kneip S, Morón MJ, Ventosa A, Bremer E, Nieto JJ (2000) Genes for the synthesis of the osmoprotectant glycine betaine from choline in the moderately halophilic bacterium *Halomonas elongata* DSM 3043. Microbiology 146:455–463

Chen ZL, Fu YN, Jiang WY, Chen RZ (2007) Characterization of the *ectC* gene and its expression product in *Halomonas* sp. Nj223 from Antarctica deep-sea sediment. Wei Sheng Wu Xue Bao 47:363–365

Chen Y-G, Zhang Y-Q, Huang H-Y, Klenk H-P, Tang S-K, Huang K, Chen Q-H, Cui X-L, Li W-J (2009) *Halomonas zhanjiangensis* sp. nov., a halophilic bacterium isolated from a sea urchin. Int J Syst Evol Microbiol 59:2888–2893

Chmura A, Shapovalova AA, van Pelt S, van Rantwijk F, Tourova TP, Muyzer G, Sorokin DY (2008) Utilization of arylaliphatic nitriles by haloalkaliphilic *Halomonas nitrilicus* sp. nov. isolated from soda soils. Appl Microbiol Biotechnol 81:371–378

Cobet AB, Wirsen C, Jones GE (1970) The effect of nickel on a marine bacterium, *Arthrobacter marinus* sp. nov. J Gen Microbiol 62:159–169

Coronado MJ, Vargas C, Kunte HJ, Galinski EA, Ventosa A, Nieto JJ (1995) Influence of salt concentration on the susceptibility of moderately halophilic bacteria to antimicrobials and its potential use for genetic transfer studies. Curr Microbiol 31:365–371

Coronado M, Vargas C, Hofemeister J, Ventosa A, Nieto JJ (2000a) Production and biochemical characterization of an α-amylase from the moderate halophile *Halomonas meridiana*. FEMS Microbiol Lett 183:67–71

Coronado MJ, Vargas C, Mellado E, Tegos G, Drainas C, Nieto JJ, Ventosa A (2000b) The α-amylase gene *amyH* of the moderate halophile *Halomonas meridiana*: cloning and molecular characterization. Microbiology 146:861–868 de la Haba RR, Arahal DR, Márquez MC, Ventosa A (2010a) Phylogenetic relationships within the family *Halomonadaceae* based on 23S and 16S rRNA comparative sequence analysis. Int J Syst Evol Microbiol 60:737–748

de la Haba RR, Sánchez-Porro C, Márquez MC, Ventosa A (2010b) Taxonomic study of the genus *Salinicola*: transfer of *Halomonas salaria* and *Chromohalobacter salarius* to the genus *Salinicola* as *Salinicola salarius* comb. nov. and *Salinicola halophilus* nom. nov., respectively. Int J Syst Evol Microbiol 60:963–971

de la Haba RR, Márquez MC, Papke RT, Ventosa A (2011) Multilocus Sequence Analysis (MLSA) of the family *Halomonadaceae*. Int J Syst Evol Microbiol doi:10.1099/ijs.0.032938-8

de la Haba RR, Sánchez-Porro C, Márquez MC, Ventosa A (2011) Taxonomy of halophiles. In: Horikoshi K (ed) Extremophiles handbook. Springer, Heidelberg, pp 255–308

Dobson SJ, Franzmann PD (1996) Unification of the genera *Deleya* (Baumann et al. 1983), *Halomonas* (Vreeland et al. 1980), and *Halovibrio* (Fendrich 1988) and the species *Paracoccus halodenitrificans* (Robinson and Gibbons 1952) into a single genus, *Halomonas*, and placement of the genus *Zymobacter* in the family *Halomonadaceae*. Int J Syst Bacteriol 46:550–558

Dobson SJ, James SR, Franzmann PD, McMeekin TA (1990) Emended description of *Halomonas halmophila* (NCMB 1971^{T}). Int J Syst Bacteriol 40:462–463

Dobson SJ, McMeekin TA, Franzmann PD (1993) Phylogenetic relationships between some members of the genera *Deleya*, *Halomonas*, and *Halovibrio*. Int J Syst Bacteriol 43:665–673

Dosta J, Nieto JM, Vila J, Grifoll M, Mata-Álvarez J (2011) Phenol removal from hypersaline wastewaters in a Membrane Biological Reactor (MBR): operation and microbiological characterisation. Bioresour Technol 102:4013–4020

Douka E, Christogianni A, Koukkou AI, Afendra AS, Drainas C (2001) Use of a green fluorescent protein gene as a reporter in *Zymomonas mobilis* and *Halomonas elongata*. FEMS Microbiol Lett 201:221–227

Duckworth AW, Grant WD, Jones BE, Meijer D, Márquez MC, Ventosa A (2000) *Halomonas magadii* sp. nov., a new member of the genus *Halomonas*, isolated from a soda lake of the East African rift valley. Extremophiles 4:53–60

Elazari-Volcani B (1940) Studies on the microflora of the Dead Sea. Ph.D. thesis, The Hebrew University of Jerusalem (in Hebrew)

Essghaier B, Fardeau ML, Cayol JL, Hajlaoui MR, Boudabous A, Jijakli H, Sadfi-Zouaoui N (2009) Biological control of grey mould in strawberry fruits by halophilic bacteria. J Appl Microbiol 106:833–846

Fallet C, Rohe P, Franco-Lara E (2010) Process optimization of the integrated synthesis and secretion of ectoine and hydroxyectoine under hyper/hypo-osmotic stress. Biotechnol Bioeng 107:124–133

Fendrich C (1988) *Halovibrio variabilis* gen. nov., sp. nov., *Pseudomonas halophila* sp. nov. and a new halophilic aerobic coccoid *Eubacterium* from Great Salt Lake, Utah, USA. Syst Appl Microbiol 11:36–43

Fernandez-Castillo R, Vargas C, Nieto JJ, Ventosa A, Ruiz-Berraquero F (1992) Characterization of a plasmid from moderately halophilic eubacteria. J Gen Microbiol 138:1133–1137

Franzmann PD, Tindall BJ (1990) A chemotaxonomic study of members of the family *Halomonadaceae*. Syst Appl Microbiol 13:142–147

Franzmann PD, Burton HR, McMeekin TA (1987) *Halomonas subglaciescola*, a new species of halotolerant bacteria isolated from Antarctica. Int J Syst Bacteriol 37:27–34

Franzmann PD, Wehmeyer U, Stackebrandt E (1988) *Halomonadaceae* fam. nov., a new family of the class *Proteobacteria* to accommodate the genera *Halomonas* and *Deleya*. Syst Appl Microbiol 11:16–19

Frillingos S, Linden A, Niehaus F, Vargas C, Nieto JJ, Ventosa A, Antranikian G, Drainas C (2000) Cloning and expression of α-amylase from the hyperthermophilic archaeon *Pyrococcus woesei* in the moderately halophilic bacterium *Halomonas elongata*. J Appl Microbiol 88: 495–503

Gadda G, McAllister-Wilkins EE (2003) Cloning, expression, and purification of choline dehydrogenase from the moderate halophile *Halomonas elongata*. Appl Environ Microbiol 69: 2126–2132

García MT, Mellado E, Ostos JC, Ventosa A (2004) *Halomonas organivorans* sp. nov., a moderate halophile able to degrade aromatic compounds. Int J Syst Evol Microbiol 54:1723–1728

García-Estepa R, Cánovas D, Iglesias-Guerra F, Ventosa A, Csonka LN, Nieto JJ, Vargas C (2006) Osmoprotection of *Salmonella enterica* serovar *Typhimurium* by N_{γ}-acetyldiaminobutyrate, the precursor of the compatible solute ectoine. Syst Appl Microbiol 29:626–633

Garriga M, Ehrmann MA, Arnau J, Hugas M, Vogel RF (1998) *Carnimonas nigrificans* gen. nov., sp. nov., a bacterial causative agent for black spot formation on cured meat products. Int J Syst Bacteriol 48:677–686

Garrity GM, Bell JA, Lilburn T (2005a) Class III. *Gammaproteobacteria* class. nov. In: Brenner DJ, Krieg NR, Staley JT, Garrity GM (eds) Bergey's manual of systematic bacteriology, vol 2, 2nd edn, (The *Proteobacteria*), part B (The *Gammaproteobacteria*). Springer, New York, p 1

Garrity GM, Bell JA, Lilburn T (2005b) Family IV. *Halomonadaceae* Franzmann, Wehmeyer and Stackebrandt 1989, 205^{VP} emend. Dobson and Franzmann 1996, 558. In: Brenner DJ, Krieg NR, Staley JT, Garrity GM (eds) Bergey's manual of systematic bacteriology, vol 2, 2nd edn, (The *Proteobacteria*), part B (The *Gammaproteobacteria*). Springer, New York, pp 300–323

Göller K, Ofer A (1998) Construction and characterization of an NaCl-sensitive mutant of *Halomonas elongata* impaired in ectoine biosynthesis. FEMS Microbiol Lett 161:293–300, Erratum in FEMS Microbiol Lett (1998) 165:379

González-Domenech CM, Béjar V, Martínez-Checa F, Quesada E (2008a) *Halomonas nitroreducens* sp. nov., a novel nitrate- and nitrite-reducing species. Int J Syst Evol Microbiol 58:872–876

González-Domenech CM, Martínez-Checa F, Quesada E, Béjar V (2008b) *Halomonas cerina* sp. nov., a moderately halophilic, denitrifying, exopolysaccharide-producing bacterium. Int J Syst Evol Microbiol 58:803–809

González-Domenech CM, Martínez-Checa F, Quesada E, Béjar V (2009) *Halomonas fontilapidosi* sp. nov., a moderately halophilic, denitrifying bacterium. Int J Syst Evol Microbiol 59: 1290–1296

González-Domenech CM, Martínez-Checa F, Béjar V, Quesada E (2010) Denitrification as an important taxonomic marker within the genus *Halomonas*. Syst Appl Microbiol 33:85–93

Grammann K, Volke A, Kunte HJ (2002) New type of osmoregulated solute transporter identified in halophilic members of the bacteria domain: TRAP transporter TeaABC mediates uptake of ectoine and hydroxyectoine in *Halomonas elongata* DSM 2581^{T}. J Bacteriol 184:3078–3085

Guan T-W, Xiao J, Zhao K, Luo X-X, Zhang X-P, Zhang L-L (2010) *Halomonas xinjiangensis* sp. nov., a halotolerant bacterium isolated from a salt lake. Int J Syst Evol Microbiol 60:349–352

Guo J, Zhou J, Wang D, Tian C, Wang P, Uddin MS (2008) A novel moderately halophilic bacterium for decolorizing azo dye under high salt condition. Biodegradation 19:15–19

Gutierrez T, Morris G, Green DH (2009) Yield and physicochemical properties of EPS from *Halomonas* sp. strain TG39 identifies a role for protein and anionic residues (sulfate and phosphate) in emulsification of n-hexadecane. Biotechnol Bioeng 103:207–216

Guzmán H, Van-Thuoc D, Martín J, Hatti-Kaul R, Quillaguamán J (2009) A process for the production of ectoine and poly(3-hydroxybutyrate) by *Halomonas boliviensis*. Appl Microbiol Biotechnol 84:1069–1077

Guzmán D, Quillaguamán J, Muñoz M, Hatti-Kaul R (2010) *Halomonas andesensis* sp. nov., a moderate halophile isolated from the saline lake Laguna Colorada in Bolivia. Int J Syst Evol Microbiol 60:749–753

He J, Huang X, Gu LF, Jiang JD, Li SP (2006) Cloning of the ectoine biosynthesis gene ectABC from *Halomonas* sp. BYS-1 and salt stressed expression in *Escherichia coli*. Wei Sheng Wu Xue Bao 46:28–32

He J, Zhen Q, Qiu N, Liu Z, Wang B, Shao Z, Yu Z (2009) Medium optimization for the production of a novel bioflocculant from *Halomonas* sp. V3a′ using response surface methodology. Bioresour Technol 100:5922–5927

Hebert AM, Vreeland RH (1987) Phenotypic comparison of halotolerant bacteria: *Halomonas halodurans* sp. nov., nom. rev., comb. nov. Int J Syst Bacteriol 37:347–350

Henrissat B, Bairoch A (1993) New families in the classification of glycosyl hydrolases based on amino acid sequence similarities. Biochem J 293:781–788

Heyrman J, Balcaen A, De Vos P, Swings J (2002) *Halomonas muralis* sp. nov., isolated from microbial biofilms colonizing the walls and murals of the Saint-Catherine chapel (Castle Herberstein, Austria). Int J Syst Evol Microbiol 52:2049–2054

Hinteregger C, Streichsbier F (1997) *Halomonas* sp. a moderately halophilic strain, for biotreatment of saline phenolic waste-water. Biotechnol Lett 19:1099–1102

Hof T (1935) An investigation of the microorganisms commonly present in salted beans. Rec Trav Bot Neerl 32:151–173

Huang H-Y, Chen Y-G, Wang Y-X, Liu J-H, Tang S-K, Peng Q, Wen M-L, Yu H, Cui X-L (2008) *Halomonas sediminis* sp. nov., a new halophilic bacterium isolated from salt-lake sediment in China. Extremophiles 12:829–835

Huang X, Shao Z, Hong Y, Lin L, Li C, Huang F, Wang H, Liu Z (2010) Cel8H, a novel endoglucanase from the halophilic bacterium *Halomonas* sp. S66-4: molecular cloning, heterogonous expression, and biochemical characterization. J Microbiol 48:318–324

Ishibashi M, Yamashita S, Tokunaga M (2005) Characterization of halophilic alkaline phosphatase from *Halomonas* sp. 593, a moderately halophilic bacterium. Biosci Biotechnol Biochem 69:1213–1216

Ishibashi M, Oda K, Arakawa T, Tokunaga M (2011) Cloning, expression, purification and activation by Na ion of halophilic alkaline phosphatase from moderate halophile *Halomonas* sp. 593. Protein Expr Purif 76:97–102

James SR, Dobson J, Franzmann PD, McMeekin TA (1990) *Halomonas meridiana*, a new species of extremely halotolerant bacteria isolated from Antarctic saline lakes. Syst Appl Microbiol 13:270–277

Jeon CO, Lim JM, Lee JR, Lee GS, Park DJ, Lee JC, Oh HW, Kim CJ (2007) *Halomonas kribbensis* sp. nov., a novel moderately halophilic bacterium isolated from a solar saltern in Korea. Int J Syst Evol Microbiol 57:2194–2198

Joshi AA, Kanekar PP, Kelkar AS, Shouche YS, Vani AA, Borgave SB, Sarnaik SS (2008) Cultivable bacterial diversity of alkaline Lonar lake, India. Microb Ecol 55:163–172

Kamekura M (1986) Production and function of enzymes of eubacterial halophiles. FEMS Microbiol Rev 39:145–150

Kawata Y, Aiba S (2010) Poly(3-hydroxybutyrate) production by isolated *Halomonas* sp. KM-1 using waste glycerol. Biosci Biotechnol Biochem 74:175–177

Kaye JZ, Márquez MC, Ventosa A, Baross JA (2004) *Halomonas neptunia* sp. nov., *Halomonas sulfidaeris* sp. nov., *Halomonas axialensis* sp. nov. and *Halomonas hydrothermalis* sp. nov.: halophilic bacteria isolated from deep-sea hydrothermal-vent environments. Int J Syst Evol Microbiol 54:499–511

Kharroub K, Jiménez-Pranteda ML, Aguilera M, Boulahrouf A, Ramos-Cormenzana A, Monteoliva-Sánchez M (2008) *Halomonas sabkhae* sp. nov., a moderately halophilic bacterium isolated from an Algerian sabkha. Int J Syst Evol Microbiol 58:40–44

Kim H, Choo YJ, Cho JC (2007a) *Litoricolaceae* fam. nov., to include *Litoricola lipolytica* gen. nov., sp. nov., a marine bacterium belonging to the order *Oceanospirillales*. Int J Syst Evol Microbiol 57:1793–1798

Kim KK, Jin L, Yang HC, Lee S-T (2007b) *Halomonas gomseomensis* sp. nov., *Halomonas janggokensis* sp. nov., *Halomonas salaria* sp. nov. and *Halomonas denitrificans* sp. nov., moderately halophilic bacteria isolated from saline water. Int J Syst Evol Microbiol 57: 675–681

Kim D, Kim SW, Choi KY, Lee JS, Kim E (2008) Molecular cloning and functional characterization of the genes encoding benzoate and p-hydroxybenzoate degradation by the halophilic *Chromohalobacter* sp. strain HS-2. FEMS Microbiol Lett 280:235–241

Kim KK, Lee KC, Oh H-M, Lee J-S (2010a) *Halomonas stevensii* sp. nov., *Halomonas hamiltonii* sp. nov. and *Halomonas johnsoniae* sp. nov., isolated from a renal care centre. Int J Syst Evol Microbiol 60:369–377

Kim MS, Roh SW, Bae JW (2010b) *Cobetia crustatorum* sp. nov., a novel slightly halophilic bacterium isolated from traditional fermented seafood in Korea. Int J Syst Evol Microbiol 60: 620–626

Kim MS, Roh SW, Bae JW (2010c) *Halomonas jeotgali* sp. nov., a new moderate halophilic bacterium isolated from a traditional fermented seafood. J Microbiol 48:404–410

König S (1998) Subunit structure, function and organisation of pyruvate decarboxylase from various organisms. Biochim Biophys Acta 1385:271–286

Kuhlmann SI, Terwisscha van Scheltinga AC, Bienert R, Kunte HJ, Ziegler C (2008) 1.55Å structure of the ectoine binding protein TeaA of the osmoregulated TRAP-transporter TeaABC from *Halomonas elongata*. Biochemistry 47:9475–9485

Kulkarni SO, Kanekar PP, Nilegaonkar SS, Sarnaik SS, Jog JP (2010) Production and characterization of a biodegradable poly (hydroxybutyrate-co-hydroxyvalerate) (PHB-co-PHV) copolymer by moderately haloalkalitolerant *Halomonas campisalis* MCM B-1027 isolated from Lonar Lake, India. Bioresour Technol 101:9765–9771

Kunte HJ, Galinski EA (1995) Transposon mutagenesis in halophilic eubacteria: conjugal transfer and insertion of transposon Tn5 and Tn1732 in *Halomonas elongata*. FEMS Microbiol Lett 128:293–299

Kushner DJ (1978) Life in high salt and solute concentrations: halophilic bacteria. In: Kushner DJ (ed) Microbial life in extreme environments. Academic, London, pp 317–368

Lang Y, Ren Y, Bai L, Zhang L (2009) Hydroxyectoine synthesis and release under osmotic shock in *Cobetia marina* CICC10367. Wei Sheng Wu Xue Bao 49:1590–1595

Lee J-C, Jeon CO, Lim J-M, Lee S-M, Lee J-M, Song S-M, Park D-J, Li W-J, Kim C-J (2005) *Halomonas taeanensis* sp. nov., a novel moderately halophilic bacterium isolated from a solar saltern in Korea. Int J Syst Evol Microbiol 55:2027–2032

Li H-B, Zhang L-P, Chen S-F (2008) *Halomonas korlensis* sp. nov., a moderately halophilic, denitrifying bacterium isolated from saline and alkaline soil. Int J Syst Evol Microbiol 58: 2582–2588

Lim J-M, Yoon J-H, Lee J-C, Jeon CO, Park D-J, Sung C, Kim C-J (2004) *Halomonas koreensis* sp. nov., a novel moderately halophilic bacterium isolated from a solar saltern in Korea. Int J Syst Evol Microbiol 54:2037–2042

Liu S, Dien BS, Cotta MA (2005) Functional expression of bacterial *Zymobacter palmae* pyruvate decarboxylase gene in *Lactococcus lactis*. Curr Microbiol 50:324–328

Liu F, Wang D, Du L, Zhu Y, Xu W (2010) Diversity of the predominant spoilage bacteria in water-boiled salted duck during storage. J Food Sci 75:317–321

Llamas Company I, Béjar V, Martínez-Checa F, Martínez-Cánovas MJ, Molina I, Quesada E (2011) *Halomonas stenophila* sp. nov., a halophilic bacterium that produces sulphate exopolysaccharides with biological activity. Int J Syst Evol Microbiol. doi: 10.1099/ijs.0.026369-0

Llamas I, Argandoña M, Quesada E, del Moral A (2000) Transposon mutagenesis in *Halomonas eurihalina*. Res Microbiol 151:13–18

Llamas I, Sánchez MJ, Argandoña M, Béjar V, Quesada E, del Moral A (2002) Analysis of the genome of the moderate halophile *Halomonas eurihalina*. Curr Microbiol 45:233–923

Llamas I, Suárez A, Quesada E, Béjar V, del Moral A (2003) Identification and characterization of the carAB genes responsible for encoding carbamoylphosphate synthetase in *Halomonas eurihalina*. Extremophiles 7:205–211

Ludwig W, Strunk O, Klugbauer S, Klugbauer N, Weizenegger M, Neumaier J, Bachleitner M, Schleifer KH (1998) Bacterial phylogeny based on comparative sequence analysis. Electrophoresis 19:554–568

Maltseva O, McGowan C, Fulthorpe R, Oriel P (1996) Degradation of 2,4-dichlorophenoxyacetic acid by haloalkaliphilic bacteria. Microbiology 142:1115–1122

Martens M, Delaere M, Coopman R, De Vos P, Gillis M, Willems A (2007) Multilocus sequence analysis of *Ensifer* and related taxa. Int J Syst Bacteriol 57:489–503

Martínez-Cánovas MJ, Béjar V, Martínez-Checa F, Quesada E (2004a) *Halomonas anticariensis* sp. nov., from Fuente de Piedra, a saline-wetland wildfowl reserve in Málaga, southern Spain. Int J Syst Evol Microbiol 54:1329–1332

Martínez-Cánovas MJ, Quesada E, Llamas I, Béjar V (2004b) *Halomonas ventosae* sp. nov., a moderately halophilic, denitrifying, exopolysaccharide-producing bacterium. Int J Syst Evol Microbiol 54:733–737

Martínez-Cánovas MJ, Quesada E, Martínez-Checa F, Béjar V (2004c) A taxonomic study to establish the relationship between exopolysaccharide-producing bacterial strains living in diverse hypersaline habitats. Curr Microbiol 48:348–353

Martínez-Checa F, Toledo FL, Vilchez R, Quesada E, Calvo C (2002) Yield production, chemical composition, and functional properties of emulsifier H28 synthesized by *Halomonas eurihalina* strain H-28 in media containing various hydrocarbons. Appl Microbiol Biotechnol 58:358–363

Martínez-Checa F, Béjar V, Martínez-Cánovas MJ, Llamas I, Quesada E (2005) *Halomonas almeriensis* sp. nov., a moderately halophilic, exopolysaccharide-producing bacterium from Cabo de Gata, Almería, south-east Spain. Int J Syst Evol Microbiol 55:2007–2011

Martínez-Checa F, Toledo FL, El Mabrouki K, Quesada E, Calvo C (2007) Characteristics of bioemulsifier V2-7 synthesized in culture media added of hydrocarbons: chemical composition, emulsifying activity and rheological properties. Bioresour Technol 98:3130–3135

Maskow T, Babel W (2001) Calorimetrically obtained information about the efficiency of ectoine synthesis from glucose in *Halomonas elongata*. Biochim Biophys Acta 1527:4–10, Erratum in Biochim Biophys Acta (2001) 1528:60

Maskow T, Kleinsteuber S (2004) Carbon and energy fluxes during haloadaptation of *Halomonas* sp. EF11 growing on phenol. Extremophiles 8:133–141

Mata JA, Martínez-Cánovas J, Quesada E, Béjar V (2002) A detailed phenotypic characterisation of the type strains of *Halomonas* species. Syst Appl Microbiol 25:360–375

Mata JA, Béjar V, Llamas I, Arias S, Bressollier P, Tallon R, Urdaci MC, Quesada E (2006) Exopolysaccharides produced by the recently described halophilic bacteria *Halomonas ventosae* and *Halomonas anticariensis*. Res Microbiol 157:827–835

McSweeney NJ, Plumb JJ, Tilbury AL, Nyeboer HJ, Sumich ME, McKinnon AJ, Franzmann PD, Sutton DC, Kaksonen AH (2011) Comparison of microbial communities in pilot-scale bioreactors treating Bayer liquor organic wastes. Biodegradation 22:397–407

Mellado E, Asturias JA, Nieto JJ, Timmis KN, Ventosa A (1995a) Characterization of the basic replicon of pCM1, a narrow-host-range plasmid from the moderate halophile *Chromohalobacter marismortui*. J Bacteriol 177:3443–3450

Mellado E, Moore ERB, Nieto JJ, Ventosa A (1995b) Phylogenetic inferences and taxonomic consequences of 16S ribosomal DNA sequence comparison of *Chromohalobacter marismortui*, *Volcaniella eurihalina*, and *Deleya salina* and reclassification of *V. eurihalina* as *Halomonas eurihalina* comb. nov. Int J Syst Bacteriol 45:712–716

Mellado E, Nieto JJ, Ventosa A (1995c) Construction of novel shuttle vectors for use between moderately halophilic bacteria and *Escherichia coli*. Plasmid 34:157–164

Mellado E, García MT, Roldán E, Nieto JJ, Ventosa A (1998) Analysis of the genome of the Gram-negative moderate halophiles *Halomonas* and *Chromohalobacter* by using pulsed-field gel electrophoresis. Extremophiles 2:435–438

Mellado E, Sánchez-Porro C, Martín S, Ventosa A (2003) Extracellular hydrolytic enzymes produced by moderately halophilic bacteria. In: Ventosa A (ed) Halophilic microorganisms. Springer, Heidelberg, pp 285–295

Menes RJ, Viera CE, Farias ME, Seufferheld MJ (2011) *Halomonas vilamensis* sp. nov., isolated from high-altitude Andean lakes. Int J Syst Evol Microbiol 61:1211–1217

Moghaieb RE, Tanaka N, Saneoka H, Murooka Y, Ono H, Morikawa H, Nakamura A, Nguyen NT, Suwa R, Fujita K (2006) Characterization of salt tolerance in ectoine-transformed tobacco plants (*Nicotiana tabacum*): photosynthesis, osmotic adjustment, and nitrogen partitioning. Plant Cell Environ 9:173–182

Mormile MR, Romine MF, Garcia MT, Ventosa A, Bailey TJ, Peyton BM (1999) *Halomonas campisalis* sp. nov., a denitrifying, moderately haloalkaliphilic bacterium. Syst Appl Microbiol 22:551–558

Muñoz JA, Pérez-Esteban B, Esteban M, de la Escalera S, Gómez MA, Martínez-Toledo MV, González-López J (2001) Growth of moderately halophilic bacteria isolated from sea water using phenol as the sole carbon source. Folia Microbiol (Praha) 46:297–302

Nakayama H, Yoshida K, Ono H, Murooka Y, Shinmyo A (2000) Ectoine, the compatible solute of *Halomonas elongata*, confers hyperosmotic tolerance in cultured tobacco cells. Plant Physiol 122:1239–1247

Ntougias S, Zervakis GI, Fasseas C (2007) *Halotalea alkalilenta* gen. nov., sp. nov., a novel osmotolerant and alkalitolerant bacterium from alkaline olive mill wastes, and emended description of the family *Halomonadaceae* Franzmann et al. 1989, emend. Dobson and Franzmann 1996. Int J Syst Evol Microbiol 57:1975–1983

Oie CS, Albaugh CE, Peyton BM (2007) Benzoate and salicylate degradation by *Halomonas campisalis*, an alkaliphilic and moderately halophilic microorganism. Water Res 41:1235–1242

Okamoto T, Taguchi H, Nakamura K, Ikenaga H, Kuraishi H, Yamasato K (1993) *Zymobacter palmae* gen. nov., sp. nov., a new ethanol-fermenting peritrichous bacterium isolated from palm sap. Arch Microbiol 160:333–337

Okamoto T, Maruyama A, Imura S, Takeyama H, Naganuma T (2004) Comparative phylogenetic analyses of *Halomonas variabilis* and related organisms based on 16S rRNA, *gyrB* and *ectBC* gene sequences. Syst Appl Microbiol 27:323–333

Ono H, Sawada K, Khunajakr N, Tao T, Yamamoto M, Hiramoto M, Shinmyo A, Takano M, Murooka Y (1999) Characterization of biosynthetic enzymes for ectoine as a compatible solute in a moderately halophilic eubacterium, *Halomonas elongata*. J Bacteriol 181:91–99

Oren A, Gurevich P, Azachi M, Henis Y (1992) Microbial degradation of pollutants at high salt concentrations. Biodegradation 3:387–398

Osman O, Tanguichi H, Ikeda K, Park P, Tanabe-Hosoi S, Nagata S (2010) Copper-resistant halophilic bacterium isolated from the polluted Maruit Lake, Egypt. J Appl Microbiol 108:1459–1470

Pastor JM, Salvador M, Argandoña M, Bernal V, Reina-Bueno M, Csonka LN, Iborra JL, Vargas C, Nieto JJ, Cánovas M (2010) Ectoines in cell stress protection: uses and biotechnological production. Biotechnol Adv 28:782–801

Peçonek J, Gruber C, Gallego V, Ventosa A, Busse H-J, Kämpfer P, Radax C, Stan-Lotter H (2006) Reclassification of *Pseudomonas beijerinckii* Hof 1935 as *Chromohalobacter beijerinckii* comb. nov., and emended description of the species. Int J Syst Evol Microbiol 56: 1953–1957

Pérez-Fernández ME, Quesada E, Gálvez J, Ruiz C (2000) Effect of exopolysaccharide V2-7, isolated from *Halomonas eurihalina*, on the proliferation in vitro of human peripheral blood lymphocytes. Immunopharmacol Immunotoxicol 22:131–141

Peters P, Galinski EA, Trüper HG (1990) The biosynthesis of ectoine. FEMS Microbiol Lett 71: 157–162

Peyton BM, Mormile MR, Petersen JN (2001) Nitrate reduction with *Halomonas campisalis*. Kinetics of denitrification at pH 9 and 12.5% NaCl. Water Res 35:4237–4242

Plisova EY, Balabanova LA, Ivanova EP, Kozhemyako VB, Mikhailov VV, Agafonova EV, Rasskazov VA (2005) A highly active alkaline phosphatase from the marine bacterium *Cobetia*. Mar Biotechnol 7:173–178

Poli A, Moriello VS, Esposito E, Lama L, Gambacorta A, Nicolaus B (2004) Exopolysaccharide production by a new *Halomonas* strain CRSS isolated from saline lake Cape Russell in Antarctica growing on complex and defined media. Biotechnol Lett 26:1635–1638

Poli A, Esposito E, Orlando P, Lama L, Giordano A, de Appolonia F, Nicolaus B, Gambacorta A (2007) *Halomonas alkaliantarctica* sp. nov., isolated from saline lake Cape Russell in Antarctica, an alkalophilic moderately halophilic, exopolysaccharide-producing bacterium. Syst Appl Microbiol 30:31–38

Prabhu J, Schauwecker F, Grammel N, Keller U, Bernhard M (2004) Functional expression of the ectoine hydroxylase gene (*thpD*) from *Streptomyces chrysomallus* in *Halomonas elongata*. Appl Environ Microbiol 70:3130–3132

Prado B, Lizama C, Aguilera M, Ramos-Cormenzana A, Fuentes S, Campos V, Monteoliva-Sánchez M (2006) *Chromohalobacter nigrandesensis* sp. nov., a moderately halophilic, Gram-negative bacterium isolated from Lake Tebenquiche on the Atacama saltern, Chile. Int J Syst Evol Microbiol 56:647–651

Qu L, Lai Q, Zhu F, Hong X, Zhang J, Shao Z, Sun X (2011) *Halomonas daqiaonensis* sp. nov., a moderately halophilic, denitrifying bacterium isolated from a littoral saltern of the Yellow Sea of China. Int J Syst Evol Microbiol. doi: 10.1099/ijs.0.026344-0

Quesada E, Valderrama MJ, Bejar V, Ventosa A, Gutiérrez MC, Ruiz-Berraquero F, Ramos-Cormenzana A (1990) *Volcaniella eurihalina* gen. nov., sp. nov., a moderately halophilic nonmotile gram-negative rod. Int J Syst Bacteriol 40:261–267

Quillaguamán J, Delgado O, Mattiasson B, Hatti-Kaul R (2004a) *Chromohalobacter sarecensis* sp. nov., a psychrotolerant moderate halophile isolated from the saline Andean region of Bolivia. Int J Syst Evol Microbiol 54:1921–1926

Quillaguamán J, Hatti-Kaul R, Mattiasson B, Alvarez MT, Delgado O (2004b) *Halomonas boliviensis* sp. nov., an alkalitolerant, moderate halophile isolated from soil around a Bolivian hypersaline lake. Int J Syst Evol Microbiol 54:721–725

Quillaguamán J, Hashim S, Bento F, Mattiasson B, Hatti-Kaul R (2005) Poly(β-hydroxybutyrate) production by a moderate halophile, *Halomonas boliviensis* LC1 using starch hydrolysate as substrate. J Appl Microbiol 99:151–157

Quillaguamán J, Doan-Van T, Guzmán H, Guzmán D, Martín J, Everest A, Hatti-Kaul R (2008) Poly(3-hydroxybutyrate) production by *Halomonas boliviensis* in fed-batch culture. Appl Microbiol Biotechnol 78:227–232

Quillaguamán J, Guzmán H, Van-Thuoc D, Hatti-Kaul R (2010) Synthesis and production of polyhydroxyalkanoates by halophiles: current potential and future prospects. Appl Microbiol Biotechnol 85:1687–1696

Raj KC, Talarico LA, Ingram LO, Maupin-Furlow JA (2002) Cloning and characterization of the *Zymobacter palmae* pyruvate decarboxylase gene (pdc) and comparison to bacterial homologues. Appl Environ Microbiol 68:2869–2876

Reddy GSN, Raghavan PUM, Sarita NB, Prakash JSS, Nagesh N, Delille D, Shivaji S (2003) *Halomonas glaciei* sp. nov. isolated from fast ice of Adelie Land, Antarctica. Extremophiles 7:55–61

Reid MF, Fewson CA (1994) Molecular characterization of microbial alcohol dehydrogenases. Crit Rev Microbiol 20:13–56

Robinson J, Gibbons NE (1952) The effects of salts on the growth of *Micrococcus halodenitrificans* n. sp. Can J Bot 30:147–154

Rodríguez-Moya J, Argandoña M, Reina-Bueno M, Nieto JJ, Iglesias-Guerra F, Jebbar M, Vargas C (2010) Involvement of EupR, a response regulator of the NarL/FixJ family, in the control of the uptake of the compatible solutes ectoines by the halophilic bacterium *Chromohalobacter salexigens*. BMC Microbiol 10:256

Rodríguez-Sáiz M, Sánchez-Porro C, De La Fuente JL, Mellado E, Barredo JL (2007) Engineering the halophilic bacterium *Halomonas elongata* to produce β-carotene. Appl Microbiol Biotechnol 77:637–643

Rohban R, Amoozegar MA, Ventosa A (2009) Screening and isolation of halophilic bacteria producing extracellular hydrolyses from Howz Soltan Lake, Iran. J Ind Microbiol Biotechnol 36:333–340

Rojas R, Miranda CD, Amaro AM (2009) Pathogenicity of a highly exopolysaccharide-producing *Halomonas* strain causing epizootics in larval cultures of the Chilean scallop *Argopecten purpuratus* (Lamarck, 1819). Microb Ecol 57:129–139

Romanenko LA, Schumann P, Rohde M, Mikhailov VV, Stackebrandt E (2002) *Halomonas halocynthiae* sp. nov., isolated from the marine ascidian *Halocynthia aurantium*. Int J Syst Evol Microbiol 52:1767–1772

Romano I, Nicolaus B, Lama L, Manca MC, Gambacorta A (1996) Characterization of a haloalkalophilic strictly aerobic bacterium, isolated from Pantelleria Island. Syst Appl Microbiol 19:326–333

Romano I, Nicolaus B, Lama L, Trabasso D, Caracciolo G, Gambacorta A (2001) Accumulation of osmoprotectants and lipid pattern modulation in response to growth conditions by *Halomonas pantelleriense*. Syst Appl Microbiol 24:342–352

Romano I, Giordano A, Lama L, Nicolaus B, Gambacorta A (2005) *Halomonas campaniensis* sp. nov., a haloalkaliphilic bacterium isolated from a mineral pool of Campania Region, Italy. Syst Appl Microbiol 28:610–618

Romano I, Lama L, Nicolaus B, Poli A, Gambacorta A, Giordano A (2006) *Halomonas alkaliphila* sp. nov., a novel halotolerant alkaliphilic bacterium isolated from a salt pool in Campania (Italy). J Gen Appl Microbiol 52:339–348

Romano I, Lama L, Orlando P, Nicolaus B, Giordano A, Gambacorta A (2007) *Halomonas sinaiensis* sp. nov., a novel halophilic bacterium isolated from a salt lake inside Ras Muhammad Park, Egypt. Extremophiles 11:789–796

Rosenberg A (1983) *Pseudomonas halodurans* sp. nov., a halotolerant bacterium. Arch Microbiol 136:117–123

Ruiz-Ruiz C, Srivastava GK, Carranza D, Mata JA, Llamas I, Santamaría M, Quesada E, Molina IJ (2011) An exopolysaccharide produced by the novel halophilic bacterium *Halomonas stenophila* strain B100 selectively induces apoptosis in human T leukaemia cells. Appl Microbiol Biotechnol 89:345–355

Sánchez-Porro C, Martín S, Mellado E, Ventosa A (2003) Diversity of moderately halophilic bacteria producing extracellular hydrolytic enzymes. J Appl Microbiol 94:295–300

Sánchez-Porro C, Tokunaga H, Tokunaga M, Ventosa A (2007) *Chromohalobacter japonicus* sp. nov., a moderately halophilic bacterium isolated from a Japanese salty food. Int J Syst Evol Microbiol 57:2262–2266

Sánchez-Porro C, de la Haba RR, Soto-Ramírez N, Márquez MC, Montalvo-Rodríguez R, Ventosa A (2009) Description of *Kushneria aurantia* gen. nov., sp. nov., a novel member of the family *Halomonadaceae*, and a proposal for reclassification of *Halomonas marisflavi* as *Kushneria marisflavi* comb. nov., of *Halomonas indalinina* as *Kushneria indalinina* comb. nov. and of *Halomonas avicenniae* as *Kushneria avicenniae* comb. nov. Int J Syst Evol Microbiol 59: 397–405

Sánchez-Porro C, Kaur B, Mann H, Ventosa A (2010) *Halomonas titanicae* sp. nov., a halophilic bacterium isolated from the RMS *Titanic*. Int J Syst Evol Microbiol 60:2768–2774

Sauer T, Galinski EA (1998) Bacterial milking: a novel bioprocess for production of compatible solutes. Biotechnol Bioeng 57:306–313

Schubert T, Maskow T, Benndorf D, Harms H, Breuer U (2007) Continuous synthesis and excretion of the compatible solute ectoine by a transgenic, nonhalophilic bacterium. Appl Environ Microbiol 73:3343–3347

Schweikhard ES, Kuhlmann SI, Kunte HJ, Grammann K, Ziegler CM (2010) Structure and function of the universal stress protein TeaD and its role in regulating the ectoine transporter *TeaABC* of *Halomonas elongata* DSM 2581^T. Biochemistry 49:2194–2204

Schwibbert K, Marin-Sanguino A, Bagyan I, Heidrich G, Lentzen G, Seitz H, Rampp M, Schuster SC, Klenk HP, Pfeiffer F, Oesterhelt D, Kunte HJ (2011) A blueprint of ectoine metabolism from the genome of the industrial producer *Halomonas elongata* DSM 2581^T. Environ Microbiol. doi: 10.1111/j.1462-2920.2010.02336.x.

Shapovalova AA, Khijniak TV, Tourova TP, Sorokin DY (2009) *Halomonas chromatireducens* sp. nov., a new denitrifying facultatively haloalkaliphilic bacterium from solonchak soil capable of aerobic chromate reduction. Microbiology (English translation of Mikrobiologiya) 78:102–111

Shrestha A, Dhamwichukorn S, Jenwitheesuk E (2010) Modeling of pyruvate decarboxylases from ethanol producing bacteria. Bioinformation 4:378–384

Shrivastav A, Mishra SK, Shethia B, Pancha I, Jain D, Mishra S (2010) Isolation of promising bacterial strains from soil and marine environment for polyhydroxyalkanoates (PHAs) production utilizing Jatropha biodiesel byproduct. Int J Biol Macromol 47:283–287

Simon-Colin C, Raguenes G, Cozien J, Guezennec JG (2008) *Halomonas profundus* sp. nov., a new PHA-producing bacterium isolated from a deep-sea hydrothermal vent shrimp. J Appl Microbiol 104:1425–1432

Sorokin DY, Tindall BJ (2006) The status of the genus name *Halovibrio* Fendrich 1989 and the identity of the strains *Pseudomonas halophila* DSM 3050 and *Halomonas variabilis* DSM 3051. Request for an opinion. Int J Syst Evol Microbiol 56:487–489

Stackebrandt E, Frederiksen W, Garrity GM, Grimont PAD, Kämpfer P, Maiden MCJ, Nesme X, Rosselló-Mora R, Swings J, Trüper HG, Vauterin L, Ward AC, Whitman WB (2002) Report of the ad hoc committee for the re-evaluation of the species definition in bacteriology. Int J Syst Evol Microbiol 52:1043–1047

Stevens DA, Hamilton JR, Johnson N, Kim KK, Lee JS (2009) *Halomonas*, a newly recognized human pathogen causing infections and contamination in a dialysis center: three new species. Medicine (Baltimore) 88:244–249

Sutherland IW (1998) Novel and established application of microbial polysaccharides. Trends Biotechnol 16:41–46

Tegos G, Vargas C, Vartholomatos G, Perysinakis A, Nieto JJ, Ventosa A, Drainas C (1997) Identification of a promoter region on the *Halomonas elongata* cryptic plasmid pHE1 employing the *inaZ* reporter gene of *Pseudomonas syringae*. FEMS Microbiol Lett 154:45–51

Tegos G, Vargas C, Perysinakis A, Koukkou AI, Christogianni A, Nieto JJ, Ventosa A, Drainas C (2000) Release of cell-free ice nuclei from *Halomonas elongata* expressing the ice nucleation gene *inaZ* of *Pseudomonas syringae*. J Appl Microbiol 89:785–792

Tetsch L, Kunte HJ (2002) The substrate-binding protein TeaA of the osmoregulated ectoine transporter TeaABC from *Halomonas elongata*: purification and characterization of recombinant TeaA. FEMS Microbiol Lett 211:213–218

Tindall BJ, Rosselló-Móra R, Busse H-J, Ludwig W, Kämpfer P (2010) Notes on the characterization of prokaryote strains for taxonomic purposes. Int J Syst Evol Microbiol 60:249–266

Van-Thuoc D, Guzmán H, Quillaguamán J, Hatti-Kaul R (2010a) High productivity of ectoines by *Halomonas boliviensis* using a combined two-step fed-batch culture and milking process. J Biotechnol 147:46–51

Van-Thuoc D, Guzmán H, Thi-Hang M, Hatti-Kaul R (2010b) Ectoine production by *Halomonas boliviensis*: optimization using response surface methodology. Mar Biotechnol (NY) 12:586–593

Vargas C, Fernández-Castillo R, Cánovas D, Ventosa A, Nieto JJ (1995) Isolation of cryptic plasmids from moderately halophilic eubacteria of the genus *Halomonas*. Characterization of a small plasmid from *H. elongata* and its use for shuttle vector construction. Mol Gen Genet 246:411–418

Vargas C, Tegos G, Drainas C, Ventosa A, Nieto JJ (1999a) Analysis of the replication region of the cryptic plasmid pHE1 from the moderate halophile *Halomonas elongata*. Mol Gen Genet 261:851–861

Vargas C, Tegos G, Vartholomatos G, Drainas C, Ventosa A, Nieto JJ (1999b) Genetic organization of the mobilization region of the plasmid pHE1 from *Halomonas elongata*. Syst Appl Microbiol 22:520–529

Vargas C, Argandoña M, Reina-Bueno M, Rodríguez-Moya J, Fernández-Aunión C, Nieto JJ (2008) Unravelling the adaptation responses to osmotic and temperature stress in *Chromohalobacter salexigens*, a bacterium with broad salinity tolerance. Saline Systems 4:14

Ventosa A, Nieto JJ (1995) Biotechnological applications and potentialities of halophilic microorganisms. World J Microbiol Biotechnol 11:85–94

Ventosa A, Gutiérrez MC, García MT, Ruiz-Berraquero F (1989) Classification of "*Chromobacterium marismortui*" in a new genus, *Chromohalobacter* gen. nov., as *Chromohalobacter marismortui* comb. nov., nom. rev. Int J Syst Bacteriol 39:382–386

Ventosa A, Nieto JJ, Oren A (1998) Biology of moderately halophilic aerobic bacteria. Microbiol Mol Biol Rev 62:504–544

Vreeland RH, Litchfield CD, Martin EL, Elliot E (1980) *Halomonas elongata*, a new genus and species of extremely salt tolerant bacteria. Int J Syst Bacteriol 30:485–495

Wang Y-N, Cai H, Chi C-Q, Lu A-H, Lin X-G, Jiang Z-F, Wu X-L (2007a) *Halomonas shengliensis* sp. nov., a moderately halophilic, denitrifying, crude-oil-utilizing bacterium. Int J Syst Evol Microbiol 57:1222–1226

Wang Y-N, Cai H, Yu S-L, Wang Z-Y, Liu J, Wu X-L (2007b) *Halomonas gudaonensis* sp. nov., isolated from a saline soil contaminated by crude oil. Int J Syst Evol Microbiol 57:911–915

Wang Y, Tang S-K, Lou K, Mao P-H, Jin X, Jiang C-L, Xu L-H, Li W-J (2008a) *Halomonas lutea* sp. nov., a moderately halophilic bacterium isolated from a salt lake. Int J Syst Evol Microbiol 58:2065–2069

Wang Y, Wu Y-H, Wang C-S, Xu X-W, Oren A, Zhu X-F, Wu M (2008b) *Halomonas salifodinae* sp. nov., a halophilic bacterium isolated from a salt mine in China. Int J Syst Evol Microbiol 58: 2855–2858

Wang Y, Tang S-K, Lou K, Lee J-C, Jeon CO, Xu L-H, Kim C-J, Li W-J (2009) *Aidingimonas halophila* gen. nov., sp. nov., moderately halophilic bacteria isolated from a salt lake. Int J Syst Evol Microbiol 59:3088–3094

Wolfe-Simon F, Blum JS, Kulp TR, Gordon GW, Hoeft SE, Pett-Ridge J, Stolz JF, Webb SM, Weber PK, Davies PC, Anbar AD, Oremland RS (2010) A bacterium that can grow by using arsenic instead of phosphorus. Science, published online December 2, 2010. doi: 10.1126/science.1197258

Wu G, Wu X-Q, Wang Y-N, Chi C-Q, Tang Y-Q, Kida K, Wu X-L, Luan Z-K (2008a) *Halomonas daqingensis* sp. nov., a moderately halophilic bacterium isolated from an oilfield soil. Int J Syst Evol Microbiol 58:2859–2865

Wu Y-H, Xu X-W, Huo Y-Y, Zhou P, Zhu X-F, Zhang H-B, Wu M (2008b) *Halomonas caseinilytica* sp. nov., a halophilic bacterium isolated from a saline lake on the Qinghai-Tibet Plateau, China. Int J Syst Evol Microbiol 58:1259–1262

Xu X-W, Wu Y-H, Zhou Z, Wang C-S, Zhou Y-G, Zhang H-B, Wang Y, Wu M (2007) *Halomonas saccharevitans* sp. nov., *Halomonas arcis* sp. nov. and *Halomonas subterranea* sp. nov., halophilic bacteria isolated from hypersaline environments of China. Int J Syst Evol Microbiol 57:1619–1624

Yanase H, Yamamoto K, Sato D, Okamoto K (2005) Ethanol production from cellobiose by *Zymobacter palmae* carrying the *Ruminocuccus albus* β-glucosidase gene. J Biotechnol 118:35–43

Yanase H, Sato D, Yamamoto K, Matsuda S, Yamamoto S, Okamoto K (2007) Genetic engineering of *Zymobacter palmae* for production of ethanol from xylose. Appl Environ Microbiol 73: 2592–2599

Yang C, Cao G, Li Y, Zhang X, Ren H, Wang X, Feng J, Zhao L, Xu P (2008) A constructed alkaline consortium and its dynamics in treating alkaline black liquor with very high pollution load. PLoS ONE 3:e3777

Yang C, Niu Y, Su H, Wang Z, Tao F, Wang X, Tang H, Ma C, Xu P (2010a) A novel microbial habitat of alkaline black liquor with very high pollution load: microbial diversity and the key members in application potentials. Bioresour Technol 101:1737–1744

Yang C, Wang Z, Li Y, Niu Y, Du M, He X, Ma C, Tang H, Xu P (2010b) Metabolic versatility of halotolerant and alkaliphilic strains of *Halomonas* isolated from alkaline black liquor. Bioresour Technol 101:6778–6784

Yoon J-H, Lee K-C, Kho YH, Kang KH, Kim C-J, Park Y-H (2002) *Halomonas alimentaria* sp. nov., isolated from jeotgal, a traditional Korean fermented seafood. Int J Syst Evol Microbiol 52:123–130

Zeigler DR (2003) Gene sequences useful for predicting relatedness of whole genomes in bacteria. Int J Syst Evol Microbiol 53:1893–1900

Zhang LH, Lang YJ, Nagata S (2009) Efficient production of ectoine using ectoine-excreting strain. Extremophiles 13:717–724

Zhou MY, Chen XL, Zhao HL, Dang HY, Luan XW, Zhang XY, He HL, Zhou BC, Zhang YZ (2009) Diversity of both the cultivable protease-producing bacteria and their extracellular proteases in the sediments of the South China sea. Microb Ecol 58:582–590

ZoBell CE, Upham HC (1944) A list of marine bacteria including descriptions of sixty new species. Bull Scripps Inst Oceanogr Univ Calif 5:239–292

Zou Z, Wang G (2010) *Kushneria sinocarnis* sp. nov., a moderately halophilic bacterium isolated from a Chinese traditional cured meat. Int J Syst Evol Microbiol 60:1881–1886

Chapter 4
The Hypersaline Lakes of Inner Mongolia: The MGAtech Project

William D. Grant, Eulyn Pagaling, M. Carmen Márquez, M. Carmen Gutiérrez, Don A. Cowan, Yanhe Ma, Brian E. Jones, Antonio Ventosa, and Shaun Heaphy

4.1 The MGAtech Project

After several years of informal collaboration based on a common interest in the microbial ecology of extreme environments, in the late 1990s, colleagues from five institutions ranging from academia to industry conceived the idea of an European Union–China project to investigate some of the many microbiologically largely unexplored thermal and saline sites in China. The partners in this ambitious venture were the University of Leicester in the UK, the University of Seville in Spain, the University of the Western Cape in South Africa, Beijing's Institute of Microbiology at the Chinese Academy of Sciences and industrial partner Genencor International in The Netherlands.

The concept of the project was that the Chinese environments were likely to harbor hitherto unaccessed microbial genetic resources that could be harnessed as sources of new and valuable products. Central to the rationale behind the bid for funding was the observation that while conventional culturing procedures were

W.D. Grant (✉) • E. Pagaling • S. Heaphy
Department of Infection, Immunity and Inflammation, University of Leicester, Leicester LE1 9HN, UK
e-mail: wdg1@leicester.ac.uk

M.C. Márquez • M.C. Gutiérrez • A. Ventosa
Department of Microbiology and Parasitology, Faculty of Pharmacy, University of Sevilla, 41012 Sevilla, Spain

D.A. Cowan
Institute for Microbial Biotechnology and Metagenomics, University of the Western Cape, Cape Town, South Africa

Y. Ma
State Key Laboratory of Microbial Resources, Institute of Microbiology, Chinese Academy of Sciences, Beijing 100101, China

B.E. Jones
Genencor International BV, Archimedesweg 30, Leiden 2333CN, The Netherlands

A. Ventosa et al. (eds.), *Halophiles and Hypersaline Environments*,
DOI 10.1007/978-3-642-20198-1_4,

often quite successful in obtaining new organisms with their concomitant genetic resource, it is known that only a small proportion (0.01–10%) of the microbial population from any environment is cultivable, and as a consequence, the majority of microbial genes are inaccessible for study using culture techniques. Accordingly, the project was designed to be innovative in that it proposed the additional use of methods to efficiently access total microbial genomic diversity (both prokaryotic and eukaryotic) in these environments through the direct acquisition of the total metagenome by on site DNA and RNA extraction (hence the project acronym MGAtech-*Multigenome Access Technology for Industrial Catalysts*).

Three preparatory meetings in Beijing took place, and in October 2002, the five partners were successful in obtaining research funding of one million euros from the European Union under Framework V. Supporting funds were provided by agencies in China and South Africa plus the industrial partner. The University of Leicester coordinated the project in view of expedition skills developed over many years in African locations. This was the first time that European Scientists had been given access to Chinese extreme environments and established an important bridge-head in microbial biotechnology cooperation between the EU and China with all the concomitant scientific and political interactions.

The objectives of the project were:

1. To obtain material from a range of unique Chinese hypersaline and thermal sites.
2. Survey the microbial biodiversity in these environments. Establish and archive a collection of novel microbes.
3. Refine methods for the efficient isolation and recovery of microbial genes from these sites.
4. Identify and produce high added value new biomolecules derived from these sites and assess these new biomolecules for commercial exploitation.

Three sampling expeditions were conducted involving around ten personnel representing the five partners. The first expedition to the geothermal areas of the Yunnan Province (the Rehai area of the Tenchong region) was completed in March 2003. The second expedition was to the hypersaline lakes of Inner Mongolia in September 2003. The third expedition to hypersaline sites in Tibet took place in August 2004. Sites were extensively described, physical parameters measured and samples taken for later culture studies and water chemistry determinations. DNA and RNA extraction and stabilization on site was undertaken, modifying procedures previously developed on African expeditions.

The project ran from October 2002 to June 2006 and during this period more than 100 hydrolase-producing thermophiles were isolated from the Tenchong sites including examples of a new genus (Xue et al. 2006). The hypersaline sites of Inner Mongolia and Tibet produced nearly 400 isolates of halophilic and haloalkaliphilic bacteria and archaea including several new genera of each. Screening produced several enzymes appropriate for biotechnological exploitation. Access to the metagenome in Yunnan and Inner Mongolian samples showed great prokaryote biodiversity, indicating great potential for isolating novel genes from gene libraries. Eukaryote biodiversity was low but the proposed technology for isolating functional eukaryote genes via cDNA generation from isolated RNA was proven to work well (Grant et al. 2006).

This chapter describes studies carried out on seven Inner Mongolian hypersaline lakes during the MGAtech project.

4.2 The Study Area

The Autonomous Region of Inner Mongolia in the northern part of the P.R. China is a large inland plateau. The area of study, around 600 km north of the Chinese capital, Beijing is located in the Xilin Gol League (Prefecture) in an area stretching roughly from the city of Xilinhot (43°57′N, 116°5′E) in the east to Erenhot (also called Erlian) (43°38′N, 111°58′E) on the border with Mongolia in the west. At an elevation of around 1,000 m the region experiences a typical temperate semi-arid continental climate with about 150–400 mm annual precipitation, mainly in the summer months, with mean maximum and minimum temperatures of 19°C and −22°C, respectively. Strong winds and high rates of evapotranspiration, often exceeding annual precipitation, are also characteristic of the region. The study area featured typical undulating steppe grassland at 1,250 m in the east descending to 900 m with [Gobi] desert scrub in the west. Two major fault lines, the Erenhot and Xilinhot faults, of tectonic (Paleozoic era) origin running SW to NE form shallow depressions in which salt lakes can form (Zheng 1991; Yu et al. 2001). The lake basins are often the sites of Quaternary precursor lakes with serial evaporite and sedimentary deposits reflecting climatic changes since the collision of the Indian and Laurasian tectonic plates. Since this is now a tectonically stable region (Zheng et al. 1993), it is aeolian activity and deflation of recent origin, not tectonic activity (in contrast to the soda lakes of the East African Rift Valley (Jones et al. 1994)) which has led to the development of the present large, shallow endorheic drainage basins in which one or more salt lakes have developed (Williams 1991). During the Holocene the climate became drier and lake waters evaporated and concentrated, creating conditions favorable for salt formation as evaporite deposits (Zheng 1991).

Many of the shallow lakes of Inner Mongolia are ephemeral or have variable morphometry. Few have been investigated limnologically and for many basic chemical data such as salinity and major ionic composition are lacking (Williams 1991). Such data that do exist (Zheng et al. 1992, 1993) suggest that many of the saline lakes along the Erenhot and Xilinhot faults are alkaline; $HCO_3^- + CO_3^{2-}$ type or Na_2SO_4 sub-type. An expedition was conducted in September 2003 (after the summer monsoon rains) to collect samples for microbiological and chemical analysis from seven lakes. It is worthy of note (since rainfall affects lake salinity) that Xilinhot weather records show that the summer of 2003 experienced 42% more rainfall than average and that the previous year had been unusually wet with almost four times the usual amounts. The lakes sampled were Ejinor, Bagaejinor, an unnamed lake west of Xilinhot on 101 Provincial Road now known to be called Bayannur (Google Maps), Shangmatala, Chahannor, Chagannor, and Erliannor.

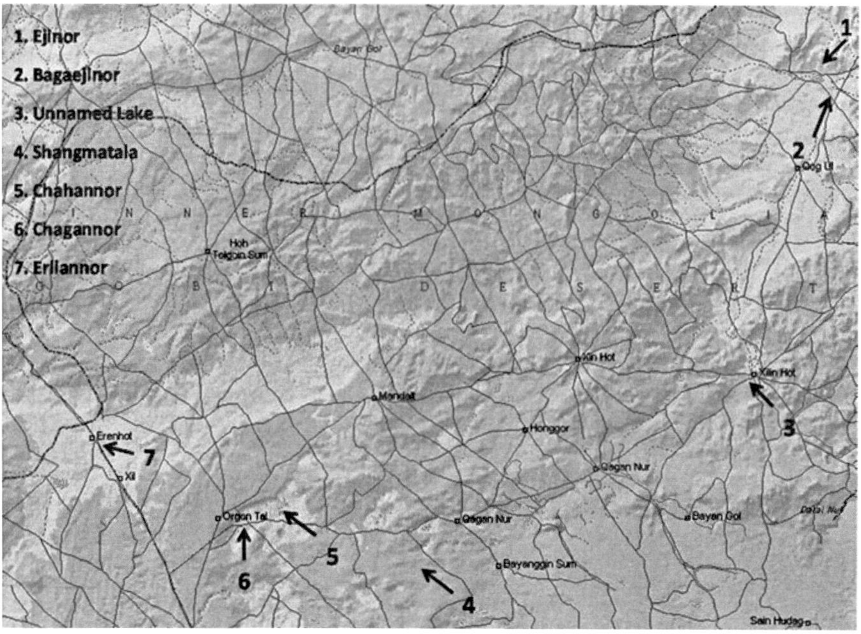

Fig. 4.1 Location of the seven hypersaline lakes studied from Inner Mongolia, northern area in P.R. China

Figure 4.1 shows the area and the location of the lakes. Samples consisted of water, sediment, salt crystals and microbial mass from filtered lake water. Secondary observations during sampling often indicated vigorous microbial activity and active mineral crystallization. Conditions during sampling were generally mild with air temperatures around 20°C, strong winds and occasional rain showers. Table 4.1 shows the features of the sampling sites in the seven lakes studied.

4.3 Description of the Lakes

Ejinor (EJ)

This is a sulfate-type salt lake with a surface area of 26 km^2. The lake is situated in the northern part of the Dabusugu tectonic basin (700 km^2) (Yu et al. 2001) and consists of an open body of water separated by a north-south causeway from a playa of salt (mainly mirabilite) sediment (15 km^2). On the north shore are a series of actively worked salterns. At the time of sampling the eastern shore consisted of a collection of dislocated evaporated pools and lagoons separated from the main body of water. The depth of brine varies with the seasons; 0.15–0.3 m in the rainy season, 0.05–0.1 m in the dry season. The lake is fed by direct precipitation (250–300 mm $year^{-1}$) and ground water. The average evaporation is 2,000 mm $year^{-1}$ (Yu et al. 2001).

Table 4.1 Features of the sampling sites from the seven lakes investigated

Sample	Coordinates	Description	T (°C)	pH	Salinity (%)	Conductivity (mS cm^{-1})
EJ1	45°14′4.52″N 116°32′4.77″E	Isolated pond	32	8.5	15.1	166
EJ2	–	Yellow colored lagoon	28	7.2	13.8	–
EJ3	45°14′7.94″N 116°32′8.52″E	Saltern of reddish water; orange flaky crystals on bottom	28	7.5	11.1	161
EJ4	–	Green scum, drainage channel	–	7.5	–	–
BJ1	45°08′5.27″N 116°36′1.67″E	Isolated pond	21	8.5	>30	147
XH1	43°55′3.55″N 115°36′7.57″E	Open water	26	9.0	–	–
XH2	43°55′1.85″N 115°37′5.20″E	Small lagoon	24	8.5	>30	185
XH3	43°55′1.70″N 115°37′.12″E	Drainage channel	21	9.5	–	287
SH1	43°12′7.51″N 114°01′3.61″E	West lagoon	20	8.5	16.7	480
SH2	43°12′7.51″N 114°01′3.61″E	East lagoon	22	8.5	–	502
CG1	43°16′1.31″N 112°55′6.36″E	South lagoon	17	10.5	18	213
CH1	43°21′5.83″N 113°08′1.93″E	Brown water with suspended silty clay	17	9.5	–	Very low
EN1	43°44′4.26″N 112°02′0.81″E	Saltern covered by white salt crust	17.9–17.2	7.5–8	–	44.9
EN2	–	Saltern covered by pink salt crust	18.4	7.5	–	–

–, No data available

Bagaejinor (BJ)

The lake lies in the SW of the Dabusu fault basin surrounded by grassland and has a surface area of 5 km^2. Zheng et al. (1992) classifies the lake as the sodium sulfate sub-type which is confirmed by our chemical analysis (Table 4.2). The lake had largely evaporated during the summer leaving a few small pools and lagoons in an extensive soft mud saline playa. The pools had a hard pink or brown crystalline base and comprised clear water surrounded by pink or brown flakes of salt crystals, probably mirabilite (Glauber's salt). There was much evidence of farm animal contamination close to the lake.

Table 4.2 Chemical composition of the seven lakes from Inner Mongolia, China

Lake	Year	Salinity (%)	pH	Na (M)	K (mM)	Ca (mM)	Mg (M)	Cl (M)	SO_4 (M)	HCO_3 (mM)	CO_3 (mM)	B_2O_3 (mM)	B (mM)	Li (mM)	Br (mM)	PO_4 (mM)	S (M)
Ejinor	1962	33.8	7.4	3.80	70.59		1.03	5.10	0.42		13.71	0.95					
	1983	33.3	7.5	4.35	31.07	5.09	0.69	5.17	0.32	6.87		1.81		0.61		0.008	
	2003 (EJ1)	15.1	8.5	2.71	38.26	10.85	0.68	3.87					2.83	0.56	16.2		0.36
	2003 (EJ3)	11.1	7.5	2.82	68.9	3.04	2.08	4.36		9.84	23.3		4.42	1.34	14.0		0.94
Bagaejinor	1962	26.9	8.2	3.77	103.0		0.32	4.15	0.23	21.68		1.67					
	1983	29.5	7.7	4.26	19.67	4.22	0.28	3.61	0.61	9.70	4.45	1.71		0.35		0.042	
	2003 (BJ1)	>30	8.5	5.32	33.2	0.77	0.35	4.61		7.40	3.30		4.25	0.33	8.05		1.07
Unnamed	2003 (XH2)	>30	8.5	5.06	53.1	3.1	0.08	5.4		13.9	1.7		2.87	0.23	7.78		0.33
Lake (Xilinhot)	2003 (XH3)		9.5	2.16	16.14	0.84	0.03	2.43					1.43	0.1	1.80		0.07
Shangmatala	1962	17.3	10.2	0.22	12.53	0.45	0.01	0.07	0.04	85.6	2.32	0.56					
	2003 (SH1)	16.7	8.5	5.38	150.0	2.62	0.26	4.69		7.4	13.0		13.0	1.93	5.18		0.81
Chagannor	1983	17.3	9.98	2.94	12.5		3.55	0.97	0.23	179.1	632.6	3.23				1.30	
	2003 (CG1)	18	10.5	2.89	14.0	0.12	0.001	1.08		360.0	410.0		3.81	0.014	1.25		0.43
Chahannor	1962	3.6	10.2	1.57	83.38			0.34	0.06	374.3	234.0	1.57					
	2003	Not recorded															
Erliannor	1962	20.3	8.2	2.80	61.38	17.09	0.27	2.91	0.26	12.3		0.67					
	1983	25.0	7.3	4.99	19.23	2.99	0.52	4.99	0.49	7.44		3.26		2.16		0.017	
	2003 (EN1)		8.2	4.2	39.0	0.12	0.86	5.33		4.1	8.3		5.24	3.3	9.57		0.48
	2003 (EN2)		7.5	3.23	61.66	4.61	1.64	5.35					10.57	6.4	15.3		0.53

Chemical composition in molarity. The elemental analysis was determined using Inductively Coupled Plasma Optical Emission Spectroscopy at the University of Leicester, UK. Carbonate and bicarbonate was determined by titration. The historical analyses are derived from Zheng et al. (1992)

Unnamed Lake West of Xilinhot (Bayannur) (XH)

No known descriptions of this lake exist; although it is close to the main 101 Provincial Road, about 40 km west of Xilinhot and it is clear from an abandoned soda factory and numerous lagoons and causeways, the former site of mineral extraction. An unusual feature of many of the lagoons was the active crystallization causing "iceberg-like" structures and a thick soda pavement. Also unusual in the colorless water was an orange colored brine shrimp. Since this alkaline water was close to the maximum pH for the survival of brine shrimp it is unclear if this is the same species of *Artemia* found in other salt lakes around the world. It may be more closely related to *Artemia monica*, a brine shrimp found in alkaline Mono Lake (California) than the more common *Artemia salina*.

Shangmatala (also known as Dermatala) (SH)

The lake is very isolated and lies in a shallow basin surrounded by low hills in a secondary depression of the Tenger fault basin at an elevation of 987 m. The surface area of lake is 2.5 km^2, with a 0.1–0.15 m depth of brine in the wet season. The lake is surrounded by grassland and vegetation almost up to the water's edge and there was much evidence of disturbance by grazing animals. The gravels and silty clays nearest the lake were soda soils covered by surface lichen. This layer was sampled. It was noted that an unpleasant odor was emitted, possibly dimethyl sulfide. Further disturbance by primitive mineral extraction was in evidence by piles of off-white "bricks," probably rock salt or Glauber's salt. The water was shallow – only a few cm deep and appeared pink due to the coloration of salt crystals. The surface of shore line mud was covered by rod-shaped crystals. The undisturbed lake sediments were stratified with a band of pink crystalline salt overlaying a green cyanobacterial layer (possibly *Microcoleus* sp.) above black anaerobic silt.

Chagannor (CG) (also known as Qagan Nur)

The lake is the site of the largest soda ash factory in China, the Xilingol Sunite Alkali Industry Company. The factory is served by a branch line of the Erenhot–Beijing railway and has several coal-fired kilns. A small town has grown up around the factory on the north side to accommodate factory workers. The lake consists of 21 km^2 of carbonate minerals mainly Pleistocene and Holocene series trona and/or natron deposits up to 30 meters deep (Zheng 1991). However, core samples reveal that this is a relict of a vast paleolake with nine different trona layers interspersed with muds reflecting the sedimentation cycles during the formation of a freshwater lake (20,000 years B.P.) and the evolution through wet and dry periods into an actively salt precipitating soda lake (Yu et al. 2001). Now there is little natural open water, only a series of peripheral [drainage] lagoons. The lake is situated in an arid area of scrubland and lies in the Erdaojing fault basin. Sampling took place on the south side well away from the industrial activity. Here the lake water was greenish and the shoreline consisted of fine, soft, grey, viscous mud making conditions for sampling precarious.

Chahannor (CH)

The lake lies on the Erdaojing–Xilaxulun faultline in a vast basin of scrubland and dried mudflats with a typical pattern of surface cracking, surrounded by low hills and escarpments. Zheng et al. (1992) indicate that the area of salt sediment covers 15–20 km^2 but is covered with a 2 meter layer of sandy clay and "there is a silt clay deposit formed by recent freshening environment in the lake surface [sic]." It appears that the original concentrated alkaline lake which was recorded 10 years previously has been covered by thick aeolian sediment from the encroaching Gobi desert. The basin had partly filled with recent meteoric water and surface runoff to form a small, probably temporary, largely freshwater lake of brown water on a bed of impervious clay. Satellite imagery confirms this picture and the changing nature of the landscape.

Erliannor (EN) (also known as Erliandabusnoer)

The lake is located a few km NE of Erenhot, a large city on the border with Mongolia and the Beijing branch of the Trans-Siberian railway. The area of the lake is 8.75 km^2, with 0.1–0.3 m depth of brine which is recorded (Zheng et al. 1992) as being the magnesium sulfate sub-type. The lake is situated in the Erlian fault basin containing alluvial and lacustrine deposits of sand and silty clays with Glauber's salt and halite (Zheng et al. 1992). Zheng et al. (1993) comment on the high levels of Br^- (300 mg l^{-1}) in the lake. However, we found four times this amount (1,220 mg l^{-1}; sample EN2) with significantly higher levels of boron and lithium compared to the other lakes sampled (Table 4.2). The natural lake is largely unrecognizable due to the small-scale salt industry which has given rise to an extensive series of new and abandoned salterns. According to the local workers there is approximately 40 km^2 of underground soda, which is harvested in the winter months. In the summer, microbial blooms turn the salterns red (presumably due to haloarchaea) and salt (halite) is harvested. At the time of sampling (9 September 2003), all the salterns were white and the water was clear and colorless with a thermal gradient.

4.3.1 Chemical Composition of the Lakes

Data for the major ionic composition of the lakes derived from the present study are presented together with historical data derived from Zheng et al. (1992) in Table 4.2. The data are not entirely consistent but the variation may reflect seasonal and/or longer term climatic effects. Probably of greater influence is the degree of disturbance due to mineral extraction. Clearly the chemical composition will differ depending upon whether the sample is from open water or from a crystallizer pond or saltern. The data indicates that these are all hypersaline lakes but suggest that only Chagannor can be considered to be a soda lake (pH > 9) with elevated $HCO_3^- + CO_3^{2-}$ and depleted Ca + Mg (Jones et al. 1994). The soda lake Chahannor no longer exists as a permanent surface feature and lies buried beneath windblown and alluvial sediment. Shangmatala may be evolving into an alkaline lake but the extensive disturbance of the natural lake into enclosed lagoons for mineral extraction make historical comparisons difficult. Our data confirm the lake as a source of Glauber's salt ($Na_2SO_4 \cdot 10H_2O$), halite (NaCl) and epsomite ($MgSO_4.7H_2O$) (Zheng et al. 1992). The data also confirm that Ejinor has the magnesium sulfate sub-type chemistry (Zheng et al. 1992) which concentrates in the crystallizer pond (EJ3) along with minor elements boron and lithium. A similar picture is seen at Erliannor which comprises almost entirely crystallizer ponds of various types where the co-concentration of Br^- is also significant.

4.4 Molecular Analysis of the Salt Lakes

An extensive study of the microbial populations in the water column of six salt lakes was done (at sampling points BJ1, CG1, EJ3, EN1, SH1, and XH2). Lake Chahannor was not included in this study due to the low salinity at the sampling time. Total genomic DNA was isolated from biomass collected at the sampling site by the GenomicPrep Cells and Tissue DNA Isolation Kit. The biomass (usually cells collected from brine filtration on site, filters being placed in ice-cold brine for transport to processing site) was pelleted in 1.5 ml eppendorf tubes by centrifugation at 13,000 rpm for 5 min and the DNA extracted according to manufacturer's protocols. These samples were then stable for at least 18 months at room temperature and were stored as such until further processing in the laboratory.

Molecular characterization involved cloning PCR amplified 16S rRNA archaeal and bacterial gene sequences from community DNA to construct 16S clone libraries. Unique clones were selected by patterns produced upon digestion with restriction endonucleases. Rarefaction curves were used to identify when enough clone sampling had taken place on each sample to guarantee extensive coverage of the diversity present. This statistical approach to determining overall biodiversity at these sites, together with the use of other biodiversity indices and richness estimators such as Chao1, Simpson's Index, the Shannon–Weaver Index and the Jaccard Index to compare sites is detailed in Pagaling et al. (2009).

4.4.1 Bacterial Diversity of the Salt Lakes

A total of 51% of the clone sequences were related to uncultured organisms. These sequences were found from a variety of sources including other salt lakes, crystallizers, salt plains and marine environments. More surprisingly are the clone sequences related to clones detected in freshwater environments, hot springs and anoxic sediments (since samples were taken from the water column). Seven clone sequences are related to those detected in Lake Chaka in Tibet, which is an environment where some of the archaeal clones were affiliated with (Jiang et al. 2006). Six are related to sequences found in hypersaline saltern ponds in Salin-de-Giraud, France (Mouné et al. 2003). Five were related to those found in the Great Salt Plains of Oklahoma (Caton et al. 2004).

The remaining 49% of sequences were related to known species belonging to the *Bacteroidetes* (such as *Salinibacter* and *Salegentibacter*), *Firmicutes* (such as *Orenia salinaria*, *Halanaerobacter chitinovorans*), *Alphaproteobacteria* (such as *Roseovarius* sp., *Paracoccus* sp.), *and Gammaproteobacteria* (the Halomonads). There are also clone sequences related to chloroplast rRNAs from eukaryotes *Dunaliella salina* and *Chlamydomonas applanata*.

Phylogenetic trees were drawn containing clone sequences and existing 16S rRNA gene sequences from all major bacterial lines of descent known to have halophilic or halotolerant members (Grant 2004), as well as the sequences of the clones' nearest neighbors. The related numbered sequences shown in the trees are unpublished sequence data base entries. The root chosen for the bacteria was *Methanospirillum hungatei*. Resulting trees distribute the clone sequences from the salt lakes into nine monophyletic assemblages: *Bacteroidetes*, Chloroplasts (*Chlorophyceae*), *Cyanobacteria*, *Aquificae*, *Actinobacteria*, *Firmicutes*, *Alphaproteobacteria*, *Deltaproteobacteria*, and *Gammaproteobacteria*. Bootstrap values above 50% are shown and clone sequences are color coded according to the lake they were detected in. Many of these groups are well supported: Chloroplasts (97%), *Cyanobacteria* (68%), *Aquificae* (67%), and *Gammaproteobacteria* (67%). There are also additional well supported lineages (80–98%) that branch between these known groups, designated Salt Lake (SL) Clusters 1–3. Two additional SL Clusters (4 and 5) are near the *Aquificae* and the *Firmicutes*, both with low bootstrap values. The largest groups are the *Proteobacteria* with 30% of the clone sequences and the *Bacteroidetes* and *Firmicutes*, both with 20% of the clone sequences.

4.4.1.1 *Bacteroidetes*

This group is shown in Fig. 4.2, which has been magnified from the original tree (see Fig. 4.4), and therefore, does not show the root. A small group of sequences related to *Salinibacter* has formed their own well-supported clade (93%) within the *Bacteroidetes*. *Salinibacter* are typical inhabitants of salt lakes and salterns. One clone sequence from EJ3 (164) is closely related to halophilic eubacterium EHB-2 (99%). This clone was in fact the most frequently occurring in the 16S library for EJ3, with a clone frequency of 24. It was originally detected in salterns in Alicante, Spain, which had total salinities between 22.4% and 37% (w/v). Studies by Antón and colleagues (2000, 2002) recognised the existence of large numbers of these

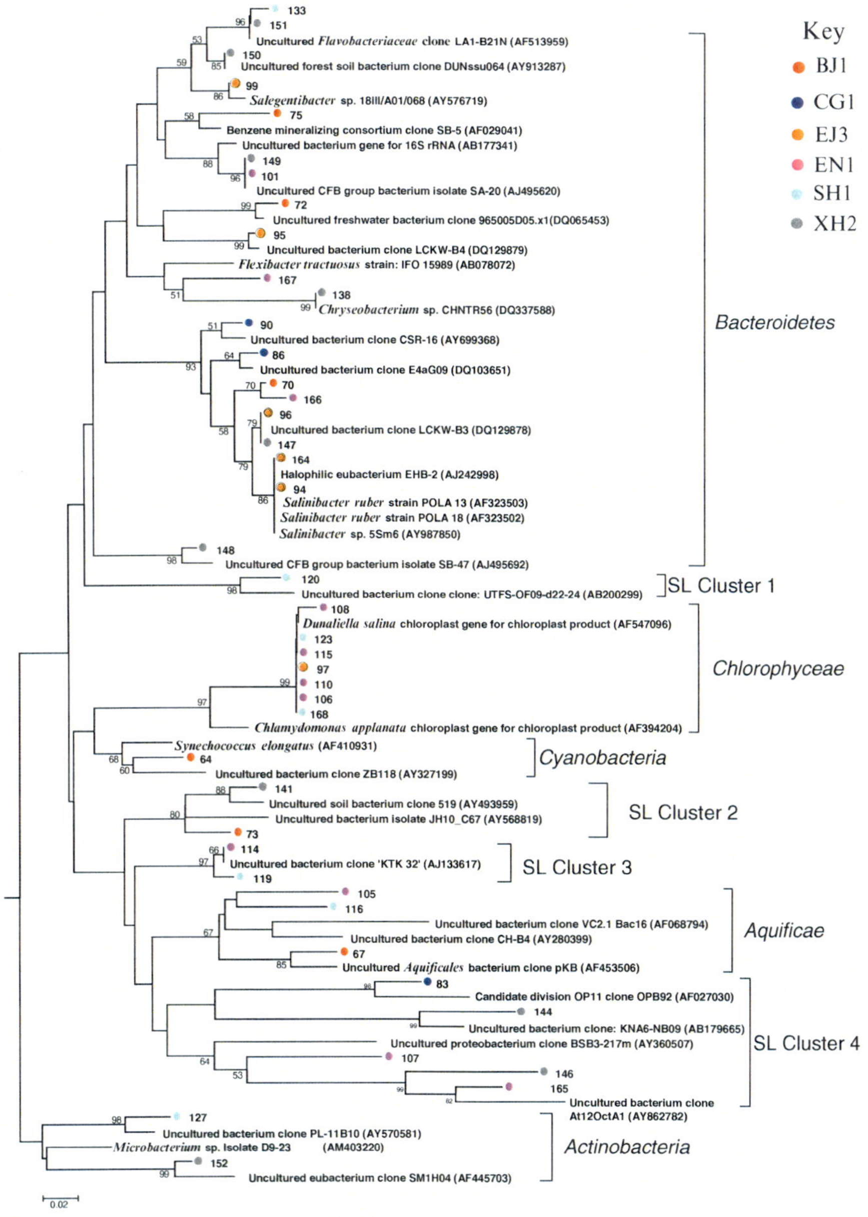

Fig. 4.2 Phylogenetic relationships showing the clone sequences with existing 16S rRNA gene sequences for *Bacteroidetes*, chloroplasts, *Cyanobacteria*, *Aquificae*, *Actinobacteria* and other uncultured isolates

Salinibacter types in hypersaline environments, making up the major bacterial constituent of these microbial ecosystems. One clone sequence from EN1 (101) and one from XH2 (149) are related to a bacterial isolate found in anoxic sediments underlying cyanobacterial mats in hypersaline saltern ponds in Salin-de-Giraud, France.

Apart from the *Salinibacter* clade, several other sequences are found within the *Bacteroidetes*. One clone sequence from SH1 (133) and XH2 (151) are related to *Flavobacteriaceae* clone LA1-B21N (both at 96%), detected in inland waters of remote Hawaiian islands, either from hypersaline Lake Laysan or from a brackish pond (Donachie et al. 2004). Several other sequences are related to unknown uncultured isolates from various environments. One clone sequence from BJ1 (75) is related to an uncultured bacterium clone: ODP1251B5.13 (93%) from deep marine sediments of the Pacific Ocean Margin (at the Peru and Cascadia Margins) that contain methane and methane hydrate. Phylogenetic analysis in that study similarly placed this clone with the *Bacteroidetes* (Inagaki et al. 2006).

4.4.1.2 Chloroplasts

Also illustrated in Fig. 4.2 are several sequences related to chloroplast genes from *Chlorophyceae*. It is likely that eukaryotes were present in the salt lakes, and the chloroplasts in them were lysed and DNA extracted along with the rest of the community DNA. The detection of these genes in the clone libraries has, therefore, given an insight into the eukaryotic diversity in the salt lakes. Eukaryotic algae, particularly *Dunaliella salina* are typical inhabitants of salt lakes, contributing to primary production, providing a source of organic compounds (Oren 1994). Several clone sequences from lakes EN1 (106, 108, 110 and 115) and SH1 (123 and 168) are related to chloroplasts in *Dunaliella salina* (97–99%). This is in fact the most frequently observed clone in the 16S libraries for both EN1 and SH1, with a clone frequency of 19 and 6, respectively. One clone sequence from EJ3 (97) is related to *Chlamydomonas applanata* (95%).

4.4.1.3 *Cyanobacteria*

Only one clone sequence from BJ1 (64) was detected, appearing in the same clade as *Synechococcus elongatus* (see Fig. 4.2) an oxygenic photosynthetic cyanobacterium also contributing to primary production in saline environments. This sequence is related to clone ZB118, showing 95% similarity. This clone was detected in a slightly saline sulfide-rich spring. Phylogenetic analysis in that study also placed clone ZB118 with the *Cyanobacteria*, with *Oscillatoria terebriformis* as its nearest neighbor, a known mat-building organism.

4.4.1.4 *Aquificae*

It is surprising to find sequences related to *Aquificales*, since representatives of this group are not halophilic or halotolerant (Grant 2004), although they have been found in deep sea hydrothermal vent systems (Reysenbach et al. 2000) so are presumably tolerant to marine conditions. Nevertheless, there appears to be one clone sequence from BJ1 (67) that is related to an uncultured *Aquificales* clone pKB (94%), which comes from a near-neutral thermal Spring in Kamchatka, Russia's largest volcanic belt flanked by the Pacific Ocean and the Sea of Okhotsk. Moreover, other sequences appear in the same clade as this *Aquificales* clone. A clone sequence from EN1 (105) is distantly related to an uncultured bacterium clone VC2.1 Bac16 (88%). This clone comes from a growth chamber deployed in a Mid-Atlantic Ridge hydrothermal vent. Another clone sequence from SH1 (116) is related to a sequence that came from a *Paralvinella sulfincola* tube-worm and the adjacent substratum on an active deep-sea vent chimney (Page et al. 2004).

4.4.1.5 *Actinobacteria*

Actinobacteria are high% G+C Gram-positives that are readily found in saline environments (Grant 2004). However, only two clone sequences were detected that appear in the same clade as *Microbacterium* sp. isolate D9–23 (Fig. 4.2), which was found in a marine aquaculture biofilter. One clone sequence from SH1 (127) is distantly related to clone PL-11B10 (91%), which was detected in the waters of a low temperature biodegraded oil reservoir. Clone sequence 152 from XH2 is distantly related to eubacterium clone SM1H04 (91%), which was detected in Angel Terrace at the Mammoth Hot Springs in Yellowstone National Park, Wyoming, USA.

4.4.1.6 *Firmicutes*

The *Firmicutes* are low% G+C Gram-positives, also readily isolated from saline environments (Grant 2004) and was in fact one of the largest groups detected in five of the six salt lakes. This group consists of several haloanaerobes: *Halanaerobacter*, *Halanaerobium*, *Orenia* and "*Natronoanaerobium*," which can be seen in Fig. 4.3. This has also been magnified from the original tree, and therefore, does not show the root. Several clone sequences from SH1 (124, 125 and 134) and one from XH2 (143) are related to *Halanaerobacter lacunaris* (96–98%). This isolate was found in the silt of the hypersaline Lake Chokrak on the Kerch Peninsula (Zhilina et al. 1992). One clone sequence from XH2 (136) was related to *Halanaerobacter chitinovorans* (98%). This isolate was found in sediment in a solar saltern in Chula Vista in Southern California (Liaw and Mah 1992). One sequence from CG1 (77) is distantly related to *Halanaerobiaceae* bacterium SLAS-1 (90%) isolated from Searles Lake, which is an alkaline salt-saturated brine,

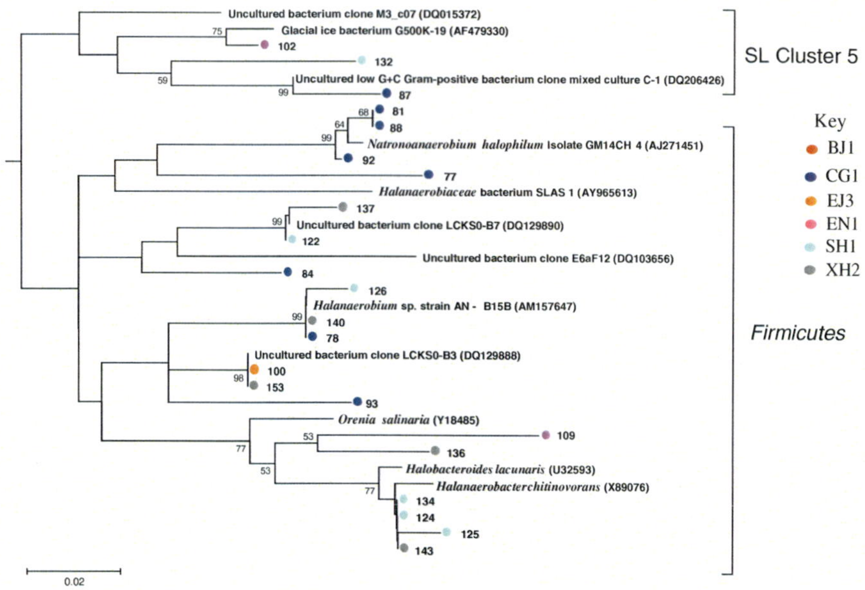

Fig. 4.3 Phylogenetic relationships showing the clone sequences with existing 16S rRNA gene sequences for *Firmicutes* and other uncultured isolates

rich in arsenic (Oremland et al. 2005). One clone sequence from CG1 (78), one from SH1 (126) and one from XH2 (140) are closely related to *Halanaerobium* sp. AN-B15B (97–98%), originally isolated from the oxic-anoxic transition of a deep sea halocline (Daffonchio et al. 2006). This clone was also one of the most dominant in the library for SH1, with a clone frequency of 6. EN1 (109) is distantly related to *Orenia salinaria* (88%). This fermentative bacterium was isolated from anoxic sediment that was rich in sulfides in Salin-de-Giraud salterns, France (Mouné et al. 2000). Three clone sequences from CG1 (81, 88, and 92) are related to "*Natronoanaerobium halophilum*" isolate G-M14CH-4 (93, 95, and 96%, respectively), which was detected in soda lakes (Jones et al. 1998).

The remaining clone sequences are related to uncultured isolates. One clone sequence from EJ3 (100), one from SH1 (122), and two from XH2 (137, 153) are related to two clones found in Lake Chaka in Tibet. One sequence from CG1 (93) is distantly related to uncultured clone E6aF12 (94%), which was detected in a hypersaline endoevaporitic microbial mat from a saltern in Israel. Phylogenetic analysis similarly placed this clone with the *Firmicutes*, with *Halocella cellulolytica* as its nearest neighbor (Sørensen et al. 2005).

4.4.1.7 *Proteobacteria*

This group comprises the largest detected from all six salt lakes. Sequences from three divisions of *Proteobacteria* were found as seen in Fig. 4.4 (4% belong to the *Alphaproteobacteria*, 6% to the *Deltaproteobacteria* and 20% to the *Gammaproteobacteria*).

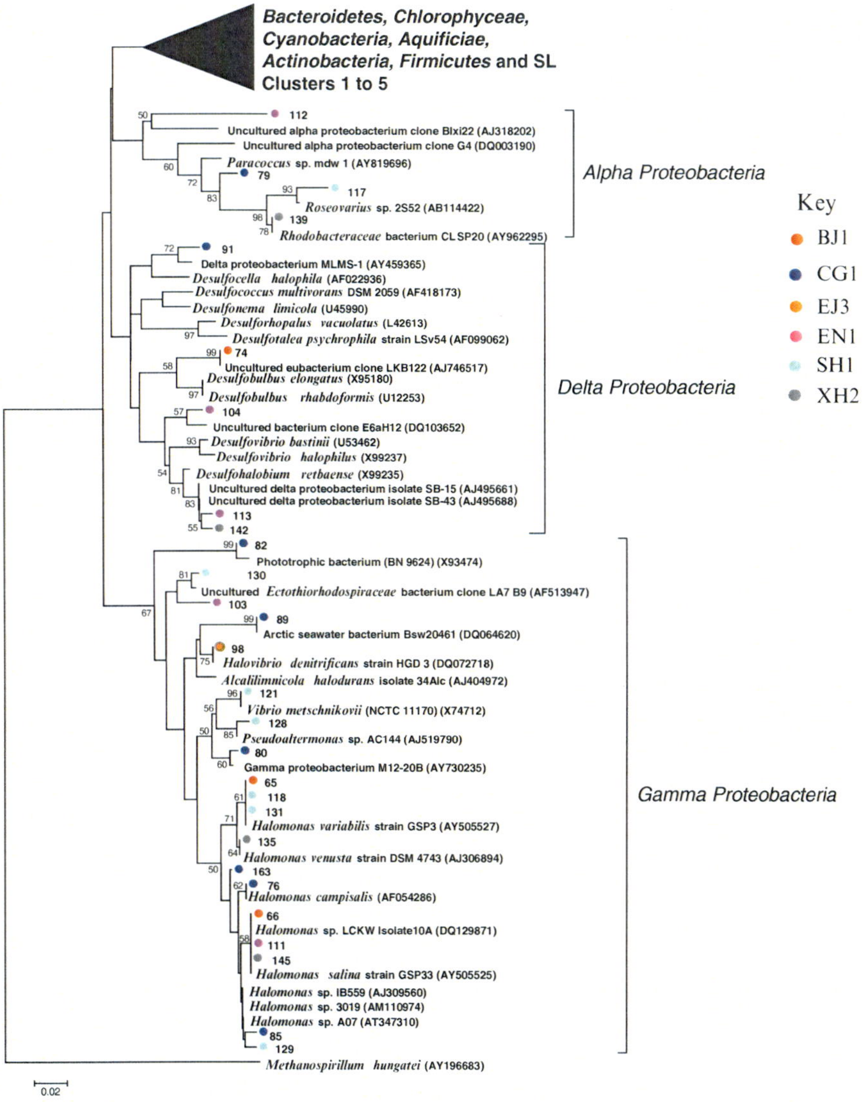

Fig. 4.4 Phylogenetic relationships showing the clone sequences with existing 16S rRNA gene sequences for *Proteobacteria* and other uncultured isolates

4.4.1.8 *Alphaproteobacteria*

A small proportion of sequences were affiliated with this group. One clone sequence from XH2 (139) is closely related to a *Rhodobacteriaceae* clone (98%), which is from a solar saltern. One clone sequence from SH1 (117) is related to *Roseovarius* sp. 2S5-2 (95%) (Fuse et al. 2003).

4.4.1.9 *Deltaproteobacteria*

Five clone sequences were found in this group of sulfate reducers. One clone sequence from CG1 (91) is distantly related to a *Deltaproteobacteria* clone MLMS-1 (92%), which was isolated from anoxic bottom water of Mono Lake, and is known to reduce arsenate using sulfide as the electron donor (Hoeft et al. 2004). One sequence from EN1 (113) and a sequence from XH2 (142) are related to uncultured isolates SB-15 and SB-43, respectively (95% and 98%, respectively). These isolates were found in anoxic sediments underlying cyanobacterial mats in Salin-de Giraud salterns in France. Phylogenetic analysis in that study also placed these clones with the *Deltaproteobacteria* near the sulfate reducers; *Desulfohalobium retbaense* their nearest neighbor (Mouné et al. 2003). Another clone sequence from EN1 (104) is distantly related to an uncultured clone E6aH12 (88%), which was detected in a hypersaline endoevaporitic microbial mat in Israel placed with the *Deltaproteobacteria*, near *Desulfocella halophila* (Sørensen et al. 2005).

4.4.1.10 *Gammaproteobacteria*

This is the largest group of *Proteobacteria* detected in the salt lakes. The majority of these sequences are Halomonads, but there are also two sequences related to Halovibrios, one to *Pseudomonas* and one to an *Ectothiorhodospiraceae* clone which are all typical of hypersaline environments, particularly the halomonads (Grant 2004).

Clone sequences from BJ1 (65) and SH1 (118 and 131) are related to *Halomonas variabilis* strain GSP 3 (96–98%). In addition, one clone sequence from BJ1 (66) and EN1 (111) are related to *Halomonas salina* strain GSP33 (99% and 97%, respectively). These isolates were found in the Great Salt Plains in Oklahoma, USA. One clone sequence from CG1 (76) is related to *Halomonas campisalis* (97%). This isolate was found in the sediments of the Salt Plain of Alkali Lake in Washington, USA (Mormile et al. 1999). One sequence from XH2 (135) is closely related to *Halomonas venusta* strain DSM 4743 (97%). This clone is the most frequently occurring clone in the library for XH2, with a clone frequency of 9.

Several clone sequences are related to other *Halomonas* species. One clone sequence from CG1 (163) is closely related to *Halomonas* sp. A-07 (99%), which was detected in soda lakes in Tanzania. Another clone sequence from CG1 (85) is closely related to *Halomonas* sp. 3019 (97%), which was detected in deep sea sediment in the East Pacific. One clone sequence from SH1 (129) is related to *Halomonas* sp. IB-559, showing 95% identity. One clone sequence from XH2 (145) is closely related to *Halomonas* sp. LCKW-Isolate10 (99%), which is another isolate found in Lake Chaka. Phylogenetic analysis in that study placed this clone with the *Gammaproteobacteria*, with *Halomonas ventosae* as its nearest neighbor (Jiang et al. 2006). Another clone sequence from EJ3 (98) is closely related to *Halovibrio denitrificans* strain HGD 3 (98%). One clone sequence from SH1 (130) is related to an uncultured *Ectothiorhodospiraceae* bacterium clone LA7-B9. *Ectothiorhodospiraceae* are halophilic anoxygenic phototrophs. A clone sequence from CG1 (82) is related to Phototrophic bacterium (BN 9624) (97%). This clone was shown to be related to *Ectothiorhodospira* species (Imhoff and Süling 1996). One sequence clone from EN1 (103) is related to *Alcalilimnicola halodurans* isolate (94%) which was found in sediments from Lake Natron in the African Rift Valley (Yakimov et al. 2001). One clone sequence from CG1 (80) is related to gammaproteobacterium M12-20B (95%) detected in Mono Lake.

4.4.1.11 Novel Lineages

These novel lineages have been called Salt Lake (SL) Clusters 1–5 (Figs. 4.2 and 4.3). SL Cluster 1 branches near to the *Bacteroidetes*. It consists of just one clone sequence from SH1 (120), which is distantly related to uncultured bacterium clone: UTFS-OF09-d22-24 (92%). This clone was detected in activated sludge.

SL Clusters 2 and 3 branch between the *Cyanobacteria* and the *Aquificae*. SL Cluster 2 consists of one sequence from BJ1 (73), which is distantly related to an uncultured bacterium isolate JH10_C67 (90%). This clone was detected in the intertidal flat of Ganghwa Island. Another sequence from XH2 (141) is distantly related to uncultured soil bacterium clone 519 (91%), which was detected in soil aggregates.

SL Cluster 3 consists of one sequence from EN1 (114) and one from SH1 (119), which are both related to an uncultured bacterium "KTK 32" (91% and 97%, respectively). This clone is from highly saline brine sediments of Kebrit Deep in the Northern Red Sea. Phylogenetic analysis in that study similarly placed this clone in a separate clade that did not belong to any of the known genus in the domain Bacteria, but was positioned between the orders *Thermotoga* and *Aquificae* (Eder et al. 1999).

SL Cluster 4 branches from the *Aquificae*, consisting of several clone sequences. One sequence from CG1 (83) is distantly related to an uncultured candidate division OP11 bacterium FL11H04 (88%). This clone was detected in a hot spring called Obsidian Pool in Yellowstone National Park, Wyoming, USA. The group, designated OP11, was positioned near the *Deinococci*. This study also demonstrated the presence of this division in a range of environments including Carolina Bay sediment, Amazonian soil and Australian deep subsurface water, thereby demonstrating the ubiquitous nature of this novel lineage in the environment (Hugenholtz et al. 1998). One clone sequence from EN1 (165) and one from XH2 (146) are related to an uncultured bacterium clone At12OctA1 (95% and 89%, respectively). This clone was detected in Salar de Atacama in Northern Chile (Demergasso et al. 2004).

SL Cluster 5 branches near the *Firmicutes*. One clone sequence from EN1 (102) is closely related to a glacial ice bacterium G500K-19, showing 98% sequence similarity (Christner 2002). One clone sequence from SH1 (132) is related to an uncultured bacterium clone M3_c07 (94%), which was detected in the gut of obese mice (Ley et al. 2005). One clone sequence from CG1 (87) is closely related to an uncultured low G+C Gram-positive bacterium clone mixed culture C-1 (98%). This clone was also detected in Mono Lake. Phylogenetic analysis in that study positioned this clone as distantly related to *Bacillus selenitireducens*, which belongs to the *Firmicutes* (Hollibaugh et al. 2006).

4.4.2 Archaeal Diversity of the Salt Lakes

The vast majority (81%) of the clone sequences obtained for the archaeal 16S rRNA gene libraries from the salt lakes were related to uncultured organisms. The related sequences were found in other salt lakes, salterns or crystallizer ponds, saline soils, saline waters of Antarctica and one sequence found in Permo-triassic rock (Walsh et al. 2005). Fifteen of the clone sequences are related to those found in crystallizer ponds in Australia (Burns et al. 2004) and a further fifteen related to those found in saline soils in British Columbia, Canada. Many of the sequences found in the British Columbian soils were closely related to those found in Permo-triassic rock. Nine clone sequences are related to those found in Lake Zabuye in Tibet (Fan et al. 2004).

The remaining 19% of clone sequences are related to known species. Three sequences are related to *Halorubrum saccharovorum*; two are related to *Halorubrum tibetense*; two are related to *Halorubrum vacuolatum*; one sequence is related to *Halorubrum xingjiangense*; and one sequence is related to *Halorubrum terrestre*. Four other sequences are related to unspecified *Halorubrum* type species. One sequence is related to *Halosimplex carlsbadense* and one sequence is related to a *Haloarcula* species.

Phylogenetic trees were drawn containing clone sequences and existing 16S rRNA gene sequences from all major archaeal lines of descent known to have halophilic or halotolerant members, as well as unpublished sequences of the clones' nearest neighbors deposited in sequence data bases. The root chosen for the archaea was *Escherichia coli*. This phylogenetic analysis shows the distribution of sequences into eight monophyletic assemblages within the order *Halobacteriales*. Many branch within the previously mentioned groups *Halorubrum*, *Halosimplex* and *Haloarcula*. Additional sequences branch with *Halobacterium*, *Halobaculum*, *Halogeometricum*, *Halomicrobium*, and the alkaliphilic group *Natronomonas*. However, there are additional lineages that form between these nodes, designated Salt Lake (SL) Clusters 1–3. This is summarized in Figs. 4.5–4.7; only bootstrap values above 50% are shown and all clone sequences are color coded according to the lake that they were detected in. All known genera within the order *Halobacteriales* are represented in the tree; however, none of the sequences in the SL Clusters affiliate closely with any of them. There appears to be good bootstrap support for the branches *Halorubrum* (82%) and *Natronomonas* (100%). All the SL Clusters are also well supported (98–100%), with the latter forming an outer group to the *Halobacteriales* (see later).

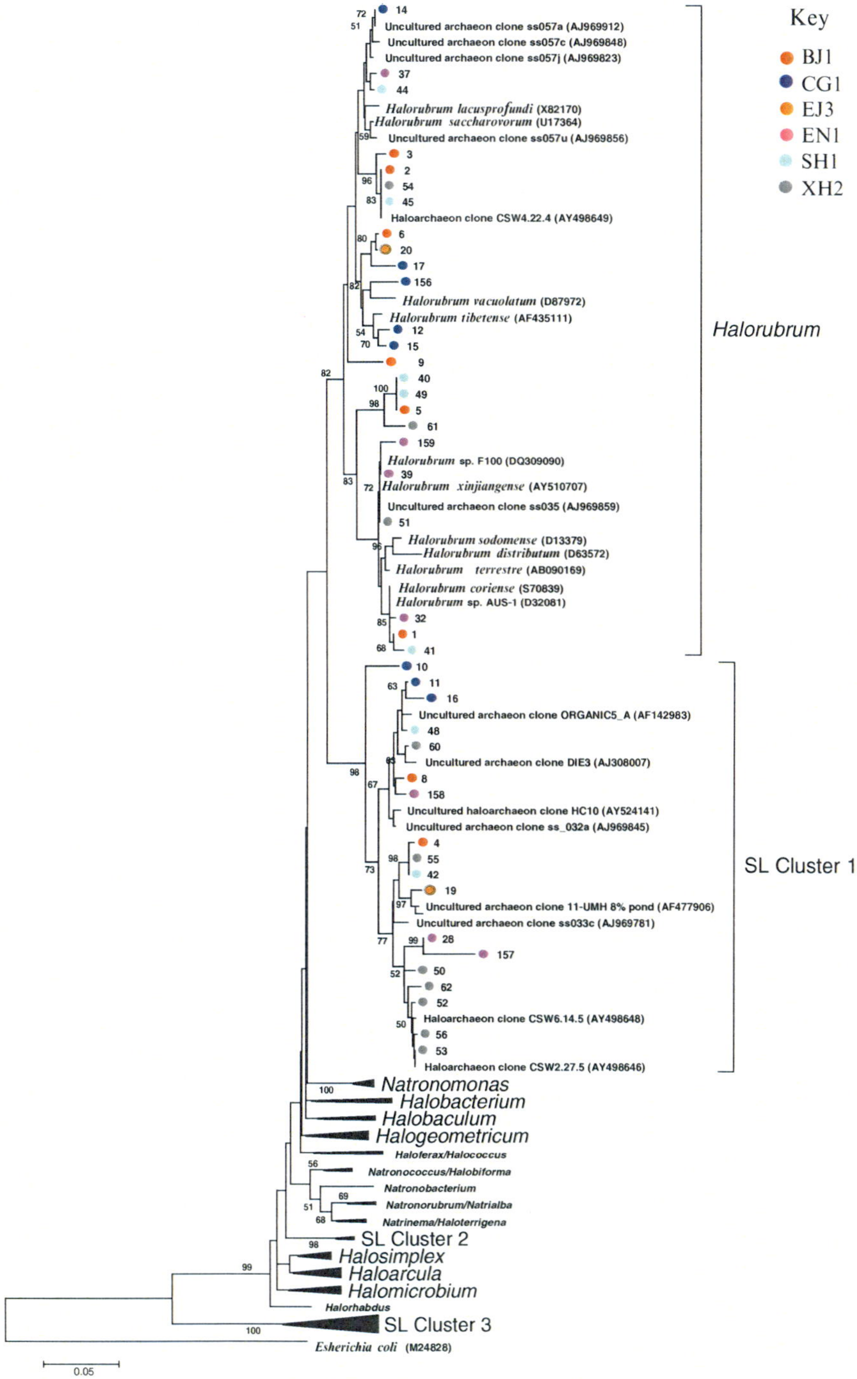

Fig. 4.5 Phylogenetic relationships showing the clone sequences with existing 16S rRNA gene sequences for *Halorubrum* and other uncultured isolates

4.4.2.1 *Halorubrum*

The 34% of clone sequences branch with representatives of *Halorubrum*, therefore, comprising the largest group detected in all six salt lakes. This group is shown in more detail in Fig. 4.5. *Halorubrum* includes extreme halophiles requiring between 1.5 and 5.2 M NaCl for growth and are strict aerobes (Grant et al. 2001). One clone sequence from BJ1 (5) and two from SH1 (40 and 49) are related to *Halorubrum saccharovorum* (all at 94%). The gene of this isolate is in fact closely related to the dominating clone found in the 16S library for SH1, showing a clone frequency of 15. The isolate was first isolated by Tomlinson and Hochstein (1976) from a mixture of mud and brine from a saltern in the southern section of San Francisco Bay. This isolate is a strict aerobe, requiring 1.5–5.2 M NaCl for growth, with an optimum concentration of 3.5–4.5 M. It also requires 0.005 M Mg^{2+}. It can grow between 30°C and 56°C, growing optimally at 50°C (Grant et al. 2001). Two clone sequences from CG1 (12 and 15) are closely related to *Halorubrum tibetense* (96% and 97%, respectively). Moreover, the gene of this isolate is closely related to the second most frequent clone in the library for lake CG1, with a clone frequency of 11. This is a haloalkaliphilic archaeon isolated from Lake Zabuye on the Tibetan Plateau, which is an alkaline chloride–sulfate salt lake with a pH of 9.4 (Fan et al. 2004). Two more clone sequences from CG1 (17 and 156) are related to *Halorubrum vacuolatum* (95% and 96%, respectively). This is also an obligate haloalkaliphilic archaeon requiring an optimum pH of 9.5 for growth (Grant et al. 2001). It was first isolated from Lake Magadi in Kenya, an alkaline soda lake with a pH of 11 (Mwatha and Grant 1993; Grant et al. 1998). One clone sequence from XH2 (51) is closely related to *Halorubrum xinjiangense* (99%). This was isolated from Xiao-Er-Kule Lake in Xinjiang, China and grows optimally at 3.1–3.4 M NaCl and pH 7.0–7.5 (Feng et al. 2004). Another clone sequence from XH2 (61) is related to *Halorubrum terrestre* (93%). Isolated from saline soils, *Halorubrum terrestre* grows optimally at 25% (w/v) NaCl and pH 7.5 (Ventosa et al. 2004). One clone sequence from EN1 (159) is closely related to *Halorubrum* sp. F100 (98%). This was found in saline lakes in Turkey. Clone sequences from BJ1 (1), EN2 (32) and SH1 (41) are closely related to *Halorubrum* AUS-1, which was isolated from a nameless clay pan in Western Australia. This isolate grows on medium containing 25% (w/v) NaCl at pH 7.4 (Mukohata et al. 1988).

The remaining clone sequences are related to uncultured isolates that cluster with *Halorubrum*. Four clone sequences from CG1, EN1 and SH1 (14, 37, 39 and 44) are related to uncultured archaeon clones ss057_a, ss057j, ss057u and ss035. These were sequences detected in saline soils in British Columbia, Canada at distances between 0 and 300 cm from the salt spring source exhibiting 7% (w/v) NaCl, and pH 7–10. The phylogenetic tree in that study similarly placed clones ss057_a, ss057j, ss057u and ss035 with *Halorubrum* (Walsh et al. 2005). Five sequences from BJ1, EJ3, SH1 and XH2 (2, 3, 20, 45 and 54) are related to haloarchaeon clone CSW4.22.4 (96–99%). This is an isolate from crystallizer ponds of Corio Bay in Victoria, Australia, which exhibited a total salt concentration

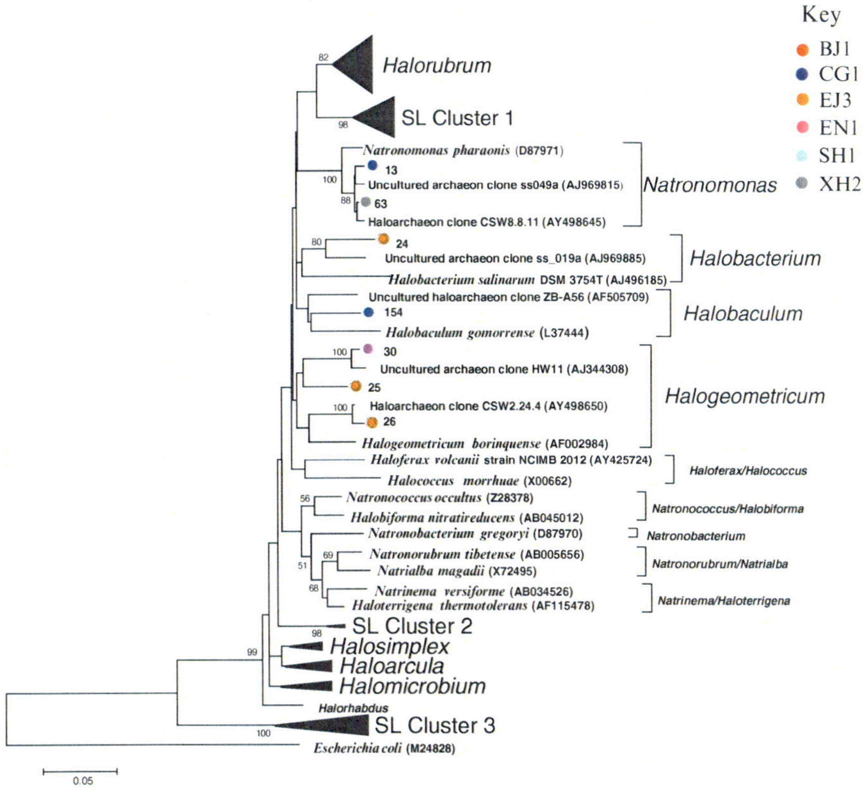

Fig. 4.6 Phylogenetic relationships showing the clone sequences with existing 16S rRNA gene sequences for *Natronomonas*, *Halobacterium*, *Halobaculum* and *Halogeometricum*

of 33% (w/v) and a pH of 8.1 (Burns et al. 2004). This clone is also the most abundant in the 16S library for lake BJ1, with a clone frequency of 22.

4.4.2.2 *Natronomonas*

This group of alkaliphiles grow at 2–5.2 M NaCl and pH 7–10 (Grant et al. 2001). Only two clone sequences from the libraries are found in this group. One sequence from CG1 (13) is closely related to an uncultured clone ss049a (97%), which is another clone detected in saline soils in British Columbia as previously described. Phylogenetic analysis in that study has shown this clone to be distantly related to *Natronomonas pharaonis* (Walsh et al. 2005). One clone sequence from XH2 (63) is closely related to an uncultured clone CSW8.8.11 (97%) from Australian crystallizer ponds as previously described. Although both are related to uncultured archaea, they are distributed within the same clade as *Natronomonas pharaonis* (Fig. 4.6). In fact, both are distantly related to *Natronomonas pharaonis*, showing 95% and 94% similarity, respectively.

4.4.2.3 *Halobacterium*

Members of this group are extremely halophilic, requiring 3.0–5.2 M NaCl, but grow optimally at 3.5–4.5 M NaCl. They also grow at temperatures between 35 and 50°C, with a pH range of 5.5–8.5 (Grant et al. 2001). Only one clone sequence from EJ3 (24) clusters within the same clade as *Halobacterium* (Fig. 4.6). It is related to an uncultured clone ss_019a, found in saline soils in British Columbia as previously described (92%). It is also distantly related to *Halobacterium salinarum* showing 87% similarity.

4.4.2.4 *Halobaculum*

These are aerobic extreme halophiles, which grow optimally in 1–2.5 M NaCl and 0.6–1 M $MgCl_2$ (Grant et al. 2001). One clone sequence from CG1 (154) is related to an uncultured haloarchaeon clone ZB-A56 (95%), which is from Lake Zabuye (as previously described). These sequences are distributed within the same clade as *Halobaculum gomorrense* (see Fig. 4.6). Sequence 154 is distantly related to *Halobaculum gomorrense*, showing 88% similarity.

4.4.2.5 *Halogeometricum*

These aerobes grow optimally at 3.5–4 M NaCl and 0.04–0.08 $MgCl_2$, and so are also extremely halophilic (Grant et al. 2001). This genus currently has two species being *Halogeometricum borinquense* the type species, which is distantly related to *Haloferax* (Montalvo-Rodríguez et al. 1998). Three clone sequences appear in the *Halogeometricum* clade (Fig. 4.6). Clone sequence 26 from EJ3 is related to clone CSW2.24.4, a sequence found in an Australian crystallizer pond as previously described. Phylogenetic analysis in that study similarly placed this clone near to, but distantly related to *Halogeometricum borinquense* (Burns et al. 2004). One clone sequence from EN1 (30) is closely related to uncultured archaeon clone HW11 (97%), which is a sequence found in Permo-Triassic rock salt. Clone sequence 25 showed a high localized similarity to *Halosimplex* but clearly full length homology places this sequence near *Halogeometricum*. Sequences 25, 26, and 30 form separate branches within the *Halogeometricum* clade with low bootstrap values, suggesting that they are only distantly related. In fact, they only show 88–89% similarity to *Halogeometricum borinquense* (AF002984).

4.4.2.6 *Haloarcula*

The genus *Haloarcula* includes a group of extremely halophilic archaea, requiring 2.0–5.2 M NaCl for growth. They are aerobic or facultative anaerobes

(Grant et al. 2001). They are shown in Fig. 4.7. One clone sequence from BJ1 (7) is related to an uncultured clone Sec16SC8 (95%), which was a sequence found in the crystallizers of an Adriatic solar saltern. This had a water activity of 0.759 at 30°C (the water activity for a saturated NaCl solution at 30°C is 0.769) and was at pH 8.0. The phylogenetic tree in that study also showed clone Sec16SC8 affiliated with *Haloarcula* (Pašić et al. 2005). One clone sequence from XH2 (58) is related to a *Haloarcula* sp. (94%). Finally, another clone sequence from EN2 (31) is related to *Haloarcula* sp. NCIMB 720 (92%).

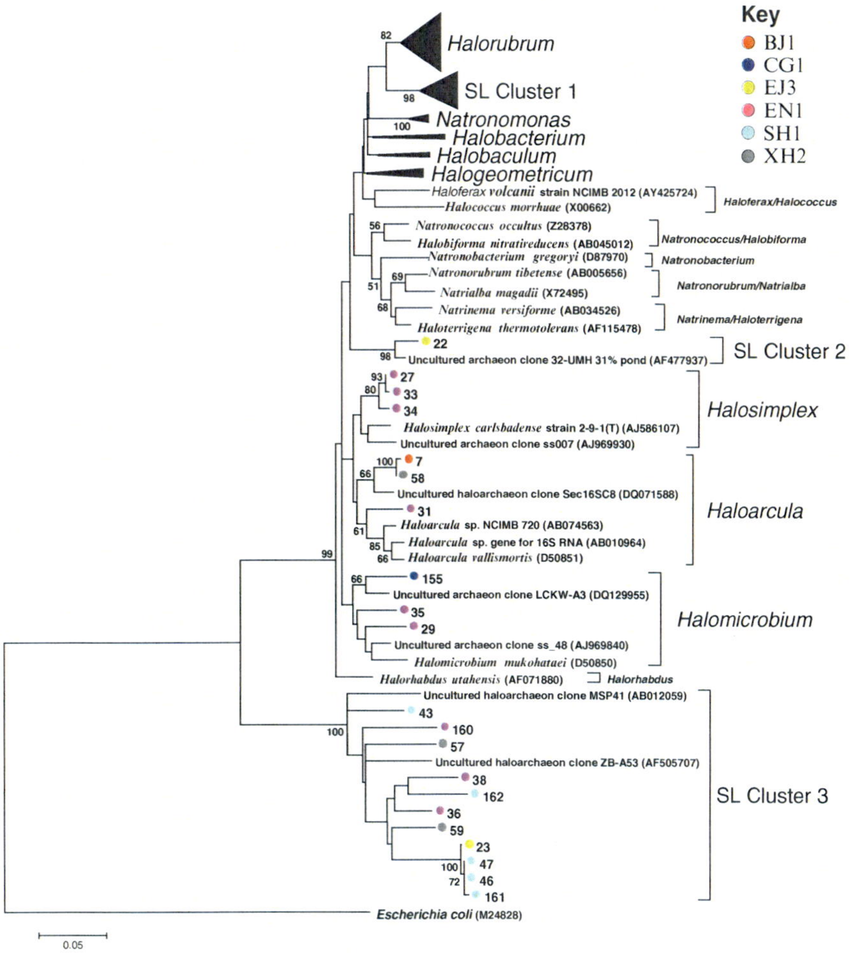

Fig. 4.7 Phylogenetic relationships showing the clone sequences with existing 16S rRNA gene sequences for *Halosimplex*, *Haloarcula*, *Halomicrobium*, and other uncultured isolates

4.4.2.7 *Halosimplex*

The genus *Halosimplex* is a deep branch distantly related to *Haloarcula* (Fig. 4.7). They are unique in that they have three different genes encoding their 16S rRNA. The strain *Halosimplex carlsbadense* was isolated from a Permian halite deposit in south eastern Mexico. It requires a defined medium for growth containing glycerol and acetate or pyruvate (Vreeland et al. 2002). Three clone sequences from EN2 (27, 33, 34) branch with the *Halosimplex* clade. Although related to an uncultured archaeon ss007, which was detected in saline soils in British Columbia (93–94%), they all show 93% similarity to *Halosimplex carlsbadense* strain 2-9-1.

4.4.2.8 *Halomicrobium*

This genus currently contains two species, *Halomicrobium mukohataei* being the type species, which was isolated from salt flats in Argentina and somewhat related to *Haloarcula*. However, analysis of its morphology, polar lipid composition and 16S rRNA gene sequence showed that it was significantly different from *Haloarcula* and was proposed as a new genus within the *Halobacteriales*. It is extremely halophilic aerobe and a facultative anaerobe in the presence of nitrate (Oren et al. 2002). Three clone sequences appear in the same clade as *Halomicrobium mukohataei* (Fig. 4.7). One clone sequence from CG1 (155) is related to an uncultured clone LCKW-A3 (93%) from Lake Chaka. This athalassohaline lake possesses a total salinity of 32.5% (w/v) with a water temperature between −8 to 4.2°C in the winter and 6 to 20°C in the summer. It also had a pH of 7.4 (Jiang et al. 2006). Two clone sequences from EN2 (29 and 35) are related to an uncultured clone ss_48 from saline soils in Canada as previously described (93% and 94%). That study also placed clone ss_48 near the *Haloarcula-Halomicrobium* branch (Walsh et al. 2005).

4.4.2.9 Novel Lineages

These novel lineages have been called Salt Lake (SL) Clusters 1–3 as shown in Figs. 4.5 and 4.7. Branching near the *Halorubrum* group, SL Cluster 1 is the second largest group comprising 27% of the clone sequences. Several clone sequences from XH2 (50, 52, 53, 56, and 62) and one from EN1 (28) are again closely related to uncultured clones from Australian crystallizer ponds as described. Haloarchaeon clone 6.14.5 is the most frequently observed clone in the library for Lake EN1, with a clone frequency of 13. Haloarchaeon clone 2.27.5 is also the dominating clone in the library for XH2, with a clone frequency of 11. Clone sequences from BJ1 (4), CG1 (16), EN2 (158), SH1 (42) and XH2 (55) are related to saline soils in Canada. Clone sequences from BJ1 (8) and CG1 (10) are related to uncultured clones ORGANIC4_A and ORGANIC5_A (95% and 94%, respectively). These sequences were found in anoxic sediment in the Vestfold Hills in Eastern Antarctica. The lakes that were sampled in this basin were at temperatures between −1.8 and 3°C, with salinities between 14 and 45 g/kg (Bowman et al. 2000). Clone ORGANIC5_A is also the most frequently occurring clone in the 16S library for Lake CG1, with a clone frequency of 12. One clone sequence from XH2 (60) is related to an uncultured archaeon DIE3 (92%). This sequence was detected in a hypersaline pond from the bottom of a slag heap of a former potassium mine in Germany. This pond had a pH of 5.6 with a total salinity above 32% (w/v). As expected, this pond had a very high potassium concentration. Phylogenetic analysis in that study also placed this clone in an unknown cluster that formed an outer group to *Halorubrum* (Ochsenreiter et al. 2002). One clone sequence from CG1 (11) and one sequence from SH1 (48) are related to a haloarchaeon clone HC10 (95% and 94%, respectively). This clone was detected in a solar saltern in San Diego, USA (Bidle et al. 2005). One clone sequence from EJ3 (19) is related to an uncultured clone 11-UHM 8% pond (97%). This was detected in a multipond solar saltern "Bras del Port" located in Alicante, Spain, showing 8% (w/v) total salinity (Benlloch et al. 2002).

SL Cluster 2 is a small group branching near the group of known haloarchaea. It consists of just one clone sequence from EJ3 (22), which is related (93%) to an uncultured archaeon from a solar saltern in Alicante, Spain, showing 31% (w/v) total salinity (Benlloch et al. 2002).

SL Cluster 3 appears to form its own separate lineage to the order *Halobacteriales*. Several sequences from EJ3 (23), EN2 (38), SH1 (46, 47, 161 and 162) and XH2 (57and 59) are related (88–91%) to sequences detected in Zabuye Lake. Two sequences from EN2 (36 and 160) and one from SH1 (43) are related to an uncultured haloarchaeon clone MSP41. This clone was detected in salt crystallizing pond at Lake Magadi, Kenya. Similarly, phylogenetic analysis in that study positioned this clone in a novel lineage on the periphery of known haloarchaea. It was suggested that this clone was a member of a deeply branching group of the *Euryarchaeota* (Grant et al. 1999).

4.5 Biogeography

The data set just described was subject to a rigorous statistical analysis (Pagaling et al. 2009). The archaeal community was found to consist of closely related lineages whereas the bacterial community showed much more diversity. This analysis was also used to determine important factors correlating with biotic composition. This showed that temperature, Na^+ and Mg^{2+} ions, and pH were the factors that drive microbial community composition in these salt lakes. In general, the closer the chemical composition of the lakes, the closer the biotic composition. Geographic distance was not a statistically significant driver but there was a positive correlation between lake proximity and biotic composition. In other words, with regard to micro-organism biogeography in closely spaced saline systems, "the milieu selects."

4.6 Haloarchaeal Viruses

Strains with 16S rRNA gene sequences 98% identical to *Halorubrum saccharovorum* were isolated from Lake Bagaejinor in Inner Mongolia. Two lytic viruses infecting these were isolated from the lake water. The host has since been designated as *Halorubrum kocurii* (Gutiérrez et al. 2008b). The BJ1 genome sequence has been determined and has very low sequence identity to any previously described virus (Pagaling et al. 2007). BJ1 has an icosahedral head and tail morphology and most likely a linear double stranded DNA genome exhibiting terminal redundancy. Its genome sequence has 42,271 base pairs with a G+C content of ~65 mol%. The genome of BJ1 is predicted to encode 70 ORFs, including one for a tRNA. Fifty of the seventy ORFs had no identity to data base entries; twenty showed sequence identity matches to archaeal viruses and to haloarchaea. ORFs possibly coding for an origin of replication complex, integrase, helicase, and structural capsid proteins were identified. Evidence for viral integration was obtained. The second virus BJ2 has been partially sequenced; 44 contigs containing a total of 97,602 base pairs with a G+C content of 51 mol%. It has no discernible sequence identity to the BJ1 virus. This virus is more fully described by Pagaling (2007).

4.7 Culture-Dependent Study of the Salt Lakes

In addition to the cultivation-independent approach described previously, culture-dependent microbiological techniques were employed to investigate the microbial community and abundance from salt lakes. Water, soil and sediment samples were collected from the athalassohaline Lakes Bagaejinor, Chagannor, Chahannor, Ejinor, Erlianor, Shangmatala, and Xilinhot (see Sect. 4.2 and 4.3 of this chapter).

The number of microorganisms in the samples collected from these environments was determined. MH medium (Ventosa et al. 1982) with different NaCl concentrations (0.5, 10, 20, and 25%) was used to determine the viable cells. The number of colony forming units (cfu) per milliliter of water ranged from 1×10^2 to 1.2×10^4 cfu/ml in MH medium with 10% NaCl and from 1.9×10^3 to 2.6×10^4 cfu/ml in the same medium with 25% NaCl. With respect to the soil/sediment samples, these values were in the range 1×10^3 to 1.4×10^6 cfu/g in media with 10% NaCl, and from 1×10^2 to 3.8×10^4 cfu/g in media with 25% NaCl. On the other hand, the determination of the total and viable cell counts in the samples collected was carried out by epifluorescence by using the Live-Dead Cell Staining kit. The total counts ranged from 6.6×10^6 to 1.3×10^7cells/ml, while the number of viable cells ranged from 1.3×10^6 to 8.4×10^6 cells/ml. The higher numbers of cells determined by direct counts are normal in comparison with the colony counts since in the last procedure only those microorganisms able to grow in culture media are detected (Castillo 2008; Carrasco 2009).

Direct plating on different media containing similar salts to those found in the salt lakes, but present in different concentration, was used to allow maximum recovery of halophilic bacterial and archaeal microorganisms from these hypersaline environments. The isolates were selected by their colony morphology, salt response and later by their 16S rRNA gene sequences (Castillo 2008; Carrasco 2009).

4.7.1 Bacterial Diversity of the Salt Lakes

A total of 179 moderately halophilic bacterial strains were isolated; however, no extremely halophilic bacteria were isolated from any of the salt lakes studied. The 16S rRNA sequence of the isolates was determined by PCR amplification of the 16S rRNA gene by using the forward primer 16F27 and the reverse primer 16R1488. The 16S rRNA gene sequence analysis was performed with the ARB software package (Ludwig et al. 2004).

The phylogenetic analysis determined that all the bacterial isolates were closely related to members of *Firmicutes* (representing 70%) and *Gammaproteobacteria* (30%). These results contrast with those obtained in other studies on saline environments, in which the larger proportion of isolates are Gram-negative bacteria (Yeon et al. 2005; Jiang et al. 2006). Within the *Firmicutes*, most isolates were closely related to endospore-forming bacteria belonging or closely related to the genera *Halobacillus* (40 strains), *Alkalibacillus* (27), *Bacillus* (21), *Oceanobacillus* (7), *Gracilibacillus* (6), *Thalassobacillus* (3), *Filobacillus* (2), *Halalkalibacillus* (2), *Virgibacillus* (2), and *Salimicrobium* (2), whereas only 10% of isolates were closely related to Gram-positive non-endospore-forming cocci: *Marinococcus* (12 strains), and *Staphylococcus* (1). The second largest group cultivated from the salt lakes was the *Gammaproteobacteria*, *Halomonas* being the most prominent genus, with 40 isolates, followed by *Halovibrio* (4), *Chromohalobacter* (3), *Pseudomonas* (3), *Alkalispirillum* (2), *Salicola* (1) and *Stenotrophomonas* (1) (Carrasco 2009). The abundance of endospore-forming microorganisms in the salt lakes studied may be due to the ability of the endospores to be transported and their remarkable capacity for resistance and dormancy, which might allow them to survive in these habitats subjected to substantial seasonal changes (Claus and Berkeley 1986).

Novel Taxa

Some of the isolates were distantly related to any previously known bacterial species and thus, a polyphasic taxonomic study, based on phenotypic, genotypic and phylogenetic characteristics, was performed in order to describe them and to determine their correct taxonomic position.

4.7.1.1 Study of Strains CG12 and CG13 (Described as *Aquisalimonas asiatica* gen. nov., sp. nov.)

The proposal of the creation of the novel genus *Aquisalimonas* was based on two strains, designated CG 12 and CG 13, which were isolated from water samples collected from Lake Chagannor (Márquez et al. 2007). This genus belongs to the *Gammaproteobacteria*, and constitutes a new phyletic sublineage within the *Alkalispirillum-Alkalilimnicola* group of the family *Ectothiorhodospiraceae*. The genus *Aquisalimonas* was defined as Gram-negative, motile, non-endospore forming, moderately halophilic and alkali-tolerant rods, occurring singly, in pairs or long chains, catalase and oxidase positive, and able to reduce nitrate to nitrite. Cellular fatty acids mainly consist of $C_{18:1}\omega7c$, $C_{16:0}$, and $C_{12:0}$. *Aquisalimonas asiatica* is the type species of the genus and strain CG12 is the type strain. The G+C content of its genomic DNA is 63.6 mol%. The polar lipids detected in this strain are phosphatidylglycerol, diphosphatidylglycerol, phosphatidylethanolamine, phosphatidylcholine, phosphoglycolipid and six different unidentified phospholipids (Márquez et al. 2007).

4.7.1.2 Study of Strain CH9d (Described as *Salsuginibacillus kocurii* gen. nov., sp. nov.)

Strain CH9d was isolated from a sediment sample of Lake Chahannor. When phylogenetic analysis was carried out, it was observed that its 16S rRNA gene sequence was very distant from those of other known taxa. *Thalassobacillus devorans* was its closest neighbor with a similarity value of only 91%, followed by other members of the genera *Bacillus, Halobacillus* and *Marinococcus*. The phylogenetic results, together with the phenotypic and chemotaxonomic differences found between the isolate and its phylogenetically related relatives led to the proposal of the creation of a novel genus, *Salsuginibacillus,* with the single species *Salsuginibacillus kocurii* (Carrasco et al. 2007b). Strain CH9d is a Gram-positive rod that forms non-pigmented colonies when grown at 37°C on alkaline, saline medium and produces ellipsoidal endospores located in central or subterminal position in swollen sporangia. However, *Thalassobacillus devorans* produces endospores located centrally without swelling of the sporangia (García et al. 2005). Conversely, species of the genus *Marinococcus* are non-spore-forming cocci characterized by their orange-pigmented colonies (Hao et al. 1984; Li et al. 2005). Strain CH9d is moderately halophilic and alkali-tolerant, whereas *Bacillus agaradhaerens* (its nearest *Bacillus* species) is a halotolerant and strictly alkaliphilic bacterium (Nielsen et al. 1995). In addition, strain CH9d differs from *T. devorans* and *B. agaradhaerens* in its inability to hydrolyze gelatin. In contrast to species of the genus *Halobacillus,* strain CH9d is negative for oxidase and is able to reduce nitrate to nitrite (Spring et al. 1996). On the other hand, the DNA base composition of strain CH9d (44.7 mol%) is similar to those of the previously described *Halobacillus* species (40–45 mol%) (Spring et al. 1996; Amoozegar et al. 2003; Yoon et al. 2003, 2004, 2005; Liu et al. 2005) but different to those described for *T. devorans* (42.4 mol%) (García et al. 2005), *B. agaradhaerens* (39.3–39.4 mol %) (Nielsen et al. 1995), and *Marinococcus* species previously described (46.4–48.5 mol%) (Hao et al. 1984; Li et al. 2005). The peptidoglycan type of the strain CH9d is A1γ based on *meso*-diaminopimelic acid, the isoprenoid quinones are MK-7 (88%) and MK-6 (12%), and its major fatty acids are anteiso-$C_{15:0}$, anteiso-$C_{17:0}$, iso-$C_{17:0}$ and iso-$C_{15:0}$. With respect to the polar lipid composition, strain CH9d contains diphosphatidylglycerol, phosphatidylglycerol, phosphatidylethanolamine and two phospholipids of unknown structure (Carrasco et al. 2007b).

4.7.1.3 Study of Strain SH4s (Described as *Aquisalibacillus elongatus* gen. nov., sp. nov.)

The genus *Aquisalibacillus* with the single species *Aquisalibacillus elongatus* was proposed on the basis of the features of strain SH4s, isolated from water of the saline Lake Shangmatala. This strain is phylogenetically related to the genera *Filobacillus*, *Piscibacillus* and *Tenuibacillus,* within the family *Bacillaceae*. The 16S rRNA gene sequence similarity values between strain SH4s and the type strains of these three genera ranged from 95.4 to 95.9%. These values are lower than those found between the genera *Tenuibacillus* and *Piscibacillus* and the genus *Filobacillus*, their most closely phylogenetically related taxon (97 and 96.9%, respectively). Strain SH4s shares some phenotypic features with these three genera but, in contrast to *Piscibacillus* and *Tenuibacillus*, it is negative for Gram-staining and oxidase activity. Besides, the isolate clearly differs from members of the three genera in the motility (cells of strain SH4s are non-motile), ability to reduce nitrate to nitrite, DNA base composition (45.9 mol%), fatty acids profile (with iso-$C_{16:0}$ and iso-$C_{15:0}$ as the major components) and, overall, the peculiar peptidoglycan composition present in its cell wall (type A4β, based on L-Orn-D-Asp) (Márquez et al. 2008).

4.7.1.4 Study of Strain EN8d (Described as *Sediminibacillus halophilus* gen. nov., sp. nov.)

Strain EN8d was isolated from a sediment sample from Erliannor salt lake. This strain is most closely related phylogenetically to the genera *Thalassobacillus* (93.6% 16S rRNA gene sequence similarity) and *Halobacillus* (95–96%), although constitutes a separate line of descent within the radiation of Gram-positive rods. Cells of strain EN8d are Gram-positive, rod-shaped that occur singly, in pairs or in short chains. This strain is motile, moderately halophilic, facultatively anaerobic, oxidase and catalase positive, and is able to reduce nitrate to nitrite. Its genomic DNA G+C content is 47.5 mol%, it has a cell-wall peptidoglycan type A1γ with *meso*-diaminopimelic acid, MK-7 as the predominant menaquinone and anteiso-$C_{15:0}$ and anteiso-$C_{17:0}$ as the major fatty acids. The polar lipids include diphosphatidylglycerol, phosphatidylglycerol, and a glycolipid. The novel genus *Sediminibacillus,* and the novel species *S. halophilus* was proposed for this strain (Carrasco et al. 2008).

4.7.1.5 Study of Strains XH-62, XH-63 and EJ-15 (Described as *Gracilibacillus orientalis* sp. nov.)

Three strains (designated XH-62, XH-63 and EJ-15) were phylogenetically related to species of the genus *Gracilibacillus*, *Gracilibacillus dipsosauri* their nearest phylogenetic neighbor (with 16S rRNA sequence similarities ranging between 95.4 and 95.8%). Strains XH-62 and XH-63 were isolated from water samples from the lake near Xilinhot (Lake Bayannur), whereas strain EJ-15 was isolated from sediments collected from Lake Ejinor. The three strains are Gram-positive, moderately halophilic rods that grow optimally at 10% NaCl. They are oxidase negative, catalase positive and strictly aerobic. They produce spherical, terminal and deforming endospores as well as colonies with a cream pigmentation. DNA G+C content of the three strains is in the range 36.1–37.1 mol%. They have *meso*-diaminopimelic acid in the cell wall peptidoglycan. The polar lipid pattern of strain XH-63, selected as representative strain of the isolates, consists of diphosphatidylglycerol, phosphatidylglycerol, phosphatidylethanolamine and a phospholipid and two amino phospholipids of unknown structure. Besides, strain XH-63 possess menaquinone of the MK-7 type and a fatty acid composition very similar to those described for other species of the genus *Gracilibacillus* (Wainø et al. 1999). All the features of the three new isolates clearly indicated that they represent a new species for which the new name *Gracilibacillus orientalis* sp. nov. was proposed, with strain XH-63 as the type strain (Carrasco et al. 2006).

4.7.1.6 Study of Strain CG-15 (Described as *Bacillus chagannorensis* sp. nov.)

Strain CG-15 was isolated from a water sample collected from Lake Chagannor. The phylogenetic analysis showed that this strain belongs to the genus *Bacillus,* and forms a phyletic group with *Bacillus saliphilus* (96.0% 16S rRNA gene sequence similarity), *Bacillus agaradhaerens* (94.0%) and *Bacillus clarkii* (93.5%). Strain CG-15 is a Gram-positive, endospore-forming motile rod, facultative anaerobe that grows over a wide range (3–20%) of salt concentrations, with optimal growth at 7% salts and thus, it can be considered as a moderately halophilic microorganism. Besides, it is an alkalitolerant bacterium that grows at pH 5.8–11.0 (optimal at pH 8.5) and 6–40°C (optimal at 37°C). The DNA G+C content of this isolate is 53.8 mol%; this value is within the range for *Bacillus* but is higher than those of *Bacillus saliphilus* (48.4 mol%) (Romano et al. 2005), *Bacillus agaradhaerens* (39.2 mol%) (Nielsen et al. 1995) and *Bacillus clarkii* (42.2 mol%) (Nielsen et al. 1995).

Strain CG-15 contains a cell wall type based on *meso*-diaminopimelic acid and MK-7 as the major menaquinone. With respect to the cellular fatty acids composition, anteiso-$C_{15:0}$, iso-$C_{15:0}$ and anteiso-$C_{17:0}$, are its major components. The polar lipids detected are diphosphatidylglycerol, phosphatidylglycerol, phosphatidylethanolamine and three different phospholipids of unknown structure. These chemotaxonomic characteristics are typical of those found in members of the genus *Bacillus* previously described (Priest et al. 1988; Heyrman et al. 2004, 2005; Wieser et al. 2005; Lim et al. 2006a, 2006b). However, strain CG-15 could be distinguish from other phylogenetically related *Bacillus* species in some features, such as cell morphology, range and optimal salt concentration for growth and optimal pH for growth as well as the genomic DNA G+C content or fatty acids composition. Therefore, on the basis of these polyphasic taxonomic data, the name *Bacillus chagannorensis* sp. nov. was proposed for the novel isolate (Carrasco et al. 2007a).

4.7.1.7 Study of Strain XH-22 (Described as *Virgibacillus salinus* sp. nov.)

A strain, related to the genus *Virgibacillus*, that was also studied in detail is strain XH-22. This isolate exhibited 16S rRNA gene sequence similarity values of 97.6 and 97.5%, with respect to *V. carmonensis* and *V. necropolis*, and values comprised between 96.9 and 94.9% with respect to the other *Virgibacillus* species. The values of DNA–DNA hybridization between strain XH-22 and the type strains of *V. carmonensis* and *V. necropolis* were 32% and 28%, respectively. These values are clearly lower than 70% cutoff generally accepted for species delineation and support the placement of strain XH-22 as a genotypically distinct species within the genus *Virgibacillus* (Wayne et al. 1987). Strain XH-22 is a Gram-positive, motile rod, with optimal growth at 10% total salts, pH 7.5 and 37°C. It has *meso*-diaminopimelic acid in the cell wall peptidoglycan and anteiso-$C_{15:0}$, $C_{16:0}$, and iso-$C_{14:0}$ as the major fatty acids. As other members of the genus *Virgibacillus* previously described, MK-7 is its predominant menaquinone; however, strain XH-22 has in addition to this component, MK-6, MK-5 and an unidentified respiratory quinone in minor amounts, whereas *Virgibacillus carmonensis* and *Virgibacillus necropolis* have MK-6 and MK-8 as minor compounds (Heyrman et al. 2003). With respect to the polar lipid composition, strain XH-22 contains diphosphatidylglycerol, phosphatidylglycerol, a glycolipid and two different phospholipids of unknown structure. The presence of diphosphatidylglycerol and phosphatidylglycerol as major polar lipids are common characteristics of *Virgibacillus* species (Heyrman et al. 2003; Wang et al. 2008) but the presence of a glycolipid had not been reported previously in any species of this genus. Strain XH-22 can be clearly distinguished from *V. carmonensis*, its nearest phylogenetic neighbor, by differences in the oxidase reaction, colony pigmentation, optimum temperature for growth and acid production from some substrates. Therefore, on the basis of the data obtained, strain XH-22 belongs to the genus *Virgibacillus*, but constitutes a novel species, for which the name *Virgibacillus salinus* sp. nov. was proposed (Carrasco et al. 2009).

4.7.2 Archaeal Diversity of the Salt Lakes

On the basis of their growth at different salt concentrations, cell morphology and colony pigmentation and some biochemical tests, 268 isolates, representing extremely halophilic archaea, were obtained. We did not isolate any haloarchaea from Lake Chahannor, probably because the salinity of this lake was too low and haloarchaea were not present. The 16S rRNA gene sequences of extremely halophilic archaea isolated were amplified by PCR using forward primer ARCHF and reverse primer ARCHR (López-García et al. 2001; Arahal et al. 1996). The sequences obtained were analyzed and aligned using the ARB software package (Ludwig et al. 2004). The result of the comparison of these sequences with those from previously described haloarchaeal species obtained from the databases indicate that most of the archaea isolated belong to the genera *Halorubrum* (68), *Haloarcula* (60), and *Haloterrigena* (51). The other halophilic archaea are related to haloarchaea belonging to the following genera: *Natrinema* (27), *Haloferax* (19), *Natronococcus* (17), *Natronobacterium* (9), *Halomicrobium* (7), *Natrialba* (6), *Halobacterium* (2), and *Natronolimnobius* (2). The hypersaline environments studied show a great haloarchaeal diversity, since representatives from at least 11 different genera belonging to the family *Halobacteriaceae* were isolated. Besides, the higher number of isolates belong to well known genera, that according to previous studies are frequently isolated from hypersaline habitats, such as the genera *Halorubrum*, *Haloarcula* or *Haloferax* (Torreblanca et al. 1986; Grant and Ross 1986; Gutierrez et al. 1989; McGenity and Grant 1995; Oren and Ventosa 1996; Ochsenreiter et al. 2002). However, a large number of isolates were related to the genera *Haloterrigena* (51) or *Natrinema* (27), which are considered as not so widely represented in hypersaline environments. On the contrary, only two strains were related to the genus *Halobacterium*, which according to previous studies is very abundant in other hypersaline habitats (McGenity et al. 1998; Ventosa et al. 1999).

4.7.2.1 Novel Taxa

Several new isolates showed a low phylogenetic relationship with the previously described haloarchaeal species and they have been studied in more detail in order to determine their taxonomic position.

4.7.2.2 Study of Strain EJ-46 (Described as *Halovivax asiaticus* gen. nov., sp. nov.)

Strain EJ-46 was isolated from sediment of the saline lake Ejinor. This organism is neutrophilic and requires at least 15% NaCl for growth. $MgCl_2$ is not required. Optimal growth is in media containing 20% NaCl at pH 7.0–7.5. Polar lipids analysis revealed the presence of phosphatidylglycerol and phosphatidylglycero-methylphosphate derived from both $C_{20}C_{20}$ and $C_{20}C_{25}$ glycerol diethers. Four glycolipids were detected, one of which may be novel. The 16S rRNA gene analysis revealed that strain EJ-46 is a member of the phylogenetic group defined by the family *Halobacteriaceae*, but there was a low degree of similarity to other members of this family. The highest 16S rRNA similarity values (94.9–94.8%) were obtained with the haloalkaliphilic species *Natronococcus occultus* and *Natronococcus amylolyticus*, respectively. The phenotypic, genotypic and phylogenetic analysis of strain EJ-46 showed that it should be classified as a new genus and species within the haloarchaea, for which the name *Halovivax asiaticus* gen. nov., sp. nov. was proposed (Castillo et al. 2006a).

4.7.2.3 Study of Strain XH-70 (Described as *Halovivax ruber* sp. nov.)

The pleomorphic, extremely halophilic archaeon strain XH-70 was isolated from the saline lake near Xilinhot. The strain required at least 2.5 M NaCl and 5 mM Mg^{2+} for growth. The 16S rRNA gene sequence analysis indicated that this strain belongs to the family *Halobacteriaceae*, showing 99.5% similarity with *Halovivax asiaticus*, and 94.7% and 94.6% similarity with *Natronococcus amylolyticus* and *Natronococcus occultus*, respectively. Polar lipid analysis supported the placement of strain XH-70 in the genus *Halovivax*. DNA–DNA hybridization studies (32% with *H. asiaticus*), as well as biochemical and physiological characterization, allowed strain XH-70 to be differentiated from *Halovivax asiaticus*. A novel species, *Halovivax ruber* sp. nov., was proposed to accommodate this strain (Castillo et al. 2007a).

4.7.2.4 Study of Strain XH-48 (Described as *Halostagnicola larseni* gen. nov., sp. nov.)

This organism is pleomorphic and exhibits optimal growth at 20% NaCl. The G+C content of its DNA is 61 mol%. TLC of polar lipids showed that strain XH-48 contained phosphatidylglycerol and phosphatidylglyceromethylphosphate derived from both $C_{20}C_{20}$ and $C_{20}C_{25}$ glycerol diethers and two unidentified glycolipids. The 16S rRNA gene phylogenetic analysis showed the position of strain XH-48 as a new distinct branch in the *Natrialba* clade, related to *Natrialba aegypticaca* (94.5%) and *Natrialba asiatica* (93.3%). Phenotypic characteristics of these two species of the genus *Natrialba* are very different from those of strain XH-48. The phenotypic, polar lipid profile and phylogenetic data based on the 16S rRNA sequence comparison support the placement of strain XH-48 in a new genus and species within the haloarchaea, for which was proposed the name *Halostagnicola larsenii* gen. nov., sp. nov. (Castillo et al. 2006b).

4.7.2.5 Study of Strain SH-6 (Described as *Halopiger xanaduensis* gen. nov., sp. nov.)

Strain SH-6 was isolated from a sediment sample from the Shangmatala salt lake. This strain was capable of growing over a wide range of NaCl concentrations, ranging from 15 to 30% NaCl. It grew optimally in the presence of 25% NaCl, as has been shown for most extremely halophilic archaea. $MgCl_2$ was not required for growth. Strain SH-6 was non-motile and pleomorphic, although long rod-shaped cells were most common. TLC of polar lipids showed that strain SH-6 contained phosphatidylglycerol and phosphatidylglyceromethylphosphate derived from both $C_{20}C_{20}$ and $C_{20}C_{25}$ glycerol diethers. The glycolipid S2-DGD-1 was also detected. The DNA G+C content of strain SH-6 is 63.1 mol%. Comparison of its 16S rRNA gene sequence with those of members of the family *Halobacteriaceae* revealed that strain SH-6 was distantly related to the other haloarchaeal genera investigated. *Natronolimnobius innermongolicus* and *Natronolimnobius baerhuensis* were shown to be the closest relatives of strain SH-6, having a 16S rRNA sequence similarity value of 94.6%. However, the similarity values were almost identical for several haloarchaea representing various genera, e.g., *Natrialba aegypticaca* (94.5%), *Natronorubrum tibetense* (93.5%) and *Natronococcus occultus* (94.1%). The phenotypic, polar lipid composition and phylogenetic data based on the 16S rRNA gene sequence comparison clearly supported the placement of strain SH-6 in a new genus and species within the haloarchaea, for which was proposed the new name *Halopiger xanaduensis* gen. nov., sp. nov. (Gutiérrez et al. 2006).

4.7.2.6 Study of Strains EJ-52, EJ-32, BG-1, EN-2 and SH-4 (Described as Four Species of the Genus *Halorubrum: Hrr. orientale* sp. nov., *Hrr. ejinorense* sp. nov., *Hrr. kocurii* sp. nov. and *Hrr. aquaticum* sp. nov.)

Five novel halophilic archaeal strains (designated EJ-52, EJ-32, BG-1, EN-2 and SH-4) were isolated from the saline lakes Ejinor, Bagaejinor, Erliannor and Shangmatala. Phylogenetic analysis based on the 16S rRNA gene sequence comparison, polar lipid composition and phenotypic characteristics indicate that these isolates belong to the genus *Halorubrum*. All strains require at least 2.5 M NaCl but $MgCl_2$ is not required. Strain EJ-52 was motile, pleomorphic and red-pigmented. Analysis of its 16S rRNA gene sequence showed that the isolate was closely related to *Halorubrum saccharovorum* (96.1% sequence similarity), *Halorubrum lacusprofundi* (95.9%), *Halorubrum tibetense* (95.2%), *Halorubrum alcaliphilum* (95.2%) and *Halorubrum vacuolatum* (95.1%). The polar lipids of strain EJ-52 were $C_{20}C_{20}$ derivatives of phosphatidylglycerol phosphate and phosphatidylglycerol phosphate methyl ester and a sulfated diglycosyl diether. On the basis of the data obtained in this study, strain EJ-52 represents a novel species, for which the name *Halorubrum orientale* sp. nov. was proposed (Castillo et al. 2006c).

On the basis of 16S rRNA gene sequence similarities, strain EJ-32 was shown to be phylogenetically related to *Halorubrum coriense* (97.9%), *Halorubrum trapanicum* (97.9%), *Halorubrum sodomense* (97.8%), *Halorubrum tebenquichense* (97.8%), *Halorubrum xinjiangense* (97.6%) and *Halorubrum terrestre* (97.4%). Strain EJ-32 was found to be non-motile and Gram-negative. The G+C content of its DNA is 64.0 mol%. Values for DNA–DNA hybridization with respect to phylogenetically related *Halorubrum* species were $\leq$ 49%, indicating that strain EJ-32 represents a novel species of the genus *Halorubrum*, for which the name *Halorubrum ejinorense* sp. nov. was proposed (Castillo et al. 2007b).

Strain BG-1 grows over a pH range from 6.0 to 9.0 with an optimum at pH 7.5. Hypotonic treatment caused cell lysis. Analysis of its 16S rRNA gene sequence positioned this isolate within the genus *Halorubrum* in the family *Halobacteriaceae*. Strain BG-1 was most closely related to *Halorubrum aidingense* (98.8% sequence similarity), *Halorubrum saccharovorum* (98.6%), *Halorubrum lacusprofundi* (98.6%), and *Halorubrum lipolyticum* (98.4%). However, DNA–DNA hybridization values between strain BG-1 and the most closely related members of the genus *Halorubrum* were below 40%. The polar lipid composition revealed the presence of mannosyl-2-sulfate-(1–4)-glycosyl-archaeol (S-DGA-3), the main glycolipid of neutrophilic species of the genus *Halorubrum*. The G+C content of its genomic DNA was 69.4 mol%. Comparison of the phenotypic characteristics between strain BG-1 and *Halorubrum* species supported the conclusion that BG-1 is a novel species within this genus, for which the name *Halorubrum kocurii* sp. nov. was proposed (Gutiérrez et al. 2008b).

Cells of strains EN-2 and SH-4 were motile rods and strictly aerobic. On the basis of 16S rRNA gene sequence analysis, these strains were closely related to *Halorubrum cibi* (97.9 and 98.0% similarity, respectively), *Halorubrum tibetense* (97.6 and 97.3%), *Halorubrum alkaliphilum* (96.8 and 97.1%), *Halorubrum luteum* (96.8 and 97.0%) and *Halorubrum lipolyticum* (96.8 and 97.0%). DNA–DNA hybridization experiments showed that strains EN-2 and SH-4 had relatedness higher than 70% but were not related to these species, with levels of DNA–DNA hybridization equal to or below 45%. Polar lipid analysis revealed that strains EN-2 and SH-4 contained phosphatidylglycerol, phosphatidylglycerol phosphate methyl ester; sulfated diglycosyl diether and several unidentified glycolipids. It was concluded that both strains represent a novel species of the genus *Halorubrum*, for which the name *Halorubrum aquaticum* sp. nov. was proposed (Gutierrez et al. 2011).

4.7.2.7 Study of Strain EJ-57 (Described as *Natrinema ejinorense* sp. nov.)

A Gram-negative, non-motile, neutrophilic, pleomorphic and extremely halophilic archaeon, strain EJ-57, was isolated from the salt lake Ejinor. Strain EJ-57 is able to grow at 25–50°C and requires at least 1.8 M NaCl for growth, with an optimum at 3.4 M NaCl, and grows over a pH range from 6.0 to 8.5, with an optimum at pH 7.0. Hypotonic treatment caused cell lysis. Analysis of its almost complete 16S rRNA gene sequence positioned the isolate within the genus *Natrinema*, in the family *Halobacteriaceae*. Strain EJ-57 was most closely related to *Natrinema versiforme* (96.2% sequence similarity), *Natrinema pallidum* NCIMB (95.9%), *Natrinema altunense* (95.8%) and *Natrinema pellirubrum* (95.5%). However, DNA–DNA hybridization experiments showed that this strain was not related to these species, with DNA relatedness equal or lower than 39%. The major polar lipids of the isolate were $C_{20}C_{20}$ and $C_{20}C_{25}$ derivatives of phosphatidylglycerol, phosphatidylglycerol phosphate methyl ester and a disulfated glycolipid S2-DGA-1. Comparative analysis of the phenotypic characteristics between strain EJ-57 and *Natrinema* species supported the conclusion that EJ-57 is a novel species within this genus, for which the name *Natrinema ejinorense* sp. nov. was proposed (Castillo et al. 2006d).

4.7.2.8 Study of Strains XH-65 (Described as *Haloterrigena salina* sp. nov.)

The extremely halophilic strain XH-65, isolated from the unnamed salt lake near Xilinhot was subjected to a polyphasic taxonomic characterization. This strain is neutrophilic, non-motile and grows optimally at 3.4 M NaCl, at pH 7.5 and at 37°C. Magnesium is not required for growth. On the basis of 16S rRNA gene sequence analysis, strain XH-65 was shown to belong to the genus *Haloterrigena* and was related to *Haloterrigena turkmenica* (98.1% sequence similarity) and other *Haloterrigena* species ($\leq$96.9%). DNA–DNA hybridization revealed 37% relatedness between strain XH-65 and *Haloterrigena turkmenica*. The polar lipid analysis revealed the presence of phosphatidylglycerol, phosphatidylglycerol phosphate methyl ester and mannose-2,6-disulfate (1 $\rightarrow$ 2)-glucose glycerol diether (S_2-DGD). The results of the DNA–DNA hybridization and physiological and biochemical tests allowed genotypic and phenotypic differentiation of strain XH-65 from the *Haloterrigena* species with validly published names. Strain XH-65 represents a novel species, for which the name *Haloterrigena salina* sp. nov. was proposed (Gutiérrez et al. 2008a).

4.7.2.9 Study of Strains CG-6 and CG-4 (Described as *Natronorubrum sediminis* sp. nov.)

Two novel haloalkaliphilic archaea, strains CG-6 and CG-4, were isolated from the sediment of Lake Chagannor. Cells of the two strains were pleomorphic, non-motile and strictly aerobic. They required at least 2.5 M NaCl for growth with an optimum at 3.4 M NaCl, and grew at a pH range from 8.0 to 11.0, with an optimum at pH 9.0. Hypotonic treatment with less than 1.5 M NaCl caused cell lysis. They had similar polar lipid composition, possessing the $C_{20}C_{20}$ and $C_{20}C_{25}$ derivatives of phosphatidylglycerol phosphate and phosphatidylglycerol phosphate methyl ester. No glycolipids were detected. Comparison of 16S rRNA sequences and morphological features placed them in the genus *Natronorubrum*. Their 16S rRNA sequence similarities with the recognized species of the genus *Natronorubrum* were 96.2–93.8%. Detailed phenotypic characterization and

DNA–DNA hybridization studies revealed that the two strains belong to a new species in the genus *Natronorubrum*, for which the name *Natronorubrum sediminis* sp. nov. was proposed (Gutiérrez et al. 2010).

Acknowledgments We thank I. J. Carrasco, A. M. Castillo and Y. Xue for their collaboration during the studies described in this chapter. This study was supported by a grant from the Quality of Life and Management of Living Resources Programme of the European Commission (project Multigenome Access Technology for Industrial Catalysts, QLK3-CT-2002-01972).

References

Amoozegar MA, Malekzadeh F, Malik KA, Schumann P, Spröer C (2003) *Halobacillus karajensis* sp. nov., a novel moderate halophile. Int J Syst Evol Microbiol 53:1059–1063

Antón J, Rosselló-Mora R, Rodriguez-Valera F, Amann R (2000) Extremely halophilic *Bacteria* in crystallizer ponds from solar salterns. Appl Environ Microbiol 66:3052–3057

Antón J, Oren A, Rodríguez-Valera F, Amann R, Rosselló-Mora R (2002) *Salinibacter ruber* gen. nov., sp. nov., a novel, extremely halophilic member of the Bacteria from saltern crystallizer ponds. Int J Syst Evol Microbiol 52:485–491

Arahal DR, Dewhirst FE, Paster BJ, Volcani BE, Ventosa A (1996) Phylogenetic analyses of some extremely halophilic archaea isolated from Dead Sea water, determined on the basis of their 16S rRNA sequences. Appl Environ Microbiol 62:3779–3786

Benlloch S, López-López A, Casamayor EO, Øvreås L, Goddard V, Daae FL, Smerdon G, Massana R, Joint I, Thingstad F, Pedrós-Alió C, Rodríguez-Valera F (2002) Prokaryotic genetic diversity throughout the salinity gradient of a coastal solar saltern. Environ Microbiol 4:349–360

Bidle K, Amadio W, Oliveira P, Paulish T, Hicks S, Earnest C (2005) A phylogenetic analysis of haloarchaea found in a solar saltern. Bios 76:89–96

Bowman JP, Rea SM, McCammon S, McMeekin T (2000) Diversity and community structure within anoxic sediment from marine salinity meromictic lakes and a coastal meromictic marine basin, Vestfold Hills, Eastern Antarctica. Environ Microbiol 2:227–237

Burns DG, Camakaris HM, Jannsen D, Dyall-Smith ML (2004) Combined use of cultivation-dependent and cultivation-independent methods indicates that members of most haloarchaeal groups in an Australian crystallizer pond are cultivable. Appl Environ Microbiol 70:5258–5265

Carrasco IJ (2009) Biodiversidad de bacterias halófilas moderadas en ambientes salinos de China. Doctoral Thesis, University of Seville

Carrasco IJ, Márquez MC, Yanfen X, Ma Y, Cowan DA, Jones BE, Grant WD, Ventosa A (2006) *Gracilibacillus orientalis* sp. nov., a novel moderately halophilic bacterium isolated from a salt lake in Inner Mongolia, China. Int J Syst Evol Microbiol 56:599–604

Carrasco IJ, Márquez MC, Xue Y, Ma Y, Cowan DA, Jones BE, Grant WD, Ventosa A (2007a) *Bacillus chagannorensis* sp. nov., a moderate halophile from a soda lake in Inner Mongolia, China. Int J Syst Evol Microbiol 57:2084–2088

Carrasco IJ, Márquez MC, Xue Y, Ma Y, Cowan DA, Jones BE, Grant WD, Ventosa A (2007b) *Salsuginibacillus kocurii* gen. nov., sp. nov., a moderately halophilic bacterium from soda-lake sediment. Int J Syst Evol Microbiol 57:2381–2386

Carrasco IJ, Márquez MC, Xue Y, Ma Y, Cowan DA, Jones BE, Grant WD, Ventosa A (2008) *Sediminibacillus halophilus* gen. nov., sp. nov., a moderately halophilic, Gram-positive bacterium from a hypersaline lake. Int J Syst Evol Microbiol 58:1961–1967

Carrasco IJ, Márquez MC, Ventosa A (2009) *Virgibacillus salinus* sp. nov., a moderately halophilic bacterium from sediment of a saline lake. Int J Syst Evol Microbiol 59:3068–3073

Castillo AM (2008) Estudio taxonómico de haloarqueas de lagos hipersalinos situados en Mongolia Interior (China). Doctoral Thesis, University of Seville

Castillo AM, Gutiérrez MC, Kamekura M, Ma Y, Cowan DA, Jones BE, Grant WD, Ventosa A (2006a) *Halovivax asiaticus* gen. nov., sp. nov., a novel extremely halophilic archaeon isolated from Inner Mongolia, China. Int J Syst Evol Microbiol 56:765–770

Castillo AM, Gutiérrez MC, Kamekura M, Ma Y, Cowan DA, Jones BE, Grant WD, Ventosa A (2006b) *Halostagnicola larsenii* gen. nov., sp. nov., an extremely halophilic archaeon from a saline lake in Inner Mongolia, China. Int J Syst Evol Microbiol 56:1515–1517

Castillo AM, Gutierrez MC, Kamekura M, Xue Y, Ma Y, Cowan DA, Jones BJ, Grant WD, Ventosa A (2006c) *Halorubrum orientale* sp. nov., a halophilic archaeon isolated from Lake Ejinor, Inner Mongolia, China. Int J Syst Evol Microbiol 56:2559–2563

Castillo AM, Gutierrez MC, Kamekura M, Xue Y, Ma Y, Cowan DA, Jones BJ, Grant WD, Ventosa A (2006d) *Natrinema ejinorense* sp. nov., isolated from a saline lake in Inner Mongolia, China. Int J Syst Evol Microbiol 56:2683–2687

Castillo AM, Gutiérrez MC, Kamekura M, Ma Y, Cowan DA, Jones BJ, Grant WD, Ventosa A (2007a) *Halovivax ruber* sp. nov., an extremely halophilic archaeon isolated from Lake Xilinhot, Inner Mongolia, China. Int J Syst Evol Microbiol 57:1024–1027

Castillo AM, Gutiérrez MC, Kamekura M, Ma Y, Cowan DA, Jones BJ, Grant WD, Ventosa A (2007b) *Halorubrum ejinorense* sp. nov., a isolated from Lake Ejinor, Inner Mongolia, China. Int J Syst Evol Microbiol 57:2538–2542

Caton TM, Caton IR, Witte LR, Schneegurt M (2004) Halotolerant aerobic heterotrophic bacteria from the Great Salt Plains of Oklahoma. Microb Ecol 48:449–462

Christner BC (2002) Recovery of bacteria from glacial and subglacial environments. PhD Thesis, Ohio State University, Columbus

Claus D, Berkeley RCW (1986) Genus *Bacillus* Cohn 1872. In: Sneath PHA, Mair NS, Sharpe ME, Holt JG (eds) Bergey's manual of systematic bacteriology, vol 2. Williams & Wilkins, Baltimore, pp 1105–1139

Daffonchio D, Borin S, Brusa T, Brusetti L, van der Wielen PW, Bolhuis H, Yakimov MM, D'Auria G, Giuliano L, Marty D, Tamburini C, McGenity TJ, Hallsworth JE, Sass AM, Timmis KN, Tselepides A, de Lange GJ, Hübner A, Thomson J, Varnavas SP, Gasparoni F, Gerber HW, Malinverno E, Corselli C, Garcin J, McKew B, Golyshin PN, Lampadariou N, Polymenakou P, Calore D, Cenedese S, Zanon F, Hoog S, Party BS (2006) Stratified prokaryote network in the oxic-anoxic transition of a deep-sea halocline. Nature 440:203–207

Demergasso CS, Casamayor EO, Chong G, Galleguillos P, Escudero L, Pedrós-Alió C (2004) Distribution of prokaryotic genetic diversity in athalassohaline lakes of the Atacama Desert, Northern Chile. FEMS Microbiol Ecol 48:57–69

Donachie SP, Bowman JP, Alam M (2004) *Psychroflextis tropicus* sp. nov., an obligately halophilic *Cytophaga-Flavobacterium-Bacteroides* group bacterium from an Hawaiian hypersaline lake. Int J Syst Evol Microbiol 54:935–940

Eder W, Ludwig W, Huber R (1999) Novel 16S rRNA gene sequences retrieved from highly saline brine sediments of Kebrit Deep, Red Sea. Arch Microbiol 172:213–218

Fan H, Xue Y, Ma Y, Ventosa A, Grant W (2004) *Halorubrum tibetense* sp. nov., a novel haloalkaliphilic archaeon from Lake Zabuye in Tibet, China. Int J Syst Evol Microbiol 54:1213–1216

Feng J, Zhou P, Liu S (2004) *Halorubrum xinjiangense* sp nov., a novel halophile isolated from saline lakes in China. Int J Syst Evol Microbiol 54:1789–1791

Fuse H, Inoue H, Murakami K, Takimura O, Yamaoka Y (2003) Production of free and organic iodine by *Roseovarius* spp. FEMS Microbiol Lett 229:189–194

García MT, Gallego V, Ventosa A, Mellado E (2005) *Thalassobacillus devorans* gen. nov., sp. nov., a moderately halophilic, phenol-degrading, Gram-positive bacterium. Int J Syst Evol Microbiol 55:1789–1795

Grant WD (2004) Life at low water activity. Phil Trans R Soc Lond B 359:1249–1267

Grant WD, Ross HNM (1986) The ecology and taxonomy of halobacteria. FEMS Microbiol Rev 39:9–15

Grant WD, Gemmell RT, McGenity TJ (1998) Halophiles. In: Horikoshi K, Grant WD (eds) Extremophiles: microbial life in extreme environments. Wiley-Liss, USA, pp 93–132

Grant S, Grant W, Jones BE, Kato C, Li L (1999) Novel archaeal phylotypes from an East African alkaline saltern. Extremophiles 3:139–145

Grant WD, Kamekura M, McGenity TJ, Ventosa A (2001) Class III. Halobacteria class nov. In: Boone DR, Castenholz R, Garrity GM (eds) Bergey's manual of systematic bacteriology, vol 1. Springer, New York, pp 294–334

Grant S, Grant WD, Cowan D, Jones BE, Ma Y, Ventosa A, Heaphy S (2006) Identification of eukaryotic open reading frames in metagenomic cDNA libraries made from environmental samples. Appl Environ Microbiol 72:135–143

Gutierrez MC, Ventosa A, Ruiz-Berraquero F (1989) DNA–DNA homology studies among strains of *Haloferax* and other halobacteria. Curr Microbiol 18:253–256

Gutierrez MC, Castillo AM, Corral P, Kamekura M, Ventosa A (2011) *Halorubrum aquaticum* sp. nov., an archaeon isolated from hypersaline lakes. Int J Syst Evol Microbiol 61:1144–1148

Gutiérrez MC, Castillo AM, Kamekura M, Ma Y, Cowan DA, Jones BJ, Grant WD, Ventosa A (2006) *Halopiger xanaduensis* gen. nov., sp. nov., an extremely halophilic archaeon isolated from a saline lake Shangmatala in Inner Mongolia, China. Int J Syst Evol Microbiol 57:1402–1407

Gutiérrez MC, Castillo AM, Kamekura M, Ventosa A (2008a) *Haloterrigena salina* sp. nov., an extremely halophilic archaeon isolated from a salt lake. Int J Syst Evol Microbiol 58:2880–2884

Gutiérrez MC, Castillo AM, Pagaling E, Heaphy S, Kamekura M, Xue Y, Ma Y, Cowan DA, Jones BJ, Grant WD, Ventosa A (2008b) *Halorubrum kocurii* sp. nov., an archaeon isolated from a saline lake. Int J Syst Evol Microbiol 58:2031–2035

Gutiérrez MC, Castillo AM, Corral P, Minegishi H, Ventosa A (2010) *Natronorubrum sediminis* sp. nov., an archaeon isolated from a saline lake. Int J Syst Evol Microbiol 60:1802–1806

Hao MV, Kocur M, Komagata K (1984) *Marinococcus* gen. nov., a new genus for motile cocci with meso-diaminopimelic acid in the cell wall; and *Marinococcus albus* sp. nov. and *Marinococcus halophilus* (Novitsky and Kushner) comb. nov. J Gen Appl Microbiol 30:449–459

Heyrman J, Logan NA, Busse H-J, Balcaen A, Lebbe L, Rodriguez-Díaz M, Swings J, De Vos P (2003) *Virgibacillus carmonensis* sp. nov., *Virgibacillus necropolis* sp. nov. and *Virgibacillus picturae* sp. nov., three novel species isolated from deteriorated mural paintings, transfer of the species of the genus *Salibacillus* to *Virgibacillus*, as *Virgibacillus maismortui* comb. nov. and *Virgibacillus salexigens* comb. nov., and emended description of the genus *Virgibacillus*. Int J Syst Evol Microbiol 53:501–511

Heyrman J, Vanparys B, Logan NA, Balcaen A, Rodriguez-Diaz M, Felske A, De Vos P (2004) *Bacillus novalis* sp. nov., *Bacillus vireti* sp. nov., *Bacillus soli* sp. nov., *Bacillus bataviensis* sp. nov. and *Bacillus drentensis* sp. nov., from the Drentse A grasslands. Int J Syst Evol Microbiol 54:47–57

Heyrman J, Rodriguez-Diaz M, Devos J, Felske A, Logan NA, De Vos P (2005) *Bacillus arenosi* sp. nov., *Bacillus arvi* sp. nov. and *Bacillus humi* sp. nov., isolated from soil. Int J Syst Evol Microbiol 55:111–117

Hoeft S, Kulp TR, Miller L, Saltikov C, Murphy J, Han S (2004) Dissimilatory arsenate and sulfate reduction in sediments of two hypersaline, arsenic-rich soda lakes: Mono and Searles lakes, California. Appl Environ Microbiol 72:6514–6526

Hollibaugh JT, Budinoff C, Hollibaugh R, Ransom B, Bano N (2006) Sulfide oxidation coupled to arsenate reduction by a diverse microbial community in a soda lake. Appl Environ Microbiol 72:2043–2049

Hugenholtz P, Pitulle C, Hershberger K, Pace N (1998) Novel division level bacterial diversity in a Yellowstone hot spring. J Bacteriol 180:366–376

Imhoff JF, Süling J (1996) The phylogenetic relationship among *Ectothiorhodospiraceae*: a reevaluation of their taxonomy on the basis of 16S rDNA analyses. Arch Microbiol 165:106–113

Inagaki F, Nunoura T, Nakagawa S, Teske A, Lever M, Lauer A (2006) Biogeographical distribution and diversity of microbes in methane hydrate-bearing deep marine sediments on the Pacific Ocean Margin. Proc Natl Acad Sci USA 103:2815–2820

Jiang H, Dong H, Zhang G, Yu B, Chapman L, Fields M (2006) Microbial diversity in water and sediment of Lake Chaka, an athalassohaline lake in northwestern China. Appl Environ Microbiol 72:3832–3845

Jones BE, Grant WD, Collins NC, Mwatha WE (1994) Alkaliphiles: diversity and identification. In: Priest FG, Ramos-Cormenzana A, Tindall BJ (eds) Bacterial diversity and systematics. Plenum, New York, pp 195–230

Jones BE, Grant WD, Duckworth A, Owenson G (1998) Microbial diversity of soda lakes. Extremophiles 2:191–200

Ley RE, Backhed F, Turnbaugh P, Lozupone C, Knight R, Gordon J (2005) Obesity alters gut microbial ecology. Proc Natl Acad Sci USA 102:11070–11075

Li W-J, Schumann P, Zhang Y-Q, Chen G-Z, Tian X-P, Xu L-H, Stackebrandt E, Jiang C-L (2005) *Marinococcus halotolerans* sp. nov., isolated from Qinghay, north-west China. Int J Syst Evol Microbiol 55:1801–1804

Liaw HJ, Mah RA (1992) Isolation and characterization of *Halanaerobacter chitinovorans* gen. nov., sp. nov., a halophilic, anaerobic, chitinolytic bacterium from a solar saltern. Appl Environ Microbiol 58:260–266

Lim J-M, Jeon CO, Kim C-J (2006a) *Bacillus taeanensis* sp. nov., a halophilic Gram-positive bacterium from a solar saltern in Korea. Int J Syst Evol Microbiol 56:2903–2908

Lim J-M, Jeon CO, Lee J-C, Ju YJ, Park D-J, Kim C-J (2006b) *Bacillus koreensis* sp. nov., a spore-forming bacterium, isolated from the rhizosphere of willow roots in Korea. Int J Syst Evol Microbiol 56:59–63

Liu WY, Zeng J, Wang L, Dou YT, Yang SS (2005) *Halobacillus dabanensis* sp. nov. and *Halobacillus aidingensis* sp. nov., isolated from salt lakes in Xinjiang, China. Int J Syst Evol Microbiol 55:1991–1996

López-García P, Moreira D, López-López A, Rodríguez-Valera F (2001) A novel haloarchaeal-related lineage is widely distributed in deep oceanic regions. Environ Microbiol 3:72–78

Ludwig W, Strunk O, Westram R, Richter L, Meier H, Yadhukumar BA, Lai T, Steppi S, Jobb G, Förster W, Brettske I, Gerber S, Ginhart AW, Gross O, Grumann S, Hermann S, Jost R, König A, Liss T, Lüßmann R, May M, Nonhoff B, Reichel B, Strehlow R, Stamatakis A, Stuckmann N, Vilbig A, Lenke M, Ludwig T, Bode A, Schleifer KH (2004) ARB: a software environment for sequence data. Nucleic Acids Res 32:1363–1371

Márquez MC, Carrasco IJ, Xue Y, Ma Y, Cowan DA, Jones BE, Grant WD, Ventosa A (2007) *Aquisalimonas asiatica* gen. nov., sp. nov., a moderately halophilic bacterium isolated from an alkaline, saline lake in Inner Mongolia, China. Int J Syst Evol Microbiol 57:1137–1142

Márquez MC, Carrasco IJ, Xue Y, Ma Y, Cowan DA, Jones BE, Grant WD, Ventosa A (2008) *Aquisalibacillus elongatus* gen. nov., sp. nov., a moderately halophilic bacterium of the family *Bacillaceae* isolated from a saline lake. Int J Syst Evol Microbiol 58:1922–1926

McGenity TJ, Grant WD (1995) Transfer of *Halobacterium saccharovorum, Halobacterium sodomense, Halobacterium trapanicum* NRC 34021 and *Halobacterium lacusprofundi* to the genus *Halorubrum* gen. nov., as *Halorubrum saccharovorum* comb nov., *Halorubrum sodomense* comb nov., *Halorubrum trapanicum* comb. nov., and *Halorubrum lacusprofundi* comb. nov. Syst Appl Microbiol 18:237–243

McGenity TJ, Gemmell RT, Grant WD (1998) Proposal of a new halobacterial genus *Natrinema* gen. nov., with two species *Natrinema pellirubrum* nom. nov. and *Natrinema pallidum* nom. nov. Int J Syst Bacteriol 48:1187–1196

Montalvo-Rodríguez R, Vreeland RH, Oren A, Kessel M, Betancourt C, Lopez-Garriga JJ (1998) *Halogeometricum borinquense* gen. nov., sp. nov., a novel halophilic archaeon from Puerto Rico. Int J Syst Bacteriol 48:1305–1312

Mormile MR, Romine MF, Garcia T, Ventosa A, Bailey T, Peyton B (1999) *Halomonas campisalis* sp. nov., a denitrifying, moderately haloalkaliphilic bacterium. Syst Appl Microbiol 22:551–558

Mouné S, Eatock C, Matheron R, Willison J, Hirschler A, Herbert R (2000) *Orenia salinaria* sp. nov., a fermentative bacterium isolated from anaerobic sediments of Mediterranean salterns. Int J Syst Evol Microbiol 50:721–729

Mouné S, Caumette P, Matheron R, Willison J (2003) Molecular sequence analysis of prokaryotic diversity in the anoxic sediments underlying cyanobacterial mats of two hypersaline ponds in Mediterranean salterns. FEMS Microbiol Ecol 44:117–130

Mukohata Y, Sugiyama Y, Ihara K, Yoshida M (1988) An Australian halobacterium contains a novel proton pump retinal protein: archaerhodopsin. Biochem Biophys Res Commun 151:1339–1345

Mwatha WE, Grant WD (1993) *Natronobacterium vacuolata* sp. nov., a haloalkaliphilic archaeon isolated from Lake Magadi, Kenya. Int J Syst Bacteriol 43:401–404

Nielsen P, Fritze D, Priest FG (1995) Phenetic diversity of alkaliphilic *Bacillus* strains: proposal for nine new species. Microbiology 141:1745–1761

Ochsenreiter T, Pfeifer F, Schleper C (2002) Diversity of Archaea in hypersaline environments characterized by molecular-phylogenetic and cultivation studies. Extremophiles 6:267–274

Oremland RS, Kulp TR, Blum J, Hoeft S, Baesman S, Miller L (2005) A microbial arsenic cycle in a salt-saturated, extreme environment. Science 308:1305–1308

Oren A (1994) The ecology of the extremely halophilic archaea. FEMS Microbiol Rev 13:415–440

Oren A, Ventosa A (1996) A proposal for the transfer of *Halorubrobacterium distributum* and *Halorubrobacterium coriense* to the genus *Halorubrum* as *Halorubrum distributum* comb. nov. and *Halorubrum coriense* comb. nov., respectively. Int J Syst Bacteriol 46:1180

Oren A, Elevi R, Watanabe S, Ihara K, Corcelli A (2002) *Halomicrobium mukohataei* gen. nov., comb. nov., and emended description of *Halomicrobium mukohataei*. Int J Syst Evol Microbiol 52:1831–1835

Pagaling E (2007) Microbial diversity of Chinese Lakes. PhD Thesis, University of Leicester. https://lra.le.ac.uk/handle/2381/7662

Pagaling E, Haigh R, Grant W, Cowan D, Jones B, Ma Y, Ventosa A, Heaphy S (2007) Sequence analysis of an archaeal virus isolated from a hypersaline lake in Inner Mongolia, China. BMC Genomics 8:410, http://www.biomedcentral.com/content/pdf/1471-2164-8-410.pdf

Pagaling E, Venables M, Wallace A, Grant WD, Cowan DA, Jones BE, Ma Y, Ventosa A, Heaphy S (2009) Microbial biogeography of six salt lakes in Inner Mongolia China and a salt lake in Argentina. Appl Environ Microbiol 75:5750–5760

Page A, Juniper SK, Olagnon M, Alain K, Desrosiers G, Querellou J (2004) Microbial diversity associated with a *Parvinella sulfincola* tube and the adjacent substratum on an active deep-sea vent chimney. Geobiology 2:225–238

Pašić L, Bartual S, Ulrih N, Grabnar M, Velikonja B (2005) Diversity of halophilic archaea in the crystallizers of an Adriatic solar saltern. FEMS Microbiol Ecol 54:491–498

Priest FG, Goodfellow M, Todd C (1988) A numerical classification of the genus *Bacillus*. J Gen Microbiol 134:1847–1882

Reysenbach AL, Longnecker K, Kirshtein J (2000) Novel bacterial and archaeal lineages from an in situ growth chamber deployed at a Mid-Atlantic Ridge hydrothermal vent. Appl Environ Microbiol 66:798–806

Romano I, Lama L, Nicolaus B, Gambacorta A, Giordano A (2005) *Bacillus saliphilus* sp. nov., isolated from a mineral pool in Campania, Italy. Int J Syst Evol Microbiol 55:159–163

Sørensen KB, Canfield DE, Teske A, Oren A (2005) Community composition of a hypersaline endoevaporitic microbial mat. Appl Environ Microbiol 71:7352–7365

Spring S, Ludwig W, Márquez MC, Ventosa A, Schleifer KH (1996) *Halobacillus* gen. nov., with descriptions of *Halobacillus litoralis* sp. nov., and *Halobacillus trueperi* sp. nov., and transfer of *Sporosarcina halophila* to *Halobacillus halophilus* comb. nov. Int J Syst Bacteriol 46:492–496

Tomlinson G, Hochstein L (1976) *Halobacterium saccharovorum* sp. nov., a carbohydrate-metabolising extremely halophilic bacterium. Can J Microbiol 22:587–591

Torreblanca M, Rodriguez-Valera F, Juez G, Ventosa A, Kamekura M, Kates M (1986) Classification of non-alkaliphilic halobacteria based on numerical taxonomy and polar lipid composition and description of *Haloarcula* gen. nov. and *Haloferax* gen. nov. Syst Appl Microbiol 8:89–99

Ventosa A, Quesada E, Rodríguez-Valera F, Ruiz-Berraquero F, Ramos-Cormenzana A (1982) Numerical taxonomy of moderately halophilic Gram-negative rods. J Gen Microbiol 128:1959–1968

Ventosa A, Gutiérrez MC, Kamekura M, Dyall-Smith ML (1999) Proposal to transfer *Halococcus turkmenicus, Halobacterium trapanicum* JCM 9743 and strain GSL-11 to *Haloterrigena turkmenica* gen. nov., comb. nov. Int J Syst Bacteriol 49:131–136

Ventosa A, Gutierrez MC, Kamekura M, Zvyagintseva I, Oren A (2004) Taxonomic study of *Halorubrum distributum* and proposal of *Halorubrum terrestre* sp. nov. Int J Syst Evol Microbiol 54:389–392

Vreeland RH, Straigh ST, Krammes J, Dougherty K, Rosenzweig W, Kamekura M (2002) *Halosimplex carlsbadense* gen. nov., sp. nov., a unique halophilic archaeon, with three 16S rRNA genes, that grows only in defined medium with glycerol and acetate or pyruvate. Extremophiles 6:445–452

Wainø M, Tindall BJ, Schumann P, Ingvorsen K (1999) *Gracilibacillus* gen. nov., with description of *Gracilibacillus halotolerans* gen. nov., sp. nov.; transfer of *Bacillus dipsosauri* to *Gracilibacillus dipsosauri* comb. nov., and *Bacillus salexigens* to the genus *Salibacillus* gen. nov., as *Salibacillus salexigens* comb. nov. Int J Syst Bacteriol 49:821–831

Walsh DA, Papke RT, Doolittle W (2005) Archaeal diversity along a soil salinity gradient prone to disturbance. Environ Microbiol 7:1655–1666

Wang C-Y, Chang C-C, Ng CC, Chen T-W, Shyu Y-T (2008) *Virgibacillus chiguensis* sp. nov., a novel halophilic bacterium isolated from Chigu, a previously commercial saltern located in southern Taiwan. Int J Syst Evol Microbiol 58:341–345

Wayne LG, Brenner DJ, Colwell RR, Grimont PAD, Kandler O, Krichevsky MI, Moore LH, Moore WEC, Murray RGE, Stackebrandt E, Starr MP, Trüper HG (1987) Report of the ad hoc committee on reconciliation of approaches to bacterial systematics. Int J Syst Bacteriol 37:463–464

Wieser M, Worliczek H, Kämpfer P, Busse H-J (2005) *Bacillus herbersteinensis* sp. nov. Int J Syst Evol Microbiol 55:2119–2123

Williams WD (1991) Chinese and Mongolian saline lakes: a limnological overview. Hydrobiologia 210:39–66

Xue Y, Zhang X, Zhou C, Zhao Y, Cowan D, Heaphy S (2006) *Caldalkalibacillus thermarum* gen. nov., sp nov., a novel alkalithermophilic bacterium from a hot spring in China. Int J Syst Evol Microbiol 56:1217–1221

Yakimov MM, Giuliano L, Chernikova T, Gentile G, Abraham W, Lunsdorf H (2001) *Alcalilimnicola halodurans gen.* nov., sp. nov., an alkaliphilic, moderately halophilic and extremely halotolerant bacterium, isolated from sediments of soda-depositing Lake Natron, East Africa Rift Valley. Int J Syst Evol Microbiol 51:2133–2143

Yeon SH, Jeong WJ, Park JS (2005) The diversity of culturable organotrophic bacteria from local solar salterns. J Microbiol 43:1–10

Yoon J-H, Kang KH, Park Y-H (2003) *Halobacillus salinus* sp. nov., isolated from a salt lake on the coast of the East Sea in Korea. Int J Syst Evol Microbiol 53:687–693

Yoon J-H, Kang KH, Oh T-K, Park Y-H (2004) *Halobacillus locisalis* sp. nov., a halophilic bacterium isolated from a marine solar saltern of the Yellow Sea in Korea. Extremophiles 8:23–28

Yoon J-H, Kang S-J, Lee C-H, Oh HW, Oh T-K (2005) *Halobacillus yeomjeoni* sp. nov., isolated from a marine solar saltern in Korea. Int J Syst Evol Microbiol 55:2413–2417

Yu G, Harrison SP, Xue B (2001) Lake status records from China: data base documentation. MPI-BGC Tech Rep 4:46–49

Zheng X (1991) Salt lakes on the Inner Mongolian plateau of China. Chin Geogr Sci 1:83–94

Zheng X, Zhang M, Dong J, Gao Z, Xu C, Han Z, Zhang B, Sun D, Wang K (1992) Salt lakes in Inner Mongolia of China. Science Press, Beijing

Zheng M, Tang J, Lui J, Zhang F (1993) Chinese saline lakes. Hydrobiologia 267:23–36

Zhilina TN, Zavarzin GA, Bulygina E, Kevbrin V, Osipov G, Chumakov K (1992) Ecology, physiology and taxonomy studies on a new taxon of *Haloanaerobiaceae, Haloincola saccharolytica* gen. nov., sp. nov. Syst Appl Microbiol 15:275–284

Chapter 5
From Genomics to Microevolution and Ecology: The Case of *Salinibacter ruber*

Arantxa Peña, Hanno Teeling, Jaime Huerta-Cepas, Fernando Santos, Inmaculada Meseguer, Marianna Lucio, Philippe Schmitt-Kopplin, Joaquín Dopazo, Ramon Rosselló-Móra, Margarete Schüler, Frank Oliver Glöckner, Rudolf Amann, Toni Gabaldón, and Josefa Antón

5.1 Introduction

Salinibacter ruber is an extremely halophilic bacterium of the *Bacteroidetes* phylum that inhabits hypersaline environments, such as crystallizer ponds from solar salterns. Haloarchaea usually dominate the microbial communities in such

A. Peña • F. Santos • J. Antón (✉)
Departamento de Fisiología, Genética y Microbiología, Universidad de Alicante, Apartado 99, Alicante, Spain
e-mail: anton@ua.es

H. Teeling • M. Schüler • R. Amann
Max Planck Institute for Marine Microbiology, Celsiusstrasse 1, Bremen, Germany

J. Huerta-Cepas • T. Gabaldón
Bioinformatics and Genomics Program, Centre for Genomics Regulation (CRG), UPF, Dr. Aiguader, 83, 08003 Barcelona, Spain

I. Meseguer
Departamento de Producción Vegetal y Microbiología, Universidad Miguel Hernández. Avda. de la Universidad s/n., Elche (Alicante), Spain

M. Lucio • P. Schmitt-Kopplin
Helmholtz Zentrum München, German Research Center for Environmental Health, Institute of Ecological Chemistry, Neuherberg, Germany

J. Dopazo
Functional Genomics Node (INB), Centro de Investigación Príncipe Felipe (CIPF), Valencia, Spain

R. Rosselló-Móra
Marine Microbiology Group, Institut Mediterrani d'Estudis Avançats (CSIC-UIB), C/Miquel Marqués 21, 07190 Esporles, Spain

F.O. Glöckner
Max Planck Institute for Marine Microbiology, Celsiusstrasse 1, Bremen, Germany
and
Jacobs University Bremen GmbH, Campusring 1, Bremen, Germany

A. Ventosa et al. (eds.), *Halophiles and Hypersaline Environments*,
DOI 10.1007/978-3-642-20198-1_5,

salterns, while *S. ruber* can account from 2 to 30%. *S. ruber* is globally widespread, but nonetheless its 16S rRNA gene and 16S–23S rRNA internal transcribed spacers (ITS) sequences are highly congruent (see below). At the genomic level, however, this species presents a high level of microdiversity (Peña et al. 2005; Antón et al. 2008). Indeed, when new *S. ruber* isolates are obtained from the same salterns where the species was first retrieved and their digested genomic DNAs are analyzed by pulsed gel field electrophoresis (PFGE), a remarkably high diversity of patterns is obtained. For instance, 35 new strains isolated in 2008 provided as many new patterns, although all had identical 16S rRNA gene sequences. In addition, none of these patterns corresponded to previously retrieved strains.

To date, genome comparisons offer the most powerful way to study micro-diversity. This approach allows unveiling the diversity hidden within a species' population as well as providing clues on the ecological mechanisms that drive diversification. Until recently, such studies were for obvious reasons focused on pathogenic bacteria. In the last years, however, microdiversity has also been studied in aquatic prokaryotic species, with a focus on the characterization of ecotypes (i.e., sub-populations within the same species with distinct ecological strategies) (Cohan and Koeppel 2008). Such comparative genome analyses at the sub-species level include the studies on *Prochlorococcus* sp. (Coleman et al. 2006), and *Alteromonas macleodi* isolates (Ivars-Martínez et al. 2008), for which ecotypes have been found that can be distinguished via their 16S rRNA gene(s) or ITS. However, there is a scale of diversity that falls beyond the ecotype level, like clusters above 99% 16S rRNA similarity as described by Acinas et al. (2004) for coastal bacterioplankton. In hypersaline environments, such clusters can be found as well (Antón et al. 2008; Oh et al. 2010).

The situation of *S. ruber* represents a really short scale of differentiation, since in most cases no differences in the whole ribosome operon can even be observed although, systematically, different strains are being continuously isolated from the same environment. From an ecological perspective, the microdiversity in *S. ruber* is really intriguing, in particular in consideration of the few opportunities for micro-niche differentiation that crystallizer ponds seem to offer.

5.2 Choosing the Strains: M8 and M31

Our objective was to envisage the microdiversity within the species *S. ruber* and understand its meaning. For this purpose, we chose the two most closely related strains in order to elucidate (a) the extent of the genomic differences among these strains, (b) whether the genomic differences have any ecological benefit, and (c) the mechanisms causing microdiversity in the context of the apparent lack of heterogeneity in the crystallizers that *S. ruber* inhabits.

Among all strains within our collection from different parts of the world, M8 and M31 showed the closest relatedness. Both had identical ribosomal operons (including 5S, 16S, 23S, and ITS), a DNA–DNA hybridization value of around 90%, and had been sampled at the same time in September 1999 on the Mediterranean island Mallorca from a pond of the solar saltern of Campos. On the other hand, a genomic micro-diversity was detectable, since both strains showed distinct genomic PFGE patterns that, like random amplification of polymorphic DNA (RAPD) patterns, clustered with above 90% similarity (Peña et al. 2005). Since strains M8 and M31 were among the five isolates that had been used for the *S. ruber* species description (Antón et al. 2002), and the genome of strain M31^{T} had already been sequenced by Mongodin et al. (2005), this provided a unique opportunity for a strain level genome comparison study (Peña et al. 2010).

5.3 Differences Between the M8 and M31 Genomes

The first step of the comparative analysis was to sequence the M8 genome (a single chromosome of 3.6 Mbp, and four plasmids ranging from 11.2 to 84.34 kbp) and to analyze features such as GC content, tetranucleotide frequency or codon usage. These parameters turned out not to be homogeneous along the chromosome, since there were two areas (that we called hypervariable regions – HVRs) with deviant GC contents and tetranucleotide frequencies as well as less adapted codon usage (Fig. 5.1). The annotation of the genome indicated a higher density of transposases in these areas as well as genes related to the cell envelope (cell wall, capsules, and outer membrane, including many glycosyltransferases and sulfotransferases).

The genomes of M8 and M31 were compared with different tools. As expected, both genomes were highly similar as it could be observed from the calculations of the average amino acid identity (AAI = 94.2%), average nucleotide identity for orthologous genes (ANIo = 93.5%), and average nucleotide identity (ANIb = 98.5%), but they also displayed some remarkable differences as discussed below. The differences between ANIo and ANIb are basically due to the differences in the thresholds for their calculations. In one side, ANIo indicates the degree of identity among recognized orthologous genes shared between the strains. On the other side, ANIb is a raw indication of the identity of the genome sequences in a similar way to that given by the use of DNA–DNA hybridization assays (Richter and Rosselló-Móra 2009).

With the exception of the HVRs, high levels of gene synteny (gene order and orientation) were observed between both chromosomes (Fig. 5.2). The two HVRs of M8 aligned with two of the three "genomic islands" that were described in M31. In addition, strain-specific genes were identified along the chromosome (represented as dots in the axis of Fig. 5.2). These are genes present only in one of the two strains, and thus part of the accessory genome of the *S. ruber* species, while the remaining genes present in both strains are part of the core genome (Feil 2004). Together, core and accessory genomes form the species' pan-genome (Medini et al. 2005). In total, 325 genes were found in M8 that were absent from M31, which implies that about 10% of the M8 genes were strain-specific. Among these, 164 genes were unique to M8 and constituted new genes absent from any

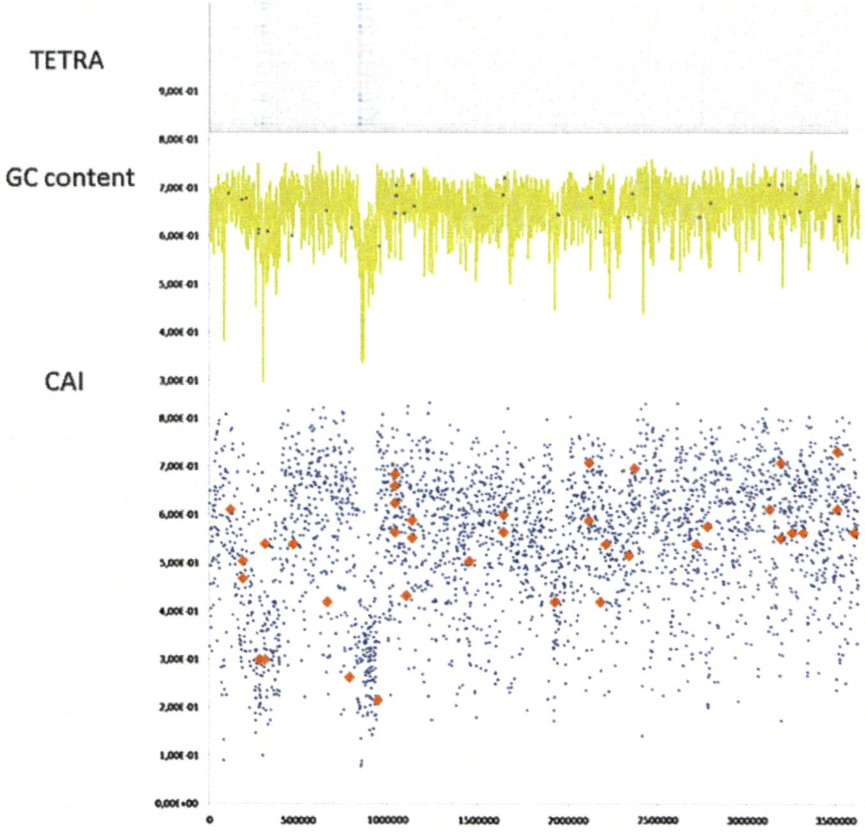

Fig. 5.1 Characteristics of the M8 genome. From *top* to *bottom*: tetranucleotide frequency, GC content and codon adaptation indexes. In *blue* in the GC plot and *red* in the CAI plot, the LGT genes described in Table 5.1. The whole-genome alignment shows the positions of orthologous genes in *S. ruber* M8 and M31. Genes present only in one of the strains appear on the axes. The location of HVR and CR in M8 are marked with *gray bars*. Modified from Peña et al. (2010)

sequence database. It is remarkable that analyzing the genome of a new strain (with a ribosomal operon identical to a known genome) reveals 164 novel genes. This points towards an extensive pan-genome, as discussed at the end of this chapter.

The core genome of *S. ruber* included genes coding for (conserved) hypothetical proteins, that were present only in M8 and M31 but not in the available databases; these genes can be considered as species-specific. Most of the genes shared by both strains displayed high levels of high amino acid identity, but 18% of the shared genes were rather divergent (AAI below 90%) with a few very divergent genes (AAI below 50%). In general, the divergent genes were distributed along the whole chromosome, albeit with a higher density in the HVRs (Peña et al. 2010).

Thus, the types of observed differentiations between M8 and M31 included: the HRVs, insertions and deletions outside of the HRVs, divergence among homologs,

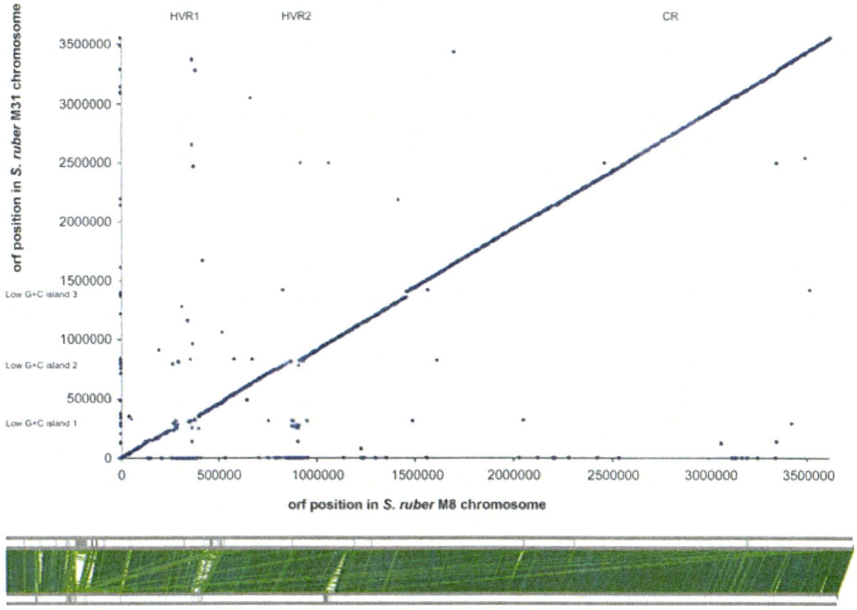

Fig. 5.2 Whole-genome alignment between M8 and M31. The *green bars* linking both genomes represent orthologs matches identified by FASTA analysis. The sequences have been aligned from the predicted replication origin. From Peña et al. (2010), with permission

and specific gene duplications in every genome. Of course, this raised the question whether these differences cause as yet undetected phenotypic strain-level differences. Therefore, we analyzed the strain's metabolomes and we indeed detected differences, mainly in the extracellular fractions, that showed a higher content of sulfonated and glycosylated metabolites in M8 (Fig. 5.3). In addition, we also found that both strains were able to compete to each other: when M8 and M31 were co-cultured in a salt-saturated medium, M31 inhibited the growth of M8 (Peña et al. 2010). Therefore, the differences found between the genomes of both strains were indeed having an effect on their phenotypes and therefore cannot be considered neutral from an ecological point of view.

5.4 The Role of LGT in the Shaping of *S. ruber* Genome

Salinibacter shares its habitat with haloarchaea (family *Halobacteriaceae*). In most, but not all (Oh et al. 2010) of the hypersaline environments analyzed so far, the most abundant prokaryote in the community is the square archaeon *Haloquadratum walsbyi*, while *S. ruber* is the most abundant bacterium. In other cases, *Halorubrum* and *Halobacterium* representatives dominate the community and the gammaproteobacterium *Salicola* spp. dominates the bacterial assemblage.

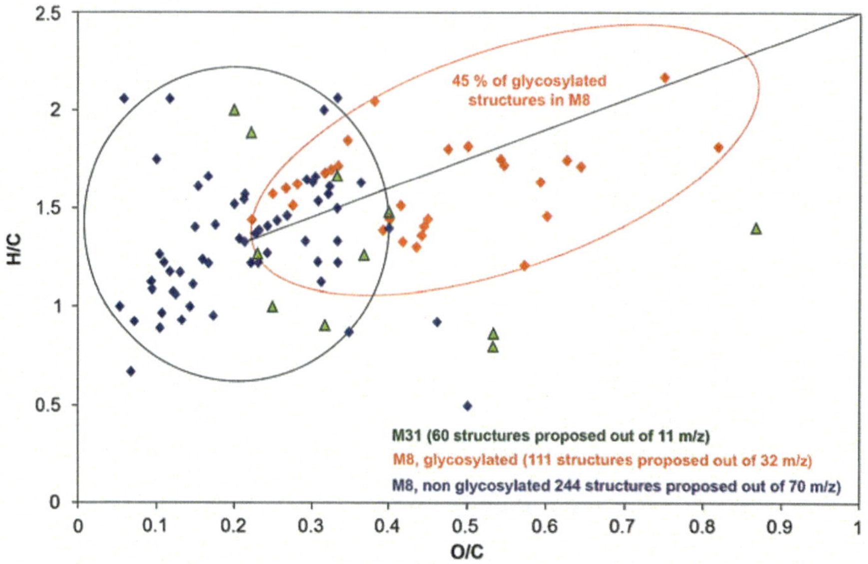

Fig. 5.3 Van Krevelen diagram structural visualization of the discriminative masses only of M8 versus M31 showing the importance of glycosylated structures in M8 as obtained after annotation of the experimental exact masses in the KEGG database (http://www.MassTRIX.org)

Whatever the case, *Salinibacter* always shares its habitat with Archaea, which provides the possibility of lateral gene transfer between these two organisms or, in other words, inter-domain (Archaea-Bacteria) gene transfer. In this context, it is noteworthy that *Salinibacter* and haloarchaea share many phenotypic traits that might be resulting either from convergent evolution or LGT: both are extremely halophilic, aerobic, heterotrophic, pigmented, maintain high intracellular potassium concentrations, have a high genomic GC content (with the exception of *Hqr. walsbyi*), and have retinal proton pumps in their membranes. The sequencing of the *S. ruber* type strain (Mongodin et al. 2005) revealed that the extent of archaeal genes in the *Salinibacter* genome was significant, albeit lower than anticipated. In our work (Peña et al. 2010), we used different approaches to identifying genes involved in inter-domain LGT in *Salinibacter* genomes and reached the same conclusion. In addition, we investigated whether these genes (hereafter referred to as LGT or archaeal genes) had been recently incorporated into the *S. ruber* genome, assuming that they were of pristine archaeal origin and not bacterial genes once transferred to archaeal genomes and now mimicking an archaeal origin, and whether these LGT genes were constitutively expressed and thus had become part of the *S. ruber* core genome.

Most of the putative LGT genes found in the M8 genome (Table 5.1) were located outside the hypervariable regions and did not display anomalous GC contents or unusual codon usage (Fig. 5.1), which would have been expected for recent acquisitions. In addition, only six of the 40 LGT genes in M8 were part of the accessory genome (i.e., not present in M31). Thus, assuming that the identified LGT

Table 5.1 Analysis of the expression of genes candidate of inter-domain LGT along the growth curve

ORF	Annotation	Closest relative[a]	Expression[b] M8/M31
94	HTR-like protein	*Halobacterium salinarum* NRC-1	++++/++++
150	Putative light-and oxygen-sensing transcription regulator	*Halobacterium salinarum* NRC-1	−+++/−++−
169	Adaptive-response sensory-kinase	*Haloarcula marismortui* DSM 3752	+++−/++−−
226	**Conserved hypothetical protein containing PIN domain (putative nucleic acid-binding protein)**	***Haloarcula marismortui* DSM 3752**	**++++/NA**
227	**Conserved hypothetical protein containing DUF0175 domain**	***Haloarcula marismortui* DSM 3752**	**++++/++++**
274	**Conserved hypothetical protein**	***Haloarcula marismortui* DSM 3752**	**−−−−/NA**
393	Cytochrome c oxidase subunit II	*Haloarcula marismortui* DSM 3752	++++/++++
538	Deoxycytidine triphosphate deaminase	*Halobacterium salinarum* NRC-1	−++−/−−+−
655	**Glycosyl transferase, group 1**	***Haloarcula marismortui* DSM 3752**	**−−−−/NA**
831	**Conserved hypothetical protein containing PIN domain (putative nucleic acid-binding protein)**	***Pyrococcus kodakaraensis* OT3**	**−−−−/NA**
911	**Probable exodeoxyribonuclease VII small subunit**	***Haloarcula marismortui* DSM 3752**	**++++/NA**
917	Ferredoxin	*Natronomonas pharaonis* DSM 2160	++++/−+++
918	Conserved hypothetical protein, membrane	*Natronomonas pharaonis* DSM 2160	+++−/++−−
919	Bacteriorhodopsin related protein	*Natronomonas pharaonis* DSM 2160	++−+/−−−−
920	Phytoene dehydrogenase	*Haloarcula marismortui* DSM 3752	++++/++++
964	Probable cell division protein	*Methanobacterium thermoautotrophicum* strain Delta H	++++/++++
995	Translation initiation factor eIF-2B alpha subunit	*Haloarcula marismortui* DSM 3752	++++/++++
1011	Conserved hypothetical protein, secreted	*Haloarcula marismortui* DSM 3752	++++/+−−−
1308	**Uncharacterized ACR**	***Archaeoglobus fulgidus* DSM 4304**	**+++−/NA**
1444	Trk system potassium uptake protein trkA	*Haloarcula marismortui* DSM 3752	++++/−++−
1448	D-lactate dehydrogenase, putative	*Natronomonas pharaonis* DSM 2160	−−−−/−−−−
1707	Conserved hypothetical protein	*Haloarcula marismortui* DSM 3752	++++/−−++
1709	Putative zinc-binding dehydrogenase	*Halobacterium salinarum* NRC-1	+−++/−++−
1859	Anion permease	*Halobacterium salinarum* NRC-1	−+++/++++
1860	Phosphate transporter	*Halobacterium salinarum* NRC-1	++++/−−+−
1903	Conserved hypothetical protein	*Haloarcula marismortui* DSM 3752	+−+−/−−+−
1924	Multi antimicrobial extrusion protein MatE	*Haloarcula marismortui* DSM 3752	−−−−/−−−−
2026	Sodium/calcium exchanger protein, membrane	*Natronomonas pharaonis* DSM 2160	+++−/+−+−
2049	Conserved hypothetical protein containing UPF0047	*Methanospirillum hungatei* DSM 864	−−−−/−−−−
2365	Putative bacterio-opsin	*Haloarcula marismortui* DSM 3752	++++/++++
2415	Heme exporter protein C	*Pyrobaculum aerophilum* DSM 7523	++−−/−−−−
2493	Trk potassium uptake system protein	*Haloarcula marismortui* DSM 3752	++++/++++
2657	Glycerol-3-phosphate dehydrogenase subunit A	*Haloarcula marismortui* DSM 3752	+++−/++++
2720	Conserved hypothetical protein, membrane	*Methanothermobacter thermautotrophicus* ΔH	++++/++++
2733	Sensory rhodopsin I (SR-I)	*Haloarcula marismortui* DSM 3752	+++−/+++−

(*continued*)

Table 5.1 (continued)

ORF	Annotation	Closest relative[a]	Expression[b] M8/M31
2780	Putative small solutes transporter	*Halobacterium salinarum* NRC-1	+++−/+++−
2798	Sensory rhodopsin I (SR-I)	*Natronomonas pharaonis* DSM 2160	++++/++++
2993	Sodium:proline symporter	*Haloarcula marismortui* DSM 3752	+−−−/−−−+
2994	Halorhodopsin	*Natronomonas pharaonis* DSM 2160	++++/++++
3079	Conserved hypothetical protein	*Natronomonas pharaonis* DSM 2160	++++/++++

Modified from Peña et al. (2010)
In boldface, LTG candidates specific of M8
Shaded in light gray, genes in HVRs; in dark gray and white font, genes in CR
[a]Closest relative according to phylogenetic trees
[b]Expression of ORFs along the growth curve as measured by RT-PCR amplification with specific primers. For every culture, 4 points were analyzed corresponding to the early, mid and late exponential phase and to stationary phase. For every point + or − is indicated. *NA* not analyzed (these ORFs were present only in M8 genome and could thus not be analyzed in M31)

genes were acquired from archaeal genomes, this interchange should have happened before the separation of the M8 and M31 lineages. This is also in agreement with the LGT gene's average to high CAI and standard GC values. Further analyses (Peña et al., manuscript in preparation) indicated that most of the LGT genes identified in M8 are also present in strains from different geographical sources. Thus, although modest, LGT has played a role in shaping the *Salinibacter* genome.

In order to see whether these 40 LGT genes were needed for the normal growth of *S. ruber*, their expression together with 8 of the 13 LGT genes present only in M31 was analyzed along the growth curve. Pure cultures of M8 and M31 were grown under standard conditions and samples were taken in early, mid, and late exponential as well as in the stationary phase. After RNA extraction, the expression of each LGT gene was detected by RT-PCR with primers specifically designed for every gene (Peña et al. 2010).

Both strains showed different LGT gene expression patterns, with a large majority of the core genome LGT genes expressed at some point in the growth curve, and a much lower proportion of expressed strain-specific LGT genes (Table 5.1; Fig. 5.4). These data indicate that most of the core genome inter-domain LGT candidates are expressed during normal growth and most likely represent functional genes needed for growth. This is in agreement with the CAI and GC traits discussed above. On the contrary, the strain-specific inter-domain LGT candidates are seldom expressed which, together with their GC and CAI characteristics, indicate that they have probably been acquired recently and did not have time to amend to their genomic environments. In any case, they do not seem necessary for growth under standard conditions.

Although beyond the goal of these experiments, the results shown in Table 5.1 deserve additional comments. The cultures were grown in the dark; however, the light-dependent retinal binding protein halorhodopsin and bacteriorhodopsin (Oren 2002) were constitutively expressed for both strains along the entire growth curve. Sensory rhodopsin (Oren 2002) was also expressed by both strains during the complete exponential phase. This is intriguing, since the expression of these

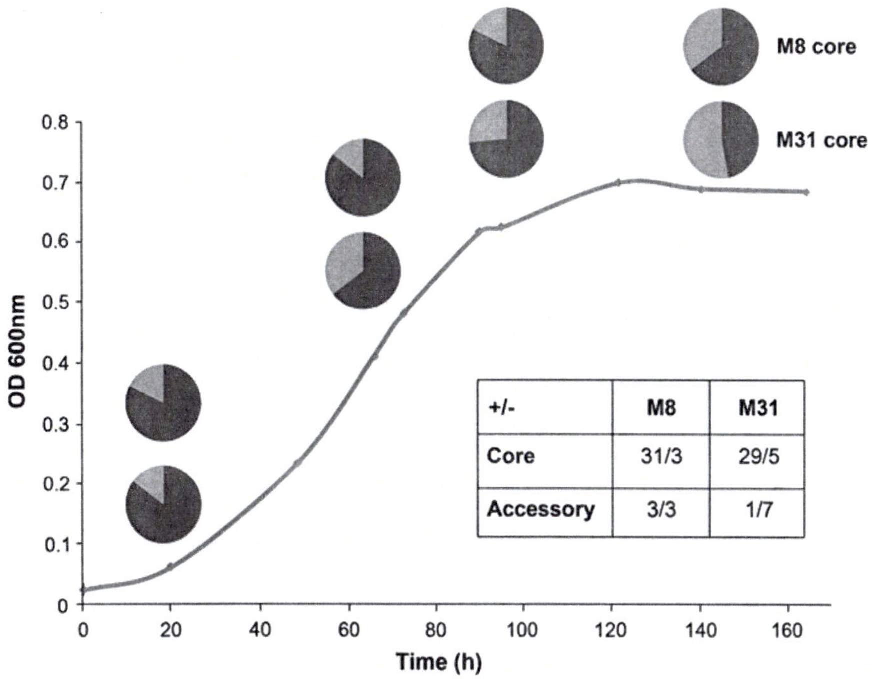

Fig. 5.4 Expression of candidate LGT genes along the growth curve. Pure cultures of M8 and M31 were set up and samples taken at four points along the growth. For every point, the proportion of core-genome LGT expressed genes for M8 and M31 (*dark gray*) are represented in *circles*. The insert shows the number of genes whose expression could (+) or could not be detected (−)

proteins should be light-dependent. Further experiments are needed to address this question and find the physiological role of these proteins, for which the activity has not been demonstrated in *S. ruber*.

5.5 *Salinibacter* and the Phages

As discussed above, many of the M8-specific genes in the HVRs were related to cell-surface properties that could be involved in phage recognition and evasion. This was also the case for the three islands in M31 (Mongodin et al. 2005). In a recent work, Pašić et al. (2009) compared the M31 genome to metagenomic fragments recovered from climax saltern crystallizers with 454 pyrosequencing technology. The three M31 islands were only scarcely represented in the metagenome and thus appear to vary among co-occurring *S. ruber* cells. Furthermore, the islands showed evidence of extensive genomic corruption with atypically low GC contents, low coding densities, and high numbers of pseudogenes and short hypothetical proteins. The annotable proteins in the islands were largely involved in

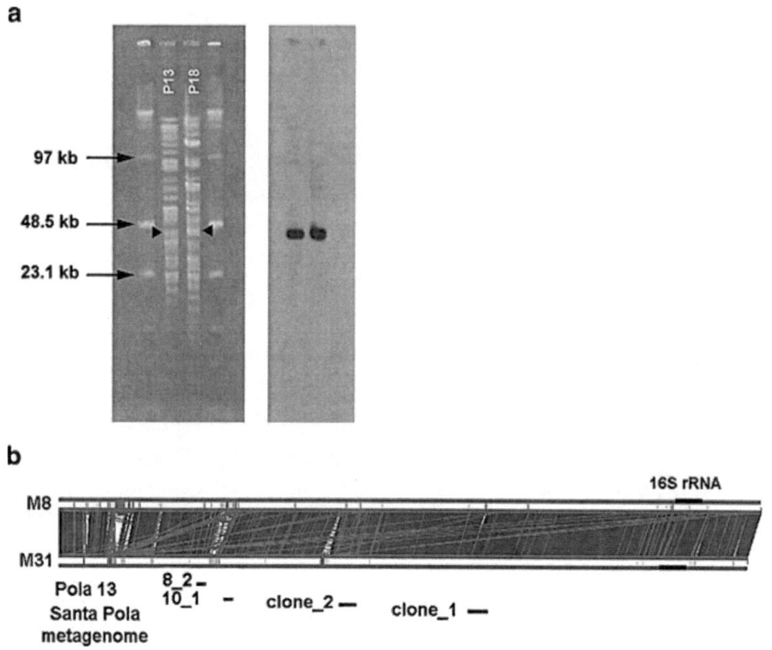

Fig. 5.5 (**a**) *Xba*I digestion of agarose embedded genomic DNA from *S. ruber* strains Pola 13 (P13) and Pola 18 (P18) (*left*). *Red arrows* point to the bands that were excised from the gel and cloned in fosmids. These bands contained the 16S rRNA gene, as hybridization with a digoxigenin-labeled probe (*right*) indicated. (**b**) Alignment of the P13 and Santa Pola metagenomics clones against M8 and M31 whole genomes. The position of the 16S rRNA gene in the genome of both microorganisms is marked in *blue*

variable cell surface characteristics. The authors proposed that "this variation... probably reflects a global strategy of bacteria to escape phage predation."

Metabolome analyses also showed differences in the exposed components of M8 and M31. Therefore, it seems reasonable to expect different phage susceptibilities for the two strains. In order to check whether this was indeed the case, we tried to isolate phages infecting M8 and M31 using different salterns waters as a source for viruses. When cultures of either strain were infected with water from the crystallizer of their isolation, we could easily isolate phages infecting M31 but not a single plaque was observed for M8. Then, we used waters from two ponds with salinities of 23.2 and 34.2% from the Santa Pola saltern (near Alicante, Spain) as source of viruses. For the low salinity water we could get a similar amount of phages for M8 and M31 ($4.47 \times 10^2 \pm 61.1$ $6.33 \times 10^2 \pm 65.1$ PFU per ml of natural sample, respectively). The high salinity water yielded a two-order magnitude increase for M8, while no plaques were formed when M31 was infected under the assayed conditions. These data indicate that, as indicated by their genomic differences, M8 and M31 indeed have different phage infection susceptibilities. This can have a profound impact on the ecology of both strains, if we consider that crystallizer ponds from solar salterns show one of the highest numbers of viruses reported for planktonic systems with values of up to 2×10^9 virus-like particles per ml. In addition, in the salterns were *Salinibacter* thrives, no bacteriovory has been detected at salinities above 25% (Guixa-Boixareu et al. 1996), which leaves viruses as a major factor for mortality. Therefore, compounds mediating interaction with phages (normally associated with cellular envelopes) should be under strong selection. This has been suggested also for the extremely halophilic archaeon *Hqr. walsbyi* (Cuadros-Orellana et al. 2007), which is present at very high abundances in the Santa Pola and Mallorca salterns. Indeed, metagenomic analyses (Santos et al. 2010, 2011) indicate that the most abundant viruses are those infecting this particular prokaryotic group.

Other prokaryotes exposed to high viral predation, such as the marine bacteria *Prochlorococcus* (Coleman et al. 2006) or "Pelagibacter" representatives (Wilhelm et al. 2007), also harbor genomic island(s) or hypervariable regions enriched in surface-related proteins. For these reasons, phages have been proposed as "drivers" of micro-diversification within prokaryote species (Rodríguez-Valera et al. 2009), not only because of their direct selection on lineages carrying specific sets of exposed cellular components, but also because of their role as gene transfer vehicles.

5.6 The Pangenome Grows

In addition to the two strains M8 and M31 from Mallorca salterns, two additional strains, Pola 13 and Pola 18 (isolated by Prof. Aharon Oren from the Santa Pola salterns) were also used for the *S. ruber* species description. To get insights into the genomic differences of Pola and Mallorca strains, we cloned and sequenced PFGE fragments containing the 16S rRNA genes of the strain Pola 13. We identified a genomic fragment containing the marker gene (Fig. 5.4a), cloned it into a fosmid vector, then sub-cloned into plasmids. Subsequent sequencing and assembly (Peña et al., unpublished results) yielded 18 contigs covering 25 kb that were aligned with the M8 and M31 genomes (Fig. 5.4b). The annotation of these contigs indicated that 19 out of the 37 predicted ORFs encoded hypothetical proteins (and thus had no homologs with M8 and M31); in addition only two ORFs (annotated as pyridoxamine phosphate oxidase and acriflavine resistance protein, respectively) shared homology with respective M8 and M31 genes. The rest of the annotated genes (with homologs in sequence databases) did not align with the M8 and M31 genomes. Thus, the analysis of only 25 kb of a new strain provided 19 completely new genes and expanded the pangenome by no less than 35 genes.

Finally, to get further insights into the extent of *S. ruber* genomic diversity, we performed a preliminary characterization of metagenomic clones containing 16S rRNA gene sequences of this bacterium (Peña et al., unpublished results). For this purpose, we constructed a metagenomic fosmid library from DNA directly retrieved from a crystallizer pond of the Santa Pola saltern from which the original Pola 13 and Pola 18 strains were isolated. In this library of 2,000 clones we could detect three clones containing the 16S marker gene. End-sequencing of the fosmids provided again an extension of *S. ruber* pangenome, since only two of the six fosmid ends were homologous to the M8 and M31 genomes, while the four remaining ones were novel and hence part of the accessory genome. These results indicate that in spite of the homogeneity between strains isolated from Santa Pola and Mallorca with respect to the ribosomal operon, their genome-level diversity is indeed very high. This corroborates the implications of the ecological differences found between M8 and M31, such as the possibility of competition, their different metabolomes and phage susceptibilities. We must keep in mind that the crystallizers in which *Salinibacter* thrives are environments with very low diversity in terms of species, but it seems that the genomic microdiversity is very high. This is the case not only for *Salinibacter*, but also for the square archaeon *Hqr. walsbyi*, which also has been shown to have a broad pangenome (Ivars-Martínez et al. 2008)

5.7 Open Questions and Future Research

The comparative study of *S. ruber* genomes in conjunction with the wet laboratory experiments summarized above have provided novel first insights into the evolution and ecology of this bacterium, and at the same time raised a plethora of further questions. Some of these questions reach far beyond studying a dedicated genus of extremophilic bacteria. We selected the following points as focus of our future work:

- Intra-specific genomic diversity. How many different genomes of *S. ruber* (or any other single species) are present in a given environment? We have isolated 35 new strains from 1 ml of crystallizer water and found 35 different genomic patterns. Sequencing of only 25 kb of a different strain yields an expansion of the pangenome by 35 genes. How many different genomes could we find at a given time? Would the genome pattern change with time for a given environment?
- The pangenome and the environment. When we compare the genomes of two extremely closely related *S. ruber* strains (M8 and M31), we find that 10% of the genes are strain-specific, which is a rather high number for such close relatives. This is also the case with other microorganisms. Which is then the size of the pangenome for these organisms? Is there a pangenome specific of every type of environment? Which is the relationship between the gene pool of an environment and the microdiversity and the "species" richness?
- Interaction between viruses and prokaryotes. Considering that closely related strains of *S. ruber* vary with respect to phage susceptibility, how do the viruses affect *Salinibacter* populations in the environment? Could the phages be mediating lateral gene transfers between *Salinibacter* and halophilic Archaea? Are the phages infecting *Salinibacter* promoting genotype recycling in the community?

As an overarching goal, we would like to unravel the units of selection in the evolution of microbial communities (strains, phage-host systems, assemblages of strains from the same species...). This may be answered for some lab model systems, but there does not seem to be a consensus for microorganisms in natural communities.

Acknowledgment This work was supported by the Spanish Ministry of Science and Innovation grants CGL2006-12714-CO2-01 and 02, and CGL2009-12651-C02-01 and 0–2 (to JA and RRM) that include European Funds (FEDER), and by the grant ACOM09/155 from the Generalitat Valenciana.

References

Acinas SG, Klepac-Ceraj V, Hunt DE, Pharino C, Ceraj I, Distel DL, Polz MF (2004) Fine-scale phylogenetic architecture of a complex bacterial community. Nature 430:551–554

Antón J, Oren A, Benlloch S, Rodríguez-Valera F, Amann R, Rosselló-Mora R (2002) *Salinibacter ruber* gen. nov., sp. nov., a novel, extremely halophilic member of the Bacteria from saltern crystallizer ponds. Int J Syst Evol Microbiol 52:485–491

Antón J, Peña A, Santos F, Martínez-García M, Schmitt-Kopplin P, Rosselló-Mora R (2008) Distribution, abundance and diversity of the extremely halophilic bacterium *Salinibacter ruber*. Saline Systems 4:15

Cohan FM, Koeppel AF (2008) The origins of ecological diversity in prokaryotes. Curr Biol 18: R1024–R1034

Coleman ML, Sullivan MB, Martiny AC, Steglich C, Barry K, DeLong EF, Chisholm SW (2006) Genomic islands and the ecology and evolution of *Prochlorococcus*. Science 311:1768–1770

Cuadros-Orellana S, Martín-Cuadrado AB, Legault B, D'Auria G, Zhaxybayeva O, Papke RT, Rodríguez-Valera F (2007) Genomic plasticity in prokaryotes: the case of the square haloarchaeon. ISME J 1:235–245

Feil EJ (2004) Small change: keeping pace with microevolution. Nat Rev Microbiol 2:483–495

Guixa-Boixareu N, Caledrón-Paz JI, Heldal M, Bratbak G, Pedrós-Alió C (1996) Viral lysis and bacterivory as prokaryotic loss factors along a salinity gradient. Aquat Microb Ecol 11: 215–227

Ivars-Martínez E, Martín-Cuadrado AB, D'Auria G, Mira A, Ferriera S, Johnson J, Friedman R, Rodríguez-Valera F (2008) Comparative genomics of two ecotypes of the marine planktonic copiotroph *Alteromonas macleodii* suggests alternative lifestyles associated with different kinds of particulate organic matter. ISME J 2:1194–1212

Medini D, Donati C, Tettelin H, Masignani V, Rappuoli R (2005) The microbial pangenome. Curr Opin Genet Dev 15:589–594

Mongodin EF, Nelson KE, Daugherty S, DeBoy RT, Wister J, Khouri H, Weidman J, Walsh DA, Papke RT, Sánchez-Pérez G, Sharma AK, Nesbø CL, MacLeod D, Baptese E, Doolittle WF, Charlebois RL, Legault B, Rodríguez-Valera F (2005) The genome of *Salinibacter ruber*. Convergence and gene exchange among hyperhalophilic Bacteria and Archaea. Proc Natl Acad Sci USA 102:18147–18152

Oh D, Porter K, Russ B, Burns D, Dyall-Smith M (2010) Diversity of *Haloquadratum* and other haloarchaea in three, geographically distant, Australian saltern crystallizer ponds. Extremophiles 14:161–169

Oren A (2002) Halophilic microorganisms and their environments. Kluwer Academic Publishers, Dordrecht

Pašić L, Rodríguez-Mueller B, Martín-Cuadraro AB, Mira A, Rohwer F, Rodríguez-Valera F (2009) Metagenomic islands of halophiles: the case of *Salinibacter ruber*. BMC Genomics 10:570

Peña A, Valens M, Santos F, Buczolits S, Antón J, Kämpfer P, Busse HJ, Amann R, Rosselló-Mora R (2005) Intraspecific comparative analysis of the species *Salinibacter ruber*. Extremophiles 9:151–161

Peña A, Teeling H, Huerta-Cepas J, Santos F, Yarza P, Brito-Echevarría J, Lucio M, Schmitt-Kopplin P, Meseguer I, Schenowitz C, Dossat C, Barbe V, Dopazo J, Rosselló-Mora R, Schüler M, Glöckner FO, Amann R, Gabaldón T, Antón J (2010) Fine-scale evolution: genomic, phenotypic and ecological differentiation in two coexisting *Salinibacter ruber* strains. ISME J 4:882–895

Richter M, Rosselló-Móra R (2009) Shifting the genomic gold standard for the prokaryotic species definition. Proc Natl Acad Sci USA 106:19126–19131

Rodríguez-Valera F, Martín-Cuadrado AB, Rodríguez-Brito B, Pašić L, Thingstad TF, Rohwer F, Mira A (2009) Explaining microbial population genomics through phage predation. Nat Rev Microbiol 7:828–836

Santos F, Yarza P, Briones C, Parro V, Antón J (2010) The metavirome of a hypersaline environment. Environ Microbiol 12:2965–2976

Santos F, Moreno M, Meseguer I, López C, Rosselló-Mora R, Parro V, Antón J (2011) Metatranscriptomic analysis of extremely halophilic viral communities. ISME J (submitted for publication) doi:10.1038/ismej.2011.34

Wilhelm LJ, Tripp HJ, Givan SC, Smith DP, Giovannoni SJ (2007) Natural variation in SAR11 marine bacterioplankton genomes inferred from metagenomic data. Biol Direct 2:19

Chapter 6
Impact of Lipidomics on the Microbial World of Hypersaline Environments

Patrizia Lopalco, Simona Lobasso, Maristella Baronio, Roberto Angelini, and Angela Corcelli

6.1 From Lipid Biochemistry to Lipidomics

Classical lipid biochemistry has given important contributions to the characterization of the extremely halophilic microorganisms inhabiting saltern crystallizer ponds worldwide. The pioneering work of Morris Kates showed that membrane lipids of halophilic Archaea, organisms abundant in coastal salterns and continental salt lakes, contain diphytanylglycerol diether lipids, in contrast with Eukarya and Bacteria, which contain mainly diacylglycerol-derived membrane lipids and some monoacylmonoalkyl-glycerol-derived lipids (Kates et al. 1965). The diether core lipid that forms the basis for most polar lipid structures present in halophilic archaeal microorganisms is 2,3-di-*O*-phytanyl-*sn*-glycerol (C_{20}, C_{20}), also called archaeol. The archaeol core is considered a chemotaxonomic marker of Archaea. A number of reviews have examined the lipid composition of halophilic Archaea in detail, their biosynthetic pathways and the analytical approaches used in their characterization (Sprott 1992; Kates 1993; Koga and Morii 2005; Corcelli and Lobasso 2006).

In the late 1990s, the general renewed interest in lipid biology and the technical innovation in mass spectrometry gave rise to the development of lipidomics as a natural evolution of lipid biochemistry. Today there is increasing awareness of the multiple dynamic roles of lipids in cell life and, in parallel, technologies

P. Lopalco
Institute for Microelectronics and Microsystems National Research Council (IMM-CNR), Lecce, Italy

S. Lobasso • M. Baronio • R. Angelini
Department of Medical Biochemistry , Biology and Physics, University of Bari, Bari, Italy

A. Corcelli (✉)
Department of Medical Biochemistry , Biology and Physics, University of Bari, Bari, Italy
and
Institute for Chemical-Physical Processes, National Research Council (IPCF-CNR), Bari, Italy
e-mail: a.corcelli@biologia.uniba.it

A. Ventosa et al. (eds.), *Halophiles and Hypersaline Environments*,
DOI 10.1007/978-3-642-20198-1_6,

able to detect even minor lipid components with short lifetimes are continuously improving. A great increase in the ability to detect and analyze lipids of all structures and sizes, even complex large molecules such as cardiolipins, has already been obtained by mass spectrometry analysis. Consequently, the way to deal with issues such as lipid–protein interactions, changes in the membrane lipid distribution and functions of bioactive lipids has changed in the lipidomics era. For example, detailed studies of membrane lipid changes in response to internal and external stimuli were technically challenging until the approach based on ESI-MS/MS was developed. Now only simple sample preparation and small samples are required to identify and quantify lipid molecular species. The recent application of MALDI-TOF/MS in lipid analysis allows rapid acquisition of data not only from the lipid extract of membranes and cells (Sun et al. 2008), but also directly from intact membranes (Angelini et al. 2010) and tissues, introducing the possibility of imaging tissues, single cell and revealing the location of endogenous lipids (Jackson and Woods 2009; Ferreira et al. 2010).

In the following section, we will show some examples illustrating the impact of novel mass spectrometry techniques on the knowledge of lipid biology of extremely halophilic Archaea and Bacteria. Microbial lipidomics allows not only a more detailed and precise knowledge of the lipid composition of individual species, but also represents an innovative tool in the study of microbial ecology.

6.2 New Lipids from the Salterns

Dense communities of halophilic microorganisms can impart a bright red color to environments such as hypersaline lakes and salterns. The existence of profuse microbial life in hypersaline environments raises a number of fundamental questions relating to the modes of osmoadaptation. The abundant Archaea together with Bacteria and halotolerant Eukarya constitute the biomass of salterns. The lipid content of microorganisms serves as a basis for chemotaxonomy and as a tool for the study of microbial ecology. In the last ten years, thank to the mass spectrometry analytical approach, new lipid molecules have been found in extreme halophilic microorganisms. The studies of the molecular structures of new lipids required an experimental approach which entailed hydrolysis of the polar head groups, selective cleavage of the chain linkages, identification of the pieces thus produced, and reconstruction of the starting molecule.

6.2.1 The Cardiolipins of Extreme Halophilic Archaea and Their Role in Osmoadaptation

Halobacterium salinarum is considered as a model organism of extremely halophilic Archaea. An interesting characteristic of *Hbt. salinarum* is represented by its ability to convert light into chemical energy for the metabolic requirements. This feature is due to the presence of a specialized membrane domain, the purple membrane (PM), containing as only protein bacteriorhodopsin, a photoactivated proton pump, which sustains the ATP synthesis.

Although at the end of 1990s it was believed that the biochemistry of both protein and lipid components of PM was completely known, new findings on PM lipids were made possible by advances in lipid mass spectrometry analyses.

Today it is known that the main lipid component in the lipid extract of the PM of *Hbt. salinarum* is a complex phosphosulfoglycolipid, glycocardiolipin, (3′-HSO_3)-Gal*p*-β-1,6Man*p*-α-1,2-Glc*p*α-1,1-[*sn*-2,3-di-*O*-phytanylglycerol]-6-[phospho-*sn*-2,3-di-*O*-phytanylglycerol] (Corcelli et al. 2000).

This membrane lipid is dimeric, being constituted by two lipid units: the glycolipid S-TGD-1 and the phosphatidic acid; it is therefore also named S-TGD-1-PA (Fig. 6.1). Having two diphytanylglycerol moieties (and four chains) in its molecule, the structure is analogous to that of cardiolipin (in particular a glycosylated cardiolipin).

The glycosylated cardiolipin (i.e., glycocardiolipin) of PM can be readily detected in a lipid extract by the characteristic ESI-MS negative signals at *m*/*z* 966 and 1,934, (bicharged and monocharged molecule, respectively) (Corcelli et al. 2000). Recently, we have shown that the ionized glycocardiolipin molecule can also easily be revealed by MALDI TOF/MS technique, as a monocharged ion both in lipid extracts and in intact membranes (Angelini et al. 2010). The presence of glycocardiolipin within the PM of *Hbt. salinarum* was discovered in our laboratory in the course of the study of the properties of two "delipidated" bacteriorhodopsin fractions (BRI and BRII) isolated by means of phenylsepharose CL-4B chromatography (Lopez et al. 1999). Mass spectrometry analyses of residual lipids associated to BRI and BRII showed a main bicharged peak at *m*/*z* 966 that was also present in the ESI-MS lipid profile of PM but never described before.

In the course of collaborative studies with Morris Kates, besides the glycocardiolipin, another novel lipid component was found, having the same backbone of eukaryal and bacterial cardiolipin (Corcelli et al. 2000). Figure 6.2 compares the structure of archaeal cardiolipin (or bisphosphatidylglycerol, BPG) with that of the bacterial cardiolipin present in *Salinibacter ruber*; the diagnostic signals for BPG in mass spectrometry analyses are at *m*/*z* 760 and 1,521 for the bicharged and monocharged ion, respectively; the bacterial cardiolipin of *Salinibacter* can also give rise to bicharged and/or monocharged ions at *m*/*z* 685.5 and 1,373, respectively.

Soon after others glycosylated cardiolipins were discovered in other halophilic Archaea. By means of ESI-MS and NMR analyses it was shown that the glycosylated cardiolipin analogue of *Halorubrum* genera has the structure of (2′-HSO_3)-Man*p*-α−1,2-Glc*p*α-1,1-[*sn*-2,3-di-*O*-phytanylglycerol]-6-[phospho-*sn*-2, 3-di-*O*-phytanylglycerol], i.e., of a phospholipid dimer constituted of the glycolipid S-DGD-5 esterified to phosphatidic acid, namely S-DGD-5-PA (see structure in Fig. 6.3). Clearly the glycolipids S-TGD-1 and S-DGD-5 are precursors of the glycosylated cardiolipins S-TGD-1-PA and S-DGD-5-PA, respectively (Lobasso

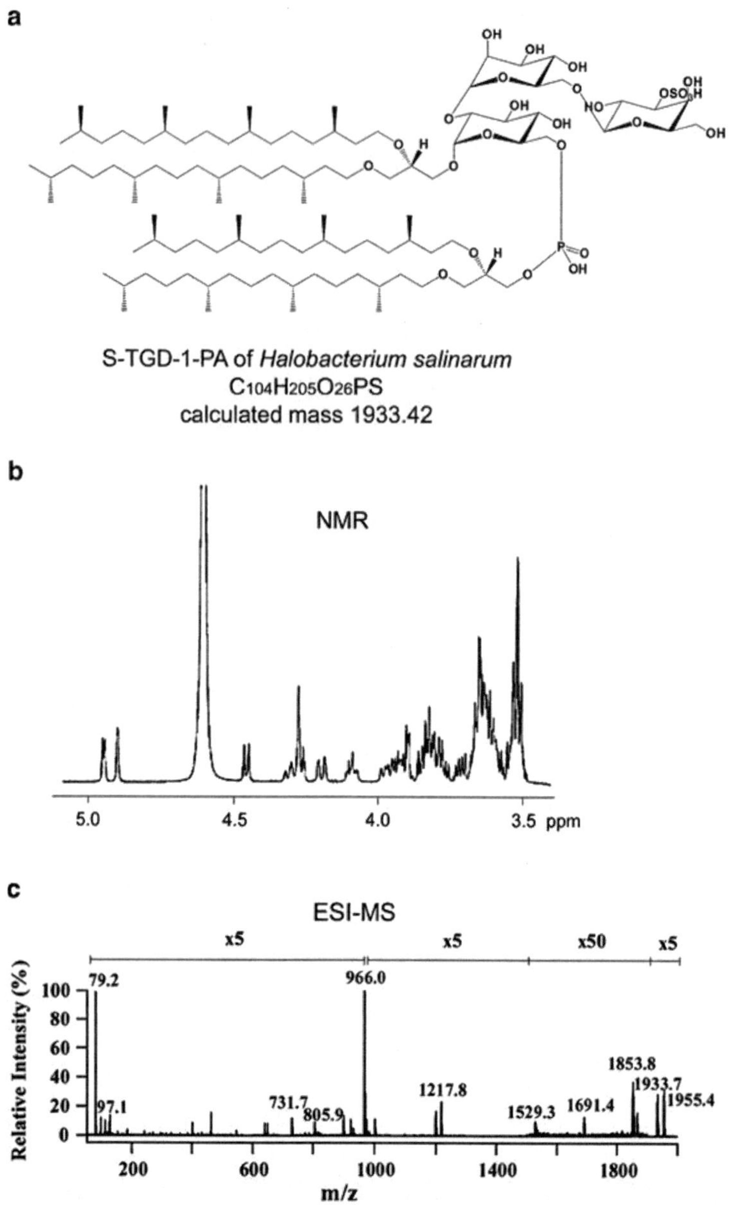

Fig. 6.1 The glycosylated cardiolipin S-TGD-1-PA (glycocardiolipin) of the purple membrane of *Halobacterium salinarum*: structure (**a**); NMR spectrum (**b**); ESI-MS analysis (**c**). The figure combines data previously shown in Corcelli et al. (2000). The structure in (**a**) (and all the followings) have been drawn as in http://www.lipidmaps.org

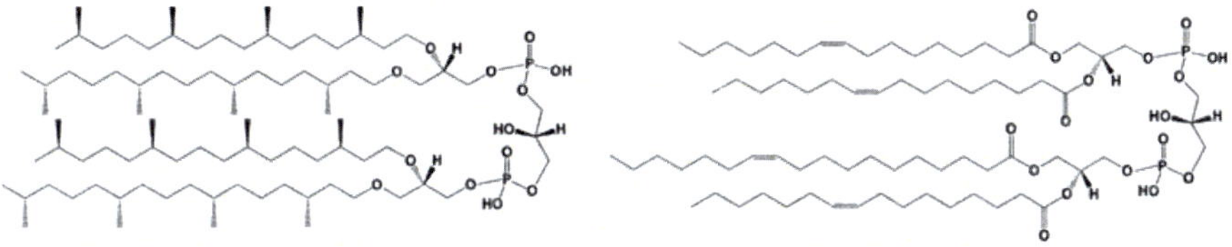

archaeal cardiolipin (BPG)
$C_{89}H_{182}O_{13}P_2$
calculated mass 1521.3

cardiolipin of *Salinibacter ruber*
$C_{75}H_{138}O_{17}P_2$
calulated mass 1372.9

Fig. 6.2 Structures of archaeal bisphosphatidylglycerol (BPG) and of the bacterial cardiolipin of *Salinibacter ruber*. Mass spectrometry signals: *m/z* 760 and 1,521 for BPG, *m/z* 685.5 and 1,373 for bacterial cardiolipin

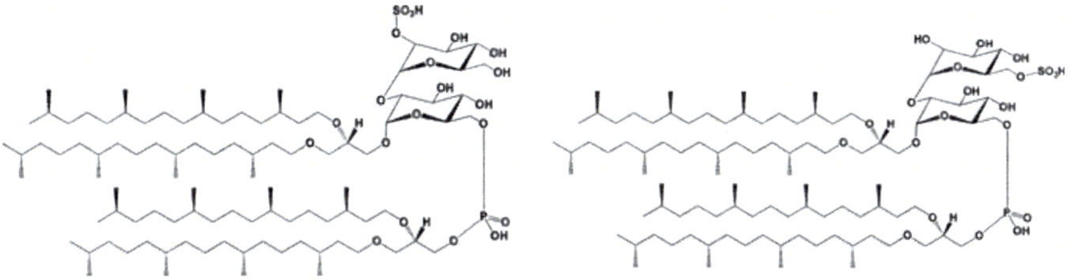

S-DGD-5-PA of *Halorubrum* sp
$C_{98}H_{195}O_{21}PS$
calculated mass 1771.36

S-GL-2 of *Haloferax volcanii*
$C_{98}H_{195}O_{21}PS$
calculated mass 1771.36

Fig. 6.3 Glycosylated cardiolipins of the *Halobacteriaceae* family. S-DGD-5-PA is constituted by the specific glycolipid S-DGD-5 of *Halorubrum* linked to phosphatidic acid (PA); S-GL-2 is constituted by the specific glycolipid S-DGD-1 of *Haloferax* linked to PA. Being structural isomers, both glycosylated cardiolipins give rise to mass spectrometry signals at *m/z* 885 and 1,770 (bicharged and monocharged ions, respectively)

et al. 2003; Lopalco et al. 2004). Furthermore, the glycosylated cardiolipin analogue S-GL-2, (6′-HSO_3)-D- Man*p*-α1,2-D-Glc*p*α1,1-[*sn*-2,3-di-*O*-phytanylglycerol]- 6-[phospho-*sn*-2,3-di-*O*-phytanylglycerol], constituted by the sulfoglycolipid S-GL-1 (i.e., S-DGD-1) esterified to phosphatidic acid, was found in the membranes of *Haloferax volcanii* (Fig. 6.3) (Sprott et al. 2003). As S-DGD-5-PA and S-DGD-1-PA (i.e., S-GL-2) are structural isomers, they have the same molecular weight and give rise to the same diagnostic signals in ESI-MS and/or MALDI-TOF/MS analyses; the bicharged and monocharged ions of these glycosylated cardiolipins are at *m/z* 885 and 1,770, respectively. In general, archaeal bisphosphatidylglycerol is abundantly present together with genus specific glycosylated cardiolipins in the different genera of *Halobacteriaceae*. *Haloquadratum walsbyi* represents an exception in this context. Lipidomic studies and recent detailed lipid analyses of residual lipids associated to solubilized BR of *Haloquadratum* showed that BPG is present in traces only and that glycosylated cardiolipin is absent (Lobasso et al. 2008).

The functional roles of archaeal cardiolipins have been described in a number of studies. As regards lipid–protein interactions, S-TGD-1-PA seems to play an important role in stability and maintaining the trimeric structure of bacteriorhodopsin (Lopez et al. 1999). The S-TGD-1-PA/BR stoichiometry in the PM has been determined by analyzing an unprocessed lipid extract of the PM with both of ^{31}P-NMR and ^{1}H-NMR spectroscopy. The molar ratio of S-TGD-1-PA to BR, in standard PM preparation, is one and appears to be consistent with its possible location in the intra-trimer space (Corcelli et al. 2002). Until now S-TGD-1-PA (or glycocardiolipin) has been found only in the extreme halophiles and in association with bacteriorhodopsin (Lattanzio et al. 2002). It is extremely difficult to remove S-TGD-1-PA from the annulus of bacteriorhodopsin in the course of solubilization with detergents, and even after lipid extraction with denaturing organic solvents a residual amount of S-TGD-1-PA still remains associated with denatured bacteriorhodopsin (Catucci et al. 2004).

Interestingly, it has been shown that there is also a tight association between cardiolipin and cytochrome c oxidase in halophilic Archaea (Corcelli et al. 2007a). The association of terminal oxidases with cardiolipins, regardless of the phylogenetic context in which this occurs, suggests that specific cardiolipins are required for the function of the respiratory chain.

Finally, some aspects regarding the regulation of the lipid biosynthesis in the archaeal extreme halophiles have been described (Lobasso et al. 2003; Lopalco et al. 2004; Corcelli et al. 2007a, b; Corcelli 2009). With the exception of *Haloquadratum walsbyi* (Lobasso et al. 2008), when the archaeal extreme halophiles of the *Halobacteriaceae* family are exposed to low salt conditions, relevant changes in the cell membrane lipid composition occur; during the adaptation of extreme halophilic microorganisms to hypo-osmolarity the membrane undergoes significant biochemical and structural changes, especially in terms of an increase in the content of cardiolipin species.

During hypoosmotic shock the level of S-TGD-1-PA in *Hbt. salinarum* increases at the expenses of the glycolipid S-TGD-1 and the phospholipid phosphatidyl glycerol phosphate methyl ester, PGP-Me (Lobasso et al. 2003). Therefore, S-TGD-1-PA is almost absent in the lipid extract of cells of *Hbt. salinarum*, while it is abundant in the PM, which are typically isolated after cells disruption by osmotic shock. In the microorganism MdS1 of the *Halorubrum* genus, it has been demonstrated that both BPG and S-DGD-5-PA are synthesized de novo at the expense of phospholipid precursors during hypoosmotic shock (Lopalco et al. 2004).

In general, the modifications in lipid composition could affect the physical chemical properties of the membrane, such as bilayer thickness, membrane fluidity and transport properties. It has been suggested that a higher cardiolipin cell content could protect the cell from lysis, acting as a glue to stabilize protein complexes within the membrane (Corcelli et al. 2007b; De Leo et al. 2009). Modulation of cardiolipin synthesis by hypoosmotic shock may represent the physiological response of extremely halophilic Archaea to the dilution of salt saturated water during rainfalls (Lobasso et al. 2003).

6.2.2 *Novel Unusual Sphingoids in Extremely Halophilic Bacteria*

The rod shaped red-colored halophilic bacterium *Salinibacter ruber* has been found among the inhabitants of saltern crystallizer ponds (Antón et al. 2000, 2002). As this organism balances the external high salt concentration by accumulating potassium ions inside, the general mechanism of adaptation to salinity appears similar to that of the extremely halophilic Archaea. However, differently from most of Archaea, the level of cardiolipin does not rise under hypoosmotic shock (Lattanzio et al. 2009).

Recently, another halophilic microorganism distantly related to *Salinibacter* has been characterized from a mesocosm experimental system filled with a mixture of Dead Sea and Red Sea water. The organism has been described as a new species belonging to new genus, *Salisaeta longa* (Vaisman and Oren 2009).

The membrane lipids of *Salinibacter ruber* and *Salisaeta longa* are typical for the bacterial domain, with glycerophospholipids containing ester-linked fatty acyl chains and not ether-linked phytanyl chains characteristic of the archaeal halophiles (Lattanzio et al. 2009). ESI-MS analyses of the total lipid extract of *Salinibacter ruber* and *Salisaeta longa* showed a prominent peak at *m*/*z* 660.6 as a negative ion that could not be assigned to previous known lipid structures. This lipid, 2-carboxy-2-amino-3,4-hydroxy-17-methyloctadec-5-ene-1-sulfonic acid (SL1), is positive to Azure A and has the structure of a sulfonated sphingoid base linked to a fatty acid, until now found in *Salinibacter* and *Salisaeta* only.

Recently, another compound positive to Azure A staining has been found in the total lipid extract of *Salisaeta longa* (Baronio et al. 2010). After isolation and purification by preparative TLC, the novel sulfonolipid (SL2) was analyzed by ESI-MS. The ESI-MS spectrum (negative ion) showed a molecular ion peak $[M\text{-}H]^-$ at 676.4 *m*/*z* that differed from the molecular ion $[M\text{-}H]^-$ at 660.6 by 16 amu, compatible with the presence of an additional OH^- residue in the novel sulfonolipid. In Fig. 6.4, the ESI-MS/MS spectra of both sulfonolipids with their molecular structures are shown. The MS/MS spectra of SL1 and SL2 contain the same *m*/*z* 418 peak corresponding to the common sphingoid base, while the peaks of the fatty acid fragments are different and occur at *m*/*z* 241 and 257 in SL1 and SL2, respectively. SL1 and SL2 appear to be structural variants of the sulfonolipids, collectively called capnoids, found as main components of the cell envelope of gliding bacteria of the genus *Cytophaga* and closely related genera (Godchaux and Leadbetter 1984) and diatoms (Anderson et al. 1978); the long chain aminosulfonate alcohol (sphingoid base) in common with SL1 and SL2 has been named halocapnine to differentiate it from the capnine of non-halophilic members of the Bacteroidetes. Figure 6.5 compares the structures of capnine and halocapnine.

Sulfonate sphingoids of members of the Bacteroidetes are considered unusual bioactive compounds (Godchaux and Leadbetter 1980, 1984; Murakami et al. 2002). Experiments are in progress to elucidate the role of sulfonolipids in the cellular physiology of *Salinibacter* and *Salisaeta*.

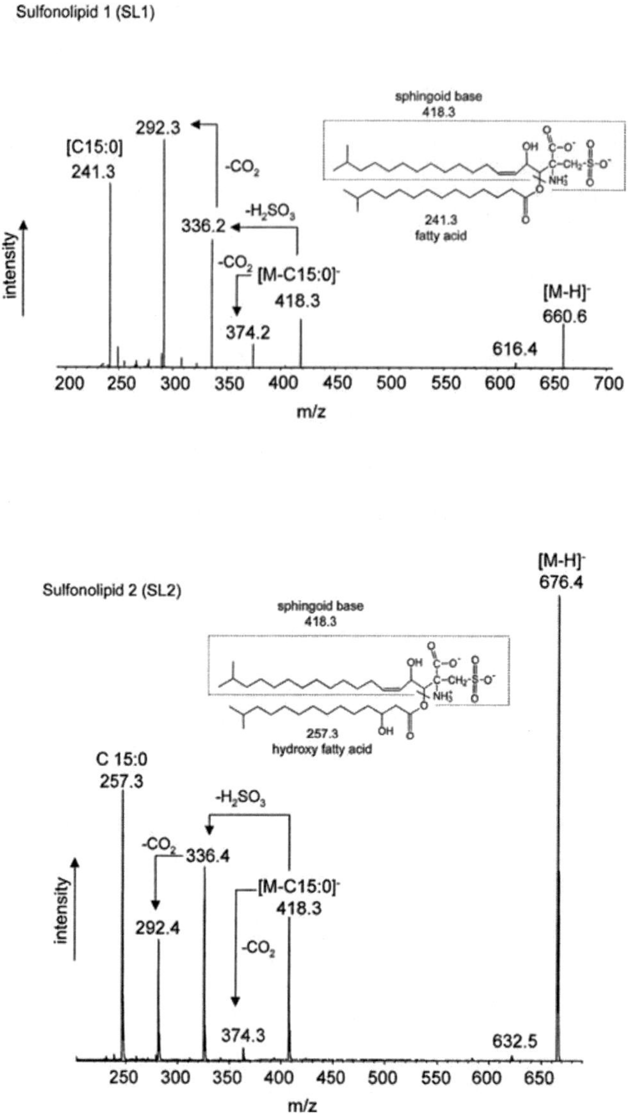

Fig. 6.4 MS/MS analyses of the isolated and purified sulfonolipids SL1 and SL2 of the halophilic Bacteroidetes. SL1 has been found both in *Salinibacter ruber* and *Salisaeta longa*, SL2 only in *Salisaeta longa*. The common peak at *m/z* 418 represents the fragment corresponding to the sulfonate sphingoid base, which is an analogue of capnine. The capnine of extremely halophilic Bacteroidetes has the specific name of halocapnine. The figure combines data previously shown in Corcelli et al. (2004) and Baronio et al. (2010)

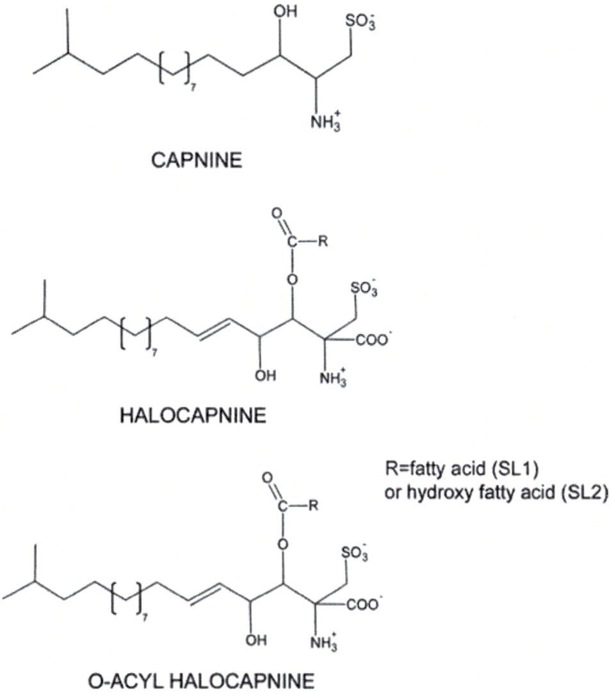

Fig. 6.5 Structures of the sulfonate sphingoid bases and of the acylated derivatives of *Salisaeta longa* and *Salinibacter ruber*

6.3 Lipid Profile of Saltern Biomass

Salterns may differ in nutrient levels, and the size of the biomass, of both algae and prokaryotes, may vary greatly. Mass spectrometry analysis of the lipid extract of saltern biomass represents a powerful tool to obtain information on the types of halophilic microorganisms that inhabit saltern brines. It is possible to detect intact polar ether lipids of Archaea, as well as polar lipids of bacteria, by ESI-MS analyses of biomass lipid extract (Lattanzio et al. 2002; Corcelli et al. 2004). Despite the complexity of the mass spectra of the lipid extracts of biomass constituted by Archaea, Bacteria and algae, the peaks of abundant ions offer a first view on the microbial life in salterns. For example, both the glycolipid isomers S-DGD-5 and S-DGD-1 give rise to a peak at *m*/*z* 1,056, which is diagnostic for the presence of the genera *Haloferax*, *Halorubrum* and *Haloquadratum*. Previous comparative analyses of lipids of biomass of Margherita di Savoia (Italy) and Eilat (Israel) showed that, while the large peak at *m*/*z* 1,056 was present both in Margherita di Savoia and in Eilat, the peak at 660 diagnostic of *Salinibacter* (and possibly *Salisaeta*), was absent in the Eilat salterns (Lattanzio et al. 2002). We here present an example of the use of MALDI-TOF/MS technique to analyze lipid extracts of saltern biomass. Figure 6.6 shows the MALDI-TOF/MS profiles of the total lipid

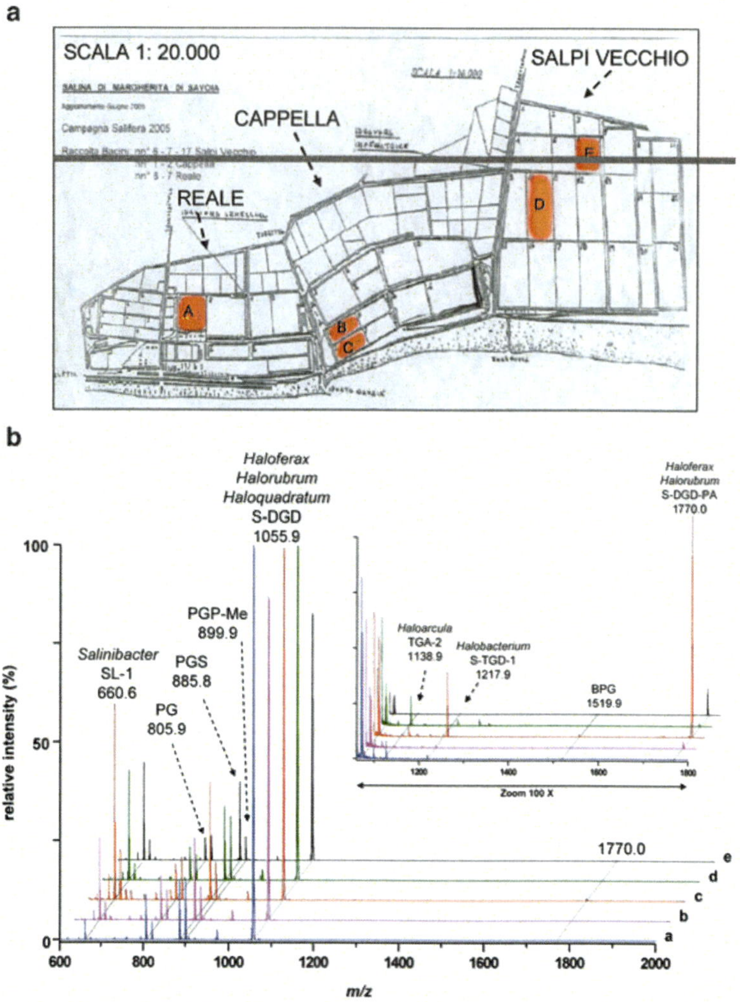

Fig. 6.6 MALDI-TOF/MS analyses of the lipid extracts of the biomass collected from different salt ponds in the salterns of Margherita di Savoia in Italy. Panel (**a**) shows a map of the salterns, where the ponds of sampling (A-B-C-D-E) are colored in *red*. In all examined samples, salinity was 27°–28° Baumé. About 5 L of *red* water were collected from each pond. Biomass was concentrated by centrifugation and lipids extracted and analyzed by following previous described procedures (Lattanzio et al. 2002; Angelini et al. 2010). Panel (**b**) shows results of mass spectrometry analyses; the inset shows a Y-axis enlargement (Zoom 100×) of the MALDI-TOF mass spectrum to visualize the minor peaks. Legend of peaks: *m/z* 660.6 corresponds to the sulfonolipid (SL1) of *Salinibacter*; *m/z* 805.9 to the archaeal phosphatidylglycerol (PG); *m/z* 885.8 to the archaeal phosphatidylglycerol sulfate (PGS); *m/z* 899.9 to the archaeal phosphatidylglycerol phosphate methyl ester (PGP-Me); *m/z* 1055.9 to the sulfated diglycosyl diether lipids (S-DGD-5 and/or S-DGD-1); in the inset, *m/z* 1138.9 to the diglycosyl diether lipid (TGD-2); *m/z* 1,218 to the sulfated triglycosyl diether lipid (S-TGD-1); *m/z* 1519.9 to the archaeal cardiolipin (BPG); *m/z* 1770.0 to the glycosylated cardiolipins (S-DGD-PA). PG, PGS and PGP-Me are ubiquitous phospholipids. S-DGD-5 is the glycolipid of *Halorubrum*. S-DGD-1 is the glycolipid of *Haloferax*; an S-DGD glycolipid has been also found in *Haloquadratum*

extracts obtained from the biomass collected from different crystallizer ponds of the salterns of Margherita di Savoia (Italy). Three different sections of the salterns have been considered for sampling: the "Cappella" (a), "Reale" (b, c) and "Salpi Vecchio" (d, e) section. Again, the main peaks in the spectrum can be attributed to the molecular ions of different kinds of polar lipids (phospholipids and glycolipids) present in the membranes of various halophilic microorganisms. Results are similar to those obtained by ESI-MS analyses; the peaks at *m/z* 660 and *m/z* 1,056 are dominant. Minor glycolipid peaks are indicative for the presence of other genera of the *Halobacteriaceae* family; the peaks in the inset of Fig. 6.6 are diagnostic for the presence in the biomass of microorganisms representative of the genera *Halobacterium* (peak at *m/z* 1,218, S-TGD-1) and *Haloarcula* (peak at *m/z* 1,138, TGA-2) as minor components. Last but not least, the peak at *m/z* 660 is present with different intensities into the examined saltern ponds. In particular, it is less intense in sample a (blue line) of "Cappella" than in "Reale" or "Salpi Vecchio" sections. Clearly, mass spectrometry data here shown should be integrated with further investigations to evaluate the precise contribution of halophilic bacteria to biomass. In the near future, as we have shown that MALDI-TOF/MS lipid analyses can also be performed in isolated intact membranes of extremely halophilic Archaea (Angelini et al. 2010), it might be possible to directly analyze lipids of intact cells within the saltern biomass.

References

Anderson R, Kates M, Volcani BE (1978) Identification of the sulfolipids in the non-photosynthetic diatom *Nitzschia alba*. Biochim Biophys Acta 528:89–106

Angelini R, Babudri F, Lobasso S, Corcelli A (2010) MALDI-TOF/MS analysis of archaebacterial lipids in lyophilized membranes dry-mixed with 9-aminoacridine. J Lipid Res 51: 2818–2825

Antón J, Rosselló-Mora R, Rodríguez-Valera F, Amann R (2000) Extremely halophilic *Bacteria* in crystallizer ponds from solar salterns. Appl Environ Microbiol 66:3052–3057

Antón J, Oren A, Benlloch S, Rodríguez-Valera F, Amann R, Rosselló-Mora R (2002) *Salinibacter ruber* gen. nov., sp. nov., a novel extreme halophilic member of the Bacteria from saltern crystallizer ponds. Int J Syst Evol Microbiol 52:485–491

Baronio M, Lattanzio VM, Vaisman N, Oren A, Corcelli A (2010) The acylhalocapnines of halophilic bacteria: structural details of unusual sulfonate sphingoids. J Lipid Res 51: 1878–1885

Catucci L, Lattanzio VMT, Lobasso S, Agostiano A, Corcelli A (2004) Role of endogenous lipids in the chromophore regeneration of bacteriorhodopsin. Bioelectrochemistry 63:111–115

Corcelli A (2009) The cardiolipin analogues of Archaea. Biochim Biophys Acta 1788:2101–2106

Corcelli A, Lobasso S (2006) Characterization of lipids of halophilic archaea. In: Rainey FA, Oren A (eds) Methods in microbiology, vol 35, Extremophiles. Elsevier/Academic, Amsterdam, pp 585–613

Corcelli A, Colella M, Mascolo G, Fanizzi FP, Kates M (2000) A novel glycolipid and phospholipid in the purple membrane. Biochemistry 39:3318–3326

Corcelli A, Lattanzio VMT, Mascolo G, Papadia P, Fanizzi FP (2002) Lipid-protein stoichiometries in a crystalline biological membrane: NMR quantitative analysis of the lipid extract of the purple membrane. J Lipid Res 43:132–140

Corcelli A, Lattanzio VMT, Mascolo G, Babudri F, Oren A, Kates M (2004) A novel sulfonolipid in the extremely halophilic bacterium *Salinibacter ruber*. Appl Environ Microbiol 70: 6678–6685

Corcelli A, Lobasso S, Palese LL, Saponetti MS, Papa S (2007a) Cardiolipin is associated with the terminal oxidase of an extremely halophilic archaeon. Biochem Biophys Res Commun 354: 795–801

Corcelli A, Lobasso S, Saponetti MS, Leopold A, Dencher NA (2007b) Glycocardiolipin modulates the surface interaction of the proton pumped by bacteriorhodopsin in purple membrane preparations. Biochim Biophys Acta 1768:2157–2163

De Leo V, Catucci L, Ventrella A, Milano F, Agostiano A, Corcelli A (2009) Cardiolipin increases in chromatophores isolated from *Rhodobacter sphareroides* after osmotic stress: structural and functional roles. J Lipid Res 50:256–264

Ferreira CR, Saraiva SA, Catharino RR, Garcia JS, Gozzo FC, Sanvido GB, Santos LFA, Lo Turco EG, Ponte JHF, Basso AC, Bertolla RP, Sartori R, Guardieiro MM, Perecin F, Meirelles FV, Sangalli JR, Eberlin MN (2010) Single embryo and oocyte lipid fingerprinting by mass spectrometry. J Lipid Res 251:1218–1227

Godchaux W III, Leadbetter ER (1980) *Capnocytophaga* spp. contain sulfonolipids that are novel in procaryotes. J Bacteriol 144:592–602

Godchaux W III, Leadbetter ER (1984) Sulfonolipids of gliding bacteria. J Biol Chem 259: 2982–2990

Jackson SN, Woods AS (2009) Direct profiling of tissue lipids by MALDI-TOF/MS. J Chromatogr B 877:2822–2829

Kates M (1993) Membrane lipids of Archaea. In: Kates M, Kushner DJ, Mateson AT (eds) The biochemistry of Archaea (Archaebacteria). Elsevier, Amsterdam, pp 261–295

Kates M, Jengoyan LS, Sastry PS (1965) A diether analog of phosphatidyl glycerophosphate in *Halobacterium cutirubrum*. Biochim Biophys Acta 98:252–268

Koga Y, Morii H (2005) Recent advances in structural research on ether lipids from archaea including comparative and physiological aspects. Biosci Biotechnol Biochem 69:2019–2034

Lattanzio VM, Corcelli A, Mascolo G, Oren A (2002) Presence of two novel cardiolipins in the halophilic archaeal community in the crystallizer brines from the salterns of Margherita di Savoia (Italy) and Eilat (Israel). Extremophiles 6:437–444

Lattanzio VMT, Baronio M, Oren A, Russell NJ, Corcelli A (2009) Characterization of polar membrane lipids of the extremely halophilic bacterium *Salinibacter ruber* and possible role of cardiolipin. Biochim Biophys Acta 1791:25–31

Lobasso S, Lopalco P, Lattanzio VMT, Corcelli A (2003) Osmotic shock induces the presence of glycocardiolipin in the purple membrane of *Halobacterium salinarum*. J Lipid Res 44: 2120–2126

Lobasso S, Lopalco P, Mascolo G, Corcelli A (2008) Lipids of the ultrathin square archaeon *Haloquadratum walsbyi*. Archaea 2:177–183

Lopalco P, Lobasso S, Babudri F, Corcelli A (2004) Osmotic shock stimulates de novo synthesis of two cardiolipins in an extreme halophilic archaeon. J Lipid Res 45:194–201

Lopez F, Lobasso S, Colella M, Agostiano A, Corcelli A (1999) Light-dependent and biochemical properties of two different bands of bacteriorhodopsin isolated on phenyl-sepharose CL-4B. Photochem Photobiol 69:599–604

Murakami C, Yamazaki T, Hanashima S, Takahashi S, Ohta K, Yoshida H, Sugawara F, Sakaguchi K, Mizushina Y (2002) Structure-function relationship of synthetic sulfoquinovosyl-acylglycerols as mammalian DNA polymerase inhibitors. Arch Biochem Biophys 403:229–236

Sprott GD (1992) Structures of archaeabacteria membrane lipids. J Bioenerg Biomembr 24: 555–566

Sprott GD, Larocque S, Cadotte N, Dicaire CJ, McGee M, Brisson JR (2003) Novel polar lipids of halophilic eubacterium *Planococcus* H8 and archaeon *Haloferax volcanii*. Biochim Biophys Acta 1633:179–188

Sun G, Yang K, Zhao Z, Guan S, Han X, Gross RW (2008) Matrix assisted laser desorption/ionisation time of flight mass spectrometric analysis of cellular glycerophospholipids enabled by multiplexed solvent dependent analyte-matrix interactions. Anal Chem 80:7576–7585

Vaisman N, Oren A (2009) *Salisaeta longa* gen. nov., sp. nov., a red, halophilic member of the Bacteroidetes. Int J Syst Evol Microbiol 59:2571–2574

Chapter 7
Molecular Mechanisms of Adaptations to High Salt Concentration in the Extremely Halotolerant Black Yeast *Hortaea werneckii*

Ana Plemenitaš and Nina Gunde-Cimerman

7.1 Introduction

As crystalline salt (NaCl) is generally considered hostile to most forms of life, it has been used for centuries as a food preservative. However, halophilic and halotolerant microorganisms can contaminate food that is preserved with salt, and they also inhabit natural hypersaline environments around the world, such as salt lakes and solar salterns. These microorganisms can adapt to extreme concentrations of NaCl, and often to high concentrations of other ions as well (Samson et al. 2004). The great majority of studies on halophilic and halotolerant microorganisms have been dedicated to halophilic Bacteria and Archaea (Brock 1979; Ramos-Cormenzana 1991; Ventosa et al. 1998), and to only one eukaryotic species, the alga *Dunaliella salina* (Oren 2005), as it has been considered that other eukaryotic organisms cannot adapt to these extreme conditions.

These considerations have been based on the fact that halophilic and halotolerant fungi were – until a decade ago – considered extremely rare. They were known only to food microbiologists, who occasionally isolated them from food preserved with high concentrations of salt or sugar, using highly selective media. Since the few known halotolerant fungal species did not show any particular preference for the chemical nature of the solute that was lowering the water potential of the medium used for their isolation (Andrews and Pitt 1987; Hocking 1993; Pitt and Hocking 1997), they were considered xerophiles. The decisive criterion for defining their

A. Plemenitaš
Institute of Biochemistry, Faculty of Medicine, University of Ljubljana, Ljubljana, Slovenia

N. Gunde-Cimerman (✉)
Department of Biology, Biotechnical Faculty, University of Ljubljana, Ljubljana, Slovenia
and
Centre of Excellence for Integrated Approaches in Chemistry and Biology of Proteins (CIPKeBiP), Jamova 39, SI-1000 Ljubljana, Slovenia
e-mail: nina.gunde-cimerman@bf.uni-lj.si

A. Ventosa et al. (eds.), *Halophiles and Hypersaline Environments*,
DOI 10.1007/978-3-642-20198-1_7,

xerophilic properties was the ability to grow on a medium which included 17% NaCl or 50% glucose, corresponding to an a_w of $\leq$0.85 (Northolt et al. 1995).

This conviction remained unchallenged until 2000, when melanized yeast-like fungi were first reported as active inhabitants of brine in solar salterns (Gunde-Cimerman et al. 2000). Later studies revealed that hypersaline environments around the globe also harbor related filamentous melanized fungi and many new or rare fungal species with previously unknown natural ecology (Zalar et al. 1999, 2005a, b, 2007, 2008a, b; Butinar et al. 2005a, b). A decade after the first discoveries of fungi in hypersaline waters, we know that the "hypersaline mycobiota" is represented by phylogenetically unrelated groups of fungi, which appear in different geographic locations at salinities higher than 17% NaCl with relatively consistency. Our understanding of the complex microbial processes in natural hypersaline environments was enhanced by including fungi as additional saprotrophic members of the hypersaline communities.

The description of halophilic and halotolerant fungi in naturally hypersaline environments has provided more appropriate and diverse model organisms for the study of eukaryotic adaptations to hypersaline conditions than the previously used salt-sensitive species such as *Saccharomyces cerevisiae*. We now know that fungi which inhabit natural hypersaline environments have a halophilic behavior that is different from that of the majority of halophilic prokaryotes: with few exceptions, they do not require salt for viability, as they can grow and adjust to the whole salinity range, from freshwater to almost saturated NaCl solutions (Plemenitaš et al. 2008). The term "extremely halotolerant" best describes this versatile type of ecological response, as in the most investigated fungal model organism, the extremely halotolerant black yeast *Hortaea werneckii*, described in this chapter.

7.2 Black Yeasts

Black yeasts or melanized yeast-like fungi or meristematic ascomycetes (Sterflinger et al. 1999) belong to orders *Capnodiales*, *Dothideales*, and *Chaetothuriales* (Crous et al. 2009; Schoch et al. 2009). These orders contain a large number of extremotolerant species that can tolerate extreme temperatures (Wollenzien et al. 1995; Sterflinger 1998; Gunde-Cimerman et al. 2003; Ruibal et al. 2009), high salt concentrations (Gunde-Cimerman et al. 2000; Butinar et al. 2005b), desiccation (Gorbushina et al. 2008; Gueidan et al. 2008), variations in pH and nutrient deficiency (Selbmann et al. 2005), and UV and ionizing radiation (Dadachova et al. 2007). The term "polyextremotolerant" can be used in many instances to describe the remarkable ability of these fungi to colonize a wide variety of different environments and endure a broad range of conditions for multiple ecological parameters, which enable them to colonize even domestic household appliances (Gostinčar et al. 2010, 2011).

This interesting group of organisms has been known since the end of the nineteenth century (de Hoog et al. 1999), although difficulties in their morphological identification together with their slow growth and low competitive ability frequently hindered their isolation and identification.

Although their unusual stress-tolerance is conferred by various specific mechanisms or pre-adaptations, some general traits characteristic for the entire group can be determined. The main characteristic is their black color, which is the result of thick and heavily melanized cell walls. Melanin shields them against adverse conditions and environmental stresses of various kinds; it contributes to virulence in pathogenic species and is a scavenger of reactive oxygen species (van Baarlen et al. 2007). Another important characteristic is the ability of black yeasts to grow meristematically at extreme environmental conditions and form microcolonies (Kogej et al. 2007), in which recolonization of old cells by new ones can occur, resulting in multilayered cell walls and additional protection of newly formed cells (Gorbushina et al. 2003). At less extreme conditions they can often shift between filamentous and yeast-like growth (Slepecky and Starmer 2009), enabling colonization of water and solid substrates. They also often produce general stress-protecting extracellular polysaccharides (Selbmann et al. 2005), which enable formation of biofilms and adherence to biotic and living surfaces. In nutrient limiting conditions, cells can be released from the biofilm, spread and colonize new habitats (van Baarlen et al. 2007).

This particular black yeast ecotype enabled successful colonization of hypersaline environments. The dominant group of fungi in hypersaline waters of the salterns are black yeasts from the ordo Capnodiales (Crous et al. 2009). In hypersaline waters at salinities above 17% NaCl, *H. werneckii* is the dominant species (Gunde-Cimerman et al. 2000).

7.3 *Hortaea werneckii*

The morphology of *H. werneckii* is typical for extremophilic species within the *Capnodiales* (Crous et al. 2009). *H. werneckii* is heavily melanized, has a thick cell wall, and is characteristically polymorphic (de Hoog 1993; Wollenzien et al. 1995; Sterflinger et al. 1999; Zalar et al. 1999). At the highest salinities it exhibits an isodiametric type of thallus expansion, which results in highly resistant, meristematic cell clumps with endogenous conidiation (Sterflinger 1998). At lower salinities it grows as yeast cells or as hyphae, depending on the substrate (Plemenitaš et al. 2008). The main morphological characteristics are presented in Fig. 7.1. Thanks to these adaptive abilities it can grow, albeit extremely slowly, in nearly saturated salt solutions, as well as completely without salt, with a broad growth optimum from 6 to 14% NaCl (Gunde-Cimerman et al. 2000; Plemenitaš et al. 2008).

Due to this polymorphic morphology, *H. werneckii* has received many designations in the past (Plemenitaš and Gunde-Cimerman 2005). Nowadays, its identification is additionally based on molecular characteristics, particularly sequencing of the ITS rDNA region and RFLP markers from SSU rDNA and ITS rDNA regions (de Hoog et al. 1999).

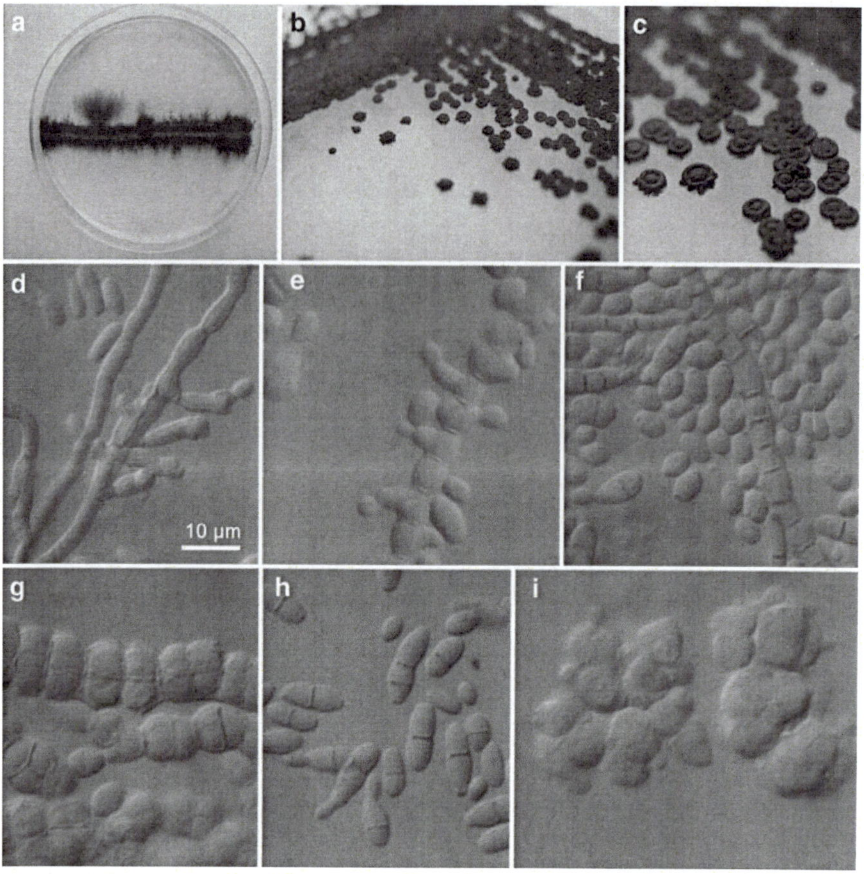

Fig. 7.1 *Hortaea werneckii* in culture: (**a**) Pure culture on MEA with 10% NaCl; (**b**, **c**) Colonies on YNB; (**d**–**f**) Proliferating hyphae on MEA with 5%, 10% and 20% NaCl, respectively; (**g**) Meristematic hyphae on MEA with 20% NaCl; (**e**). Yeast cells on MEA with 5% NaCl; (**i**) Meristematic clumps on MEA with 25% NaCl. The scalebars marked on picture (**d**) is valid also for pictures (**e**–**i**). The pictures were taken by Cene Gostinčar, Rok Miklavčič and Polona Zalar

H. werneckii is at present the best described eukaryotic halophilic model organism. Before it became evident that it is the dominant species inhabiting hypersaline waters, it attracted scientific interest as the causative agent of human *tinea nigra*. This superficial black colonization of the salty human hand, characteristically limited to the greasy stratum corneum, can be removed by vigorous scrubbing (de Hoog and Guého 1998; Göttlich et al. 1995). Nevertheless, there were many speculations about its natural ecology due to its infection potential. More than 10 years after its first discovery in the Slovenian salterns (Gunde-Cimerman et al. 2000) we know that its ecology reflects its halophilic character. *H. werneckii* can be occasionally found as a spoiling agent of food preserved with high salt, it can be isolated with low frequency from seawater-related environments (Zalar et al. 1999), from wood immersed in hypersaline waters (Zalar et al. 2005b) and from surface layers of tropical microbial mats in salterns (Cantrell et al. 2006). There is new evidence that it even inhabits spider webs in Atacama desert caves, together with a new *Dunaliella* species (Azúa-Bustos et al. 2010). However, its primary natural ecological niche is precrystallization and crystallization ponds in eutrophic salterns around the world (Butinar et al. 2005b). It appears in seasonal peaks, which correlate primarily with high environmental nitrogen and phosphorus values. At environmental salinities above 20% NaCl, *H. werneckii* usually represents between 85 and 90% of all isolated fungi, while it can be detected only occasionally when NaCl concentrations are below 10% (Gunde-Cimerman et al. 2000; Butinar et al. 2005b).

7.4 Physiological Responses in *H. werneckii*

We have identified intracellular potassium and sodium content, compatible solutes and membrane characteristics as the key physiological responses in extremely salt-tolerant black yeast *H. werneckii*.

H. werneckii maintains low intracellular Na^+ concentration in spite of the extremely high external salinity. Although the Na^+ to K^+ ratio increased with increased NaCl concentration in the medium, this increase did not follow the increase in external Na^+ concentration (Kogej et al. 2005). These data suggested that *H. werneckii* is well adapted to changes in external ion concentrations, either due to its ability to effectively extrude Na^+ or/and to prevent influx of Na^+.

Second, we found that glycerol is the main compatible solute (Petrovič et al. 2002; Kogej et al. 2005) In *H. werneckii*, intracellular glycerol concentrations correlated well with increases in salinity up to 1.5 M NaCl, whereas at higher salinities intracellular glycerol increased only slightly. We hypothesized that this is due to the rearrangements of the melanin granules on the outer parts of the cell wall of *H. werneckii*, which at lower salinities form a distinct layer, thereby reducing the permeability of cell wall to glycerol. At higher salinities on the other hand, melanization is reduced, resulting in the escape of glycerol from the cell (Kogej et al. 2007). These data suggested the presence of other compatible solutes. Besides glycerol, other low-molecular-weight organic compounds such as polyols erythritol, arabitol, and mannitol, and amino acid derivatives mycosporines were identified as compatible solutes in *H. werneckii* (Kogej et al. 2007). It should be stressed that amino acid derivatives, mycosporines, which accumulate steeply up to 1.0 M NaCl and decrease at higher salinity, have not been previously referred as compatible solutes in fungi (Oren and Gunde-Cimerman 2007).

Salt-induced changes in cellular membranes in *H. werneckii* differ considerably from responses previously observed in the cellular membranes of salt-sensitive *S. cerevisiae* (Turk et al. 2004, 2007). While the sterol-to-phospholipid ratio increases in *S. cerevisiae* with increased salt concentration, in *H. werneckii* this ratio did not alter much. On the other hand, salt stress induced an increase in fatty acid unsaturation and length. Increased salinity was accompanied by a decrease in C16:0 together with an increase in *cis*-C18:$2^{\Delta 9,12}$ (Turk et al. 2004). Such adaptations were earlier proposed to raise membrane fluidity at high salt concentrations (Russell et al. 1995). Indeed, in *H. werneckii*, changes in the fatty acid composition resulted in high plasma membrane fluidity over a wide range of NaCl concentrations. These results indicated high intrinsic salt stress tolerance, and were in good agreement with ecophysiological data and the dominance of *H. werneckii* in hypersaline waters of salterns.

7.5 Molecular Adaptations in *H. werneckii*

7.5.1 Differential Gene Expression

Fluctuation in external salinity forces eukaryotic cells to counteract the osmotic pressure by protective biochemical processes of their protein products, which is first reflected in changes in gene expression. In accordance with physiological responses to increased NaCl concentrations in *H. werneckii,* expression of genes involved in these responses at the level of membranes, ion homeostasis and glycerol synthesis were studied.

An overall increase in fatty acid chain length and unsaturation at increased salinity was observed in *H. werneckii,* with a decrease in C16:0 fatty acids and an increase in *cis*-C18:$2^{\Delta 9,12}$ fatty acids. In accordance with these results was the identification of genes encoding Δ^9-, Δ^{12}-desaturases, and elongases. We found that the genome of *H. werneckii* contains two copies of elongase *HwELO1* and two copies of desaturase *HwODE12*. The expression of these genes was upregulated with an increase in NaCl concentration, with the exception of one isoform of elongase (*HwELO1B*) (Gostinčar et al. 2009).

In line with the assumption that *H. werneckii* maintains low Na^+ due to effective Na^+ exclusion, we have identified an ENA-like ATPases. As revealed by the phylogenetic analysis, the HwENA proteins belong to the group of fungal P-type ATPases, which are phylogenetically older than the Na^+/K^+ ATPases of the salt-sensitive *S. cerevisiae*, and of the salt-tolerant *Debaryomyces hansenii*. The genome of *H. werneckii* contains two genes, *HwENA1* and *HwENA2*, encoding ENA-like ATPases. The expression of both identified genes is responsive to increased salt concentration and to increased pH, both characteristic of the natural environment of *H. werneckii* (Gorjan and Plemenitaš 2006). These results suggested that HwENA ATPases might be involved in the mechanism of the adaptation of *H. werneckii* to its natural environment.

Accumulation of glycerol in a cell is a combination of endogenous synthesis and glycerol uptake from the medium. Like in *S. cerevisiae*, where increased synthesis of glycerol is regulated by the increased expression of glycerol-phosphate dehydrogenase, which produces glycerol-phosphate from dihydroxyacetone-P, we also found salt-dependent increased expression of the gene coding for a putative glycerol-P dehydrogenase in *H. werneckii*. Extremely salt-tolerant *H. werneckii* codes for two salt-induced *GPD1* genes, with similar gene transcription regulation and with 98% amino-acid sequence identity between these paralogs. Both genes are expressed differentially at increased salinity (Lenassi et al. 2011).

Changes in external osmolarity in *S. cerevisiae* as well as in other fungi are sensed by special sensory proteins, which transmit the signal through the HOG signal transduction pathway and regulate the transcription of osmoresponsive genes. In *H. werneckii*, we have identified key components of the HOG pathway: soluble histidine kinase HwHHK7 (Lenassi and Plemenitaš 2007) and transmembrane HwSho1 (Fettich et al. 2011), both putatively involved in sensing high salinity, and three kinases of the MAP kinase module: MAPKKKHwSte11, MAPKK HwPbs2 (data not published), and terminal MAP kinase HwHog1 (Turk and Plemenitaš 2002; Lenassi et al. 2007). Furthermore, we demonstrated that genes like *HwENA1* and *HwGPD1*, both involved in physiological responses to increased salinity in *H. werneckii*, are controlled by activated MAP kinase HwHog1 (Vaupotič and Plemenitaš 2007a).

While searching for novel osmoresponsive genes, we assessed the differential gene expression of extremely halotolerant black yeast *H. werneckii* at the global level. Suppression subtractive hybridization (SSH) is a powerful technique to create a library of differentially expressed genes even from an organism without sequenced genome. We therefore constructed a cDNA subtraction library of *H. werneckii* adapted to hypersaline environment at 3 M and 4.5 M NaCl and identified an uncommon osmoprotective set of 95 differentially expressed genes, presented in Fig. 7.2. Their majority of has not been previously connected to hypersaline adaptation in *S. cerevisiae*. Novel genes putatively connected with extreme osmoadaptability of *H. werneckii* were identified. The analysis of transcriptional responses in hypersaline-adapted and hypersaline-stressed cells revealed that only a few genes responded to acute salt-stress, whereas all were differentially expressed in adapted cells (Vaupotič and Plemenitaš 2007a). It is also of interest that the majority of genes were upregulated at increased NaCl concentrations (Fig. 7.2), while only few, mostly connected to the cell cycle and cell wall biosynthesis, were downregulated (Fig. 7.2). As shown in Fig. 7.2, a high proportion of the differentially expressed genes is involved in metabolism and many of them are connected to the mitochondria. A proteomic study of the mitochondria further revealed preferential accumulation of energy metabolism enzymes in the hypersaline environment (Vaupotič et al. 2008a).

We were particularly interested in genes which are regulated by the MAP kinase HwHog1. By using a chromatin immunoprecipitation (CHIP) assay we demonstrated that more than one third of the differentially expressed genes were associated with the MAP kinase HwHog1. Based on our data on identification and characterization of the components of the HOG signaling pathway in *H. werneckii* and the identified genes that are regulated by the activated HwHog1, we propose the model shown in Fig. 7.3.

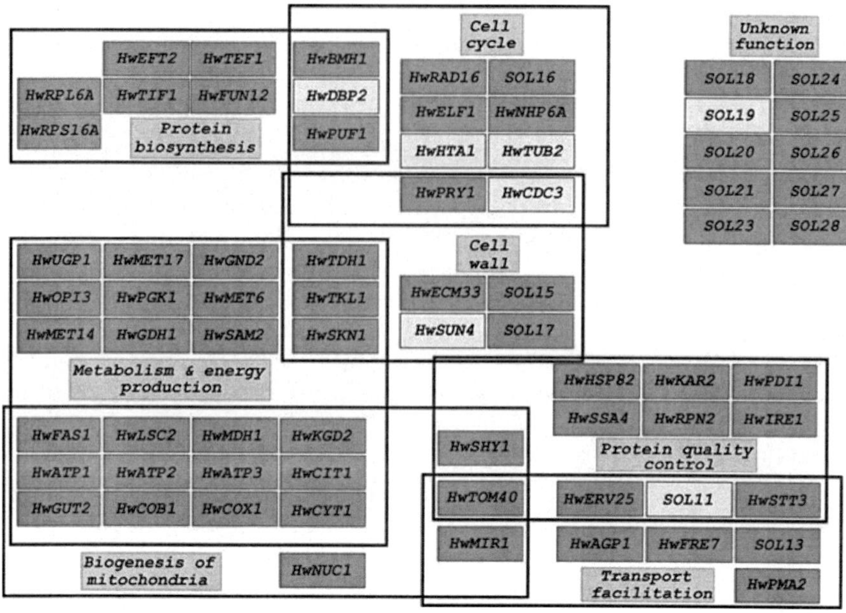

Fig. 7.2 Differentially expressed genes in *H. werneckii* grouped according to their respective functions. *Dark gray squares* represent up-regulated genes, *light gray squares* indicate down-regulated genes

7.5.2 *Proteins as Determinants of Halotolerance in* H. werneckii

At the protein level, we have identified two proteins in *H. werneckii* that are particularly responsive to increased NaCl concentration: hydroxy-methylglutaryl coenzyme A reductase (HwHMG R) and 3′-phosphoadenosine-5′-phosphatase (HwHal2), which is encoded by the *HAL2* gene.

7.5.2.1 HwHMG R

When studying the regulation of hydroxy-methylglutaryl coenzyme A reductase (HMG R), a key regulatory enzyme in the biosynthesis of sterols, we found a specific regulation pattern of this enzyme in *H. werneckii*. The activity and level of HMG R were investigated in *H. werneckii* and in other selected halophilic fungi isolated from solar saltpans. Representative fungi from the orders *Capnodiales Dothideales*, *Eurotiales*, and *Wallemiales* showed every a similar pattern of HMG R regulation, which differs from the pattern observed in salt-sensitive and moderately salt-tolerant yeasts. In all halophilic fungi studied, the HMG R amounts and activities were the lowest at optimal growth salinity and increased under hyposaline

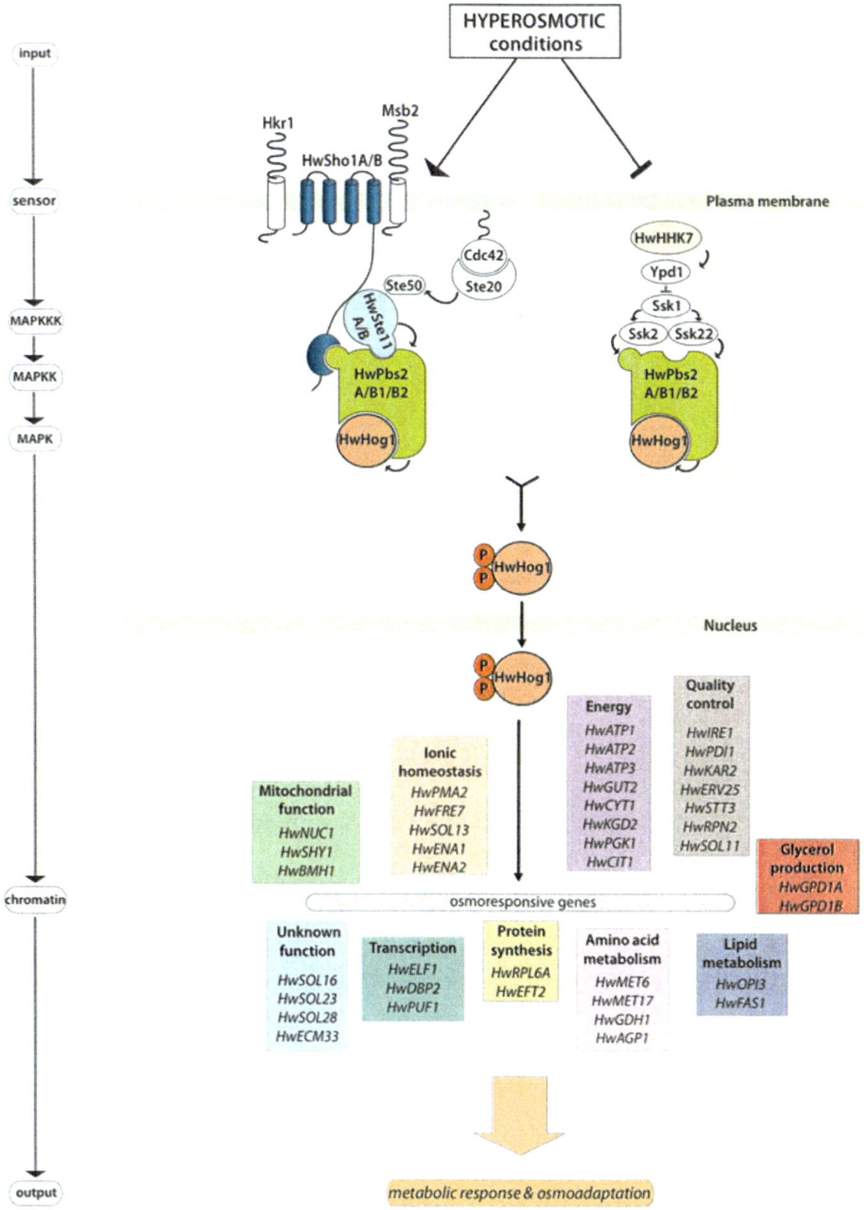

Fig. 7.3 HwHog1 MAP kinase of the HOG signal transduction pathway regulates the transcription of many osmoresponsive genes in *H. werneckii*

and hypersaline conditions. We therefore propose that HMG R could serve as one of the hallmarks in distinguishing between halophilic and non-halophilic fungal species (Vaupotič et al. 2008b).

Our studies also revealed that *H. werneckii* contains two HMGR isoenzymes located in the mitochondria (HwHmg1) and the endoplasmic reticulum (HwHmg2). We demonstrated that the activity of the microsomal isoenzyme, but not of the mitochondrial one, depends on environmental salinity. The activity of the microsomal HwHmg2 was the highest in hypo-saline and extremely hyper-saline environments, whereas it was down-regulated under optimal growth conditions. This was due to intensive ubiquitination and proteasomal degradation of the HwHmg2. The activity of the truncated mitochondrial HwHmg1 was constant in different growth conditions, suggesting an osmoadaptation-directed fate for mevalonate utilization in *H. werneckii*. As shown by the analysis of prenylated proteins, the salt-dependent HwHMG R activity profile mirrors protein prenylation rather than cellular sterol content, which does not change significantly with changing salinity (Vaupotič and Plemenitaš 2007b).

7.5.2.2 HwHal2

The 3-phosphoadenosine-5-phosphatase encoded by the HAL2 gene is a ubiquitous enzyme required for the removal of 3′-phosphoadenosine-5′-phosphate (PAP) produced during sulfur assimilation in eukaryotes. Inhibition of Hal2 by Na^+ or Li^+ results in PAP accumulation, which is toxic for yeasts. Increased amounts of PAP inhibit enzymes like sulphotransferases (Albert et al. 2000), RNA processing enzymes (Dichtl et al. 1997) and nucleoside diphosphate kinase (Schneider et al. 1998).

Two novel HAL2-like genes, HwHAL2A and HwHAL2B, have been cloned from the saltern-inhabiting extremely halotolerant black yeast *H. werneckii* (Vaupotič et al. 2007). Both HwHAL2 isoforms were inducible upon addition of salt. We clearly demonstrated halotolerance of their respective protein products compared to a homologous protein from salt-sensitive yeast. When HwHal2 proteins were expressed in a salt-sensitive strain of *S. cerevisiae*, an increase in halotolerance to up to 1.8 M NaCl was achieved, that has never been observed before with the HAL2 products of other species. HwHal2 proteins have unique structural features, possibly involved in the evolution of salt resistance. Modeling revealed two loops, META and ANA, which have not been previously identified in other Hal2 proteins. Using genetic and biochemical validation, it was demonstrated that the META sequence motif has the most evident effect on the HwHal2B-dependent salt tolerance (Vaupotič et al. 2007). The identification of HwHal2B, which significantly increases halotolerance in yeast, could provide a promising transgene for improving halotolerance in crop improvement strategies.

7.5.3 *Gene Duplications as a Global Response in* H. werneckii

Gene duplication in general is an already accepted mechanism of adaptation to various stresses. In yeast, for example, most of the duplicated genes code for membrane transporters and proteins involved in stress responses (Kondrashov et al. 2002).

From our studies it appears that gene duplication is also an important mechanism to combat stress due to fluctuations in environmental salinity in *H. werneckii*. Duplications were demonstrated in all genes involved in so far observed physiological responses in *H. werneckii*. Starting from the components of the HOG signal transduction pathway, in which gene duplications are not common in other fungi, we found gene duplications in putative sensory proteins, soluble histidine kinase HwHHK7 and transmembrane HwSho1, as well as in MAPKKK HwSte11 and MAPKK HwPbs2 (our unpublished data). While the expression of the *HwHHK7A* gene, coding for histidine kinase, increased only slightly with increased NaCl concentration, the expression of the *HwHHK7B* gene was highly salt-responsive in salt-adapted *H. werneckii* cells, as well as in stressed cells. Moreover, the expression profile of the *HwHHK7B* gene after exposure of the cells to hypersaline and hyposaline stress indicated the existence of two types of responses: an early response to hyposaline stress and a late response to the hypersaline stress (Lenassi and Plemenitaš 2007). The genome of *H. werneckii* also contains two copies of the *HwSHO1*genes (Fettich et al. 2011). Although the expression level of these genes does not change significantly with increased salt concentration, it is of importance to stress that *H. werneckii* is to our knowledge the first reported fungus with SHO1 gene duplication, as thus far only one copy of this gene (or none) has been found in the genomes of sequenced fungi (Krantz et al. 2006).

Adaptations at the level of membrane fluidity, ion homeostasis, as well as compatible solutes synthesis is also accompanied by duplications of genes coding for proteins involved in these responses in *H. werneckii*. We demonstrated gene duplications in fatty acid elongases and desaturases, involved in salt-dependent changes in fatty acids structure that correlated well with the salt-dependent changes in membrane fluidity (Turk et al. 2004; Gostinčar et al. 2009). It has been suggested earlier that fatty acid modification serves to raise membrane fluidity at high salt concentrations; thus, the duplication of the desaturases may be part of the adaptations of *H. werneckii* to saline environments, compared to the salt-sensitive *S. cerevisiae* which has only one Δ^9-desaturase and is incapable of Δ^{12} desaturation (Gostinčar et al. 2009).

We also identified two gene copies of the *HwENA1* genes that are differentially expressed and are involved with ion homeostasis. While the transcription of *HwENA1* was induced at higher level when the cells were exposed to salt stress, the expression of the *HwENA2* gene was higher in the adapted cells, suggesting their different roles in ion homeostasis. The genome of *H. werneckii* contains two copies of the *HwGPD1* gene, responsible for glycerol production. As glycerol is a key compatible solute in *H. werneckii*, duplication of the *GPD1* gene, which codes for a key regulatory enzyme in glycerol synthesis, might also be an advantage in the adaptation mechanisms of *H. werneckii*.

Gene duplications were also found in genes coding for proteins that could well be defined as markers of halotolerance, HwHal2A and B and HwHmg1 and 2. While expression of both the *HwHAL2* genes is salt-responsive, in case of HMG R, only the microsomal version (HwHmg2) is regulated by NaCl (Vaupotič and Plemenitaš 2007b; Vaupotič et al. 2007).

Given all of the above examples of gene duplications, it is tempting to speculate that gene duplications in *H. werneckii* provide evolutionary benefit for life in environments with fluctuating salt concentrations.

7.6 Conclusions

The discovery of the black yeast *H. werneckii* as the dominant fungal species in hypersaline waters enabled the introduction of a new model organism to study the mechanisms of salt tolerance in eukaryotes. Studies on the extremely halotolerant *H. werneckii* have revealed its superior mechanisms for adapting to an extremely broad range of salinities, from freshwater to almost saturated NaCl concentrations. Comparisons with salt-sensitive or moderately salt-tolerant fungi showed novel, intricate mechanisms to combat fluctuating salinity. We described novel physiological adaptations on the level of cell wall, melanization, membranes, ion homeostasis, and compatible solute strategy. On the molecular level we observed gene duplications, novel genes that are differentially expressed at different NaCl concentrations and regulated by the key MAP kinase of the HOG signaling pathway HwHog1, and we identified proteins as appropriate markers of halotolerance. Studies on *H. werneckii* also revealed a new insight into specialized evolutionary mechanisms of adaptation to high environmental salinity and other extremes in the kingdom of fungi.

Acknowledgment The scientific work was partly financed via operation "Centre of excellence for integrated approaches in chemistry and biology of proteins" number OP13.1.1.2.02.0005, and financed by European regional development fund, by Slovenian Ministry of higher education, science, and technology (15% share of financing), and Slovenian Research Agency (85% share of financing).

References

Albert A, Yenush L, Gil-Mascarell MR, Rodriguez PL, Patel S, Martinez-Ripoll M, Blundell TL, Serrano R (2000) X-ray structure of yeast Hal2p, a major target of lithium and sodium toxicity, and identification of framework interactions determining cation sensitivity. J Mol Biol 295: 927–938

Andrews S, Pitt JJ (1987) Further studies on the water relations of xerophilic fungi, including some halophiles. J Gen Microbiol 133:233–238

Azúa-Bustos A, González-Silva C, Salas L, Palma RE, Vicuña R (2010) A novel subaerial *Dunaliella* species growing on cave spiderwebs in the Atacama Desert. Extremophiles 14: 443–452

Brock TD (1979) Ecology of saline lakes. In: Shilo M, Hirsch P (eds) Strategies of microbial life in extreme environments: report of the Dahlem Workshop on Strategy of Life in Extreme Environments, Berlin, 1978, November 20–24. Verlag Chemie, Weinheim, pp 29–47

Butinar L, Santos S, Spencer-Martins I, Oren A, Gunde-Cimerman N (2005a) Yeast diversity in hypersaline habitats. FEMS Microbiol Lett 244:229–234

Butinar L, Sonjak S, Zalar P, Plemenitaš A, Gunde-Cimerman N (2005b) Melanized halophilic fungi are eukaryotic members of microbial communities in hypersaline waters of solar salterns. Bot Mar 1:73–79

Cantrell SA, Casillas L, Molina M (2006) Characterization of fungi from hypersaline environments of solar salterns using morphological and molecular techniques. Mycol Res 110: 962–970

Crous PW, Schoch CL, Hyde KD, Wood AR, Gueidan C, de Hoog GS, Groenewald JZ (2009) Phylogenetic lineages in the *Capnodiales*. Stud Mycol 64:17–47

Dadachova E, Bryan RA, Huang X, Moadel T, Schweitzer AD, Aisen P, Nosanchuk JD, Casadevall A (2007) Ionizing radiation changes the electronic properties of melanin and enhances the growth of melanized fungi. PLoS One 2:e457

de Hoog GS (1993) Evolution of black yeasts: possible adaptation to the human host. Antonie Leeuwenhoek 63:105–109

de Hoog GS, Guého E (1998) Agents of white piedra, black piedra and tinea nigra. In: Kreier J, Wakelin D, Cox F (eds) Topley & Wilson's microbiology and microbial infections. Hodder Arnold, London, pp 100–112

de Hoog GS, Zalar P, Urzi C, de Leo F, Yurlova NA, Sterflinger K (1999) Relationships of dothideaceous black yeasts and meristematic fungi based on 5.8S and ITS2 rDNA sequence comparison. Stud Mycol 43:31–37

Dichtl B, Stevens A, Tollervey D (1997) Lithium toxicity in yeast is due to the inhibition of RNA processing enzymes. EMBO J 16:7184–7195

Fettich M, Lenassi M, Veranič P, Gunde-Cimerman N, Plemenitaš A (2011) Identification and characterization of putative osmosensors, HwSho1A and HwSho1B, from the extremely halotolerant black yeast *Hortaea werneckii*. Fungal Genet Biol 48(5):475–484

Gorbushina AA, Whitehead K, Dornieden T, Niesse A, Schulte A, Hedges J (2003) Black fungal colonies as units of survival: hyphal mycosporines synthesized by rock dwelling microcolonial fungi. Can J Bot 81:131–138

Gorbushina AA, Kotlova ER, Sherstneva OA (2008) Cellular responses of microcolonial rock fungi to long-term desiccation and subsequent rehydration. Stud Mycol 61:91–97

Gorjan A, Plemenitaš A (2006) Identification and characterization of ENA ATPasesHwENA1 and HwENA2 from the halophilic black yeast *Hortaea werneckii*. FEMS Microbiol Lett 265:41–50

Gostinčar C, Turk M, Plemenitaš A, Gunde-Cimerman N (2009) The expressions of D9-, D12-desaturases and an elongase by the extremely halotolerant black yeast *Hortaea werneckii* are salt dependent. FEMS Yeast Res 9:247–256

Gostinčar C, Grube M, de Hoog GS, Zalar P, Gunde-Cimerman N (2010) Extremotolerance in fungi: evolution on the edge. FEMS Microbiol Ecol 71:2–11

Gostinčar C, Grube M, Gunde-Cimerman N (2011) Indoor potential of black fungi: enrichment or evolution? Fungal Biol (in press)

Göttlich E, de Hoog GS, Yoshida S, Takeo K, Nishimura K, Miyaji M (1995) Cell surface hydrophobicity and lipolysis as essential factors in human tinea nigra. Mycoses 38:489–494

Gueidan C, Villasenor CR, de Hoog GS, Gorbushina AA, Untereiner WA, Lutzoni F (2008) A rock-inhabiting ancestor for mutualistic and pathogen-rich fungal lineages. Stud Mycol 61: 111–119

Gunde-Cimerman N, Zalar P, de Hoog S, Plemenitaš A (2000) Hypersaline waters in salterns – natural ecological niches for halophilic black yeasts. FEMS Microbiol Ecol 32:235–240

Gunde-Cimerman N, Sonjak S, Zalar P, Frisvad JC, Diderichsen B, Plemenitaš A (2003) Extremophilic fungi in Artic ice: a relationship between adaptation to low temperature and water activity. Phys Chem Earth 28:1273–1278

Hocking AD (1993) Responses of xerophilic fungi to changes in water activity. In: Jennings DH (ed) Stress tolerance of fungi. Marcel Dekker, New York, pp 233–243

Kogej T, Ramos J, Plemenitaš A, Gunde-Cimerman N (2005) The halophilic fungus *Hortaea werneckii* and the halotolerant fungus *Aureobasidium pullulans* maintain low intracellular cation concentrations in hypersaline environments. Appl Environ Microbiol 71:6600–6605

Kogej T, Stein M, Volkmann M, Gorbushina AA, Galinski EA, Gunde-Cimerman N (2007) Osmotic adaptation of the halophilic fungus *Hortaea werneckii*: role of osmolytes and melanization. Microbiology 153:4261–4273

Kondrashov FA, Rogozin IB, Wolf YI, Koonin EV (2002) Selection in the evolution of gene duplications. Genome Biol 3:RESEARCH0008

Krantz M, Becit E, Hohmann S (2006) Comparative genomics of the HOG-signalling system in fungi. Curr Genet 49:137–151

Lenassi M, Plemenitaš A (2007) Novel group VII histidine kinase HwHhk7B from the halophilic fungi *Hortaea werneckii* has a putative role in osmosensing. Curr Genet 51:393–405

Lenassi M, Vaupotič T, Gunde-Cimerman N, Plemenitaš A (2007) The MAP kinase HwHog1 from the halophilic black yeast *Hortaea werneckii*: coping with stresses in solar salterns. Saline Syst 3:3

Lenassi M, Zajc J, Gostinčar C, Gorjan A, Gunde-Cimerman N, Plemenitaš A (2011) Adaptation of the glycerol-3-phosphate dehydrogenase Gpd1 to high salinities in the extremely halotolerant *Hortaea werneckii* and halophilic *Wallemia ichthyophaga*. Fungal Biol (in press)

Northolt MD, Frisvad JC, Samson RA (1995) Occurrence of food-borne fungi and factors for growth. In: Samson RA, Hoekstra ES, Frisvad JC, Filtenborg O (eds) Introduction to food-borne fungi. CBS, Delft, pp 243–250

Oren A (2005) A hundred years of *Dunaliella* research: 1905–2005. Saline Syst 1:2

Oren A, Gunde-Cimerman N (2007) Mycosporines and mycosporine-like amino acids: UV protectants or multipurpose secondary metabolites? FEMS Microbiol Lett 269:1–10

Petrovič U, Gunde-Cimerman N, Plemenitaš A (2002) Cellular responses to environmental salinity in the halophilic black yeast *Hortaea werneckii*. Mol Microbiol 45:665–672

Pitt JI, Hocking AD (1997) Fungi and food spoilage. Blackie Academic & Professional, London

Plemenitaš A, Gunde-Cimerman N (2005) Cellular reponses in the halophilic black yeast *Hortaea weneckii* to high environmental salinity. In: Gunde-Cimerman N, Oren A, Plemenitaš A (eds) Adaptation to life at high salt concentrations in Archaea Bacteria and Eukarya. Springer, Dordrecht, pp 455–470

Plemenitaš A, Vaupotič T, Lenassi M, Kogej T, Gunde-Cimerman N (2008) Adaptation of extremely halotolerant black yeast *Hortaea werneckii* to increased osmolarity: a molecular perspective at a glance. Stud Mycol 61:67–75

Ramos-Cormenzana A (1991) Halophilic organisms and the environment. In: Rodriguez-Valera F (ed) General and applied aspects of halophilic microorganisms. Plenum, New York, pp 15–31

Ruibal C, Sakayaroj J, Sano T, Selbmann L, Shearer CA, Shirouzu T, Slippers B, Suetrong S, Tanaka K, Volkmann-Kohlmeyer B, Wingfield MJ, Wood AR, Woudenberg JHC, Yonezawa H, Zhang Y, Spatafora JW (2009) A class-wide phylogenetic assessment of *Dothideomycetes*. Stud Mycol 64:1–15

Russell NJ, Evans RI, Steeg PF, Hellemons J, Verheul A, Abee T (1995) Membranes as a target for stress adaptation. Int J Food Microbiol 28:255–261

Samson RA, Hoekstra ES, Frisvad JC (2004) Introduction to food- and airborne fungi. Centraalbureau voor Schimmelcultures, Utrecht

Schneider B, Xu YW, Janin J, Veron M, Deville-Bonne D (1998) 3-O-Phosphorylated nucleotides are tight binding inhibitors of nucleoside diphosphate kinase activity. J Biol Chem 273: 28773–28778

Schoch CL, Crous PW, Groenewald JZ, Boehm EWA, Burgess TI, de Gruyter J, de Hoog GS, Dixon LJ, Grube M, Gueidan C, Harada Y, Hatakeyama S, Hirayama K, Hosoya T, Huhndorf SM, Hyde KD, Jones EBG, Kohlmeyer J, Kruys Å, Li YM, Lücking R, Lumbsch HT, Marvanová L, Mbatchou JS, McVay AH, Miller AN, Mugambi GK, Muggia L, Nelsen MP, Nelson P, Owensby CA, Phillips AJL, Phongpaichit S, Pointing SB, Pujade-Renaud V, Raja HA, Rivas Plata E, Robbertse B, Ruibal C, Sakayaroj J, Sano T, Selbmann L, Shearer CA, Shirouzu T, Slippers B, Suetrong S, Tanaka K, Volkmann-Kohlmeyer B, Wingfield MJ, Wood AR, Woudenberg JHC, Yonezawa H, Zhang Y, Spatafora JW (2009) A class-wide phylogenetic assessment of *Dothideomycetes*. Stud Mycol 64:1–15.

Selbmann L, de Hoog GS, Mazzaglia A, Friedmann EI, Onofri S (2005) Fungi at the edge of life: cryptoendolithic black fungi from Antarctic desert. Stud Mycol 51:1–32

Slepecky RA, Starmer WT (2009) Phenotypic plasticity in fungi: a review with observations on *Aureobasidium pullulans*. Mycologia 101:823–832

Sterflinger K (1998) Temperature and NaCl-tolerance of rock-inhabiting meristematic fungi. Antonie Leeuwenhoek 74:271–281

Sterflinger K, de Hoog GS, Haase G (1999) Phylogeny and ecology of meristematic ascomycetes. Stud Mycol 43:5–22

Turk M, Plemenitaš A (2002) The HOG pathway in the halophilic black yeast *Hortaea werneckii*: isolation of the HOG1 homolog gene and activation of HwHog1p. FEMS Microbiol Lett 216: 193–199

Turk M, Méjanelle L, Šentjurc M, Grimalt JO, Gunde-Cimerman N, Plemenitaš A (2004) Salt-induced changes in lipid composition and membrane fluidity of halophilic yeast-like melanized fungi. Extremophiles 8:53–61

Turk M, Abramović Z, Plemenitaš A, Gunde-Cimerman N (2007) Salt stress and plasma-membrane fluidity in selected extremophilic yeasts and yeast-like fungi. FEMS Yeast Res 7: 550–557

van Baarlen P, van Belkum A, Summerbell RC, Crous PW, Thomma BPHJ (2007) Molecular mechanisms of pathogenicity: how do pathogenic microorganisms develop cross-kingdom host jumps? FEMS Microbiol Rev 31:239–277

Vaupotič T, Plemenitaš A (2007a) Differential gene expression and Hog1 interaction with osmoresponsive genes in the extremely halotolerant black yeast *Hortaea werneckii*. BMC Genomics 8:280

Vaupotič T, Plemenitaš A (2007b) Osmoadaptation-dependent activity of microsomal HMG-CoA reductase in the extremely halotolerant black yeast *Hortaea werneckii* is regulated by ubiquitination. FEBS Lett 581:3391–3395

Vaupotič T, Gunde-Cimerman N, Plemenitaš A (2007) Novel 3′-phosphoadenosine-5′-phosphatases from extremely halotolerant *Hortaea werneckii* reveal insight into molecular determinants of salt tolerance of black yeasts. Fungal Genet Biol 44:1109–1122

Vaupotič T, Veranič P, Jenoe P, Plemenitaš A (2008a) Mitochondrial mediation of environmental osmolytes discrimination during osmoadaptation in the extremely halotolerant black yeast *Hortaea werneckii*. Fungal Genet Biol 45:994–1007

Vaupotič T, Veranič P, Petrovič U, Gunde-Cimerman N, Plemenitaš A (2008b) HMG-CoA reductase is regulated by environmental salinity and its activity is essential for halotolerance in halophilic fungi. Stud Mycol 61:61–66

Ventosa A, Nieto JJ, Oren A (1998) Biology of moderately halophilic aerobic bacteria. Microbiol Mol Biol Rev 62:504–544

Wollenzien U, de Hoog GS, Krumbein WE, Urzì C (1995) On the isolation of microcolonial fungi occurring on and in marble and other calcareous rocks. Sci Total Environ 167:287–294

Zalar P, de Hoog GS, Gunde-Cimerman N (1999) Ecology of halotolerant dothideaceous black yeasts. Stud Mycol 43:38–48

Zalar P, de Hoog GS, Schroers H-J, Frank JM, Gunde-Cimerman N (2005a) Taxonomy and phylogeny of the xerophilic genus *Wallemia* (Wallemiomycetes and Wallemiales, cl. et ord. nov.). Antonie Leeuwenhoek 87:311–328

Zalar P, Kocuvan MA, Plemenitaš A, Gunde-Cimerman N (2005b) Halophilic black yeasts colonize wood immersed in hypersaline water. Bot Mar 48:323–326

Zalar P, de Hoog GS, Schroers H-J, Crous PW, Groenewald JZ, Gunde-Cimerman N (2007) Phylogeny and ecology of the ubiquitous saprobe *Cladosporium sphaerospermum*, with descriptions of seven new species from hypersaline environments. Stud Mycol 58:157–183

Zalar P, Gostinčar C, de Hoog GS, Uršič V, Sudhadham M, Gunde-Cimerman N (2008a) Redefinition of *Aureobasidium pullulans* and its varieties. Stud Mycol 61:21–38

Zalar P, Frisvad JC, Gunde-Cimerman N, Varga J, Samson RA (2008b) Four new species of *Emericella* from the Mediterranean region of Europe. Mycologia 100:779–795

Chapter 8
Viruses from the Hypersaline Environment

Elina Roine and Hanna M. Oksanen

8.1 Introduction

Organisms thriving in hypersaline environments represent all three domains of life: Archaea, *Bacteria*, and *Eukarya* (Oren 2008). It has been shown, however, that the higher the salt concentration of the hypersaline environment, the more dominant are the haloarchaeal species (Benlloch et al. 2002; Casamayor et al. 2002; Pedrós-Alió et al. 2000; Burns et al. 2004a; Santos et al. 2010). The number of virus-like particles (VLPs) is also very high (Guixa-Boixareu et al. 1996; Oren et al. 1997; Pedrós-Alió et al. 2000; Maturrano et al. 2006), and it has been suggested that in hypersaline environment, viruses may be the only predators (Guixa-Boixareu et al. 1996).

Studies on organisms living in extreme environments started already in the late nineteenth century (Oren 2002; Rainey and Oren 2006) and the earliest studies on extreme halophilic environments such as Great Salt Lake and the Dead Sea were reported soon after the first reports (Daniels 1917; Wilkansky 1936). The first haloarchaeal virus was described in 1974 (Torsvik and Dundas 1974).

Although a common theme in early studies on halophilic viruses seems to be the discovery of a head–tail virus as a result of a spontaneous lysis of a haloarchaeal culture (Schnabel et al. 1982b; Vogelsang-Wenke and Oesterhelt 1988; Witte et al. 1997) or a discovery of a virus in a flagellar preparation (Torsvik and Dundas 1974), active searches for halophilic viruses were also conducted (Wais et al. 1975; Pauling 1982). All the isolated viruses were of head–tail morphotype and they were studied for the purpose of finding molecular tools for haloarchaea as well as for studying molecular mechanisms of haloarchaeal transcription (Reiter et al. 1988). Unfortunately, many of these early viruses are most probably not available anymore. Prokaryotic viruses have now returned to the spotlight as major players in

E. Roine (✉) • H.M. Oksanen
Department of Biosciences and Institute of Biotechnology, University of Helsinki, P.O. Box 56 (Viikinkaari 5), 00014 Helsinki, Finland
e-mail: elina.roine@helsinki.fi

A. Ventosa et al. (eds.), *Halophiles and Hypersaline Environments*,
DOI 10.1007/978-3-642-20198-1_8,

the biogeochemical cycles and as examples of new viral architectures (Prangishvili et al. 2006; Suttle 2007). This has also activated the search for new haloviruses. Until this day, only approximately 20 haloarchaeal viruses are characterized and most of them represent the head–tail morphology (e.g., ΦH and ΦCh1); and an icosahedral virus (SH1) as well as enveloped viruses (HRPV-1, HHPV-1, His1 and His2) have also been described. These viruses do not show as high diversity in the morphotypes as do the crenarchaeal viruses (Prangishvili et al. 2006), but they show unexpected variability, especially in the nature of the genome and its structure, even between related viruses (Witte et al. 1997; Roine et al. 2010; Vogelsang-Wenke and Oesterhelt 1988). Microscopic examination of environmental samples has also shed light into the potential diversity of the haloarchaeal VLPs, suggesting that although most of the isolated viruses are head–tail viruses, the majority of haloarchaeal viruses are of other morphotypes (Guixa-Boixareu et al. 1996; Oren et al. 1997; Diez et al. 2000; Santos et al. 2007; Sime-Ngando et al. 2010).

In this review, we give a short introduction to the isolated haloarchaeal viruses. The relatively few descriptions of the head–tail halophages (Calvo et al. 1988; Kauri et al. 1991; Kukkaro and Bamford 2009) are not included. We then provide more in-depth information on the structurally well-characterized icosahedral virus SH1 and the group of enveloped viruses exemplified by HRPV-1 and HHPV-1. The detailed structural studies of SH1 have shown evolutionary connections that cannot be revealed by studying the nucleotide or amino acid sequences alone.

8.2 Approaches to Studying Viruses in Hypersaline Environment

As the studies of other environmental samples, the ones on the micro-organisms in extreme environments can methodologically be divided into three different types of approaches. Microscopy of environmental samples gives us an idea of the number of the VLPs leaving the infectivity of these particles an open question (Weinbauer 2004). Viruses having novel particle morphology can also easily pass unrecognized. Environmental sequencing (metagenomics) is a powerful tool in generating new data on the nucleotide sequence space of a sample. As is the case also with the other approaches, it is not devoid of bias due to the sample preparation and analysis of data (Tringe and Rubin 2005; Kunin et al. 2008; Wooley et al. 2010). The third approach, i.e., isolation and cultivation of organisms from the hypersaline environments has been relatively easy, but we face the same problems as in the isolation of organisms from other sources. It has been estimated that only 1–10% of all organisms of an environment can be cultivated in laboratory conditions. An exceptional achievement was the long-lasting project of isolating the most abundant haloarchaeal organism in many hypersaline sources, the square haloarchaeon *Haloquadratum walsbyi*: It took 24 years from the first microscopic identification by Walsby (1980) to the first reports of an axenic culture (Bolhuis et al. 2004; Burns et al. 2004b). For prokaryotic viruses also the ones with temperate life cycles will easily go unnoticed in the direct screening of viruses from an environmental sample (Stedman et al. 2009).

8.3 Viral Life Cycles in the Hypersaline Environment

The life cycles of haloarchaeal viruses seem to fall in the basic modes of the bacteriophage life cycles (Weinbauer 2004). Most of the head–tail viruses of haloarchaea are temperate, replicating their genome either as part of the host genome (e.g., ΦCh1) or as independent episomes (e.g., ΦH). The increasing NaCl concentrations have shown to change the mode of Hs1 and S5100 infection from lytic to persistent (Torsvik and Dundas 1980; Daniels and Wais 1990). Although this behavior can be easily explained as a strategy of the virus to utilize the cells' resources for virus production before cell lysis due to dilution of the salt concentration (Daniels and Wais 1990), a critical step-by-step analysis of the interactions at the molecular level will be needed in order to elucidate the causes and the effects. The effect of salt concentration on the adsorption of haloarchaeal viruses and different phages in a gradient of salt was systematically tested in the study by Kukkaro and Bamford (2009). As also reported previously, they concluded that there is no commonly shared correlation between the changes in adsorption rates and a range of salt concentrations used.

The two extreme ends of the viral life cycles are represented by SH1, a virulent haloarchaeal virus, and the enveloped haloarchaeal viruses HRPV-1 and HHPV-1, which extrude continuously from the cells affecting the host cell growth only minimally (Porter et al. 2005; Pietilä et al. 2009; Roine et al. 2010). The difference in these life cycles can be seen in the one-step experiments where a culture of synchronized cells (i.e., cells that are in the same exponential growth phase) is infected at high multiplicity of infection (MOI) in order to get most of the cells infected. The virulent virus SH1 that lyses the cells in the end of the infection cycle causes a dramatic drop in the turbidity of the culture (Fig. 8.1a). In a similar set-up, the enveloped viruses, on the other hand, cause only minor decrease in host cell culture turbidity (Fig. 8.1b). The changes in turbidity also correlate with the amount of living and dividing cells in the culture: The viable cell count of a culture infected with the virulent SH1 has decreased dramatically whereas the cells infected with enveloped viruses can still divide and produce colonies on the plate. In the case of HRPV-1 or HHPV-1, the infected cells form smaller colonies which produce viruses when re-cultured (Pietilä et al. 2009; Roine et al. 2010). In conclusion, the viruses are released from the cells without lysis, but viral infection slows down the division rate of the infected cells.

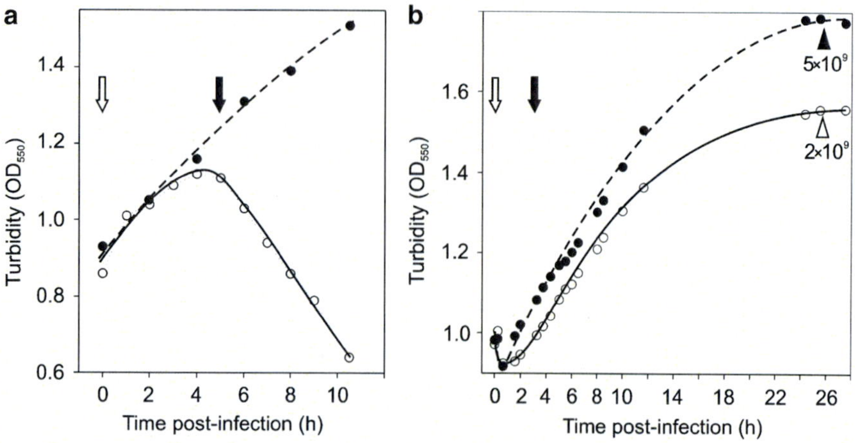

Fig. 8.1 The one-step growth curves of halophilic euryarchaeal (**a**) icosahedral virus SH1 and (**b**) pleomorphic virus HHPV-1 infecting *Haloarcula hispanica*. The turbidity of uninfected (*closed circles*) and infected (*open circles*) cultures are shown. The exponentially growing hosts were infected (*white arrows*) using multiplicity of infection (MOI) of 40 (in **a**, SH1) and MOI of 15 (in **b**, HHPV-1). The free HHPV-1 virions were removed by centrifugation after 1 h post-infection. The beginning of the release of the progeny viruses is marked by *black arrows*. (**a**) The lysis of *Haloarcula hispanica* occurs concomitantly with the SH1 release 5 h post-infection, and the maximum virus yield is approximately 1×10^{12} pfu/ml. (**b**) HHPV-1 viruses are released continuously from the cells by budding. Virus release starts 3–5 h post-infection and it reaches its maximum (~1×10^{12} pfu/ml) around 25 h post-infection. The cell densities (colony forming units/ml) around 25 h post-infection are indicated by *arrowheads*: uninfected (*black*) and infected (*white*)

8.4 Head–Tail Archaeal Viruses

There are a number of extensive reviews on the research of the head–tail haloviruses with their properties listed and we refer to these for further information (Reiter et al. 1988; Dyall-Smith et al. 2003; Porter and Dyall-Smith 2006; Porter et al. 2007, 2008b). Most of the head–tail viruses were originally reported to infect *Halobacterium halobium* or *Halobacterium curtirubrum* (now *Halobacterium salinarum*) (Porter et al. 2008b; Reiter et al. 1988). Although the status of those strains in terms of the detailed characterization for classification is unclear, many of them are now classified into *Halobacterium salinarum* (Ventosa and Oren 1996). All the isolated head–tail viruses fall into the groups of myo- and siphoviruses, and no podoviruses have been isolated. Out of the 16 isolated head–tail viruses, only the complete genome sequences of ΦCh1, HF1, HF2 and BJ1 have been published (Pagaling et al. 2007; Porter et al. 2008b). On the basis of this information it is clear, however, that the genomes show similar type of modular structure as the head–tail bacteriophages (Hendrix et al. 2000). The connection between the haloarchaeal and bacterial head–tail viruses is also clear on the basis of gene homologues (Tang et al. 2002; Pagaling et al. 2007; Krupovič et al. 2010). Three dimensional structural modeling of the haloarchaeal major capsid protein gpE of ΦCh1 resulted in a protein fold similar to the bacteriophage HK97 major capsid protein fold showing relatedness also at the higher structural level (Krupovič et al. 2010).

Still to date, the temperate myovirus ΦH isolated from spontaneously lysed cells of *Halobacterium halobium* R_1 (*Halobacterium salinarum* R1) in 1982 (Schnabel et al. 1982b) is one of the most studied among haloarchaeal viruses. Yet, only partial genome sequence of this virus is available. The distinctive feature of ΦH genome was the observed frequent rearrangements in virus population (Schnabel et al. 1982a, b). The rearrangements were caused by a varying number of copies of the insertion element ISH1.8 as well as insertion of ISH50 (Schnabel et al. 1982a). The variants of the virus containing two copies of ISH1.8 in inverted orientation were shown to experience two types of frequent recombination events: An inversion of the genomic region flanked by the two copies of ISH1.8, the so called L-segment, as well as circularization of the L-segment into a 12-kb plasmid pΦHL (Schnabel 1984; Schnabel et al. 1982a). The pΦHL was shown to confer partial resistance against the ΦH infection (Schnabel 1984). The rest of the research on ΦH deals mostly with the regulation of transcription and super-infection immunity (Schnabel et al. 1984; Gropp et al. 1989, 1992; Stolt and Zillig 1992, 1993b). A ΦH repressor was discovered that forms unusually stable complexes with DNA (Ken and Hackett 1991); also the first archaeal antisense transcript was reported that takes part into the super-infection immunity (Stolt and Zillig 1993a).

Another temperate myovirus ΦCh1 was isolated in 1997 by Angela Witte and her colleagues after spontaneous lysis of the haloalkaliphilic archaeon *Natrialba magadii* (Witte et al. 1997). ΦCh1 is the only archaeal virus known to contain both DNA and RNA in its virion. The virion-associated RNA is mainly of host-origin varying in length from 100 to 800 nucleotides (Witte et al. 1997). ΦCh1 ORFs that could be assigned to a homologue in the sequence databases were the ORFs of *Hbt. salinarum* head–tail virus ΦH (Klein et al. 2002). The close relationship between these viruses is surprising as their hosts are phylogenetically distant and in terms of pH they live in totally different environments. As in ΦH, genomic rearrangements have been shown to take place in the ΦCh1 genome. These rearrangements occur in the region encoding a putative site-specific recombinase Int1 (Rössler et al. 2004). These genomic rearrangements are not as versatile as in the ΦH involving mainly an inversion of the *int1* gene region that affects the amino acid sequences of two structural proteins of the virion (Rössler et al. 2004). It was suggested that this inversion affects the specificity of the host recognition by changing the putative tail fiber proteins (Rössler et al. 2004). As discussed above for ΦH, variations of the genome are caused by insertion elements. These rearrangements were suggested to reflect the extremely instable nature of the host genome as shown for the *Hbt. halobium* plasmid pHH1 (Pfeifer et al. 1981) as well as for the whole genome (Sapienza et al. 1982; Simsek et al. 1982; Reiter et al. 1988). Genomic instability comparable to one found in *Hbt. salinarium* has not been reported for *Nab. magadii.*

Although there are several other viruses e.g., Hs1, Ja.1, S41, S50.2, S4100, S5100, ΦN, Hh-1, Hh-3, and S45 (Porter et al. 2008b) reported to infect *Hbt. salinarum*, such genetic divergence of the viral genomes have not been studied.

Indirect evidence, however, suggests such a phenomenon for S41 and S50.2 viruses (Daniels and Wais 1998) and genomic rearrangements were also reported for the BJ1 head–tail virus infecting *Halorubrum saccharovorum* (Pagaling et al. 2007).

In addition to isolated haloviruses, the "environmental halophage 1" (EHP-1) has been identified using a metagenomic approach (Santos et al. 2007). Most of the predicted ORFs with matches to database were similar to predicted gene products of ΦCh1. The low G + C content of the 35 kb genome as well as the codon usage were similar to those of *Haloquadratum walsbyi* (Santos et al. 2007). Using the combination of a metagenomic approach and environmental microscopy, another saline environment, the Lake Retba (Senegal) was analyzed for viral and halophage content (Sime-Ngando et al. 2010). Microscopy of the concentrated salt water sample showed great diversity in the morphotypes of VLPs with less than 1% representing head–tail viruses. Also sequence data revealed very few similarities to the known sequences (Sime-Ngando et al. 2010). The majority of matches found in databases, however, represented the sequences of head–tail viruses BJ1, ΦCh1 and ΦH and a couple of sequences had matches to the enveloped virus HRPV-1 and His1 genomes.

8.5 The Icosahedral Virus SH1

SH1 was isolated in 2005 (Porter et al. 2005) from a salt lake in Western Australia. To date, SH1 is the only described euryarchaeal virus with icosahedral morphology. Under the icosahedral protein capsid there is a membrane enclosing the genome. SH1 is a virulent virus infecting *Haloarcula hispanica*. As discussed above (Fig. 8.1a), the infected host cells lyse between 5 and 6 h post infection releasing progeny particles (Porter et al. 2005).

8.5.1 The SH1 Genome

The genome is a linear dsDNA molecule of 30,898 bp, and contains 309-bp inverted repeats and 5′ terminal proteins (Bamford et al. 2005b; Porter and Dyall-Smith 2008). Thus, the genome is most probably replicated by a protein-primed mechanism. This replication strategy has been described in great detail for phage Φ29 (Salas 1984). However, the SH1 genome does not appear to have any ORF with canonical DNA polymerase motifs, and most likely uses a host enzyme (Bamford et al. 2005b). The G + C content of the genome is 68.4%, which is close to that of the host *Har. hispanica* (63.5%). The genome contains 56 predicted ORFs, which are arranged in at least seven polycistronic operons with tightly regulated promoters and containing in most cases a potential 5′ ribosome binding site (Bamford et al. 2005b; Porter et al. 2008a). Late in the infection cycle the SH1 transcription seems to be characterized by extensive counter-transcription for most of the coding transcripts that may have a function in translational regulation (Porter et al. 2008a).

Among the SH1 ORFs only few sequences are similar to previously identified sequences in the databases (Bamford et al. 2005b). This reflects the fact that only a limited number of archaeal sequences have been determined, and that archaeal genes seem to be significantly different from those of bacterial and eukaryotic origin. Recently, a proviral element (IHP) was found to be integrated in the genome of *Haloarcula marismortui* (Jalasvuori et al. 2009). The IHP element contains genes similar to the SH1 genes encoding the major capsid proteins and the ATPase (see below). The same genetic module is also found in the archaeal plasmid pHH205 of *Hbt. salinarum* as well as in bacterial viruses (phage P23-77 and IN93) revealing an evolutionary relationship between bacterial and archaeal genetic elements (Jalasvuori et al. 2009). Additionally, the ORF 17 of SH1 encodes a putative protein with ATPase motifs including classical Walker A and B motifs (Walker et al. 1982) as well as a motif common to internal membrane-containing dsDNA viruses e.g., bacteriophages PRD1 and PM2, eukaryotic viruses *Paramecium bursaria* Chlorella virus 1 (PBCV-1) and Chilo Iridescent Virus (CIV), and archaeal virus *Sulfolobus* turreted icosahedral virus (STIV) (Bamford et al. 2005a; Strömsten et al. 2005).

8.5.2 SH1 Virion Structure

Production of large amounts of highly purified viral material has enabled us to study the structure of SH1 by quantitative dissociation studies as well as by electron cryo-microscopy (cryo-EM) and image reconstruction at 9.6-Å resolution (Kivelä et al. 2006; Jäälinoja et al. 2008). The icosahedral SH1 virion has a diameter of 78 nm between facets and is composed of approximately 15 virally encoded structural protein species (VP1-VP15), of which 11 have been identified as virion proteins by protein chemistry (Fig. 8.2; Bamford et al. 2005b; Kivelä et al. 2006). Most of the structural protein species have been localized within the virions and the proteins can be divided into those forming the protein capsid, the vertex complex, and those being associated with the viral membrane (Fig. 8.2; Bamford et al. 2005b; Kivelä et al. 2006; Jäälinoja et al. 2008).

The SH1 spikes are horn-like structures projecting almost 20 nm outwards from the capsid surface (Fig. 8.2). These two-fold symmetric structures, composed of at least two proteins VP3 and VP6, are most probably involved in receptor recognition (Jäälinoja et al. 2008). Also protein VP2 is associated with the SH1 vertex structure. The glycine- and serine-rich VP2 protein with heptapeptide repeat patterns characteristic for coiled-coil proteins suggest that it could be an elongated fiber-like protein appropriate for forming a flexible structure (Bamford et al. 2005b).

In SH1, the membrane lying under the capsid follows its icosahedral shape and there are several interactions between the capsomers and the membrane (Jäälinoja et al. 2008; Fig. 8.2). The viral membrane is composed of neutral lipids and three

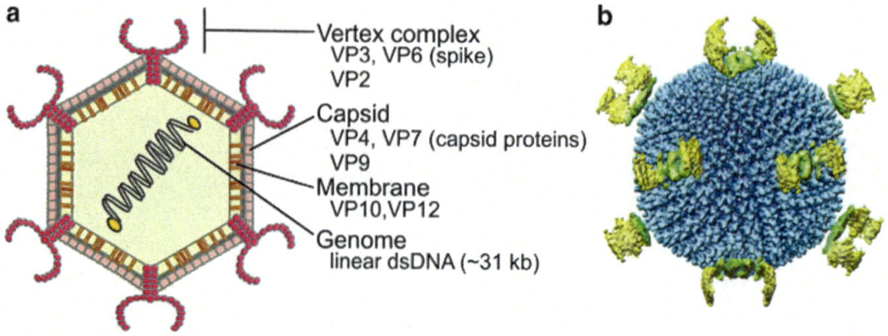

Fig. 8.2 Halophilic icosahedral membrane-containing virus SH1. (**a**) Schematic picture of the virion architecture showing the location of the major structural proteins (VPs). (**b**) Cryo-electron microscopy based structure of the SH1 virion viewed down twofold axis of symmetry. Part (**b**) is reproduced from Jäälinoja et al. (2008). Copyright (2008) National Academy of Sciences, USA

major archaeal phospholipids: phosphatidylglycerol, phosphatidylglycerosulfate, and phosphatidylglycerosulfate methyl ester (Bamford et al. 2005b). The viral lipid composition is different from that of its host cell indicating that the phospholipids are selectively acquired from the host cytoplasmic membrane during virus maturation. The overall structure of SH1 virion resembles the virion morphology of archaeal *Sulfolobus* turreted icosahedral viruses STIV and STIV-2 (Happonen et al. 2010; Khayat et al. 2005) and bacteriophages PRD1, Bam35, PM2 and P23-77 (Abrescia et al. 2004, 2008; Laurinmäki et al. 2005; Jaatinen et al. 2008) as well as eukaryotic *Paramecium bursaria* Chlorella virus 1 (PBCV-1; Nandhagopal et al. 2002). They all have a dsDNA genome encapsidated into a proteinaceous membrane residing in an icosahedral protein capsid. The capsid architecture of SH1 is most similar to that of the bacteriophage P23-77 infecting *Thermus thermophilus* (Jaatinen et al. 2008). Structurally related viruses having hosts from different domains of life support the hypothesis that viruses are ancient and these two viruses might share a common ancestry dating back to the time before the separation of the three domains of life (Bamford 2003; Krupovič and Bamford 2008).

8.6 Enveloped Viruses Infecting Haloarchaea

The characterization of the new group of pleomorphic viruses was a result of our efforts in looking for new viruses from the hypersaline environments. The first member of this group, the *Halorubrum* pleomorphic virus 1 (HRPV-1) is also the first archaeal virus reported to have a single-stranded DNA (ssDNA) genome (Pietilä et al. 2009). The group includes at the moment also a dsDNA virus, *Haloarcula hispanica* pleomorphic virus 1 (HHPV-1, Fig. 8.3a) as well as two putative proviruses, the pHK2, a genetic element previously reported to be a plasmid (Holmes et al. 1995) and a provirus in the genome of *Haloferax volcanii*

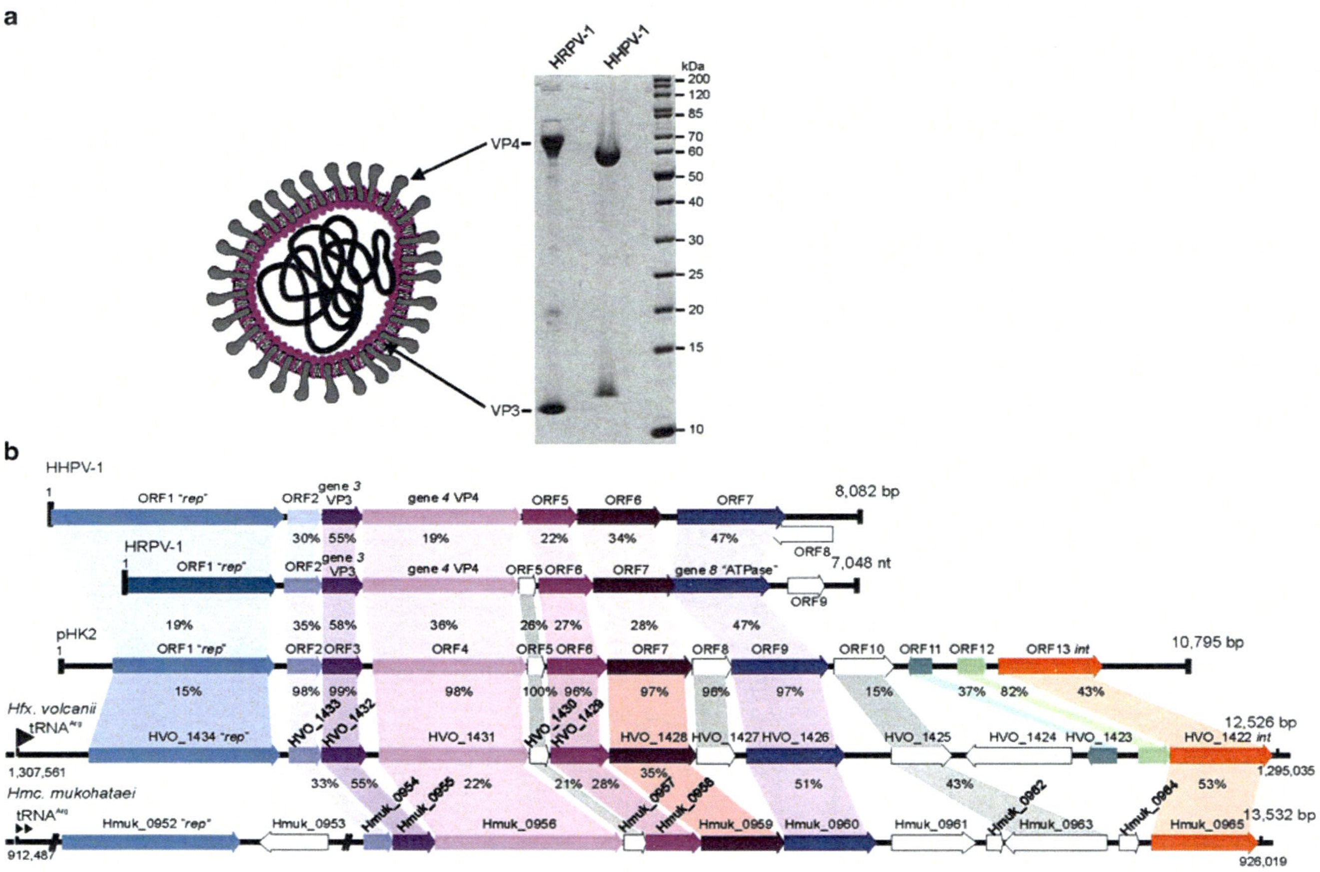

Fig. 8.3 The virion architecture and genome alignment of the new group of pleomorphic viruses. (**a**) A schematic of the pleomorphic virus architecture pointing out the predicted orientation of the two major structural proteins (*left*). A sodium dodecyl sulfate polyacrylamide gel electrophoresis (SDS-PAGE) analysis of the purified HRPV-1 and HHPV-1 virions is shown on the right. Molecular weight standards (kDa) are shown on the *right*. (**b**) Genomic

(Roine et al. 2010, Fig. 8.3b). The genomes of HRPV-1 and HHPV-1 are circular molecules of 7,048 nt and 8,082 bp, in size, respectively. Previously identified spindle-shaped viruses His1 and His2 (genus *Salterprovirus*; Bath et al. 2006) are flexible particles containing approximately 15 kb linear dsDNA genomes. The chloroform sensitivity as well as the low buoyant density of His1 and His2 suggest they belong to enveloped viruses (Bath et al. 2006). Although the genome types of pleomorphic viruses and spindle-shaped viruses are different, limited homology is shared at the amino acid sequence level between His2 virus and HHPV-1 and HRPV-1 (see below; Bath et al. 2006). It is worth emphasizing that even though the two pleomorphic viruses are closely related (see below), they contain different types of genomes: HRPV-1 contains a ssDNA genome whereas the HHPV-1 contains a dsDNA genome (Pietilä et al. 2009; Roine et al. 2010). These results conflicts with current viral classification that is based on the genome types and the mode of replication (Baltimore 1971), because in the absence of the sequence data these two viruses having two different genome types would be classified as unrelated viruses (Roine et al. 2010).

8.6.1 Genomic Comparison

Similar protein patterns of HRPV-1 and HHPV-1 with two major structural proteins already suggested relatedness (Fig. 8.3a). At the nucleotide sequence level these two genomes only show homology along a very short stretch. However, the genomes are co-linear along the whole genomes and at the amino acid sequence level the highest identity can be found between the small major structural protein VP3 (Fig. 8.3b). The co-linearity can also be found between the viruses and the putative proviruses (Fig. 8.3b; Roine et al. 2010). This region includes ORFs 1, 2, 6 (HHPV-1 ORF5), 7 (HHPV-1 ORF6) and 9 (HHPV-1 ORF7 and HRPV-1 gene *8*). ORFs 1 encode putative replication initiation proteins showing rather low identity between each other. They are predicted, however, to serve the same function (Roine et al. 2010). The region starting from ORF2 and ending in ORF9 (in pHK2 and in the *Hfx. volcanii* genomic region) show amino acid sequence identity higher than 20%. In the genomic alignment, we can see small ORFs that may have been lost in HHPV-1. HRPV-1 still seems to have a homolog of ORF5, but has lost ORF8 that can be found in pHK2 and *Hfx. volcanii* proviral elements. At the genomic level,

Fig. 8.3 (Continued) comparison of the viruses HRPV-1 and HHPV-1 as well as the putative proviral elements pHK2 and *Hfx. volcanii* genomic region. The newly identified genomic region in *Hmc. mukohataei* (DSM12286) shows the same features as the *Hfx. volcanii* proviral region with a $tRNA^{Arg}$ in the 5′ region followed by the ORF1 and a putative integrase in the end of the proviral element. The $tRNA^{Arg}$ in *Hmc. mukohataei* is a split-tRNA. The percentages show the identity of the amino acid sequences of the pairwise alignments. The Hmuk_0952 contains the DUF1424 conserved domain common to proteins involved in replication. It also shows 30% identity with the pHK2 ORF1 which is not indicated in the picture

Table 8.1 Homology of theVP4 major structural proteins and predicted amino acid sequences of the putative ORF4 products[a]

	HRPV-1 VP4	HHPV-1 VP4	His2 gp29	His1 ORF27	pHK2 ORF4	HVO_1431	HVO_0271	Hmuk_0956
HRPV-1 VP4	–	19	26	12	36	36	20	23
HHPV-1 VP4	19	–	20	12	18	18	22	17
His2 gp29	26	20	–	12	25	26	23	26
His1 ORF27	12	12	12	–	14	13	11	13
pHK2 ORF4	36	18	25	14	–	98	23	24
HVO_1431	36	18	26	13	98	–	22	22
HVO_0271	20	22	23	11	23	22	–	21
Hmuk_0956	23	17	26	13	24	22	21	–

[a]Identity percentages of the pair-wise alignments determined by CLUSTALW at the NPS@ Web server (http://npsa-pbil.ibcp.fr/) using the default settings

HRPV-1 seems to be a closer relative to pHK2 proviral element than to HHPV-1 as judged by the higher number of homologous ORFs as well as by the higher identity of the proteins VP3 and VP4 sequences (Fig. 8.3b, Table 8.1).

The pHK2 and *Hfx. volcanii* proviral elements are almost identical along the conserved ORF2 to ORF9 region (Fig. 8.3b). After that there have been genomic rearrangements, still preserving the last three predicted ORFs. The translated polypeptides of these three ORFs show high identity. The pHK2 ORF11 and ORF13 and the corresponding ORFs in the *Hfx. volcanii* genome also encode putative polypeptides that by homology are predicted to be a ΦH-type repressor and an integrase like protein, respectively (Fig. 8.3b).

The pHK2 region spanning from ORF4 to ORF9 can also be found in the genome of the spindle-shaped virus His2 (Bath et al. 2006; Pietilä et al. 2009). The major structural protein of His2 (VP1 encoded by gp29) shows 25% identity to the pHK2 ORF4 encoded polypeptide, which is higher than the identity between the HHPV-1 protein VP4 and pHK2 ORF4 encoded protein (Table 8.1). The rest of the His2 ORFs encoded putative polypeptides show lower homology to their predicted counterparts.

In addition to the homologous genetic elements mentioned above, there are regions in almost all haloarchaeal genomes that show homology to different sets of these genes (Table 8.2). After a closer examination of the genomic databases, an almost complete proviral element was found from the genome of *Halomicrobium mukohataei* (GenBank CP001688.1). This element contains all the other homologous ORFs found in the *Hfx. volcanii* proviral element except for the ORF encoding the putative ΦH repressor like protein (Fig. 8.3b). The alignment of this element with the haloarchaeal enveloped viruses and other putative proviral elements further supports the observation that the central region containing the genes encoding the major structural proteins, are the most conserved part of the viruses (Roine et al. 2010).

Table 8.2 Examples of homologous ORFs found from different haloarchaeal genomes[a]

	HVO_1434	HVO_1433	HVO_1432	HVO_1431	HVO_1430	HVO_1429	HVO_1428	HVO_1427	HVO_1426	HVO_1425	HVO_1424	HVO_1423	ORF12	HVO_1422
Hfx. volcanii				HVO_0271		HVO_0272	HVO_0273		HVO_0274			HVO_0261		HVO_1839 HVO_0385
Hfx. lucentense		pHK2 ORF2	pHK2 ORF3	pHK2 ORF4	pHK2 ORF5	pHK2 ORF6	pHK2 ORF7	pHK2 ORF8	pHK2 ORF9	pHK2 ORF10		pHK2 ORF11	pHK2 ORF12	pHK2 ORF13
Hmc. mukohataei		Hmuk_0954	Hmuk_0955	Hmuk_0457 Hmuk_0956	Hmuk_0957	Hmuk_0958	Hmuk_0459 Hmuk_0832 Hmuk_0959		Hmuk_0460 Hmuk_0831 Hmuk_0960	Hmuk_0829 Hmuk_0963		Hmuk_0443 Hmuk_1702		Hmuk_0467 Hmuk_0155 Hmuk_0965
Hrr. lacusprofundi			Hlac_1753							Hlac_3323		Hlac_3166		
Har. marismortui				rrnAC2291 rrnAC2399			rrnAC2293 rrnAC2401		rrnAC2294 rrnAC2402/ rrnAC2403					
Hrd. utahensis	Huta_0241			Huta_0799			Huta_0801		Huta_0802	Huta_1840				Huta_0110
Nab. magadii				Nmag_0288			Nmag_0286		Nmag_0283	Nmag_4156		Nmag_4173 Nmag_3625		
Nmn. pharaonis				NP2342A NP5348A			NP2346A NP5344A		NP2348A NP5342A			NP2364A NP3896A NP5364A		NP5368A
Htg. turkmenica										Htur_3124	Htur_4526	Htur_4526 Htur_5241		Htur_1403
His2 virus				VP1 (gp29)			gp31		gp33					

[a]ORFs from the *Hfx. volcanii* (HVO) pro-viral element were used as a query in PSI-BLAST search using the cut-off value of 0.005. Abbreviations: *Hfx.*, *Haloferax*; *Hmc.*, *Halomicrobium*; *Hrr.*, *Halorubrum*; *Har*, *Haloarcula*; *Hrd.*, *Halorhabdus*; *Nab.*, *Natrialba*; *Nmn.*, *Natronomonas*; *Htg.*, *Haloterrigena*

8.6.2 *Particle Architecture and Structural Proteins*

The virion structure of the pleomorphic viruses is relatively simple with the genome enveloped by protein-containing membrane. The envelope lipids are acquired from the host cytoplasmic membrane non-selectively and thus the lipid composition is approximately the same as in the host (Pietilä et al. 2010; Roine et al. 2010). There are two major structural proteins, VP3 and VP4 (Fig. 8.3). In addition, protein VP8 has been identified as a minor structural component in HRPV-1 (Pietilä et al. 2009). The smaller major structural protein VP3 is predicted to be a membrane associated protein facing the interior of the virion (Pietilä et al. 2010) whereas VP4 is the spike protein protruding outside and being anchored to the virion envelope by a C-terminal trans-membrane domain (Pietilä et al. 2009, 2010). The HRPV-1 particles are approximately 44 × 55 nm in size and not as uniform in shape as the virions containing a rigid icosahedral protein coat. This can be seen as a very diffuse light scattering band in a rate zonal sucrose gradient, for example. However, the particles of HRPV-1 seem to be more uniform in shape than the HHPV-1 particles when samples are studied by negative staining and transmission electron microscopy (TEM). This may be explained by the fact that the HRPV-1 VP4 is glycosylated but the HHPV-1 VP4 is not (Pietilä et al. 2010; Roine et al. 2010) and thus the surface of the HHPV-1 particle is more prone to the pressure by the different negative stains. Some negative stains seem to radically reduce the infectivity of HHPV-1 (Roine et al. 2010) and at the same time change the morphology of the virion from flexible into round particles that are more uniform in shape (Kukkaro, personal communication). To conclude, the definition of a pleomorphic particle, in our opinion, describes best this virus morphology.

The large major structural proteins (VP4) show very low sequence similarity to each other (Table 8.1). We have hypothesized that the VP4 protein is responsible for the receptor recognition and fusion of the envelope with the host cytoplasmic membrane (Pietilä et al. 2009; Roine et al. 2010). The proteins recognizing the host receptors are usually very diverse due to the selection pressure of the host (Lubbers et al. 1995; Brüssow and Desiere 2001; Saren et al. 2005). HRPV-1 and HHPV-1 VP4 polypeptides are both N-terminally processed and during the morphogenesis of the viral particle, the protein is translocated to the extracellular face of the host membrane (Pietilä et al. 2010). The VP4 molecules also share some putative structural features such as a predicted C-terminal membrane anchor preceded by predicted coiled-coil domain. As coiled-coil domains are usually involved in multimeric interactions, it is probable that the VP4 is multimeric at some stage of host recognition or entry. It is highly probable that VP4 protein interacts with VP3, which contains four predicted trans-membrane domains (Pietilä et al. 2009). Two of these have been experimentally verified suggesting a role as a matrix forming protein (Pietilä et al. 2009, 2010). The interaction of VP3 with the genomic DNA was not excluded, although the experimental results did not indicate any strong interaction to occur (Pietilä et al. 2010). Whether the VP3 proteins act solely as structural components or are active also in the fusion process needs to be determined.

Another predicted protein product showing relatively high homology between the identified viruses and the proviral elements are the ORF9 encoded products, the VP8 protein of HRPV-1 and the putative ORF33 of His2 virus. Many of these show the conserved motifs of P-loop NTPases (Bath et al. 2006; Pietilä et al. 2009), i.e., the Walker A and Walker B elements (Walker et al. 1982). Experimental evidence, however, is needed to show that they are functional. We propose that they may function in the assembly of the viral particles.

As discussed above, the putative ORF1 encoded proteins are all predicted to be replication initiation proteins. The conserved motifs 2 and 3 (Ilyina and Koonin 1992) found in plasmids and viruses replicating their genomes by rolling circle replication (RCR) can also be found in all of them (Roine et al. 2010). This is in good agreement with the fact that the HRPV-1 and HHPV-1 genomes are circular and that pHK2 was initially described as a haloplasmid.

8.7 Conclusions

Haloarchaeal viruses are among the least studied viruses comprising currently only around 20 characterized isolates. Our systematic approach for screening new host isolates and subsequently viruses for them has shown that there are still new types of viruses to be found. Although not as versatile in the particle architecture as the crenarchaeal viruses (Prangishvili et al. 2006), the description of the new haloarchaeal viruses have significantly contributed to the general knowledge of the viral universe. This is true especially in the case of the pleomorphic viruses HRPV-1 and HHPV-1 that are closely related, but have different genome types. The higher order viral classification divides viruses according to the genome type and replication strategy (Baltimore 1971). The pleomorphic viruses thus represent a deviation from this scheme and points to a reconsideration of the higher order taxonomy.

The virion architecture of the haloarchaeal pleomorphic viruses is to some extent reminiscent of the *Plasmaviridae* family of viruses L2 and L172 that infect *Acholeplasma laidlawii* (Dybvig et al. 1985). These viruses are enveloped and pleomorphic in structure and have a similar structural protein pattern between each other and also between the haloarchaeal enveloped viruses (Dybvig et al. 1985). L2 contains a circular 12 kb dsDNA genome whereas the L172 genome is a circular ssDNA molecule 14 kb in size. However, only the L2 genome sequence is available (Maniloff et al. 1994) and the relatedness of the plasma viruses cannot be confirmed. The structural relatedness of the plasmaviruses and the pleomorphic viruses would not be surprising when taking into account the fact that both SH1 and the head–tail viruses of haloarchaea have relatives among the bacteriophages.

Our studies together with the ones by Mike Dyall-Smith have shown that the early isolation of only head–tail haloarchaeal viruses does not reflect the virus population in halophilic environment. The isolation of many enveloped viruses correlates better with the results of microscopical examination of environmental samples, suggesting that in terms of morphotypes there is much diversity to be found (Guixa-Boixareu et al. 1996; Oren et al. 1997; Diez et al. 2000; Santos et al. 2007; Sime-Ngando et al. 2010). Although our culture-dependent studies support the observations of the environmental microscopy, not all the VLPs seen by microscopy are necessarily viruses. Viruses are genetic elements capable of packaging their genome into a particle. They are also infective agents and consequently obey Koch's postulates. This means that the viruses produce new infective progeny highly similar to the one used for the infection. Unfortunately, this can be proven only by the culture-dependent approach. Environmental sequencing has expanded our knowledge of the sequence space to be found from the hypersaline environment. Several limitations in the preparation of the sample such as viruses that do not pass the 0.2-μm filter (Hendrix 2009) or are dissociated by chloroform, and sequences that are toxic to the cloning hosts used pose serious challenges if the entire viral fraction of the sample is to be determined (Tringe and Rubin 2005; Kunin et al. 2008; Wooley et al. 2010). Consequently, the environmental sequencing approach also has a bias that may not be any less significant than the bias in the other approaches. In addition, the sequences that show no matches to databases gain their full value only when similar sequences have been identified in a virus characterized in detail. This usually involves the isolation of the virus. As when mapping the sequences in other environmental niches, the mapping of the hypersaline environment should include all the possible types of approaches from the measurement of the bulk biological processes such as primary production and physico-chemical conditions to environmental microscopy and sequencing as well as the culture-dependent approach of isolating the infective viruses.

It is clear that prokaryotic viruses are important players in the genomic plasticity of their hosts. In hypersaline environment, it is generally agreed that the amount of viruses and their hosts is high (Guixa-Boixareu et al. 1996; Oren et al. 1997; Maturrano et al. 2006; Pedrós-Alió et al. 2000). The lysis of cells due to viral infection, however, was shown to be low (Guixa-Boixareu et al. 1996). This suggests that the popular "kill the winner" phenomenon would not fully operate in hypersaline environments. Rodríguez-Valera and others (Rodriguez-Valera et al. 2009), on the other hand, propose a "kill the winner" type of control by viruses over the dominating host population preserving the microdiversity of the microbial population and generating a pan genome comprised of numerous different, but closely related clones (Papke et al. 2004; Rodriguez-Valera et al. 2009). The existence of proviruses and the numerous remnants of them that are homologous to the pleomorphic viruses suggest a very intimate relationship between the virus and the haloarchaeal host. The proposed high number of the enveloped viruses suggests that these viruses have a great impact in generating the microdiversity using a strategy more benign than the "kill the winner" strategy is. In such an extreme environment it is conceivable that the relationship of the predator and prey are closely connected and dependent on each other. Support for this can also be found in the observed changes of the head–tail virus life cycles in which the higher salinity caused a switch from lysis into proviral mode. As seen in other prokaryotes, proviruses may have dramatic effects on the behavior of their hosts (Brüssow et al. 2004). More studies, however, are needed to elucidate the complex interactions monitoring population dynamics of the viruses and their hosts by taking advantage of the detailed molecular biology data available.

Acknowledgments This work was supported by Academy of Finland Centre of Excellence Program in Virus Research grant 11296841 (2006–2011), by the University of Helsinki Three Year Grant 2010–2012 to E.R., and by the Academy of Finland grant (1127665) to H.M.O. We acknowledge Dennis Bamford for critical comments on our manuscript and Petra Kukkaro for contributing to the studies on the transmission electron microscopy of the pleomorphic viruses.

References

Abrescia NG, Cockburn JJ, Grimes JM, Sutton GC, Diprose JM, Butcher SJ, Fuller SD, San Martin C, Burnett RM, Stuart DI, Bamford DH, Bamford JK (2004) Insights into assembly from structural analysis of bacteriophage PRD1. Nature 432:68–74

Abrescia NG, Grimes JM, Kivelä HM, Assenberg R, Sutton GC, Butcher SJ, Bamford JK, Bamford DH, Stuart DI (2008) Insights into virus evolution and membrane biogenesis from the structure of the marine lipid-containing bacteriophage PM2. Mol Cell 31:749–761

Baltimore D (1971) Expression of animal virus genomes. Bacteriol Rev 35:235–241

Bamford DH (2003) Do viruses form lineages across different domains of life? Res Microbiol 154:231–236

Bamford DH, Grimes JM, Stuart DI (2005a) What does structure tell us about virus evolution? Curr Opin Struct Biol 15:655–663

Bamford DH, Ravantti JJ, Rönnholm G, Laurinavičius S, Kukkaro P, Dyall-Smith M, Somerharju P, Kalkkinen N, Bamford JK (2005b) Constituents of SH1, a novel lipid-containing virus infecting the halophilic euryarchaeon *Haloarcula hispanica*. J Virol 79:9097–9107

Bath C, Cukalac T, Porter K, Dyall-Smith ML (2006) His1 and His2 are distantly related, spindle-shaped haloviruses belonging to the novel virus group, *Salterprovirus*. Virology 350:228–239

Benlloch S, López-López A, Casamayor EO, Øvreås L, Goddard V, Daae FL, Smerdon G, Massana R, Joint I, Thingstad F, Pedrós-Alió C, Rodríguez-Valera F (2002) Prokaryotic genetic diversity throughout the salinity gradient of a coastal solar saltern. Environ Microbiol 4:349–360

Bolhuis H, te Poele EM, Rodríguez-Valera F (2004) Isolation and cultivation of Walsby's square archaeon. Environ Microbiol 6:1287–1291

Brüssow H, Desiere F (2001) Comparative phage genomics and the evolution of *Siphoviridae*: insights from dairy phages. Mol Microbiol 39:213–222

Brüssow H, Canchaya C, Hardt WD (2004) Phages and the evolution of bacterial pathogens: from genomic rearrangements to lysogenic conversion. Microbiol Mol Biol Rev 68:560–602

Burns DG, Camakaris HM, Janssen PH, Dyall-Smith ML (2004a) Combined use of cultivation-dependent and cultivation-independent methods indicates that members of most haloarchaeal groups in an Australian crystallizer pond are cultivable. Appl Environ Microbiol 70: 5258–5265

Burns DG, Camakaris HM, Janssen PH, Dyall-Smith ML (2004b) Cultivation of Walsby's square haloarchaeon. FEMS Microbiol Lett 238:469–473

Calvo C, de la Paz AG, Bejar V, Quesada E, Ramos-Cormenzana A (1988) Isolation and characterization of phage F9-11 from a lysogenic *Deleya halophila* strain. Curr Microbiol 17:49–53

Casamayor EO, Massana R, Benlloch S, Øvreås L, Diez B, Goddard VJ, Gasol JM, Joint I, Rodríguez-Valera F, Pedrós-Alió C (2002) Changes in archaeal, bacterial and eukaryal assemblages along a salinity gradient by comparison of genetic fingerprinting methods in a multipond solar saltern. Environ Microbiol 4:338–348

Daniels LL (1917) On the flora of Great Salt Lake. Am Nat 51:499–506
Daniels LL, Wais AC (1990) Ecophysiology of bacteriophage S5100 infecting *Halobacterium cutirubrum*. Appl Environ Microbiol 56:3605–3608
Daniels LL, Wais AC (1998) Virulence in phage populations infecting *Halobacterium curtirubrum*. FEMS Microbiol Ecol 25:129–134
Diez B, Antón J, Guixa-Boixereu N, Pedrós-Alió C, Rodríguez-Valera F (2000) Pulsed-field gel electrophoresis analysis of virus assemblages present in a hypersaline environment. Int Microbiol 3:159–164
Dyall-Smith M, Tang SL, Bath C (2003) Haloarchaeal viruses: how diverse are they? Res Microbiol 154:309–313
Dybvig K, Nowak JA, Sladek TL, Maniloff J (1985) Identification of an enveloped phage, mycoplasma virus L172, that contains a 14-kilobase single-stranded DNA genome. J Virol 53:384–390
Gropp F, Palm P, Zillig W (1989) Expression and regulation of *Halobacterium halobium* phage ΦH genes. Can J Microbiol 35:182–188
Gropp F, Grampp B, Stolt P, Palm P, Zillig W (1992) The immunity-conferring plasmid pΦHL from the *Halobacterium salinarium* phage ΦH: nucleotide sequence and transcription. Virology 190:45–54
Guixa-Boixareu N, Calderon-Paz JI, Heldal M, Bratbak G, Pedrós-Alió C (1996) Viral lysis and bacterivory as prokaryotic loss factors along a salinity gradient. Aquat Microb Ecol 11: 215–227
Happonen LJ, Redder P, Peng X, Reigstad LJ, Prangishvili D, Butcher SJ (2010) Familial relationships in hyperthermo- and acidophilic archaeal viruses. J Virol 84:4747–4754
Hendrix RW (2009) Jumbo bacteriophages. Curr Top Microbiol Immunol 328:229–240
Hendrix RW, Lawrence JG, Hatfull GF, Casjens S (2000) The origins and ongoing evolution of viruses. Trends Microbiol 8:504–508
Holmes ML, Pfeifer F, Dyall-Smith ML (1995) Analysis of the halobacterial plasmid pHK2 minimal replicon. Gene 153:117–121
Ilyina TV, Koonin EV (1992) Conserved sequence motifs in the initiator proteins for rolling circle DNA replication encoded by diverse replicons from eubacteria, eucaryotes and archaebacteria. Nucleic Acids Res 20:3279–3285
Jäälinoja HT, Roine E, Laurinmäki P, Kivelä HM, Bamford DH, Butcher SJ (2008) Structure and host-cell interaction of SH1, a membrane-containing, halophilic euryarchaeal virus. Proc Natl Acad Sci USA 105:8008–8013
Jaatinen ST, Happonen LJ, Laurinmäki P, Butcher SJ, Bamford DH (2008) Biochemical and structural characterisation of membrane-containing icosahedral dsDNA bacteriophages infecting thermophilic *Thermus thermophilus*. Virology 379:10–19
Jalasvuori M, Jaatinen ST, Laurinavičius S, Ahola-Iivarinen E, Kalkkinen N, Bamford DH, Bamford JK (2009) The closest relatives of icosahedral viruses of thermophilic bacteria are among viruses and plasmids of the halophilic archaea. J Virol 83:9388–9397
Kauri T, Ackermann H-W, Goel U, Kushner DJ (1991) A bacteriophage of a moderately halophilic bacterium. Arch Microbiol 156:435–438
Ken R, Hackett NR (1991) *Halobacterium halobium* strains lysogenic for phage ΦH contain a protein resembling coliphage repressors. J Bacteriol 173:955–960
Khayat R, Tang L, Larson ET, Lawrence CM, Young M, Johnson JE (2005) Structure of an archaeal virus capsid protein reveals a common ancestry to eukaryotic and bacterial viruses. Proc Natl Acad Sci USA 102:18944–18949
Kivelä HM, Roine E, Kukkaro P, Laurinavičius S, Somerharju P, Bamford DH (2006) Quantitative dissociation of archaeal virus SH1 reveals distinct capsid proteins and a lipid core. Virology 356:4–11
Klein R, Baranyi U, Rössler N, Greineder B, Scholz H, Witte A (2002) *Natrialba magadii* virus ΦCh1: first complete nucleotide sequence and functional organization of a virus infecting a haloalkaliphilic archaeon. Mol Microbiol 45:851–863

Krupovič M, Bamford DH (2008) Virus evolution: how far does the double β-barrel viral lineage extend? Nat Rev Microbiol 6:941–948

Krupovič M, Forterre P, Bamford DH (2010) Comparative analysis of the mosaic genomes of tailed archaeal viruses and proviruses suggests common themes for virion architecture and assembly with tailed viruses of bacteria. J Mol Biol 397:144–160

Kukkaro P, Bamford DH (2009) Virus-host interactions in environments with a wide range of ionic strengths. Environ Microbiol Rep 1:71–77

Kunin V, Copeland A, Lapidus A, Mavromatis K, Hugenholtz P (2008) A bioinformatician's guide to metagenomics. Microbiol Mol Biol Rev 72:557–578

Laurinmäki PA, Huiskonen JT, Bamford DH, Butcher SJ (2005) Membrane proteins modulate the bilayer curvature in the bacterial virus Bam35. Structure 13:1819–1828

Lubbers MW, Waterfield NR, Beresford TP, Le Page RW, Jarvis AW (1995) Sequencing and analysis of the prolate-headed lactococcal bacteriophage c2 genome and identification of the structural genes. Appl Environ Microbiol 61:4348–4356

Maniloff J, Kampo GJ, Dascher CC (1994) Sequence analysis of a unique temperature phage: mycoplasma virus L2. Gene 141:1–8

Maturrano L, Santos F, Rosselló-Mora R, Antón J (2006) Microbial diversity in Maras salterns, a hypersaline environment in the Peruvian Andes. Appl Environ Microbiol 72:3887–3895

Nandhagopal N, Simpson AA, Gurnon JR, Yan X, Baker TS, Graves MV, Van Etten JL, Rossmann MG (2002) The structure and evolution of the major capsid protein of a large, lipid-containing DNA virus. Proc Natl Acad Sci USA 99:14758–14763

Oren A (2002) Halophilic microorganisms and their environments. Kluwer Academic Publishers, Dordrecht

Oren A (2008) Microbial life at high salt concentrations: phylogenetic and metabolic diversity. Saline Systems 4:2

Oren A, Bratbak G, Heldal M (1997) Occurrence of virus-like particles in the Dead Sea. Extremophiles 1:143–149

Pagaling E, Haigh RD, Grant WD, Cowan DA, Jones BE, Ma Y, Ventosa A, Heaphy S (2007) Sequence analysis of an Archaeal virus isolated from a hypersaline lake in Inner Mongolia, China. BMC Genomics 8:410

Papke RT, Koenig JE, Rodríguez-Valera F, Doolittle WF (2004) Frequent recombination in a saltern population of *Halorubrum*. Science 306:1928–1929

Pauling C (1982) Bacteriophages of *Halobacterium halobium*: isolated from fermented fish sauce and primary characterization. Can J Microbiol 28:916–921

Pedrós-Alió C, Calderón-Paz JI, MacLean MH, Medina G, Marrasé C, Gasol JM, Guixa-Boixereu N (2000) The microbial food web along salinity gradients. FEMS Microbiol Ecol 32:143–155

Pfeifer F, Weidinger G, Goebel W (1981) Genetic variability in *Halobacterium halobium*. J Bacteriol 145:375–381

Pietilä MK, Roine E, Paulin L, Kalkkinen N, Bamford DH (2009) An ssDNA virus infecting archaea: a new lineage of viruses with a membrane envelope. Mol Microbiol 72:307–319

Pietilä MK, Laurinavicius S, Sund J, Roine E, Bamford DH (2010) The single-stranded DNA genome of novel archaeal virus *Halorubrum* pleomorphic virus 1 is enclosed in the envelope decorated with glycoprotein spikes. J Virol 84:788–798

Porter K, Dyall-Smith M (2006) The isolation and study of viruses of halophilic microorganisms. In: Rainey FA, Oren A (eds) Methods in microbiology, vol 35, Extremophiles. Elsevier/Academic, London, pp 681–702

Porter K, Dyall-Smith ML (2008) Transfection of haloarchaea by the DNAs of spindle and round haloviruses and the use of transposon mutagenesis to identify non-essential regions. Mol Microbiol 70:1236–1245

Porter K, Kukkaro P, Bamford JK, Bath C, Kivelä HM, Dyall-Smith ML, Bamford DH (2005) SH1: a novel, spherical halovirus isolated from an Australian hypersaline lake. Virology 335:22–33

Porter K, Russ BE, Dyall-Smith ML (2007) Virus-host interactions in salt lakes. Curr Opin Microbiol 10:418–424

Porter K, Russ BE, Yang J, Dyall-Smith ML (2008a) The transcription programme of the protein-primed halovirus SH1. Microbiology 154:3599–3608

Porter K, Russ BE, Thorburn AN, Dyall-Smith ML (2008b) Viruses infecting Euryarchaea. In: Mahy BWJ, van Regenmortel MHV (eds) Encyclopedia of virology, vol 5. Elsevier, Oxford, pp 411–423

Prangishvili D, Forterre P, Garrett RA (2006) Viruses of the Archaea: a unifying view. Nat Rev Microbiol 4:837–848

Rainey FA, Oren A (2006) Extremophile microorganisms and the methods to handle them. In: Rainey FA, Oren A (eds) Methods in microbiology, vol 35, Extremophiles. Elsevier/Academic, London, pp 1–25

Reiter WD, Zillig W, Palm P (1988) Archaebacterial viruses. Adv Virus Res 34:143–188

Rodriguez-Valera F, Martin-Cuadrado AB, Rodriguez-Brito B, Pašić L, Thingstad TF, Rohwer F, Mira A (2009) Explaining microbial population genomics through phage predation. Nat Rev Microbiol 7:828–836

Roine E, Kukkaro P, Paulin L, Laurinavičius S, Domanska A, Somerharju P, Bamford DH (2010) New, closely related haloarchaeal viral elements with different nucleic acid types. J Virol 84:3682–3689

Rössler N, Klein R, Scholz H, Witte A (2004) Inversion within the haloalkaliphilic virus ΦCh1 DNA results in differential expression of structural proteins. Mol Microbiol 52:413–426

Salas M (1984) A new mechanism for the initiation of replication of Φ29 and adenovirus DNA: priming by the terminal protein. Curr Top Microbiol Immunol 109:89–106

Santos F, Meyerdierks A, Peña A, Róssello-Mora R, Amann R, Antón J (2007) Metagenomic approach to the study of halophages: the environmental halophage 1. Environ Microbiol 9: 1711–1723

Santos F, Yarza P, Parro V, Briones C, Antón J (2010) The metavirome of a hypersaline environment. Environ Microbiol 12:2965–2976

Sapienza C, Rose MR, Doolittle WF (1982) High-frequency genomic rearrangements involving archaebacterial repeat sequence elements. Nature 299:182–185

Saren AM, Ravantti JJ, Benson SD, Burnett RM, Paulin L, Bamford DH, Bamford JK (2005) A snapshot of viral evolution from genome analysis of the tectiviridae family. J Mol Biol 350: 427–440

Schnabel H (1984) An immune strain of *Halobacterium halobium* carries the invertible L segment of phage ΦH as a plasmid. Proc Natl Acad Sci USA 81:1017–1020

Schnabel H, Schramm E, Schnabel R, Zillig W (1982a) Structural variability in the genome of phage ΦH of *Halobacterium halobium*. Mol Gen Genet 188:370–377

Schnabel H, Zillig W, Pfaffle M, Schnabel R, Michel H, Delius H (1982b) *Halobacterium halobium* phage ΦH. EMBO J 1:87–92

Schnabel H, Palm P, Dick K, Grampp B (1984) Sequence analysis of the insertion element ISH1.8 and of associated structural changes in the genome of phage ΦH of the archaebacterium *Halobacterium halobium*. EMBO J 3:1717–1722

Sime-Ngando T, Lucas S, Robin A, Tucker KP, Colombet J, Bettarel Y, Desmond E, Gribaldo S, Forterre P, Breitbart M, Prangishvili D (2011) Diversity of virus-host systems in hypersaline Lake Retba, Senegal. Environ Microbiol 13: no. doi:10.1111/j.1462-2920.2010.02323.x

Simsek M, DasSarma S, RajBhandary UL, Khorana HG (1982) A transposable element from *Halobacterium halobium* which inactivates the bacteriorhodopsin gene. Proc Natl Acad Sci USA 79:7268–7272

Stedman KM, Porter K, Dyall-Smith ML (2009) The isolation of viruses infecting Archaea. In: Wilhelm SW, Weinbauer MG, Suttle CA (eds) Manual of aquatic viral ecology. American Society of Limnology and Oceanography, Waco, TX, pp 57–64

Stolt P, Zillig W (1992) In vivo studies on the effects of immunity genes on early lytic transcription in the *Halobacterium salinarium* phage ΦH. Mol Gen Genet 235:197–204

Stolt P, Zillig W (1993a) In vivo and in vitro analysis of transcription of the L region from the *Halobacterium salinarium* phage ΦH: definition of a repressor-enhancing gene. Virology 195: 649–658

Stolt P, Zillig W (1993b) Antisense RNA mediates transcriptional processing in an archaebacterium, indicating a novel kind of RNase activity. Mol Microbiol 7:875–882

Strömsten NJ, Bamford DH, Bamford JK (2005) In vitro DNA packaging of PRD1: a common mechanism for internal-membrane viruses. J Mol Biol 348:617–629

Suttle CA (2007) Marine viruses – major players in the global ecosystem. Nat Rev Microbiol 5: 801–812

Tang SL, Nuttall S, Ngui K, Fisher C, Lopez P, Dyall-Smith M (2002) HF2: a double-stranded DNA tailed haloarchaeal virus with a mosaic genome. Mol Microbiol 44:283–296

Torsvik T, Dundas ID (1974) Bacteriophage of *Halobacterium salinarium*. Nature 248:680–681

Torsvik T, Dundas ID (1980) Persisting phage infection in *Halobacterium salinarium* str. 1. J Gen Virol 47:29–36

Tringe SG, Rubin EM (2005) Metagenomics: DNA sequencing of environmental samples. Nat Rev Genet 6:805–814

Ventosa A, Oren A (1996) *Halobacterium salinarum* nom. corrig., a name to replace *Halobacterium salinarium* (Elazari-Volcani) and to include *Halobacterium halobium* and *Halobacterium cutirubrum*. Int J Syst Bacteriol 46:361

Vogelsang-Wenke H, Oesterhelt D (1988) Isolation of a halobacterial phage with a fully cytosine-methylated genome. Mol Gen Genet 211:407–414

Wais AC, Kon M, MacDonald RE, Stollar BD (1975) Salt-dependent bacteriophage infecting *Halobacterium cutirubrum* and *H. halobium*. Nature 256:314–315

Walker JE, Saraste M, Runswick MJ, Gay NJ (1982) Distantly related sequences in the α- and β-subunits of ATP synthase, myosin, kinases and other ATP-requiring enzymes and a common nucleotide binding fold. EMBO J 1:945–951

Walsby AE (1980) A square bacterium. Nature 238:69–71

Weinbauer MG (2004) Ecology of prokaryotic viruses. FEMS Microbiol Rev 28:127–181

Wilkansky B (1936) Life in the Dead Sea. Nature 138:467

Witte A, Baranyi U, Klein R, Sulzner M, Luo C, Wanner G, Kruger DH, Lubitz W (1997) Characterization of *Natronobacterium magadii* phage ΦCh1, a unique archaeal phage containing DNA and RNA. Mol Microbiol 23:603–616

Wooley JC, Godzik A, Friedberg I (2010) A primer on metagenomics. PLoS Comput Biol 6: e1000667

Chapter 9
Haloviruses of Great Salt Lake: A Model for Understanding Viral Diversity

Bonnie K. Baxter, Mihnea R. Mangalea, Smaranda Willcox, Shereen Sabet, Miriam-Nirvana Nagoulat, and Jack D. Griffith

9.1 Introduction

9.1.1 The Setting

Great Salt Lake, Utah, USA, is a large body of water, segmented by a railroad causeway into two general systems. The south arm is a moderate saline bay (9–15% sodium chloride content), while the north arm is a hypersaline bay (24–30% sodium chloride). As a terminal lake in an elevated biome, the lake elevation (average 1,280 m) fluctuates as does the temperature, affecting the salt content at any given time (reviewed in Baxter et al. 2005).

The north arm, virtually cut off from fresh water supply save precipitation, is an appropriate model for hypersaline studies. It is an ecosystem of extremes beyond the salt content, as the lake elevation, desert environment, and saline waters also result in desiccation and high ultraviolet light penetration (Baxter et al. 2007). While studies on halophiles from such environments are numerous, we know much less about the viruses that also survive these multiple extremes and infect halophile hosts.

B.K. Baxter (✉)
Great Salt Lake Institute, Westminster College, 1840 S 1300 E Salt Lake City, UT, USA
e-mail: bbaxter@westminstercollege.edu

M.R. Mangalea • S. Willcox • J.D. Griffith
Department of Immunology and Microbiology, University of North Carolina at Chapel Hill, Chapel Hill, NC 27599, USA

S. Sabet • M.-N. Nagoulat
Department of Biology, La Sierra University, 4500 Riverwalk Pkwy, Riverside, CA 92515, USA

A. Ventosa et al. (eds.), *Halophiles and Hypersaline Environments*,
DOI 10.1007/978-3-642-20198-1_9,

9.1.2 *Halovirus Isolation Studies*

In this chapter, our use of the term "halovirus" is to imply a virus that infects a halophile, the host being either a bacterial or archaeal species. While in isolation studies the host may be known, in other diversity studies discussed below, the host is unknown.

With the goal of examining halovirus diversity, a number of various strategies have been employed by other investigators. While all of the following techniques have been successful to a certain extent in isolating and visualizing hypersaline viruses, there are limitations that have resulted in an incomplete analysis of the total halovirus morphotypes that may exist. Some techniques outlined here, in particular the cultivation methods, are given in more detail in a recent review (Porter and Dyall-Smith 2006).

9.1.2.1 Isolation by Cultivation

A number of haloviruses have been isolated from hypersaline lakes using traditional cultivation methodology. The advantage of an isolated phage/host pair is the ability to replicate the virus and produce particles in sufficient number for more in-depth study. From this type of work, we can assess similarities and differences of structure, genomics, and biochemistry. Some reviews have addressed the question of diversity primarily summarizing the work to date on various characterized strains (e.g., Dyall-Smith et al. 2003; Prangishvili et al. 2006).

However, there are caveats to working with laboratory strains as hosts. First, direct plating only results in lytic viruses, which means that temperate viruses go undetected and unisolated. Second, optimal lytic conditions for each halovirus can be variable and also may be different than the optimal culture conditions of the host (Porter and Dyall-Smith 2006). In addition, if cultivation of environmental microorganisms is as limited as estimated, then only a small fraction of host strains may be lab-friendly. The haloviruses that infect them, then, are uncultivatable as well. Also, as other authors have noted, virus isolations are biased by use of standard lab host strains that might grow well in culture but be underrepresented in the environment (Dyall-Smith et al. 2003).

9.1.2.2 Isolation of Genome by Pulsed-Field Gel Electrophoresis

Recently, the complete genome of EHP-1, an uncultivated environmental halovirus, was sequenced (Santos et al. 2007). The whole genome of this virus was cloned and subsequently sequenced from a 37-kb band detected via pulsed-field gel electrophoresis (PFGE) analysis of an environmental brine sample. This strategy seems to be promising, yet there are caveats here as well. In this case, the halovirus is known, but not its host. So, although valuable genome sequence information can still be utilized, both virus morphology and host-virus dynamics remain a mystery until the host is discovered. Furthermore, while some uncultured viruses may be discovered through the direct cloning of whole environmental genomes, there is no guarantee that the same procedure will be successful for all viruses. Other, earlier groups attempted but failed to clone and sequence complete viral genomes from the environment, presumably due to the presence of viral genes that are toxic to cells (Forest Rohwer, personal communication). Ultimately, diversity of haloviruses is therefore not well understood from such standard and nonstandard methods as those described above.

9.1.2.3 Isolation by Phase Partition

A more detailed physical and biochemical characterization of individual phage species present in the hypersaline water would necessitate steps that provide larger amounts of phage, concentrated and at least partially purified based on parameters such as size or buoyant density. The phase partition method of Per Åke Albertsson (1960, 1986) employs a solution of polyethylene glycol in moderate salt (0.5–1.0 M NaCl) to selectively precipitate phage from solution. Precipitation can be carried out at $1 \times g$ and thus large initial volumes can be processed for subsequent concentration and analysis. In the past, one of us (J.D.G.) used this procedure to concentrate M13 phage from 200 L of broth in which M13 infected *Escherichia coli* cells had been removed by slow speed centrifugation. This large-scale concentration can then be coupled with additional methods, including cesium chloride density banding, to further purify and separate different phage species.

9.1.3 Halovirus Community Studies

Population studies of hypersaline virioplankton have employed various methods, including filtration, PFGE, TEM, and metagenomic analyses. While such methods are limited in the type of biological information acquired about haloviruses, they nonetheless contribute to providing an overall view of what exists in the halovirus community.

9.1.3.1 Filtration

Large-scale filtration and concentration techniques, such as tangential flow filtration (TFF), have been employed to collect haloviruses from environmental water samples. The obvious advantage of TFF is the ability to simultaneously filter and concentrate relatively large volumes (e.g., 20 L), resulting in a high virus concentration. However, reports of TFF with hypersaline water samples without a prior pre-filtration step show high viral loss compared to other filtration methods, such as positive pressure filtration (Diez et al. 2000). While the authors of this study concluded that viral diversity was not compromised through their use of TFF, the reduction in virus number as a result of direct and exclusive TFF use could presumably also reduce representation of the halovirus community.

9.1.3.2 Pulsed-Field Gel Electrophoresis

Diez and co-workers (2000) reported observing fusiform viruses in their Spanish saltern samples. In addition, PFGE and DNA hybridization analyses of the viral community's nucleic acid showed that different viral assemblages existed in different hypersaline ponds at the same field site, and that there was less diversity than in marine or haloalkaliphile environments, with a maximum number of eight bands observed via PFGE (Diez et al. 2000). PFGE and DNA hybridization analyses of the moderately hypersaline Mono Lake virioplankton showed that viral genome sizes were similar to those in marine and estuarine environments with the most common-sized genome being ~35 kbp, and that approximately 27 bands could be discerned via PFGE (Jiang et al. 2004). Furthermore, there were temporal and spatial differences in the viral assemblages as some viral genomes were only present at certain depths and at certain times of the year (Jiang et al. 2004; Sabet et al. 2006).

9.1.3.3 Transmission Electron Microscopy

Water samples can also be ultracentrifuged first to pellet the hypersaline viruses before adsorption onto transmission electron microscopic (TEM) grids (e.g., Guixa-Boixareu et al. 1996; Oren et al. 1997). This method seems to be the most logical strategy to investigate halovirus diversity, as it is culture-independent and does not involve time-consuming filtration protocols.

Using this method, the first filamentous virus-like particles (VLPs) were reported from Spanish salterns, along with icosahedral, tailed, and lemon-shaped viruses; furthermore, the observation was made that the number of lemon-shaped haloviruses increased with salinity (Santos et al. 2007). To visualize VLPs adsorbed to the TEM supports, concentrations in the range of 10^8–10^9/ml are required and it has been argued that different viruses have different adsorptive and diffusive properties (Grieg Steward, personal communication; Wommack and Colwell 2000). Furthermore, some viruses, especially those that are not tailed capsids, might be too fragile for direct ultracentrifugation (Porter and Dyall-Smith 2006). Collectively, these limitations could result in an incomplete representation of the hypersaline virioplankton.

Halovirus populations have been successfully visualized from water samples by directly pelleting them onto TEM grids via the airfuge, a specialized ultracentrifuge. This method is ideal as it is cultivation-independent, does not require filtration, and does not depend on a high concentration of viruses for successful adsorption. This technique has recently yielded TEM images of very diverse halovirus morphotypes discovered in Lake Retba, Senegal (Sime-Ngando et al. 2010). According to the authors, 46% of the VLPs viewed by TEM were spindle-shaped (4% with a tail and 42% without a tail); 35% were spherical (7% icosahedral and 28% non-icosahedral); 13% were linear; only 1% were head-tailed (i.e., tailed capsids such as the *Siphoviridae* and *Myoviridae*); and, surprisingly, 5% of the total viruses from the environment were unidentified or of unknown shapes including hairpin-shaped particles, bacilliform particles with and without an appendage, chains of small globules, hook-shaped particles, tadpole-shaped particles, reed-shaped particles, and complex particles with branched filaments and spherical units. These exciting new forms indicate our current categories for haloviruses are too narrow.

Another study using TEM to evaluate the morphotype diversity in the stratified water column of Mono Lake showed that "exotic" shapes did not exist, only VLPs with hexagonal capsids, prolate capsids, and capsids with single-axis symmetry. There were three different viral assemblages in three different layers. The abundance of tailed viruses was highest in the oxic layer (epilimnion) and decreased with increasing depth. Large untailed capsids, >150 nm, were most numerous in the middle layer (metalimnion) and least numerous in the oxic layer. Furthermore, tail length increased with depth, with Podoviruses existing exclusively at the epilimnion and absent in the metalimnion and hypolimnion depths; Myoviruses being most abundant in the epilimnion, then in the metalimnion, and least abundant in the hypolimnion; and Siphoviruses existing in the epilimnion, but then increasing with depth being most numerous in the hypolimnion (Brum and Steward 2010).

9.1.3.4 Metagenomics

A metagenomics approach has only very recently been applied to hypersaline aquatic habitats. One group invented a metagenomics tool called Genome relative Abundance and Average Size, or GAAS, which delivers more accurate results when undertaking genome database searches (Angly et al. 2009). By analyzing ten different hypersaline viral metagenomes, this methodology revealed that the average genome length of haloviruses was between 51 and –263 kbp. In addition, the size range of hypersaline viral genomes was relatively narrow compared to the size range of hypersaline microbial genomes (Angly et al. 2009), indicating that there is no trend between viral and microbial genome lengths and that they vary independently of each other. Another study showed that, while dominant microbial and viral communities in low-, medium-, and high-salinity aquatic environments maintained stability over time, thereby confirming that a stable geochemistry leads to a stable biology, a "fine-grained" analysis of viral genotypes and microbial strains actually reflected temporal fluctuations in both the microbiome (metagenome of microbes) and virome (metagenome of viruses) from samples collected between 1 day and 1 year of each other from a San Diego, USA, saltern system (Rodriguez-Brito et al. 2010). These data are interpreted to support the "kill-the-winner" theory that explains how dominant microbial strains in an environment grow in abundance until a certain threshold is reached when viruses then act as predators and very quickly reduce the dominant, virus-sensitive strain allowing less dominant, virus-resistant strains to predominate (Wommack and Colwell 2000).

A metagenomics analysis of a Spanish saltern showed that 75–80% of the metavirome contained hypothetical proteins, many conserved, while there were some matches to known halovirus isolates, as well as some surprising hits to the San Diego, USA, salterns metavirome (Santos et al. 2010). Single-nucleotide polymorphisms (SNPs) existed along this metavirome, the majority of which were neutral, signifying the presence of variants. There was also a detection of high mutation rates that reflected high intra-species diversity, meaning that similar hosts are infected by closely related viruses. One especially interesting result of this metagenomics study was the observation that there were low numbers of integrases within the metavirome, implying that lysogeny may not be a prevalent form of infection for haloviruses (Santos et al. 2010). This result seems to agree with the observation made that cultured haloviruses typically cause chronic infections in their host, more so than strictly lytic or strictly lysogenic infections (Porter et al. 2007). In other words, chronic infection may be the preferred choice of haloviruses.

In conjunction with more standard methodologies, metagenomics is able to contribute to our understanding of halovirus ecology. Valuable insight is being provided by such analyses that allow us to know the halovirus community more generally; however, metagenomics is unable to address the question of morphology at this time.

9.1.4 *Halovirus Diversity*

So what do we know about halovirus diversity? If it is possible to look at viruses with a "macro" lens, we can look at the shape and structure of a virus and begin some level of classification. When we get down to the genome level, we may find that our classification was ineffective, but it is a place to start. Of course, the more comprehensive our approach in addressing a question, the better our answer will be. A true picture of diversity will come from coupling methods and from exploring new hypersaline systems.

From the published studies that utilize EM, at least three major halovirus morphotypes are known to exist in significant numbers and they are described below.

9.1.4.1 Fusiform Haloviruses

Spindle-shaped viruses have been observed exclusively in the Archaea, including methanogens, thermophiles, and halophiles (reviewed in Prangishvili et al. 2006).

In a number of studies of hypersaline ecosystems, fusiforms appear to be the dominant virus type (e.g., Guixa-Boixareu et al. 1996; Oren et al. 1997). Two well-characterized fusiform haloviruses are His1 and His2, both of which infect *Haloarcula hispanica*. These have been classified in the *Salterprovirus* genus (Bath et al., 2006). These virions have their lemon shape and size range (44 × 67–77 nm) in common, and they are sometimes seen with short tails, but they are composed of dissimilar structural proteins. Both viruses are lytic and have no integrase activity for integration into the genome of the host. They are distantly related genetically in that they have similar genome structure, linear dsDNA with terminal proteins, and related DNA polymerase genes; however, they do not show homology to the other fusiform genera. For this reason, Dyall-Smith and colleagues suggest that they should be classified in a distinct family (Dyall-Smith et al. 2003).

9.1.4.2 Head–Tail Haloviruses

The members of this broad category of haloviruses are characterized by icosahedral heads and a variety of tail structures. Families are grouped by their tail structures: *Siphoviridae* viruses have long, non-contractile tails, *Myoviridae* have contractile tails, and *Podoviridae* which have very short tails (Ackermann 2007). Several hypersaline-dwelling viruses make good models for this morphotype including ΦH (Gropp et al. 1992), ΦCh1 (Klein et al. 2002; Witte et al. 1997), HF1 and HF2 (Nuttall and Dyall-Smith 1993, 1995; Tang et al. 2002). HF1/2 have been especially well-described; both are lytic phage which can infect a range of genera including *Haloferax*, *Halobacterium*, *Haloarcula*, *Natrialba*, and *Halorubrum*.

9.1.4.3 Spherical Haloviruses

Icosahedral-shaped particles without tails comprise another group of observed haloviruses. The isolated SH1, a lytic virus which infects the genera *Haloarcula* and *Halorubrum*, has an internal lipid layer beneath the protein coat (Porter et al. 2005). Though similar in appearance, the SH1 linear genome has no relationship to that of another spherical archaeal virus SSIV (Prangishvili et al. 2006), which infects *Sulfulobus* (Porter et al. 2005). The structure of SH1 is well-described in a cryo-EM study which reveals several unique structural features, especially unusual capsid proteins and protruding spikes (Jäälinoja et al. 2008). This category of viruses is difficult to assess with TEM studies in that it may be representative of head–tail species that have lost their tails or of pleomorphic viruses that may assume a spherical shape (Porter et al. 2005; Dyall-Smith, personal communication).

9.2 Results

The hypersaline north arm of Great Salt Lake and its accompanying salt production ponds, provide a vast resource for halovirus diversity exploration. There are a number of microniches, including petroleum seep areas, freshwater spring inputs, and desiccated regions. Since the lake is in an elevated desert biome, the seasonal variation is extreme, and annual precipitation is also variable (Baxter et al. 2005). In fact, Bacteria/Archaea diversity along this north meander line varies by location, by season and by year (Litchfield and Baxter, unpublished). One would then anticipate that the halovirus diversity also fluctuates. Studies here should take both a temporal and a spatial approach.

9.2.1 *Cell and Viral Counts*

In sampling twice a season over a 3-year period (2007–2010), the cell counts representing bacteria and archaea were relatively stable, a range of 4–10 $\times$ 10^7 cells/ml in the spring/summer months, and a range of 1–3 $\times$ 10^6 cells/ml in fall/winter months.

In July of 2009, we counted the phage with epifluorescence microscopy (EFM) using a standard protocol that stains the viral nucleic acid to allow counting of the particles (Patel et al. 2007) (Table 9.1). EFM is currently the most widely used approach for estimating the total abundance of virus particles in aquatic systems (Suttle 2007). Unfixed water samples were processed within several hours after collection to prevent loss of viral counts (Wen et al. 2004). Counts of haloviruses indicated values at all four locations are in the range of 3–6 $\times$ 10^9 particles/ml. Cell counts were also done for site 4, which revealed 6.2 $\times$ 10^7 cells/ml. Therefore, the viral load on the ecosystem was observed to be approximately 100-fold greater than that of the cells in Great Salt Lake. We have not repeated the viral counts in fall/winter sampling as of yet; temporal data will be forthcoming.

Table 9.1 EFM counts of virus particles in Great Salt Lake

Site	Viruses/ml
1	6.20×10^9
2	3.75×10^9
3	3.19×10^9
4	4.59×10^9

Sites 1–3 are samples from salterns along the north arm of the lake, while sample 4 is from the north arm water column

9.2.2 *Purification of the Virus Community from Great Salt Lake Brine*

In June 2007, we began a project to examine the diversity of the lake's haloviruses. Our goal was to eliminate the bias of cultivatable hosts and to find a method that would produce an overall picture of the viruses present at the gross morphology level. We envisioned a procedure that would allow one to more quickly assess the population and to potentially isolate individual haloviruses for which the host was unknown. Unlike the airfuge study (Sime-Ngando et al. 2010), this would require no specialized equipment beyond an ordinary ultracentrifuge. The use of phase partition provided a means of processing large volumes of hypersaline brine.

Forty liters of hypersaline brine were collected from the north arm of the lake. Cells were removed by slow speed centrifugation, and the supernatant was diluted from approximately 5 M to 1 M NaCl with water and then mixed with polyethylene glycol 8000 to 10%. This slurry was left in the cold room to precipitate over several days and then the supernatant was siphoned off and the flocculent precipitate fraction was spun down at 10,000 × *g* for 10 min and finally resuspended in a small volume of a buffer of 1 M NaCl, 10 mM Tris, and 2 mM $MgCl_2$. No VLPs were present in the supernatant. This precipitate thus represented the entire viral community present in the brine.

For electron microscopy analysis, 10 μl of this concentrated halovirus suspension was adsorbed to a carbon-covered copper grid (glow-charged just before application) and then stained with a 2% uranyl acetate solution for 2–3 min. After letting the grid air-dry, it was analyzed in a Tecnai 12 instrument. The analysis revealed the presence of expected morphotypes as described above (Fig. 9.1), and in addition, large amounts of a morphotype which should constitute a new halovirus category, as described below.

9.2.2.1 Fusiform Viruses

Though not the most numerous of VLP forms in our Great Salt Lake samples, many lemon-shaped particles were observed (Fig. 9.2). These virus forms were

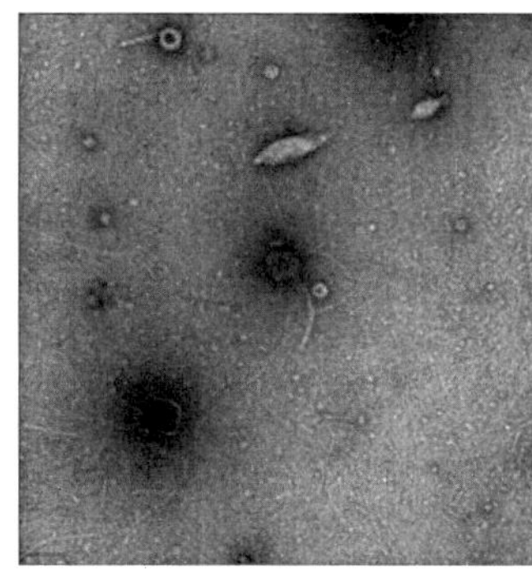

Fig. 9.1 TEM of precipitated VLPs from north arm Great Salt Lake brine showing various morphotypes. The bar measures 50 nm

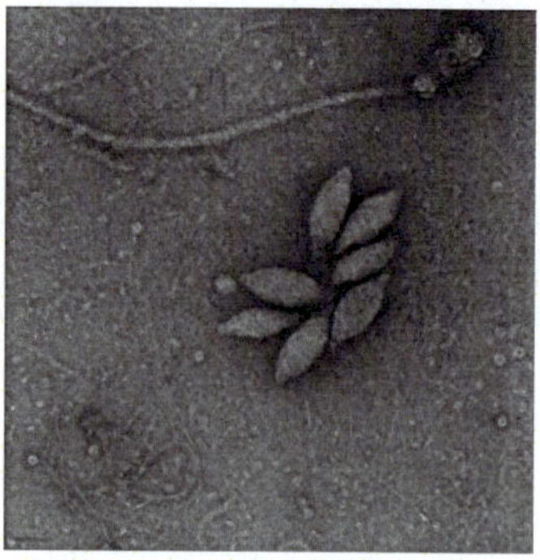

Fig. 9.2 Fusiform VLPs in Great Salt Lake samples. The bar measures 50 nm

~50–120 nm in length, appear to have a membrane enclosure, and have a small tail structure at one end. The size diversity may indicate the presence of more than a single species of fusiform halovirus present in GSL brine. However, it is important to note that purified His1 also showed a broad size distribution (Bath and Dyall-Smith 1998).

9.2.2.2 Head–Tail Viruses

Hypersaline VLPs exhibiting an icosahedral head and filamentous tail were the most numerous forms that we encountered in the TEM analysis. Perhaps this group also exhibited the most variety since the head size and tail size varied as well as the morphology of the components. Attached tails varied widely, from simple filaments ending in a point to stiff rods with a complex attachment structure at the end (Fig. 9.3). In some of the phage, the tail sheaths had contracted resulting in a thick segment closest to the head and then a long protruding thin filament extending from the thick segment (Fig. 9.3c). Some tails were long and sometimes curved (Fig. 9.3d). Examples of these forms are prevalent in the published literature, the best-described species being *E. coli* bacteriophage such as T4, which has an icosahedral head, stiff tail and complex attachment structure at its end, and phage lambda, whose tail is long, often curved and comes to a point.

Since these VLPs were so diverse, we measured the head structures observed on the micrographs to evaluate not only the range but also the distribution (Fig. 9.4). There is a clear cluster between 50 and 64 nm in head diameter. For comparison, the Senegal study reported head size ranges of 50–130 nm (Sime-Ngando et al. 2010).

9.2.2.3 Spherical Viruses

Spherical VLPs with icosahedral structure, electron dense centers, and no tail were observed throughout the Great Salt Lake samples (Fig. 9.5). These forms varied in size from ~40 to ~70 nm in diameter. Some others may have a short tail structure

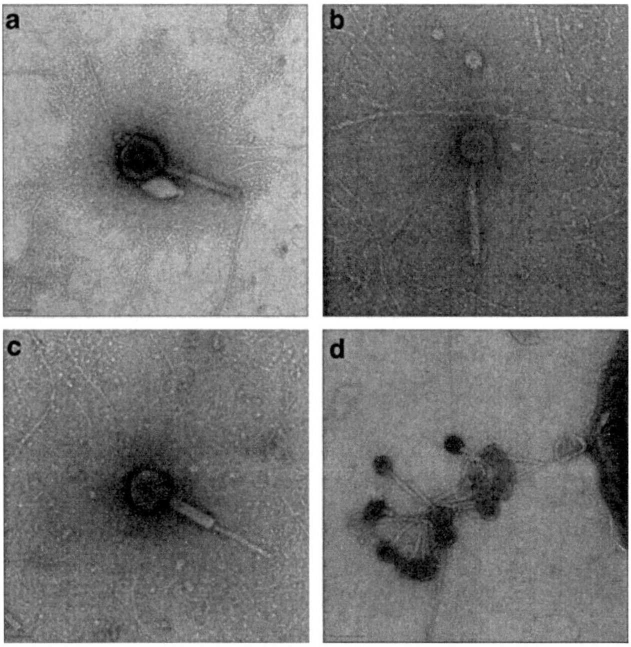

Fig. 9.3 A variety of Great Salt Lake head–tail VLP morphotoypes are observed, displaying variation in head size and tail length and structure. Panels (**a–c**) have a bar measuring 50 nm, and panel (**d**) has a bar measuring 100 nm

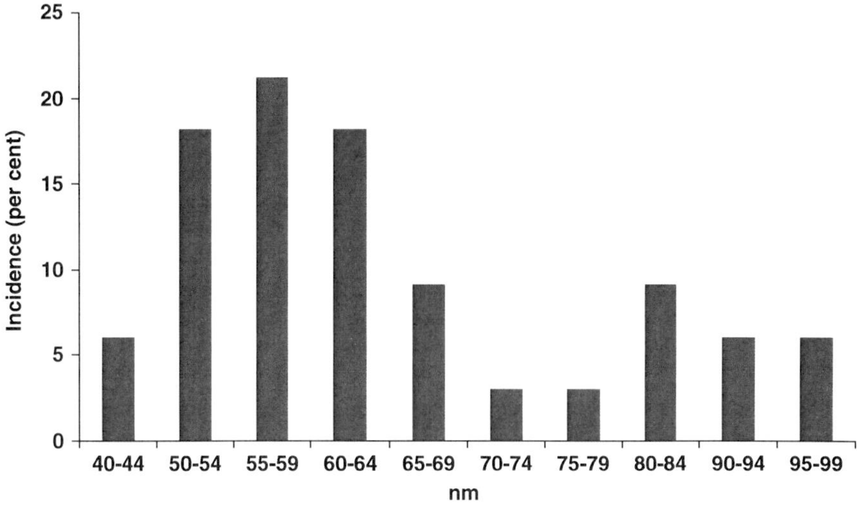

Fig. 9.4 Distribution of head width measurements in head–tail viruses from Great Salt Lake. The X-axis corresponds to head size while the Y-axis shows the incidence in percent of particles counted

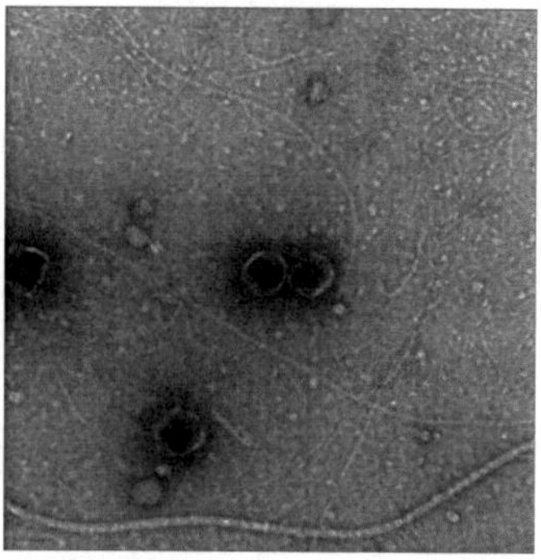

Fig. 9.5 Spherical VLPs from Great Salt Lake. The bar measures 50 nm

not easily visualized. These would mostly resemble coli phage T7. Smaller homogeneous-looking spherical particles were frequently observed in the background ranging from 20 to 40 nm in diameter. However, their small size, lack of a clear icosahedral shape, and range in sizes precluded a clear identification of them as phage in contrast to other biological structures that could have been present. Indeed, *E. coli* bacteriophage such as ΦX174 would have fallen into this category of scoring due to their small size and homogenous appearance in the EM.

9.2.2.4 A New Major Category of Haloviruses: Filamentous and Rod-Shaped Viruses

Though there are a few scattered reports of filamentous VLPs in hypersaline waters (e.g., Diez et al. 2000; Sime-Ngando et al. 2010), there has been too little data to create another category. In the negative stained images, we frequently observed many different filamentous forms. Some appeared very similar to bacterial fimbriae, pili, and flagella, being extremely stiff, variable in length, and having diameters in the range of 20 nm. In addition, however, there were also thinner, more flexible filaments and they frequently measured ~1 μm in length (Fig. 9.6a). These resemble both in dimensions and appearance, the well-known *E. coli* bacteriophages M13 and fd which are 5 nm in width and ~1 μm in length. Occasionally, wider, shorter filamentous particles with a clear central channel were observed (Fig. 9.6b) and these were strikingly similar to the contracted forms of M13 reported previously (Manning et al. 1981).

9.2.3 Coupled Fractionation and TEM

To further purify the viral precipitate, we designed a cesium chloride density gradient that would enable separation of species or morphotypes. We selected the

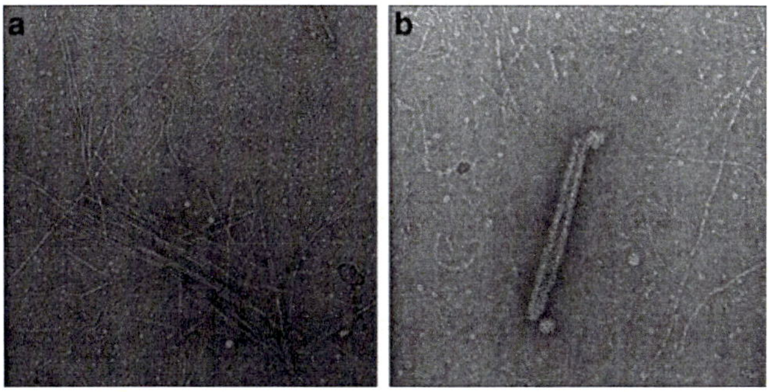

Fig. 9.6 Filamentous VLP from Great Salt Lake. Panel (**a**) has a bar measuring 50 nm, and panel (**b**) has a bar measuring 500 nm

Table 9.2 Some density values of isolated virus species used to standardize the cesium chloride gradient

Virus (host)	Density (g/ml)	Form	References
ΦH (haloarchaea)	1.72	Head–tail	Schnabel et al. (1982)
Ja.1 (haloarchaea)	1.55	Head–tail	Wais et al. (1975)
Lambda (*E. coli*)	1.50	Head–tail	Kaiser (1966)
ΦX-174 (*E. coli*)	1.40	Spherical	Vinograd and Hearst (1962)
T4 (*E. coli*)	1.5	Head–tail	Leibo and Mazur (1966)
M13 (*E. coli*)	1.29	Filament	Salivar et al. (1964)
His1 (haloarchaea)	1.28	Fusiform	Bath et al. (2006)
SSVI (thermoarchaea)	1.24	Spherical	Martin et al. (1984)

density range (1.2–1.6 g/ml) based on published viral density data, with a focus on haloviruses (Table 9.2).

The precipitate was resuspended in a buffer of 1 M NaCl, 10 mM Tris, 2 mM $MgCl_2$, and 1.35 g/ml cesium chloride to form a density gradient over 24 h ultracentrifugation at 60,000 rpm in a Beckman 70Ti rotor. The fractions formed by the cesium chloride gradient were collected at various densities as indicated in Fig. 9.7. This procedure yielded VLP-rich fractions that could be examined for structure, size, and diversity using TEM.

Analysis of VLP-rich fractions provided some intriguing results. Though there were no fractions that contained exclusively a single morphotype, there were fractions that showed significant enrichment.

For example, fraction 6 (Fig. 9.7), at a cesium chloride density of 1.28 g/ml, was enriched with fusiform VLPs. This is consistent with the density of purified haloviruses, His1, at 1.28 g/ml, and His 2, at 1.30 g/ml (see Table 9.2). However, the measurements of the Great Salt Lake fusiforms demonstrated that they were on average larger than the mean size for His1: Bath and colleagues report a 74-nm length for His1 (Bath et al. 2006) while the haloviruses in Great Salt Lake were larger, measuring 100–138 nm with a mean value of 118 nm. Another example, fractions 12–15 (Fig. 9.7), at a cesium chloride density of 1.33–1.38 g/ml, were

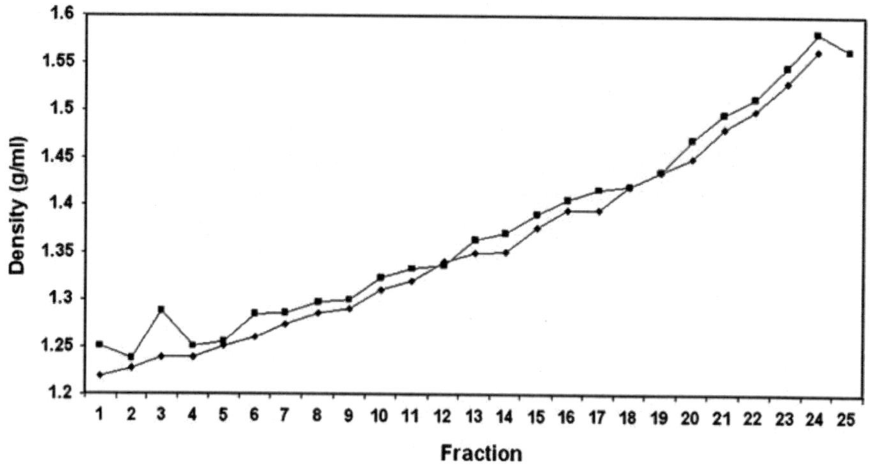

Fig. 9.7 Cesium chloride fractionation of Great Salt Lake VLP precipitate. The X-axis represents fraction number which is correlated by the density value on the Y-axis

enriched with head–tail phage, consistent with the density and size of the well-studied bacteriophages such as T4 (see Table 9.2).

Attempts at cultivation from the cesium chloride fractions with lab-isolated Great Salt Lake halophiles were unsuccessful, underscoring the fact that many host strains may be uncultivatable. However, this approach demonstrates the feasibility of large-scale phage purification from hypersaline brine and the possibility in the future of further purification of single phage species at a scale that would allow nucleic acid sequencing.

9.3 Discussion

9.3.1 Virus-Host Ratios

In hypersaline ecosytems like Great Salt Lake, it seems the predators (viruses) are in greater numbers than the prey (bacteria/archaea). We find a 100-fold difference in numbers, which is supported by other studies that show similar ratios, between 10 and 100-fold (Oren et al. 1997; Dyall-Smith et al. 2003; Jiang et al. 2004; Suttle 2007; Sime-Ngando et al. 2010). The literature suggests that as the number of bacteria/archaea increase or decrease along a salinity gradient, so do the viral counts (Guixa-Boixareu et al. 1996; Oren et al. 1997). The number of haloviruses relative to cells in the brine generally goes against the notion in ecology that predators are limited. That said, common laboratory bacteriophage have been found to be most stable in high salt solutions. Thus, these haloviruses may have a very long lifetime in the brine, and their population numbers may resist fluctuation.

Even if the Great Salt Lake virus-to-cell counts are representative of hypersaline systems, simple counts cannot reveal the actual ratio between a specific host/virus pair. Many viruses produce high particle to plaque-forming units (pfu) ratios, meaning that the majority of virions produced are non-infectious for one reason or another. Furthermore, some halophages have been shown to have a broad host range, infecting not just different species of the same genus, but even being able to infect different genera (e.g., Kauri et al. 1991; Nuttall and Dyall-Smith 1993; Porter et al. 2005). These factors and others may ultimately impact the observed high virus-to-host counts in brine.

9.3.2 Methodology

As we explore the various methods used to access viral load on a system, it seems clear that TEM is the only method that provides data on both abundance and morphology. This, combined with EfM data and fractionation by density gradient, can provide a broad overview of the morphotypes present and a way to enrich for these forms for future studies.

Most significant in our approach, fractionation of environmental samples can be used to enrich specific morphotypes. The assay presented here can be modified in that fractions can be sub-fractionated to increase the separation, for example. Though we did not succeed in isolating single species, perhaps this is because of the great diversity present, which suggests enrichment can be achieved for morphotypes, but not individual species, in an environment with such diversity.

9.3.3 A Call for New Categories of Haloviruses

Morphotypes may suggest similarities among haloviruses. As we obtain genomic data, however, we may find that shape means little. For example, His1 and His2 are fusiform, like SSV1 which infects a thermophile, *Sulfolobus*, but they do not share significant genetic homology with SSV1 (Porter et al. 2005). Nonetheless, it is a starting place to access diversity.

In the Senegal study (Sime-Ngando et al. 2010), 5% of the total VLPs are forms distinct from the categories of fusiform, spherical and head–tail shapes. The researchers observed hairpin-shaped, filamentous, chain-shaped, hook-shaped, tadpole-shaped, reed-shaped and other complex particles. Oren et al. (1997) reported various morphotypes including an unusual star-shaped morphology. Though these represent a minor component of the overall diversity, it suggests that we should broaden the halovirus morphotype categories. This suggestion comes with an understanding that until one such halovirus is isolated together with its host, we cannot truly distinguish a viral structure from a structure from another source (e.g., sub-cellular structures free in the brine).

Filamentous viruses may infect uncultivatable strains by and large, and so may be under-represented in the halophile literature. In Great Salt Lake, we see filamentous virus-like particles that appear distinct from pili or flagella. Pili and flagella have large diameters (~20 nm) and a clear helical repeat along their length as seen by EM. They are also very stiff. The classic filamentous phage on the other hand are thin, ~5 nm, and relatively flexible with a helical repeat that can be seen only under the most ideal conditions. They are also long, on the order of 1 μm. These features were present in the filamentous particles observed in the samples examined from the Great Salt Lake water. Linear or filamentous particles appear to be the predominant form of archaeal viruses in hyperthermophile communities (Rice et al. 2001; Rachel et al. 2002). However, according to the current literature, this morphotype appears to be missing from hypersaline communities save a few brief notations (e.g., Diez et al. 2000; Sime-Ngando et al. 2010).

In the history of the identification of *E. coli* phage, the filamentous phage (M13, fd) were some of the very last to be discovered using TEM. While T4 phage can be readily identified even with a TEM of minimal resolution and marginal staining methods, due to its large size and distinct appearance, the filamentous phage may be easily dismissed as background protein filaments and further, their very thin diameter requires higher resolution instruments and optimal negative staining methods. The one clear distinguishing characteristic is their uniform length (~1 μm). However, deletion mutants of M13 readily appear upon cultivation that generate shorter particles (Griffith and Kornberg 1974). Therefore, a clear proof that filamentous phage are present in these Great Salt Lake samples will ultimately require a high enough purification so that length determinations can be made and ideally single-stranded circular DNA purified from the phage. These studies are in progress.

Thus, there is strong evidence that the "filamentous" form should now be considered a new halovirus category. The diversity from a genetic and biochemical point of view within this category is yet to be determined and will depend on isolated haloviruses of this form. Findings like the ones presented here that underscore new categories point to the significance of coupled fractionation/TEM work.

References

Ackermann HW (2007) 5500 phages examined in the electron microscope. Arch Virol 152:227–243

Albertsson PA (1960, 1986) Partition of cell particles and macromolecules. Wiley, New York

Angly FE, Willner D, Prieto-Davo A, Edwards RA, Schmieder R, Vega-Thurber R, Antonopoulos DA, Barott K, Cottrell MT, Desnues C, Dinsdale EA, Furlan M, Haynes M, Henn MR, Hu Y, Kirchman DL, McDole T, McPherson JD, Meyer F, Miller RM, Mundt E, Naviaux RK, Rodriguez-Mueller B, Stevens R, Wegley L, Zhang L, Zhu B, Rohwer F (2009) The GAAS metagenomic tool and its estimations of viral and microbial average genome size in four major biomes. PLoS Comput Biol 5:e1000593

Bath C, Dyall-Smith ML (1998) His1, an archaeal virus of the Fuselloviridae family that infects *Haloarcula hispanica*. J Virol 72:9392–9395

Bath C, Cukalac T, Porter K, Dyall-Smith ML (2006) His1 and His2 are distantly related, spindle-shaped haloviruses belonging to the novel virus group, Salterprovirus. Virology 350:228–239

Baxter BK, Litchfield CD, Sowers K, Griffith JD, DasSarma PA, DasSarma S (2005) Great Salt Lake microbial diversity. In: Gunde-Cimerman N, Oren A, Plemenitaš A (eds) Adaptation to life in high salt concentrations in archaea bacteria, and eukarya. Springer, Dordrecht, pp 11–25

Baxter BK, Eddington B, Riddle MR, Webster TN, Avery BJ (2007) Great Salt Lake halophilic microorganisms as models for astrobiology: evidence for desiccation tolerance and ultraviolet radiation resistance. In: Hoover RB, Levin GV, Rozanov AY, Davies PCW (eds) Instruments, methods, and missions for astrobiology X, 6694. SPIE, Bellingham WA, document ID 669415

Brum JR, Steward GF (2010) Morphological characterization of viruses in the stratified water column of alkaline, hypersaline Mono Lake. Microb Ecol 60:636–643

Diez B, Antón J, Guixa-Boixereu N, Pedrós-Alió C, Rodríguez-Valera F (2000) Pulsed-field gel electrophoresis analysis of virus assemblages present in a hypersaline environment. Int Microbiol 3:159–164

Dyall-Smith M, Tang SL, Bath C (2003) Haloarchaeal viruses: how diverse are they? Res Microbiol 154:309–313

Griffith J, Kornberg A (1974) Mini M13 bacteriophage: circular fragments of M13 DNA are replicated and packaged during normal infections. Virology 59:139–152

Gropp F, Grampp B, Stolt P, Palm P, Zillig W (1992) The immunity-conferring plasmid p ΦHL from the *Halobacterium salinarium* phage ΦH: nucleotide sequence and transcription. Virology 190:45–54

Guixa-Boixareu N, Calderon-Paz JI, Heldal M, Bratbak G, Pedrós-Alió C (1996) Viral lysis and bacterivory as prokaryotic loss factors along a salinity gradient. Aquat Microb Ecol 11:215–227

Jäälinoja HT, Roine E, Laurinmäki P, Kivelä HM, Bamford DH, Butcher SJ (2008) Structure and host-cell interaction of SH1, a membrane-containing, halophilic euryarchaeal virus. Proc Natl Acad Sci USA 105:8008–8013

Jiang S, Steward G, Jellison R, Chu W, Choi S (2004) Abundance, distribution, and diversity of viruses in alkaline, hypersaline Mono Lake, California. Microb Ecol 47:9–17

Kaiser AD (1966) On the internal structure of bacteriophage lambda. J Gen Physiol 1:171–178

Kauri T, Ackerman HW, Goel U, Kushner DJ (1991) A bacteriophage of a moderately halophilic bacterium. Arch Microbiol 156:435–438

Klein R, Baranyi U, Rössler N, Greineder B, Scholz H, Witte A (2002) *Natrialba magadii* virus ΦCh1: first complete nucleotide sequence and functional organization of a virus infecting a haloalkaliphilic archaeon. Mol Microbiol 45:851–863

Leibo SP, Mazur P (1966) Effect of osmotic shock and low salt concentration on survival and density of bacteriophages T4B and $T4Bo_1$. Biophys J 6:747–772

Manning M, Chrysogelos S, Griffith J (1981) Mechanisms of phage M13 contraction: intermediate structures trapped at low temperature. Biophys J 37:28–31

Martin A, Yeats S, Janekovic D, Reiter WD, Aicher W, Zillig W (1984) SAV 1, a temperate U.V.-inducible DNA virus-like particle from the archaebacterium *Sulfolobus acidocaldarius* isolate B12. EMBO J 3:2165–2168

Nuttall SD, Dyall-Smith ML (1993) HF1 and HF2: novel bacteriophages of halophilic archaea. Virology 197:678–684

Nuttall SD, Dyall-Smith ML (1995) Haloviruses HF2: genome organization and replication strategy. J Virol 69:2322–2327

Oren A, Bratbak G, Heldal M (1997) Occurrence of virus-like particles in the Dead Sea. Extremophiles 1:143–149

Patel A, Noble RT, Steele JA, Schwalbach MS, Hewson I, Fuhrman JA (2007) Virus and prokaryote enumeration from planktonic aquatic environments by epifluorescence microscopy with SYBR Green I. Nat Protoc 2:269–276

Porter K, Dyall-Smith M (2006) The isolation and study of viruses of halophilic microorganisms. In: Rainey F, Oren A (eds) Methods in microbiology, vol 35, Extremophiles. Elsevier/Academic, London, pp 669–690

Porter K, Kukkaro P, Bamford JKH, Bath C, Kivelä HM, Dyall-Smith ML, Bamford DH (2005) SH1: a novel, spherical halovirus isolated from an Australian hypersaline lake. Virology 335:22–33

Porter K, Russ BE, Dyall-Smith ML (2007) Virus-host interactions in salt lakes. Curr Opin Microbiol 10:418–424

Prangishvili D, Forterre P, Garrett RA (2006) Viruses of the Archea: a unifying view. Nat Rev Microbiol 4:837–848

Rachel R, Bettstetter M, Hedlund BP, Häring M, Kessler A, Stetter KO, Prangishvili D (2002) Remarkable morphological diversity of viruses and virus-like particles in hot terrestrial environments. Arch Virol 147:2419–2429

Rice G, Stedman K, Snyder J, Wiedenfeft B, Willits D, Brumfield S, McDermott T, Young MJ (2001) Viruses from extreme thermal environments. Proc Natl Acad Sci USA 98:13341–13345

Rodriguez-Brito B, Li L, Wegley L, Furlan M, Angly F, Breitbart M, Buchanan J, Desnues C, Dinsdale E, Edwards R, Felts B, Haynes M, Liu H, Lipson D, Mahaffy J, Martin-Cuadrado AB, Mira A, Nulton J, Pašić L, Rayhawk S, Rodriguez-Mueller J, Rodriguez-Valera F, Salamon P, Srinagesh S, Thingstad TF, Tran T, Thurber RV, Willner D, Youle M, Rohwer F (2010) Viral and microbial community dynamics in four aquatic environments. ISME J 4:739–751

Sabet S, Chu W, Jiang SC (2006) Isolation and genetic analysis of haloalkaliphilic bacteriophages in a North American soda lake. Microb Ecol 51:543–554

Salivar WO, Tzagoloff H, Pratt D (1964) Some physical-chemical and biological properties of the rod-shaped coliphage M13. Virology 24:359–371

Santos F, Meyerdierks A, Peña A, Rosselló-Móra R, Amann R, Antón J (2007) Metagenomic approach to the study of haloviruses: the environmental halophage 1. Environ Microbiol 9:1711–1723

Santos F, Yarza P, Parro V, Briones C, Antón J (2010) The metavirome of a hypersaline environment. Environ Microbiol 12:2965–2976

Schnabel H, Zillig W, Pfäffle M, Schnabel R, Michel H, Delius H (1982) *Halobacterium halobium* phage ΦH. EMBO J 1:87–92

Sime-Ngando T, Lucas S, Robin A, Tucker KP, Colombet J, Bettarel Y, Desmond E, Gribaldo S, Forterre P, Breitbart M, Prangishvili D (2010) Diversity of virus-host systems in hypersaline Lake Retba, Senegal. Environ Microbiol (in press)

Suttle CA (2007) Marine viruses – major players in the global ecosystem. Nat Rev Microbiol 5:801–812

Tang SL, Nuttall S, Ngui K, Fisher C, Lopez P, Dyall-Smith M (2002) HF2: a double-stranded DNA tailed haloarchaeal virus with a mosaic genome. Mol Microbiol 44:283–296

Vinograd J, Hearst JE (1962) Equilibrium sedimentation of macromolecules and viruses in a density gradient. Fortschr Chem Org Naturst 20:372–422

Wais AC, Kon M, McDonald RE, Stollar BD (1975) Salt-dependent bacteriophage infecting *Halobacterium cutirubrum* and *Halobacterium halobium*. Nature 256:314–315

Wen K, Ortmann AC, Suttle CA (2004) Accurate estimation of viral abundance by epifluorescence microscopy. Appl Environ Microbiol 70:3862–3867

Witte A, Baranyi U, Klein R, Sulzner M, Luo C, Wanner G, Krüger DH, Lubitz W (1997) Characterization of *Natronobacterium magadii* phage ΦCh1, a unique archaeal phage containing DNA and RNA. Mol Microbiol 23:603–616

Wommack KE, Colwell RR (2000) Virioplankton: viruses in aquatic ecosystems. Microbiol Mol Biol Rev 64:69–114

Chapter 10
Initiation and Regulation of Translation in Halophilic Archaea

Jörg Soppa

10.1 Introduction

Translation is an important step in the multilevel process of the expression of a gene into a phenotype of a cell or organism. The process of translation consists of the phases of initiation, elongation, and termination. Initiation is the rate-limiting step and thus determines the rate of protein formation. Therefore, regulation of translation typically acts at the initiation step. Several mechanisms of translation initiation are known. The first mechanism depends on the so-called Shine Dalgarno (SD) motif, which has a consensus sequence of eight nucleotides and is localized a short distance upstream of open reading frames (Fig. 10.1a). The SD motif can base-pair with the 3′-end of the 16S rRNA and thereby place the small subunit of the ribosome onto the mRNA so that the start codon becomes localized in the P site. After recruitment of the charged initiator, tRNA and the large subunit of the ribosome translation can start. In bacteria, three initiation factors are involved in the process. This mechanism was long thought to be the default and with very few exceptions the only mechanism for bacteria (reviews: Marintchev and Wagner 2004; Laursen et al. 2005). It enables the existence of operons that are transcribed into polycistronic transcripts with several translation initiation sites. However, a bioinformatic analysis of bacterial and archaeal genomes revealed that only 50% of all genes are preceded by a SD motif, indicating that one or several additional mechanisms are likely to operate in prokaryotes (Chang et al. 2006).

The second mechanism is the so-called scanning mechanism and this is the default initiation mechanism of eukaryotes (Fig. 10.1b). It begins with the binding of initiation factor eIF4E to the cap structure at the 5′-end of a transcript. The small subunit is then recruited via protein–protein interactions, involving several additional initiation factors. The small subunit subsequently scans linearly along the

J. Soppa (✉)
Institute for Molecular Biosciences, Goethe-University, Biocentre, Max-von-Laue-Strasse 9, D-60438 Frankfurt, Germany
e-mail: soppa@bio.uni-frankfurt.de

A. Ventosa et al. (eds.), *Halophiles and Hypersaline Environments*,

DOI 10.1007/978-3-642-20198-1_10,

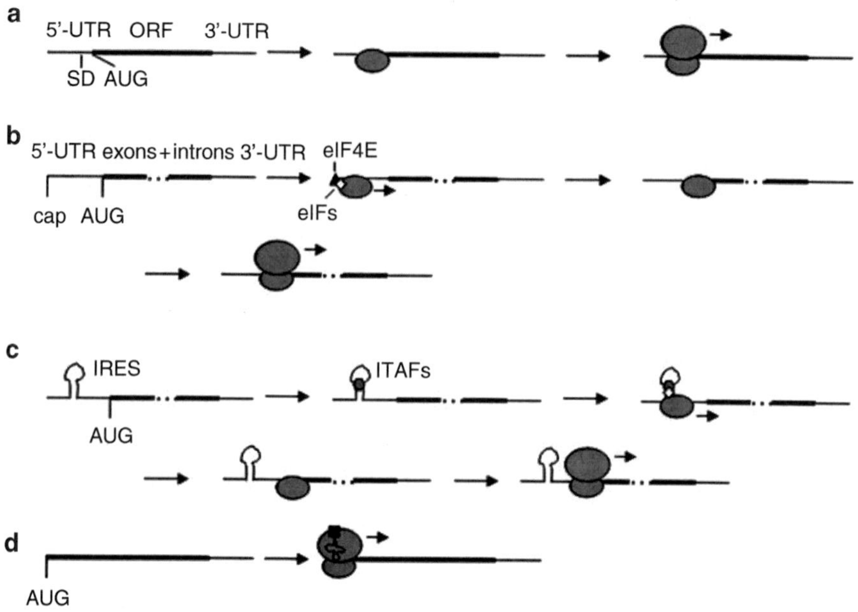

Fig. 10.1 Schematic overview of four different mechanisms of translation initiation. (**a**) The Shine Dalgarno (SD) motif-dependent mechanism in prokaryotes. (**b**) The scanning mechanism (default mechanism of most eukaryotes). (**c**) The IRES-dependent mechanism of eukaryotes. *IRES* internal ribosome entry site; *ITAF* IRES binding transacting factor. (**d**) The mechanism acting on leaderless transcripts, which is distributed in archaea, bacteria and eukaryotes

transcript until the first AUG codon is reached. Translation can begin after recruitment of the charged initiator tRNA and the large subunit (review: Kozak 2002).

A few years ago it was recognized that about 10% of eukaryotic transcripts can be efficiently translated under conditions when the scanning mechanism does not work. It is thought that in many or all of these cases so-called "*i*nternal *r*ibosome *e*ntry *s*ites" (IRES) are involved (Fig. 10.1c). IRES are folded structures within the 5′-UTRs (untranslated regions) of transcripts that can specifically be recognized by proteins called ITAFs (IRES trans-acting factors). The small subunit of the ribosome is then recruited to the transcript and can scan along the transcript until the start codon is reached. The remaining steps are identical to the scanning mechanism. The IRES mechanism is confined to eukaryotes, which have considerably larger 5′-UTRs than prokaryotes.

The fourth initiation mechanism acts on mRNAs lacking a 5′-UTR, which are called leaderless transcripts (Fig. 10.1d). The un-dissociated 70S (prokaryotes) or 80S (eukaryotes) ribosome and the charged initiator tRNA are required and form an initiation complex at the 5′-end of the transcript (O'Donnell and Janssen 2002; Udagawa et al. 2004; Andreev et al. 2006). The leaderless pathway is believed to be the evolutionary oldest mechanism, because leaderless transcripts exist in all three domains of life and leaderless transcripts can be translated in in vitro translation systems of all three domains of life (Moll et al. 2002). Furthermore, it was found to be common in several archaeal species (*Sulfolobus*: Tolstrup et al. 2000, *Pyrobaculum*: Slupska et al. 2001) and it is the default pathway in the lower eukaryote *Giardia lamblia* (Li and Wang 2004).

Figure 10.1 gives an overview of decisive steps of the four mechanisms. Several recent reviews give an insight into different aspects of translation initiation in archaea, bacteria and/or eukaryotes (Marintchev and Wagner 2004; Laursen et al. 2005; Londei 2005; Lopez-Lastra et al. 2005; Pisarev et al. 2005; Algire and Lorsch 2006; Allen and Frank 2007; Benelli and Londei 2009).

The current chapter gives an overview of initiation as well as regulation of translation in halophilic archaea, with a specific emphasis on *Haloferax volcanii*. *Hfx. volcanii* has been developed into an archaea model species, because many molecular genetic, biochemical, microbiological and cell biological techniques are available for working with this species (review: Soppa et al. 2008). The genome sequence is known (Hartman et al. 2010), a curated genome database is available (http://www.halolex.mpg.de; Pfeiffer et al. 2008), transcriptome, proteome and metabolome analyses have been established (e.g., Zaigler et al. 2003; Karadzic and Maupin-Furlow 2005; Sisignano et al. 2010), the construction of genomic mutants has become a routine procedure (Allers et al. 2004), a halo-tolerant GFP version can be applied (Reuter and Maupin-Furlow 2004), etc. Furthermore, many biological processes in addition to translation are under active investigation, including replication, DNA repair, transcriptional regulation, protein folding, protein transport, and posttranslational protein modification.

10.2 Initiation of Translation

10.2.1 Most Haloarchaeal Transcripts are Leaderless

The first 5′-ends of haloarchaeal transcripts were determined more than 25 years ago, and surprisingly it turned out that they were devoid of 5′-UTRs (e.g., Dunn et al. 1981; Betlach et al. 1984). At that time it was thought that SD motifs are essential for translation initiation and therefore it was hypothesized that SD motifs might be localized within the open reading frames. Subsequently, transcripts with 5′-UTRs were detected in halophilic and other archaea and the leaderless transcripts were considered to be exceptions, like in *Escherichia coli*.

The subject of leaderless archaeal transcripts was revitalized as sequencing of 144 *Sulfolobus solfataricus* genes revealed that about half of them were distal genes in operons and were preceded by a SD motif, while the other half were monocistronic or proximal genes in operons and were not preceded by a SD motif. It was therefore proposed that in *Sulfolobus* two different mechanisms for translation initiation operate on transcripts of these two groups of genes (Tolstrup et al. 2000). This view was experimentally corroborated by the finding that the 30S small ribosomal subunit interacts differently with leadered and with leaderless transcripts, respectively (Benelli et al. 2003). The 5′-ends of ten transcripts from *Pyrobaculum aerophilum* were mapped and all of them turned out to be leaderless (Slupska et al. 2001). A bioinformatic analysis of the genome indicated that this might be true for the majority of the *P. aerophilum* genes. Very recently, high throughput sequencing was applied to characterize the transcriptome of *S. solfataricus* and it was revealed that 69% of all transcripts of protein-encoding genes are leaderless (Wurtzel et al. 2010).

These studies indicated that leaderless transcripts are not seldom at least in crenarchaeota. To clarify whether this is also true for euryarchaeota, the 5′-ends of 40 transcripts from *Halobacterium salinarum* and *Haloferax volcanii* were determined (Brenneis et al. 2007). It was revealed that about two-third of the haloarchaeal transcripts are leaderless. A bioinformatic analysis of the more than 2,000 *Hbt. salinarum* genes that are either monocistronic or are proximal genes in operons revealed that most probably the majority of them is also leaderless, based on the distance between the basal promoter elements and the translational start codon (Brenneis et al. 2007). Therefore, the leaderless pathway is the default mechanism for translation initiation in haloarchaea. Using a reporter gene system that allowed quantifying the translational efficiency it was confirmed that leaderless transcripts are efficiently translated in vivo (Brenneis et al. 2007). This was in line with the earlier observation that the mutation of the 5′-UTR of transcript encoding a gas vesicle protein led to a decrease in translational efficiency, while the total removal of the 5′-UTR, which created an artificially leaderless transcript, led to a more than tenfold increase in translational efficiency (Sartorius-Neef and Pfeifer 2004).

However, these results obtained with two species of different haloarchaeal genera cannot be generalized to all euryarchaeota. Characterization of the *Methanosarcina mazei* transcriptome by high throughput sequencing of cDNA libraries revealed that the majority of transcripts have long 5′-UTRs (Jäger et al. 2009). Therefore, methanogenic and halophilic archaea differ in the major mechanisms employed for the initiation of translation.

10.2.2 A Novel Mechanism for Translation Initiation in Haloarchaea

It was a surprise to find that only very few haloarchaeal transcripts contain an SD motif and that most leadered transcripts are devoid of an SD motif (Brenneis et al. 2007). The small average size of the 5′-UTRs of about 20 nt and the lack of detectable structures using folding algorithms in silico excluded that an IRES type mechanism could operate on these transcripts. Therefore, only ribosomal scanning remained of the known mechanisms for initiation (Fig. 10.1). Haloarchaeal transcripts are not capped and haloarchaea are devoid of a homolog to the cap-binding protein eIF4E, but an analogous mechanism involving recognition of the 5′-terminal triphosphate could have evolved and for an initiation factor of

Sulfolobus it could indeed be shown that it specifically binds to the 5′-triphosphate of transcripts (Hasenöhrl et al. 2008). The scanning mechanism makes very clear predictions for the consequences of introducing different kinds of mutations (Kozak 2002). This allowed the design of a set of experiments specifically aimed at testing whether or not a scanning type mechanism operates on leadered SD-less haloarchaeal transcripts (Hering et al. 2009). The experiments included mutation of the start codon, introduction of additional in frame and out of frame start codons as well as introduction of stable structures at the 5′-end or near the start codon. The results unambiguously revealed that a scanning mechanism can be excluded and that a fifth and novel mechanism operates on these transcripts. It could also be shown that the mechanisms operating on leaderless and on leadered SD-less transcripts in haloarchaea are disparate, e.g., leaderless transcripts must contain less than five nucleotides upstream of the start codon to allow efficient translation, while the 5′-UTR of leadered transcripts must have a length of at least 20 nt. Another difference is the start codon selectivity, i.e., only AUG is used at leaderless transcripts, while AUG, GUG and UUG are possible for leadered transcripts. Surprisingly it was found that also 20 nt 5′-UTRs of random sequence are compatible with efficient translation initiation. The molecular mechanism of this novel pathway is currently unclear and is under investigation. The five mechanisms of translation initiation and their characteristic features are summarized in Table 10.1.

As already mentioned, many bacterial genes are not preceded by a SD motif. This is species-specific, e.g., in *Bacillus* more than 90% of all genes have an

Table 10.1 Overview of translation initiation mechanisms and characteristic features

Mechanism	Prevalence in haloarchaea	Occurrence in domain	Characteristic features
Leaderless	About 2/3	Archaea, Bacteria, Eukaryotes	70S/80S ribosome and initiator tRNA required; restricted to start codon AUG; insensitive to kasugamycin
Novel	About 1/3	Archaea (Bacteria?)	5′-UTR lacking SD motif and IRES; not inhibited by mutations that block scanning efficiency influenced by sequence of 5′-UTR
SD	Seldom	Archaea, Bacteria	Base pairing of SD motif with 3′-end of 16S rRNA; fixed distance between SD motif and start codon; enables the existence of polycistronic operons
Scanning	No	Eukaryotes	5′-Cap recognition and recruitment of 40S subunit; linear scanning of 5′-UTR until the first AUG
IRES	No	Eukaryotes	Complex structured RNA motif in 5′-UTR; recognition by specific proteins (ITAFs)

adjacent SD motif, in contrast to *Bacteroides*, where this is true for only about 15% of all genes (Chang et al. 2006). Similarly, a bioinformatic analysis of the *E. coli* genome had revealed the existence of a class of transcripts that did not fit the conventional model and it was suggested that they "presumably initiate differently from the majority of sites" (Shultzaberger et al. 2001). Therefore, the 5′-ends of 14 transcripts of *E. coli* genes not preceded by a SD motif were determined and it was found that all of them contain a SD-less 5′-UTR with an average length of 25 nt (Zerulla, Schweizer, and Soppa, unpublished data). Therefore, also in *E. coli* leadered SD-less transcripts are not a seldom exception, and it remains to be discovered whether the molecular mechanism of initiation is similar in bacteria and haloarchaea. The very different sets of translation initiation factors – only three in *E. coli* and more than ten in haloarchaea – could indicate that the mechanisms might turn out to be analogous rather than homologous.

10.3 Regulation of Translation

10.3.1 Genome-Wide Quantification of Translational Regulation

Until recently, the belief was widespread that translational regulation is typical for eukaryotes, in contrast to prokaryotes, which apply predominantly transcriptional regulation. The discovery of riboswitches, RNA thermometers in 5′-UTRs, and small regulatory RNAs (sRNAs) has somewhat changed the view, but even together these examples comprise not more than 1–3% of all genes of bacterial species (review: Liu and Camilli 2010). The fraction of genes under differential control of translational efficiency has not been determined yet for any bacterial species, but recently a genome-wide analysis of translational control has been performed with two haloarchaeal species, *Hbt. salinarum* and *Hfx. volcanii* (Lange et al. 2007). An overview of the experimental design is shown in Fig. 10.2. In short, cells were lysed very gently and cell debris was removed by centrifugation. The resulting cytoplasmic extract was applied to a sucrose density gradient, which enabled the isolation of free, un-translated RNA and polysome-bound RNA that was actively translated at the time of cell harvest. Both RNA populations were transformed into labeled cDNAs and compared using a DNA microarray. This enabled the identification of transcripts with translational efficiencies that was above or below average, respectively. The analysis was performed with exponentially growing cultures as well as with cultures in the stationary phase. Transcripts were denoted as translationally regulated when their relative translational efficiencies were different in both growth phases, e.g., they were under-translated in exponential phase and normally translated in stationary phase or vice versa. Growth phase-dependent translational regulation was observed for about 20% of *Hbt. salinarum* transcripts and 10% of *Hfx. volcanii* transcripts. These numbers are much higher than anticipated and

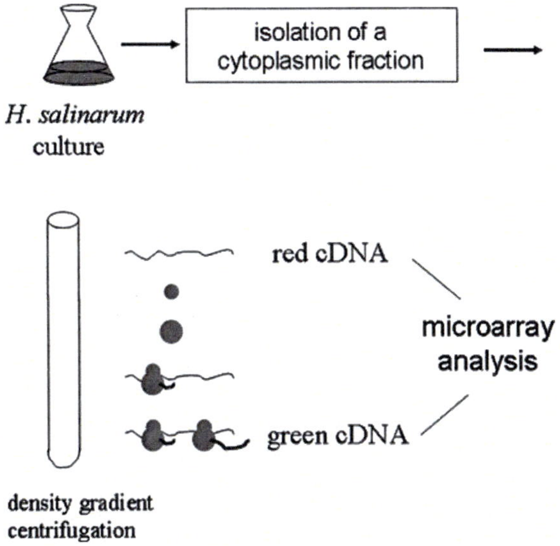

Fig. 10.2 Schematic overview of a translatome experiment. After fractionation of a cytoplasmic extract using density gradient centrifugation, RNA is isolated from the fraction containing free mRNAs and the polysomal fraction, labeled cDNA is generated and compared using a DNA microarray. For the characterization of translational regulation translatome analyses have to be performed with cultures grown under different conditions

reveal that translational control of gene expression is as important in haloarchaea as in eukaryotes. Global analyses of translational regulation have been performed with *Saccharomyces cerevisiae*, *Arabidopsis thaliana* and with human cell lines applying various conditions (e.g., Kuhn et al. 2001; Grolleau et al. 2002; Preiss et al. 2003; Branco-Price et al. 2005; Kawaguchi and Bailey-Serres 2005; Lü et al. 2006; Nicolai et al. 2006; Shenton et al. 2006). In these and further studies, translationally regulated eukaryotic genes were identified, and their fraction ranged from 1 to 25%, depending on species and conditions.

The high fraction of translationally controlled genes in species of two different haloarchaeal genera indicates that regulation of gene expression at the level of translation is older than anticipated until now and might well have been present in the common ancestor of archaea and eukaryotes. Analysis of further archaeal species and of several bacterial species is badly needed to unravel how widespread translational regulation is distributed in prokaryotes.

The analysis of growth phase-dependent translational control in *Hbt. salinarum* was paralleled with the analysis of growth phase-dependent transcriptional control. The levels of 17% of all transcripts were differentially regulated and thus this fraction is very similar to the fraction of translationally regulated genes. However, expression of only 70 genes was regulated on both levels (Lange et al. 2007). This is in contrast to yeast, in which a high fraction of homodirectional changes of transcript levels and translational efficiencies has been reported (Preiss et al. 2003; Halbeisen and Gerber 2009).

While the fraction of translationally regulated genes is high in *Hbt. salinarum* as well as in *Hfx. volcanii*, there is nearly no overlap in the identity of regulated genes in the two species. Also in eukaryotes the groups of genes under translational control vary widely in the three different species or under different conditions. It remains to be analyzed whether also in haloarchaea translational regulation is also operating under additional conditions, e.g., the application of various stresses, and whether the identity of regulated genes depends on the condition used, as has been found for eukaryotes.

10.3.2 Involvement of UTRs in Translational Regulation

The discovery that translational efficiencies of many transcripts are differentially regulated induced the aim to unravel the molecular mechanism. Two regulated transcripts were selected that are down-regulated in exponential phase and in stationary phase, respectively, to represent these two classes of genes. Six constructs were generated containing the *dhfr* (dihydrofolate reductase) reporter gene fused to either the 5′-UTR, or the 3′-UTR or both UTRs of the two translationally regulated genes. The translational efficiencies of the respective fusion transcripts were determined by quantification of the protein level using a DHFR activity test and of the transcript level using reverse transcription and Real Time PCR (qRT-PCR). It was revealed that in both cases the differential regulation of translational efficiencies could be transferred from the native transcript to the reporter transcript. However, both UTRs were necessary, the presence of 5′- or 3′-UTRs alone were not sufficient to induce translational regulation. This implies that the 5′- and the 3′-end of transcripts functionally interact in vivo, at least for transcripts that underlie translational regulation (Brenneis and Soppa 2009). The roles of 5′- and 3′-UTRs in translational regulation were investigated in further experiments. Stabilization of a predicted stem-loop structure in one of the 5′-UTRs by mutagenesis led to a total inhibition of translation, while destabilization of the predicted stem-loop structure increased translation under the non-induced condition. In both cases, translational control was lost, indicating that the role of 5′-UTRs in translational control is to down-regulate translational efficiency, which can be overcome by an element in the 3′-UTR. This hypothesis was strengthened by a UTR-swap experiment. The direction of regulation (induction in exponential versus stationary phase) remained the same when only the 5′-UTRs were exchanged; in contrast, exchange of only the 3′-UTRs led to a reversal of the direction of regulation, underscoring the importance of transcript circularization and implying that binding sites for regulatory factors are embedded in 3′-UTRs. The search for such factors, which might include regulatory proteins and/or regulatory RNAs, is currently under way.

The first results indicating circularization of eukaryotic transcripts in vivo have been published more than 10 years ago and today it is thought that it is a very general phenomenon in eukaryotes. This is based on results revealing that the polyA-binding protein (PABP) interacts with translation initiation factors, including (indirectly) the cap-binding initiation factor eIF4E (Gallie 1991; Wells et al.

1998; Svitkin et al. 2009). In this way, it can be guaranteed that translation is only initiated on transcripts that are complete, including the polyA-tail (Gallie 1998). It is also discussed that ribosomes can be re-used on the same transcript that they have just translated, enhancing the efficiency of translation. Last but not least, it has been described that this interaction is involved in translational regulation, because proteins competing with the initiation factor eIF4G for PABP binding can repress translational efficiency (e.g., Karim et al. 2006). As haloarchaea are devoid of a polyA tail at the 3′-end, the PABP, a cap at the 5′-end and its binding protein eIF4E, the mechanism of transcript circularization seems to be different in higher eukaryotes and haloarchaea. However, in both groups transcript circularization is involved in translational regulation, and it will be interesting whether this mechanism is also applied in the third domain of life, the bacteria.

10.3.3 Different Initiation Mechanisms Define "Translational Regulons"

It could be shown that haloarchaea employ three different initiation mechanisms and that the importance of translational regulation is higher than anticipated. Therefore, the question arose whether the three initiation mechanisms might be used differentially under various conditions, which would indicate that "translational regulons" exist in haloararchaea. To this end, several constructs were generated to test this hypothesis (Hering 2009). For example, the *sod* gene encoding superoxide dismutase (HVO_2913) was chosen, which is transcribed into an mRNA containing a 5′-UTR with a very good SD motif. A translational fusion with the *dhfr* reporter gene was produced, so that the fusion transcript contained two native AUG start codons. In addition, a third AUG start codon was introduced by mutation at the 5′-end, thus forming a leaderless transcript. All three start codons were in frame to the *dhfr* gene (Fig. 10.3a). Differential usage of the three potential start codons was analyzed throughout the growth curve using an anti-DHFR antibody. Figure 10.3b reveals differential usage of the start codons in exponential versus late stationary phase. Differential usage of the initiation mechanisms was also observed in cultures grown at different temperatures or salt concentrations and in cells exposed to stress conditions. Additional experiments with transcripts containing synthetic 5′-UTRs and two potential start sites exemplifying different initiation mechanisms ("leaderless" versus "novel," Table 10.1) underscored differential usage of the mechanisms under different conditions, indicating that translational regulons indeed exist in haloarchaea.

Similar to haloarchaea also eukaryotes seem to use differential usage of initiation mechanisms to define "translational regulons." Translatome analyses have shown that about 10% of all transcripts are still translated with high efficiency under conditions when cap-dependent translation is severely compromised, e.g., stress or apoptosis (Lopez-Lastra et al. 2005). It can be expected that these transcripts use IRES-dependent translation initiation at least under those conditions. As both haloarchaea and eukaryotes use mechanisms for translation initiation differentially under various

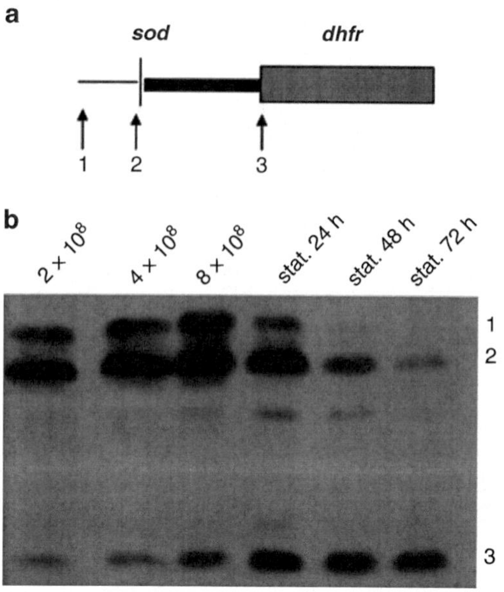

Fig. 10.3 Differential efficiencies of initiation mechanisms under various conditions. (**a**) Schematic representation of a transcript that is comprised of a translational fusion of part of the transcript of the *sod* gene including the 5′-UTR and the *dhfr* reporter gene. At the very 5′-end a third AUG start codon was generated by mutagenesis, leading to a leaderless transcript. All three AUG start codons (No. 1–3, as indicated) are in frame with the *dhfr* gene. (**b**) Western blot analysis with an anti-DHFR antibody of the three proteins translated from the fusion transcript shown in a. Aliquots were removed from a culture at different time points, the cell densities (cells ml^{-1}) and the time in stationary phase, respectively, are indicated. The numbers indicate the proteins started at the AUG start codons shown in (**a**)

conditions, it is tempting to speculate that such translational regulons also exists in bacteria. A very specific case has indeed been recently reported: the treatment of *E. coli* with kasugamycin induces the formation of specialized ribosomes (61S ribosomes) that can translate leaderless but not leadered transcripts (Kaberdina et al. 2009).

10.4 Probable Factors Involved in Regulation

10.4.1 sRNAs Exist in Haloarchaea and are Probably also Involved in Translational Regulation

The importance of small non-coding, regulatory RNAs for translational regulation in eukaryotes and in bacteria has become evident in recent years. In eukaryotes, it is estimated that thousands of small RNAs (miRNAs, siRNAs, piRNAs) exist that influence translational efficiencies of target transcripts either directly or via regulating transcript half lives. The essential roles of an increasing number of important physiological functions have been discovered (review: Liu and Paroo 2010). In bacteria, the roles of sRNAs have most intensively been studied in *E. coli*, although the characterization in many additional species is under way. Bacteria contain up to a few hundred sRNAs, which have been shown to be involved in stress response, but also in additional regulatory pathways (review: Liu and Camilli 2010). Differences between eukaryotic and bacterial sRNAs are e.g., (1) that eukaryotic sRNAs are much shorter and have a very defined size around 20 nt, while the sizes of bacterial sRNAs range from about 30 nt to several hundred nucleotides, and (2) that eukaryotic sRNAs typically bind to the 3′-UTRs of their target mRNAsm, while bacterial sRNAs typically bind to the 5′-UTRs of their targets.

The existence of sRNAs in archaea has been discovered a few years ago using bioinformatic approaches as well as experimental RNomics. All investigated species contained sRNAs, e.g., *Archaeoglobus fulgidus, Methanocaldococcus jannaschii, Methanosarcina mazei, Pyrococcus furiosus, and Sulfolobus solfataricus* (review: Schmitz-Streit et al. 2011). The numbers and size distributions resemble that of bacterial sRNAs, but neither targets nor biological functions have been unraveled as yet.

The existence of sRNAs in *Hfx. volcanii* has been investigated using in silico comparative genomics as well as several experimental approaches, including RNomics and High Throughput sequencing (Soppa et al. 2009; Straub et al. 2009; Babski et al. in press; Fischer et al. 2011). More than 200 sRNAs were discovered, ranging in size from about 30 nt to more than 200 nt. (Differential) expression of selected sRNA genes was verified using DNA microarray analysis, Northern blots, and reporter gene assays. Construction of deletion mutants and their phenotypic analysis revealed that haloarchaeal sRNAs play important roles not only in stress responses, but also in metabolic regulation and even in cellular behavior. A protein of the LSM protein family was shown to bind sRNAs in vivo and the severe and pleiotropic phenotype of a *lsm* deletion mutant indicates that the haloarchaeal LSM protein is important for the function of many sRNAs (Fischer et al. 2010), similar to homologs in eukaryotes and bacteria. As for the sRNAs of other archaea, targets of haloarchaeal sRNAs are not yet known, their molecular mode of action has not yet been investigated and it has not even been proven that they are involved in regulating translation. However, due to the universal role of regulating translational efficiencies in eukaryotes and bacteria, this can be assumed to be also true for (halo)archaeal sRNAs.

10.4.2 Translation Initiation Factors Do Not Seem to be Essential in Haloarchaea

Haloarchaea contain more than ten translation initiation factors, similar to eukaryotes and in stark contrast to bacteria, which harbor only three initiation factors (reviews: Londei 2005; Benelli and Londei 2009). It will be interesting to unravel whether some of these factors are specific for one of the three initiation mechanisms co-existing in haloarchaea and thereby can regulate the observed differential usage. To test this idea, conditional depletion mutants of all factors were generated (Schultz and Soppa, unpublished data). Surprisingly, all mutants were viable after factor depletion and only one of them revealed a growth defect. Therefore, currently deletion mutants are being generated. Several deletion mutants have been successfully constructed, indicating that at least several translation initiation factors in haloarchaea are either not essential or are redundantly encoded. The combination of deletion mutants with the reporter gene system described above, which includes three in frame AUG start codons, will allow unraveling the general or specific role of initiation factors for the different initiation mechanisms.

10.5 Conclusions and Outlook

It has been revealed that haloarchaea use three mechanisms for translation initiation simultaneously. Most transcripts are leaderless, about one-third belong to a novel group of leadered transcripts that are devoid of a SD motif, and only a few transcripts use the SD mechanism, which until recently was thought to be the default mechanism for prokaryotes. The three mechanisms are used with differential efficiency under various conditions and thus haloarchaea apply translational regulons for global regulation of gene expression. In two haloarchaeal species, it was shown that 20 and 10% of all transcripts, respectively, are subject to growth phase-dependent regulation of translational efficiencies. For selected genes it was shown that the 5′- and 3′-UTRs are sufficient to transfer the translational regulation to another, reporter transcript. The data imply that 5′- and 3′- end of transcripts have to functionally interact in vivo and transcript circularization is thus not confined to eukaryotes, but is also important in haloarchaea.

Current and future research will concentrate on the molecular details of the three initiation mechanisms and the pathways of translational control. Open issues are e.g., the characterization of cis-acting RNA motifs important for the novel initiation pathway, the presumed differential involvement of at least some translation initiation factors in the three initiation mechanisms, and the identity and role of regulatory factors binding to the 3′-UTRs of regulated genes. Furthermore, it is assumed but not yet proven that archaeal sRNAs are involved in translational regulation, like eukaryotic and bacterial sRNAs. To this end, target mRNAs of selected of the more than 200 haloarchaeal sRNAs have to be identified and it has to be revealed whether they typically bind to the 5′-end, like in bacteria, or to the 3′-UTR, like in eukaryotes. Last but certainly not least, results obtained with haloarchaea should induce new projects with bacteria, e.g., transcript circularization has been shown to occur in eukaryotes as well as in haloarchaea, but it is totally unclear whether bacteria make use of this mechanisms which holds potential for ensuring transcript integrity as well as translational regulation. The recent results showed that initiation and regulation of translation exhibit more variability than recognized a few years ago and it has been hypothesized that more variability can be anticipated and will be uncovered in the future (Malys and McCarthy 2011).

References

Algire MA, Lorsch JR (2006) Where to begin? The mechanism of translation initiation codon selection in eukaryotes. Curr Opin Chem Biol 10:480–486

Allen GS, Frank J (2007) Structural insights on the translation initiation complex: ghosts of a universal initiation complex. Mol Microbiol 63:941–950

Allers T, Ngo HP, Mevarech M, Lloyd RG (2004) Development of additional selectable markers for the halophilic archaeon *Haloferax volcanii* based on the *leuB* and *trpA* genes. Appl Environ Microbiol 70:943–953

Andreev DE, Terenin IM, Dunaevsky YE, Dmitriev SE, Shatsky IN (2006) A leaderless mRNA can bind to mammalian 80S ribosomes and direct polypeptide synthesis in the absence of translation initiation factors. Mol Cell Biol 26:3164–3169

Babski J, Tjaden B, Voss B, Jellen-Ritter A, Marchfelder A, Hess WR, Soppa J (2011) Bioinformatic prediction and experimental verification of sRNAs in the haloarchaeon *Haloferax volcanii*. RNA Biol (in press)

Benelli D, Londei P (2009) Begin at the beginning: evolution of translational initiation. Res Microbiol 160:493–501

Benelli D, Maone E, Londei P (2003) Two different mechanisms for ribosome/mRNA interaction in archaeal translation initiation. Mol Microbiol 50:635–643

Betlach M, Friedman J, Boyer HW, Pfeifer F (1984) Characterization of a halobacterial gene affecting bacterio-opsin gene expression. Nucleic Acids Res 12:7949–7959

Branco-Price C, Kawaguchi R, Ferreira RB, Bailey-Serres J (2005) Genome-wide analysis of transcript abundance and translation in *Arabidopsis* seedlings subjected to oxygen deprivation. Ann Bot 96:649–660

Brenneis M, Soppa J (2009) Regulation of translation in haloarchaea: 5′- and 3′-UTRs are essential and have to functionally interact in vivo. PLoS ONE 4:e4484

Brenneis M, Hering O, Lange C, Soppa J (2007) Experimental characterization of cis-acting elements important for translation and transcription in halophilic archaea. PLoS Genet 3:e229

Chang B, Halgamuge S, Tang SL (2006) Analysis of SD sequences in completed microbial genomes: non-SD-led genes are as common as SD-led genes. Gene 373:90–99

Dunn R, McCoy J, Simsek M, Majumdar A, Chang SH, Rajbhandary UL, Khorana HG (1981) The bacteriorhodopsin gene. Proc Natl Acad Sci USA 78:6744–6748

Fischer S, Benz J, Spath B, Maier LK, Straub J, Granzow M, Raabe M, Urlaub H, Hoffmann J, Brutschy B, Allers T, Soppa J, Marchfelder A (2010) The archaeal Lsm protein binds to small RNAs. J Biol Chem 285:34429–34438

Fischer S, Benz J, Späth B, Jellen-Ritter A, Heyer R, Dörr M, Maier LK, Menzel-Hobeck C, Lehr M, Jantzer K, Babski J, Soppa J, Marchfelder A (2011) Regulatory RNAs in *Haloferax volcanii*. Biochem Soc Trans 39:159–162

Gallie DR (1991) The cap and poly(A) tail function synergistically to regulate mRNA translational efficiency. Genes Dev 5:2108–2116

Gallie DR (1998) A tale of two termini: a functional interaction between the termini of an mRNA is a prerequisite for efficient translation initiation. Gene 216:1–11

Grolleau A, Bowman J, Pradet-Balade B, Puravs E, Hanash S, Garcia-Sanz JA, Beretta L (2002) Global and specific translational control by rapamycin in T cells uncovered by microarrays and proteomics. J Biol Chem 277:22175–22184

Halbeisen RE, Gerber AP (2009) Stress-dependent coordination of transcriptome and translatome in yeast. PLoS Biol 7:e105

Hartman AL, Norais C, Badger JH, Delmas S, Haldenby S, Madupu R, Robinson J, Khouri H, Ren Q, Lowe TM, Maupin-Furlow J, Pohlschröder M, Daniels C, Pfeiffer F, Allers T, Eisen JA (2010) The complete genome sequence of *Haloferax volcanii* DS2, a model archaeon. PLoS One 5:e9605

Hasenöhrl D, Lombo T, Kaberdina V, Londei P, Bläsi U (2008) Translation initiation factor a/eIF3 (−gamma) counteracts 5′ to 3′ mRNA decay in the archaeon *Sulfolobus solfataricus*. Proc Natl Acad Sci USA 105:2146–2150

Hering O (2009) Initiation der Translation in halophilen Archaea. PhD thesis, Goethe-University, Frankfurt, Germany

Hering O, Brenneis M, Beer J, Suess B, Soppa J (2009) A novel mechanism for translation initiation operates in haloarchaea. Mol Microbiol 71:1451–1463

Jäger D, Sharma CM, Thomsen J, Ehlers C, Vogel J, Schmitz RA (2009) Deep sequencing analysis of the *Methanosarcina mazei* Gö1 transcriptome in response to nitrogen availability. Proc Natl Acad Sci USA 106:21878–21882

Kaberdina AC, Szaflarski W, Nierhaus KH, Moll I (2009) An unexpected type of ribosomes induced by kasugamycin: a look into ancestral times of protein synthesis? Mol Cell 33:227–236

Karadzic IM, Maupin-Furlow JA (2005) Improvement of two-dimensional gene electrophoresis proteome maps of the haloarchaeon *Haloferax volcanii*. Proteomics 5:354–359

Karim MM, Svitkin YV, Kahvejian A, Crescenzo G, Costa-Mattioli M, Sonenberg N (2006) A mechanism of translational repression by competition of Paip2 with e'IF4G for poly(A) binding protein (PABP) binding. Proc Natl Acad Sci USA 103:9494–9499

Kawaguchi R, Bailey-Serres J (2005) mRNA sequence features that contribute to translational regulation in *Arabidopsis*. Nucleic Acids Res 33:955–965

Kozak M (2002) Pushing the limits of the scanning mechanism for initiation of translation. Gene 299:1–34

Kuhn KM, DeRisi JL, Brown PO, Sarnow P (2001) Global and specific translational regulation in the genomic response of *Saccharomyces cerevisiae* to a rapid transfer from a fermentable to a nonfermentable carbon source. Mol Cell Biol 21:916–927

Lange C, Zaigler A, Hammelmann M, Twellmeyer J, Raddatz G, Schuster S, Oesterhelt D, Soppa J (2007) Genome-wide analysis of growth phase-dependent translational and transcriptional regulation in halophilic archaea. BMC Genomics 8:415

Laursen BS, Sørensen HP, Mortensen KK, Sperling-Petersen HU (2005) Initiation of protein synthesis in bacteria. Microbiol Mol Biol Rev 69:101–123

Li L, Wang CC (2004) Capped mRNA with a single nucleotide leader is optimally translated in a primitive eukaryote, *Giardia lamblia*. J Biol Chem 279:14656–14664

Liu JM, Camilli A (2010) A broadening world of bacterial small RNAs. Curr Opin Microbiol 13:18–23

Liu Q, Paroo Z (2010) Biochemical principles of small RNA pathways. Annu Rev Biochem 79:295–319

Londei P (2005) Evolution of translational initiation: new insights from the archaea. FEMS Microbiol Rev 29:185–200

Lopez-Lastra M, Rivas A, Barria MI (2005) Protein synthesis in eukaryotes: the growing biological relevance of cap-independent translation initiation. Biol Res 38:121–146

Lü X, de la Peña L, Barker C, Camphausen K, Tofilon PJ (2006) Radiation-induced changes in gene expression involve recruitment of existing messenger RNAs to and away from polysomes. Cancer Res 66:1052–1061

Malys N, McCarthy JE (2011) Translation initiation: variations in the mechanism can be anticipated. Cell Mol Life Sci 68:991–1003

Marintchev A, Wagner G (2004) Translation initiation: structures, mechanisms and evolution. Q Rev Biophys 37:197–284

Moll I, Grill S, Gualerzi CO, Blasi U (2002) Leaderless mRNAs in bacteria: surprises in ribosomal recruitment and translational control. Mol Microbiol 43:239–246

Nicolai M, Roncato MA, Canoy AS, Rouquie D, Sarda X, Freyssinet G, Robaglia C (2006) Large-scale analysis of mRNA translation states during sucrose starvation in *Arabidopsis* cells identifies cell proliferation and chromatin structure as targets of translational control. Plant Physiol 141:663–673

O'Donnell SM, Janssen GR (2002) Leaderless mRNAs bind 70S ribosomes more strongly than 30S ribosomal subunits in *Escherichia coli*. J Bacteriol 184:6730–6733

Pfeiffer F, Broicher A, Gillich T, Klee K, Mejia J, Rampp M, Oesterhelt D (2008) Genome information management and integrated data analysis with HaloLex. Arch Microbiol 190: 281–299

Pisarev AV, Shirokikh NE, Hellen CU (2005) Translation initiation by factor-independent binding of eukaryotic ribosomes to internal ribosomal entry sites. C R Biol 328:589–605

Preiss T, Baron-Benhamou J, Ansorge W, Hentze MW (2003) Homodirectional changes in transcriptome composition and mRNA translation induced by rapamycin and heat shock. Nat Struct Biol 10:1039–1047

Reuter CJ, Maupin-Furlow JA (2004) Analysis of proteasome-dependent proteolysis in *Haloferax volcanii* cells, using short-lived green fluorescent proteins. Appl Environ Microbiol 70: 7530–7538

Sartorius-Neef S, Pfeifer F (2004) *In vivo* studies on putative Shine-Dalgarno sequences of the halophilic archaeon *Halobacterium salinarum*. Mol Microbiol 51:579–588

Schmitz-Streit R, Jäger D, Jellen-Ritter A, Babski J, Soppa J, Marchfelder A (2011) Archaea employ small RNAs as regulators. In: Marchfelder A, Hess WR (eds) Regulatory RNAs in prokaryotes. Springer, Heidelberg

Shenton D, Smirnova JB, Selley JN, Carroll K, Hubbard SJ, Pavitt GD, Ashe MP, Grant CM (2006) Global translational responses to oxidative stress impact upon multiple levels of protein synthesis. J Biol Chem 281:29011–29021

Shultzaberger RK, Bucheimer RE, Rudd KE, Schneider TD (2001) Anatomy of *Escherichia coli* ribosome binding sites. J Mol Biol 313:215–228

Sisignano M, Morbitzer D, Gätgens J, Oldiges M, Soppa J (2010) A 2-oxoacid dehydrogenase complex of *Haloferax volcanii* is essential for growth on isoleucine but not on other branched chain amino acids. Microbiology 156:521–529

Slupska MM, King AG, Fitz-Gibbon S, Besemer J, Borodovsky M, Miller JH (2001) Leaderless transcripts of the crenarchaeal hyperthermophile *Pyrobaculum aerophilum*. J Mol Biol 309: 347–360

Soppa J, Baumann A, Brenneis M, Dambeck M, Hering O, Lange C (2008) Genomics and functional genomics with haloarchaea. Arch Microbiol 190:197–215

Soppa J, Straub J, Brenneis M, Jellen-Ritter A, Heyer R, Fischer S, Granzow M, Voss B, Hess WR, Tjaden B, Marchfelder A (2009) Small RNAs of the halophilic archaeon *Haloferax volcanii*. Biochem Soc Trans 37:133–136

Straub J, Brenneis M, Jellen-Ritter A, Heyer R, Soppa J, Marchfelder A (2009) Small RNAs in haloarchaea: Identification, differential expression and biological function. RNA Biol 6:281–292

Svitkin YV, Evdokimova VM, Brasey A, Pestova TV, Fantus D, Yanagiya A, Imataka H, Skabkin MA, Ovchinnikov LP, Merrick WC, Sonenberg N (2009) General RNA-binding proteins have a function in poly(A)-binding protein-dependent translation. EMBO J 28:58–68

Tolstrup N, Sensen CW, Garrett RA, Clausen IG (2000) Two different and highly organized mechanisms of translation initiation in the archaeon *Sulfolobus solfataricus*. Extremophiles 4: 175–179

Udagawa T, Shimizu Y, Ueda T (2004) Evidence for the translation initiation of leaderless mRNAs by the intact 70 S ribosome without its dissociation into subunits in eubacteria. J Biol Chem 279:8539–8546

Wells SE, Hillner PE, Vale RD, Sachs AB (1998) Circularization of mRNA by eukaryotic translation initiation factors. Mol Cell 2:135–140

Wurtzel O, Sapra R, Chen F, Zhu Y, Simmons BA, Sorek R (2010) A single-base resolution map of an archaeal transcriptome. Genome Res 20:133–141

Zaigler A, Schuster SC, Soppa J (2003) Construction and usage of a onefold-coverage shotgun DNA microarray to characterize the metabolism of the archaeon *Haloferax volcanii*. Mol Microbiol 48:1089–1105

Chapter 11
Protein Transport Into and Across Haloarchaeal Cytoplasmic Membranes

Kieran Dilks, Maria Ines Gimenez, Manuela Tripepi, and Mechthild Pohlschröder

11.1 Introduction

Like other prokaryotic cytoplasmic membranes, haloarchaeal cytoplasmic membranes are decorated with a large number of distinct proteins that are involved in processes that range from energy conversion to signal transduction and substrate trafficking into or out of the cytoplasm. In addition to protein insertion into the cell membrane, protein transport pathways are used by cells to secrete a wide array of proteins across cytoplasmic membranes (Yuan et al. 2010). Secreted haloarchaeal substrates comprise a wide variety of proteins, including halocins (a family of protein toxins that kill either closely or distantly related archaea), polymer-degrading enzymes of possible commercial interest (such as laccases, amylases and proteases) and substrate-binding proteins (Kobayashi et al. 1994; O'Connor and Shand 2002; Gimenez et al. 2007; De Castro et al. 2008; Uthandi et al. 2010). In addition, surface structure subunits such as pilins, flagellins, and the S-layer glycoprotein must be transported across the hydrophobic membranes prior to incorporation into extracytoplasmic structures (Alam and Oesterhelt 1984; Sumper et al. 1990; Szabo et al. 2007; Tripepi et al. 2010).

While protein transport has been studied extensively in bacteria, related research in archaea had been lagging. However, in recent years, a clearer understanding of archaeal protein transport has emerged through studies on model archaea, most notably *Haloferax volcanii* (Pohlschröder et al. 2005a; Calo and Eichler 2010). These studies have revealed that while some aspects of protein transport are universally conserved, others are apparently unique to their respective prokaryotic domain. Moreover, protein transport in halophilic archaea (haloarchaea) significantly

K. Dilks • M. Tripepi • M. Pohlschröder (✉)
Department of Biology, University of Pennsylvania, Philadelphia, PA 19104–6018, USA
e-mail: pohlschr@sas.upenn.edu

M.I. Gimenez
Instituto de Investigaciones Biológicas, Facultad de Ciencias Exactas y Naturales, Universidad Nacional de Mar del Plata, Funes 3250 4 degrees Nivel, Mar del Plata 7600, Argentina

A. Ventosa et al. (eds.), *Halophiles and Hypersaline Environments*,
DOI 10.1007/978-3-642-20198-1_11,

differs from transport in non-halophilic archaea, perhaps due to the high cellular and extracellular salt concentrations in the environments that haloarchaea inhabit (Rose et al. 2002; Gimenez et al. 2007; Kwan et al. 2008; Storf et al. 2010).

In this chapter, the mechanisms that facilitate and regulate the transport of proteins across the cytoplasmic membrane of *Hfx. volcanii* are discussed. In particular, it focuses on the universally conserved Sec pathway and the Twin-arginine transport (Tat) pathway, which is employed primarily by prokaryotes but is also found in chloroplasts and some protists (Driessen and Nouwen 2008; Yuan et al. 2010). In addition, the machinery and substrates of the archaeal pathways are compared and contrasted to those of the same pathways in other organisms. Several hypothetical explanations for the differences observed between the substrates of these pathways in halophilic and non-halophilic prokaryotes, as well as between halophilic bacteria and haloarchaea, are examined. Finally, this chapter highlights how the unique features of haloarchaeal pathways enabled the development of computational tools that permit studies that should promote a better understanding of protein transport in general.

11.2 The Sec Transport Pathway

All sequenced species, eukaryotes and prokaryotes alike, appear to employ the Sec protein transport pathway as a means by which to insert proteins into the cytoplasmic membrane in addition to secreting proteins into the extracytoplasmic environment (Pohlschröder et al. 2005b). Understandably, given the roles it plays, this pathway is essential in all species thus far tested, including the haloarchaeon *Hfx. volcanii* (Rose and Pohlschröder 2002; Haddad et al. 2005). Transport via this pathway involves several components including the signal recognition particle (SRP), a ribonucleoprotein complex that initiates the targeting of many Sec substrates to a membrane-embedded Sec pore (Zwieb and Bhuiyan 2010).

11.2.1 The Signal Recognition Particle

SRPs, which form a close association with ribosomes, either recognize an N-terminal transmembrane segment or a highly hydrophobic stretch in the signal peptide of the targeted protein as the nascent peptide chain extrudes from the ribosome during translation (see below). The interactions among the SRP, ribosome, and nascent chain complex temporarily halt translation of the Sec substrate. The complex is targeted to the membrane-embedded Sec pore via an SRP receptor that is associated with the Sec pore. Proper targeting relieves the translational pause and results in co-translational transport of the newly formed protein across the

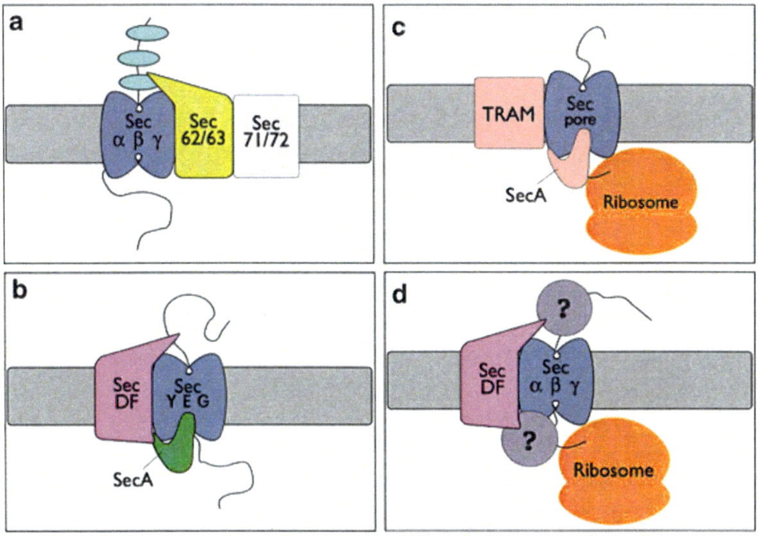

Fig. 11.1 Variations on Sec substrate translocation energetics. The driving force behind unidirectional protein translocation can differ based upon both the organism and the translocated substrate. (**a**) In eukaryotes, post-translationally translocated proteins are not energetically "pushed" through the Sec pore, but rather the ER lumenal protein Bip (teal ovals) binds the substrate and prevents retrotranslocation. In this way, Bip binding acts like a ratchet. See text regarding accessory proteins Sec61/62 and Sec72/73. (**b**) In bacterial post-translational translocation, the cytoplasmic motor protein SecA couples ATP hydrolysis to propelling the substrate through the Sec channel. See text regarding accessory proteins SecD and SecF. (**c**) In organisms from all domains, it is believed that for some proteins translation and translocation are coupled. In this scenario, the protein is extruded from the ribosome directly through the Sec pore as translation occurs, and therefore, an additional source of energy may not be required. SecA (in bacteria) and TRAM (in eukaryotes) play roles in the co-translational transport of at least some substrates (see text). (**d**) While every sequenced bacterial and eukaryotic organism encodes a Bip or SecA homolog, archaeal species do not. Thus, the energetics of archaeal translocation may be solely reliant upon translation, or unidentified proteins regulate this process

cytoplasmic membrane (Natale et al. 2008; Yuan et al. 2010; Zimmermann et al. 2011) (Fig. 11.1a).

The eukaryotic SRP and its receptor are more complex than the analogous prokaryotic structures and the archaeal and bacterial SRP receptors are more similar to each other than to the eukaryotic receptor (Grudnik et al. 2009). However, the archaeal SRP54 component, which is essential in *Hfx. volcanii*, most closely resembles the component of the eukaryotic SRP (Rose and Pohlschröder 2002), just one example of the complex evolutionary relationships that exist among the three domains of life.

A subset of Sec substrates is transported in an SRP-independent manner and is consequently translocated through the Sec pore post-translationally. Cytoplasmic caperones maintain these fully translated proteins in a secretion-competent state and translocation-specific ATPases, including the bacterial cytoplasmic ATPase SecA and the ER-lumenal ATPase Bip in eukaryotes, drive their transport across the cytoplasmic membrane (Driessen and Nouwen 2008; Yuan et al. 2010) (Fig. 11.1b, c). Interestingly, archaea lack homologs of either SecA or Bip (see below).

11.2.2 The Translocon

The Sec translocase generally contains three core components: SecYEG in bacteria and Sec61αβγ in archaea and eukaryotes. Although SecY/Sec61α and SecE/Sec61γ are highly evolutionarily conserved across all three domains, the nomenclature reflects the higher degree of total sequence similarity between the archaeal and eukaryotic core components (Cao and Saier 2003; Pohlschröder et al. 2005b; Yuan et al. 2010).

Analyses of the Sec pore crystal structures of the archaeon *Methanocaldococcus jannaschii* and *Pyrococcus furiosus* suggest that it is the core of a protein-conducting channel that forms a clamshell-like structure within the membrane (van den Berg et al. 2004; Egea and Stroud 2010). While the eukaryotic and archaeal Sec61α sequences are most similar, the bacterial *secY* and archaeal *sec61*α are transcriptionally co-regulated with ribosomal genes, underlining the ancient origin of the Sec pathway as well as its essential nature (Cao and Saier 2003; Pohlschröder et al. 2005b; Yuan et al. 2010).

A second component, Sec61γ in eukaryotes and SecE in eukaryotes, shares very limited homology. In fact, the co-regulation of many archaeal *sec61*γ homologs and bacterial *secE* genes with the gene encoding the transcription factor *nusG* provided the clue that ultimately led to conclusive evidence that these prokaryotic genes are related to each other and to the eukaryotic gene *sec61*γ (Hartmann et al. 1994).

Consistent with the close relationship between the archaeal and eukaryotic Sec61α and Sec61γ homologs, the third pore component, Sec61β, is also conserved between archaea and eukaryotes. Conversely, no clear homology exists between Sec61β and the third bacterial core component, SecG. Since this component is less critical to pore function than the other components of the pore, it may or may not be distantly related; however, based on currently available sequence and genome localization data, a relationship has not been established.

11.2.3 Sec Accessory Components

Successful transport of substrates through the Sec pore requires additional accessory components. For co-translational transport, eukaryotes require the presence of pore-associated translocation-associated complex membrane components (TRAM) while post-translational transport involves a membrane complex formed by Sec62/63, and in some cases a Sec71/72-complex (Osborne et al. 2005) (Fig. 11.1b). The exact function(s) of these complexes are unknown; however, the cytoplasmic region of Sec62 can act as a chaperone, and its luminal aspect facilitates binding by the ATPase Bip. Bip, which is homologous to the Hsp70 chaperone, is required for post-translational protein transport via the Sec pathway. It binds secreted proteins as they exit the pore, physically preventing unintentional retro-translocation as proteins enter the endoplasmic reticulum (Zimmermann et al. 2011). Since it is not required for in vitro co-translational transport of the few substrates tested thus far, a co-translational transport role for Bip is uncertain at best. ATP is not available to prokaryotes on the extracytoplasmic side of the membrane; hence, an extracellular or periplasmic Bip homolog would be inactive. Therefore, it is not surprising that prokaryotic genomes do not encode a Bip homolog.

In bacteria, Sec substrate post-translational, and some co-translational transport, requires the cytoplasmic ATPase, SecA, which is peripherally associated with the membrane (Fig. 11.1c). SecA provides the motive force that drives proteins across the membrane, and it is also probably critical to targeting post-translationally transported proteins to the Sec pore (Driessen and Nouwen 2008; Yuan et al. 2010). The membrane protein complex SecDF may play a role in stabilizing SecA, and might also provide a chaperone function that facilitates efficient folding of transported proteins on the extracytoplasmic face of the membrane (Matsuyama et al. 1993; Duong and Wickner 1997).

While the lack of a Bip homolog in archaea is understandable, it is curious that archaea also lack a SecA homolog (Cao and Saier 2003; Pohlschröder et al. 2005a; Yuan et al. 2010). Without a translocation ATPase, the mechanism that generates the energy necessary for archaeal protein transport remains a mystery (Fig. 11.1d). Although some reports have suggested that post-translational transport can occur in archaea (Ortenberg and Mevarech 2000; Eichler and Adams 2005), the lack of a translocation ATPase may indicate that archaeal protein transport via the Sec pathway is primarily co-translational. Conversely, most archaea, including *Hfx. volcanii*, have SecDF homologs, and *Hfx. volcanii secDF* deletion mutants exhibit phenotypes similar to those of analogous *Escherichia coli* mutants. While the stabilization of SecA is apparently not among the functions of SecDF, preliminary data suggest that this membrane complex can act as a chaperone (Hand et al. 2006). Archaea also lack homologs of the eukaryotic TRAM, Sec62/63 and Sec71/72 accessory components.

11.2.4 Sec Signal Peptides and Peptidases

The regions of signal peptides that are critical to targeting secreted proteins to the Sec pore include a charged N terminus and a hydrophobic stretch of amino acid residues (Paetzel et al. 2002; Natale et al. 2008). While the signal peptide can serve as an N-terminal membrane anchor for some Sec substrates, it usually contains sequences that target it for cleavage by a signal peptidase (SPase I) (Tuteja 2005). The universally conserved SPase I associates with three additional proteins to form a complex that can process the signal peptide as a substrate enters the ER (Zimmermann et al. 2011). The bacterial SPase I homolog, which is thought to be responsible for processing most bacterial Sec signal peptides, has unique catalytic sites not found in its eukaryotic counterpart and it acts as a monomer. Data mining of archaea suggests that the archaeal SPase I also acts as a monomer, as it lacks homologs of the components that interact with the eukaryotic SPase I (Tuteja 2005). Interestingly, the primary sequence of the archaeal SPase I suggests it is a chimera, resembling the bacterial SPase I in some regions and the eukaryotic SPase I in others (Ng et al. 2007). *Hfx. volcanii*, but not all haloarchaea, expresses two SPase I paralogs; however, the significance of this is unclear (Eichler 2002; Fine et al. 2006; Fink-Lavi and Eichler 2008).

If a signal peptide contains an SPase I processing site following the hydrophobic stretch of amino acid residues, the processed substrate is released from the cytoplasmic membrane unless it is anchored to the membrane via a C-terminal transmembrane segment or via interactions with other membrane-associated proteins (Fig. 11.2).

Conversely, the bacterial SPase II recognizes a processing site, subsequent to the hydrophobic stretch, known as a lipobox, and processes signal peptides containing this motif upon acylation of an invariant cysteine located at the C-terminal end of the lipobox (Fig. 11.2) (Sankaran and Wu 1994; Paetzel et al. 2002; Hutchings et al. 2009; Tschumi et al. 2009). The acylated cysteine anchors the protein to the membrane and is the most N-terminal amino acid residue remaining in the processed substrate (Fig. 11.2). Although archaea, particularly

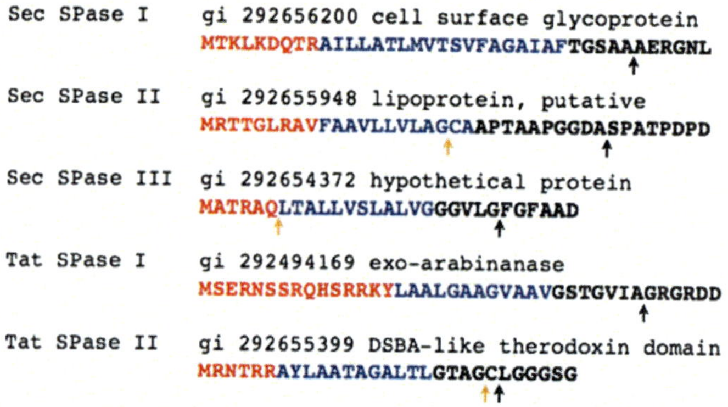

Fig. 11.2 Signal peptides predicted by Phobius vary in translocation and processing pathways. Phobius identifies a large number of signal peptides; however, it is poor at identifying non-Sec signal peptides, fails to differentiate the translocation pathway and does not consider different peptidase cleavage sites. Five signal peptides from *Hfx. volcanii* that were identified by Phobius to have a signal peptide are shown, with the predicted N-terminal and hydrophobic regions of the signal shown in *red* and *blue*, respectively. The Phobius-suggested processing site of each signal is indicated by a *black arrow*. Further analysis using LipoP, TatFind and FlaFind reveals that each of these Phobius positive proteins belongs to a unique class of secretory protein: Sec SPase I, Sec SPase II, Sec SPase III, Tat SPase I, and Tat SPase II. For SPase II and SPase III signal peptides, the more-likely processing site predicted by LipoP and FlaFind is indicated by an *orange arrow*

euryarchaea, express proteins containing lipobox motifs (see below), neither an archaeal SPase II nor an acylase homolog has yet been identified (Gimenez et al. 2007; Storf et al. 2010).

Finally, archaeal and bacterial SPase III each recognize a processing site located between the charged region and the hydrophobic stretch of the signal peptide (Paetzel et al. 2002; Szabo et al. 2007). These signal peptides target subunits of type IV pilus-like surface structures to the Sec translocase (Francetic et al. 2007), and its hydrophobic stretch is critical for multimerization of pilins or pilin-like subunits, which allows the formation of an α-helical core that serves as a scaffold for the assembly of a surface structure (Fig. 11.2) (Craig and Li 2008; Pelicic 2008; Albers and Pohlschröder 2009). While some archaea and bacteria express multiple paralogs of SPase III, each targeting specific subsets of pilin-like proteins, to date, all sequenced haloarchaeal genomes encode only a single paralog (Szabo et al. 2007; Albers and Pohlschröder 2009).

11.3 The Twin Arginine Transport Pathway

The Twin arginine transport (Tat) pathway is a mode of protein transport that is limited to prokaryotes, chloroplasts and a few protists (Natale et al. 2008; Yuan et al. 2010). It transports proteins after they have folded in the cytoplasm, and hence its substrates have distinct sizes and conformations. Since the Sec pore cannot accommodate the translocation of substrates exceeding a diameter of approximately 20 Å, the transport of pre-folded proteins necessitates components and mechanisms that are fundamentally different from those employed by the Sec pathway.

11.3.1 Tat Components

Three distinct integral membrane components of the Tat machinery have been identified: TatA, TatB and TatC. At least one TatA and one TatC homolog are required for the Tat pathway to function properly in prokaryotes (Natale et al. 2008; Yuan et al. 2010). Negative stain electron microscopy has revealed that TatA is situated in ring-like structures in *E. coli*, forming a pore ranging in size from 30 to 70 Å through which Tat substrates might be translocated (Gohlke et al. 2005). Multimers of *E. coli* TatB and TatC form a separate complex that may interact directly with substrates (Cline and Mori 2001; Gerard and Cline 2006).

Haloarchaea are the only group of related species known to employ the Tat pathway for nearly half of their secreted proteins (see below) (Rose et al. 2002; Storf et al. 2010). Tat machineries in haloarchaea are comprised of four components: TatAo, TatAt, TatCo and TatCt. The haloarchaeon *Hfx. volcanii* was the first organism in which the Tat pathway was shown to be essential (Dilks et al. 2005). TatAt is essential in this species, but TatAo deletion mutants remain viable, suggesting that if there are proteins transported in a strict TatAo-dependent manner, they are not necessary for *Hfx. volcanii* viability when grown under standard laboratory conditions (Dilks et al. 2005).

The haloarchaeal TatCt is predicted to have 14 TM segments, a unique feature among known TatC paralogs, which typically possess 6 TM segments (Dilks et al. 2005). Sequence analysis has revealed that the six TM segments nearest the N terminus and the six TM segments nearest the C terminus of *Hfx. volcanii* TatCt have significant homology to other TatC paralogs, suggesting that the structure of this protein is the result of a duplication event. The seventh and eighth TM segments lack homology to any protein thus far sequenced (Dilks et al. 2005). Although both TatC paralogs are essential for growth, *tatCt* can complement a *tatCo* deletion, but *tatCo* does not complement a *tatCt* deletion. When expressed at levels similar to the wild-type gene, truncated proteins lacking either the final six or eight TM segments of TatCt, TatCtII/6 and TatCtII/8, respectively, do not complement a chromosomal deletion of *tatCt*. However, overexpression of either of these truncated proteins from a plasmid construct does allow growth of this deletion mutant, with the rescued mutant lacking a detectable phenotype. Conversely, overexpressed TatCtII/6 or TatCtII/8 do not complement a chromosomal deletion of *tatCt* (Gimenez and Pohlschröder, unpublished data).

Interestingly, the TatC paralog of *Salinibacter ruber*, a halophilic bacterium that like haloarchaea accumulates KCl in its cytoplasm, does not contain duplicated TM segments (Mongodin et al. 2005). Moreover, this bacterium uses its Tat pathway to transport only a very limited number of substrates (see below). Hence, while the duplication in the *Hfx. volcanii* TatCt might represent an adaptation to high salt environments, a more likely explanation is that it facilitates the extensive use of the Tat pathway observed in haloarchaea (see below).

As noted above, it is currently thought that Tat protein translocation is mediated through substrate signal peptide interactions with the TatBC complex that initiate recruitment of TatA, causing a major conformational change in the TatBC complex and leading to passage of the substrate through the TatA pore (Natale et al. 2008; Yuan et al. 2010). Given the prominent role of TatC in Tat protein transport, it is possible that by increasing the number of available substrate recognition and binding sites in TatC, a duplication of the TM segments might facilitate more efficient protein transport. This may be the reason that when only the six TM segments nearest the N terminus of TatCt (TatCtI) are expressed, very high levels of expression are required to support extensive use of this pathway in *Hfx. volcanii*. On the other hand, subtle sequence differences between TatCtI and TatCtII might be the reason it does not rescue the *tatCt* deletion mutation, even when expressed at high levels (Gimenez and Pohlschröder, unpublished; Fig. 11.3). This truncated version of TatCt may fail to bind substrates efficiently enough to meet the needs of haloarchaeal protein transport, or alternatively, it might not support complex formation or transport in the absence of the N terminus.

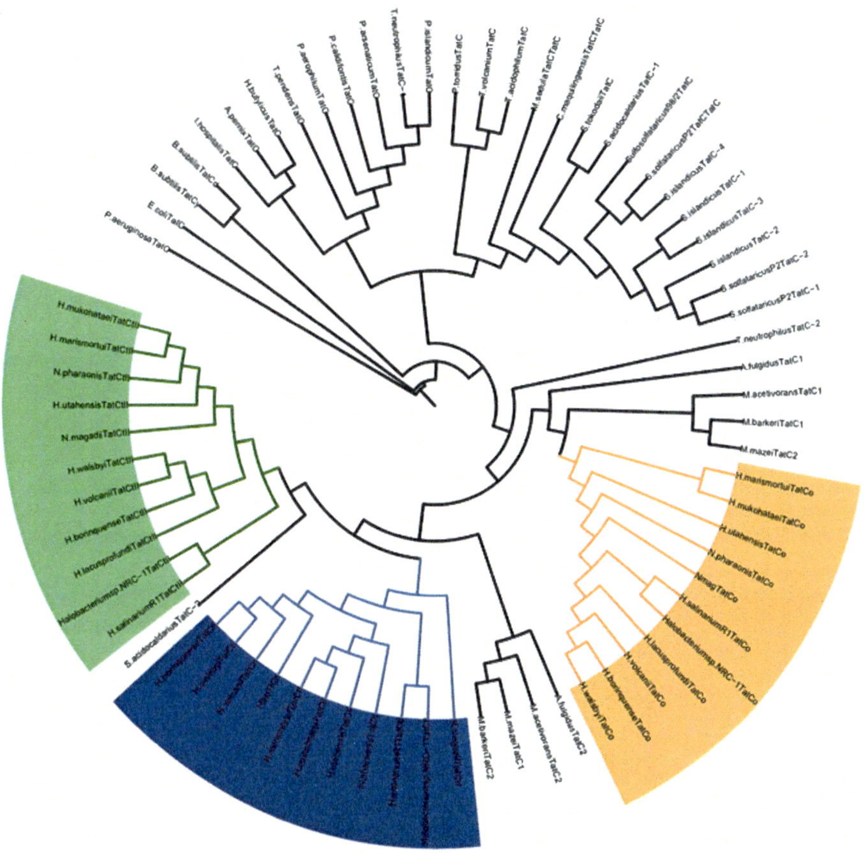

Fig. 11.3 Haloarchaeal TatCt internal paralogs (TatCtI and TatCtII, respectively) cluster with distinct TatC subgroups. Multiple sequence alignment (MSA) of TatC protein sequences obtained from NCBI was carried out by MUSCLE (Edgar 2004). The phylogenetic tree was constructed by maximum likelihood method implemented in PHYML (Anisimova and Gascuel 2006) program with 100 non-parametric bootstraps

11.3.2 Tat Signal Peptides and Peptidases

Although they exhibit subtle differences, Tat signal peptides and Sec signal peptides are structurally similar (see below) (Natale et al. 2008; Yuan et al. 2010). They each contain an N-terminal charged region, a hydrophobic stretch and a signal peptidase-processing site. While most prokaryotic Tat substrates are predicted to be processed by SPase I homologs, some signal peptides also contain a lipobox. Unlike Tat substrates in other prokaryotes, the majority of haloarchaeal Tat substrates contain predicted lipoboxes (see below and Fig. 11.4).

	Tat			Sec			
	Secreted Soluble	C-term Anchor	Lipid Anchor	Secreted Soluble	C-term Anchor	Lipid Anchor	Flagella / Pili
H. halophila	17	0	1	244	8	61	17
S. ruber M8	13	1	2	398	3	112	3
B. halodurans	4	0	0	224	15	136	4
Hfx. volcanii	41	4	77	92	23	18	37

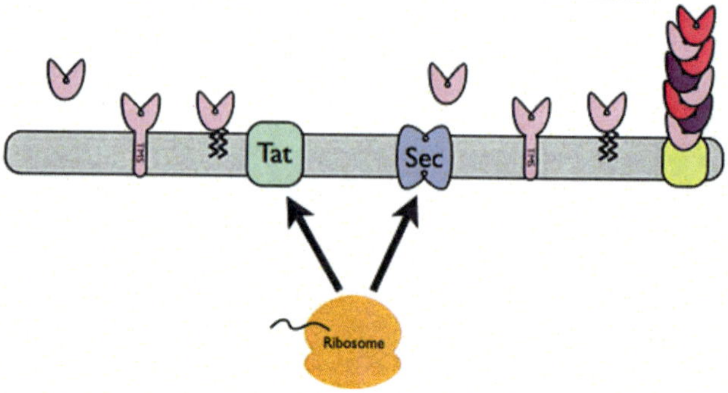

Fig. 11.4 Haloarchaea exhibit an atypical protein secretion profile, even when compared to other halophiles. The secreted proteins of *Haloferax volcanii* along with three other halophilic organisms (*Halorhodospira halophila SL1, Salinibacter ruber M8,* and *Bacillus halodurans C-125*) were identified and segregated into classes using several computational programs. In brief, all ORF sequences for each organism were obtained from http://www.ncbi.nlm.nih. The ORF sequences were analyzed using TatFind, Phobius. LipoP, and FlaFind/PilFind. By comparing the output of the four analyses, seven classes of proteins were predicted: Tat secreted soluble proteins, Tat C-terminally anchored proteins, Tat lipoproteins, Sec secreted soluble proteins, Sec C-terminally anchored proteins, Sec lipoproteins, and Sec flagella/pili. If a protein was mutually predicted by TatFind and Phobius, then it was defined as a Tat substrate as Phobius commonly predicts Tat substrates as Sec substrates. Any protein identified by any program that was also predicted by Phobius to have an internal transmembrane domain or have more than one transmembrane domain was excluded from being termed a secreted protein

11.4 Prediction of Subcellular Localization of Haloarchaeal Proteins

In silico predictions of transmembrane segments and signal peptides provide valuable information as to whether a protein will remain in the cytoplasm, inserts into the cytoplasmic membrane, or be secreted across the cytoplasmic membrane. These predictions also shape our understanding of how extensively specific transport pathways are used in various organisms as well as the signal peptidase that processes a specific substrate. The signal peptidase that processes a substrate is indicative of whether it will be secreted into the extracellular environment, remain associated with the membrane via a C-terminal or lipid anchor, or be incorporated into a surface structure (Fig. 11.4). Therefore, accurate prediction of the class of signal peptide sequence sheds light on the role of proteins of unknown function.

In eukaryotes, the vast majority of extracytoplasmic proteins is inserted into or are transported across the endoplasmic reticulum membrane via the Sec pore. Sec substrates are targeted to this universally conserved pore via an N-terminal hydrophobic domain located in either the first transmembrane segment of a membrane protein or in the signal peptide (Zimmermann et al. 2011). During or after translocation, the signal peptide of a eukaryotic Sec substrate is removed by SPase I. The software program Phobius predicts the presence of transmembrane segments and SPase I processing recognition sites in Sec signal peptides, and also predicts whether the processed substrate is inserted into the endoplasmic reticulum (ER) membrane or is transported across it (Kall et al. 2004). Proteins not predicted by Phobius are likely to remain in the cytoplasm.

In contrast, prokaryotic Sec substrates, as noted above, can possess signal peptides cleaved by SPase I, SPase II or SPase III, respectively (Fig. 11.2). Moreover, in many prokaryotes, structurally similar signal peptides may differentially target the protein for secretion via the Sec or Tat pathways, and may be processed by either SPase I or SPase II (see above, Figs. 11.2 and Figs. 11.4). Hence, to make accurate predictions concerning the subcellular localization and pathways employed to transport specific prokaryotic proteins, Phobius must be used in conjunction with programs designed to predict SPase II or SPase III processed signal peptides, as well as programs that can distinguish Tat and Sec substrates (Figs. 11.2 and Figs. 11.4).

A variety of software programs that predict specific signal peptide classes have been developed. For example, LipoP and PredLipo were designed to predict signal peptides processed by SPase II in Gram-negative and Gram-positive bacteria, respectively (Bagos et al. 2008; Juncker et al. 2003). FlaFind and PilFind predict archaeal and bacterial class III signal peptides, respectively, and TatFind and TatP are designed to predict prokaryotic Tat signal peptides (Rose et al. 2002; Bendtsen et al. 2005; Szabo et al. 2007, Imam and Pohlschröder, unpublished). These programs are complementary to some extent. For instance, a Tat-translocated lipoprotein would be correctly predicted only by using TatFind and LipoP consecutively. Due to the recent interest in this specific subgroup of proteins, the program TatLipo was designed to specifically identify haloarchaeal Tat substrates containing lipoboxes (signalfind.org) (Storf et al. 2010).

It should be noted that although they are structurally conserved, there is significant sequence divergence among the sequences of SPase I processed signal peptides, and due to these low sequence constraints, programs such as Phobius have a strong tendency to predict false positives for both Sec signal peptides and SPase I cleavage sites (Fig. 11.2). Moreover, since both the Sec and Tat pathway specific signal peptides contain a charged region followed by a hydrophobic core, and many contain an SPase I cleavage site, Phobius is inclined to misclassify many Tat signal peptides as Sec signal peptides. Thus, the use of TatFind would help eliminate false-positives predicted by Phobius. In fact, Pred-Tat, a program that combines predictions of Tat and Sec signal peptides, was recently developed (Bagos et al. 2010).

Similarly, since SPase I has a relaxed specificity, many lipoproteins cleaved by SPase II are predicted by Phobius to have SPase I processed signal peptides. Therefore, Phobius predictions of SPase I cleavage sites in predicted lipoproteins are likely to be incorrect. Clearly, applying these programs in concert, supported by in vivo data can clarify subcellular localization of processed substrates by ensuring more accurate predictions of the class of the signal peptide that each substrate contains (Fig. 11.2).

While some existing programs are trained on archaeal protein sequences, the number of confirmed archaeal secreted proteins for each signal peptide class remains very low. Pred-Signal was recently developed to specifically predict archaeal SPase I processed signal peptides. However, contrary to the authors claim, only a few substrates used to train this program have been verified (Bagos et al. 2009). Conversely, TatFind, a program initially trained on a large set of predicted Tat substrates, has been validated by in vivo verification of numerous substrates in archaea as well as bacteria, suggesting that these signal peptide sequences are similar in bacteria and archaea (Rose et al. 2002; Dilks et al. 2003; Widdick et al. 2006). Some sequence differences are present in the Tat signal peptides of various prokaryotic species. For example, a small fraction of Tat substrates tolerate replacement of one of the arginines in the twin arginine motif with a lysine, while in *Hfx. volcanii* neither arginine can be replaced (DeLisa et al. 2002; Kwan and Bolhuis 2010). Moreover, there seems to be some variability in the relative specificity of the amino acid residues allowed in the vicinity of the twin arginines, both between bacterial species and between bacteria and archaea (McDonough et al. 2008).

Interestingly, the sequences of SPase II processing sites of Tat and Sec substrates appear to be highly conserved across domains, and the majority of predicted haloarchaeal lipobox motifs contain the canonical LAGC sequence (Gimenez et al. 2007; Storf et al. 2010). As noted above, this is particularly intriguing since an archaeal SPase II homolog has not yet been identified. Conversely, there are clear differences between archaeal and bacterial SPase III processing sites, and therefore, FlaFind and PilFind were developed independently to predict archaeal and bacterial type IV pilin-like proteins, respectively (Szabo et al. 2007; see signalfind.org). While both signal peptides require a G/A, or rarely an S, at the -1 position, the positive charge at position -2 appears to be specific to archaeal SPase III recognition sites. Moreover, only a subset of archaeal species express an SPase III that recognizes a signal peptide that has a glutamate residue at the +5 position, which appears to be nearly universal in SPase III processing sites in bacteria (Szabo et al. 2007; Imam and Pohlschröder, unpublished). While the number of confirmed secreted proteins is still relatively low for haloarchaea, the computational tools now available allow us to estimate the diversity of archaeal secreted substrates and the extent to which each protein transport pathway is being used (Fig. 11.4 and signalfind.org).

11.5 Haloarchaeal Secreted Proteins

More than 20 years ago it was demonstrated that the Sec signal peptide of the *Hfx. volcanii* S-layer glycoprotein is processed at a predicted SPase I cleavage site (Sumper et al. 1990). Subsequently, it was shown that this *Hfx. volcanii* signal peptide can also target an *E. coli* Sec reporter protein to the bacterial translocon (Pohlschröder et al. 2004). While many additional putative Sec substrates have been identified in *Hfx. volcanii* by Phobius, haloarchaeal S-layer glycoproteins are the only haloarchaeal Sec substrates predicted to contain SPase I processing sites that have been confirmed in vivo. Instead, most verified translocated proteins in haloarchaea are either non-Sec or non-SPase I substrates, in stark contrast to known bacterial substrates (Gimenez et al. 2007; Storf et al. 2010; Tripepi et al. 2010).

A number of Phobius predicted haloarchaeal Sec substrates also contain predicted lipoboxes and are thus likely be processed by an SPase II analog (Fig. 11.4). Replacement mutations of the conserved lipobox cysteine render these proteins less stable, although they are still processed (Storf et al. 2010). Further analysis is required to clarify whether the lipobox is wrongly predicted or alternatively, SPase I processes the mutant protein at a cryptic signal peptidase recognition site. It should be noted that the predicted lipobox motifs are conserved in homologous proteins from a variety of haloarchaeal species, indicating that these motifs probably play an important role in protein maturation (Storf, unpublished data).

A third subset of archaeal Sec substrates are those that possess a SPase III cleavage site. In archaea, including the haloarchaea, flagellar biosynthesis is reminiscent of bacterial type IV pilus biosynthesis (Jarrell and McBride 2008). That is, the flagellin subunits are secreted via the Sec pathway rather than via a type III secretion system similar to that used to transport bacterial flagellins, and archaeal flagellin signal peptides are processed by FlaK/PibD, a SPase III homolog. The biosynthesis of archaeal flagella requires two additional proteins, an ATPase and a membrane protein that are homologous to bacterial PilB and PilC, respectively. PilB and PilC are required for bacterial type IV pilus biosynthesis (Jarrell and McBride 2008; Tripepi et al. 2010).

Taking advantage of the large number of conserved flagellin genes identified in archaeal genomes, FlaFind was developed to identify genes encoding proteins having signal peptides processed by SPase III (Szabo et al. 2007). The majority of these genes are co-regulated with homologs of the pilus biosynthesis genes *pilB* and *pilC*. Interestingly, while known bacterial and archaeal pilins are generally no larger than 300 amino acid residues, FlaFind identified several much larger archaeal proteins containing predicted SPase III processing sites. These proteins are reminiscent of *Sulfolobus solfataricus* substrate binding proteins processed by the SPase III homolog PibD, and proposed to be incorporated into a substrate-binding surface structure known as a bindosome (Zolghadr et al. 2007). In *Hfx. volcanii* and *Halobacterium* strains, in addition to sequences and sizes that vary, predicted pilin genes also appear to have distinctive transcription patterns, as revealed by microarray data from the laboratories of Jörg Soppa and Charles Daniels.

Consistent with these proteins being processed by SPase III, a *Hfx. volcanii pibD* deletion strain fails to process either flagellins or several predicted pilin-like substrates (Tripepi et al. 2010). Many predicted pilin-like proteins were not tested because they are unstable when expressed from an inducible or constitutive promotor. In fact, only those proteins detected by microarray analyses are stably expressed from plasmid constructs (Tripepi and Pohlschröder, unpublished). Other than the flagella, the functional roles played by the putative archaeal pili and pilus-like structures are unknown, however, the failure of *pibD* deletion strains to attach to a glass coverslip suggests that type IV pilus-like structures are involved in *Hfx. volcanii* surface attachment (Tripepi et al. 2010).

As noted above, the *Hfx. volcanii* Tat pathway is essential and a significant portion of its predicted secreted proteins contain a twin arginine motif (Dilks et al. 2005) (Fig. 11.4). If the twin arginines are mutated to twin lysines, the mutant substrates are not translocated, presumably due to failed substrate targeting. Using this strategy, several haloarchaeal Tat substrates have been confirmed in vivo, including several halocyanins, a DsbA-like protein, an iron-binding protein, a maltose-binding protein, an amylase and a laccase (Dilks et al. 2005; Gimenez et al. 2007; Kwan and Bolhuis 2010; Rose et al. 2002; Storf et al. 2010; Uthandi et al. 2010).

Some Tat substrates, which have signal peptides that contain SPase I processing sites, are secreted into the extracellular environment while others are bound to the membrane via a C-terminal transmembrane segment (TMS) (Dilks et al. 2005; Gimenez et al. 2007) (Fig. 11.4).

However, the majority of haloarchaeal Tat substrates contain a putative lipobox (Storf et al. 2010) (Fig. 11.4). Mutants in which the cysteine in a putative lipobox is replaced with a serine have been used to reveal that the cysteine is required for the efficient processing of the precursor and proper membrane localization (Gimenez et al. 2007; Storf et al. 2010). In some instances, the C-S replacement mutants remain membrane-associated. Since the lipobox acts as the SPase II recognition site, it follows that perturbation of the lipobox may render some of these mutants incapable of being processed. In these instances, the signal peptide likely tethers the unprocessed protein to the membrane. The large number of Tat substrates containing a lipobox is particularly striking since SPase II and acylating enzyme homologs have not been identified in archaea (see above). Archaea may require distinct enzymes to perform these functions because of differences in the cytoplasmic membranes of bacteria and archaea. While other euryarchaea also express proteins having lipobox motifs, the percentage of Tat substrates having predicted lipoboxes is significantly higher in haloarchaea than it is in other archaea (Storf et al. 2010). Similarly, the percentage of Sec substrates with lipobox motifs is similar in halophilic archaea and halophilic bacteria. However, haloarchaea are unique in the high percentage of their Tat substrates that have a predicted lipobox (Fig. 11.4).

Tat substrates having a predicted SPase III processing site have not been identified. To date, tested substrates containing confirmed SPase III processing sites are subunits of surface structures where incorporation likely requires that the protein is at least partially unfolded.

11.6 Protein Transport and Adaptation to High Salt

The unique translocation strategies haloarchaea have adopted raise an important scientific question: Why haloarchaea? Of the vast array of highly diverse prokaryotic species found in moderate and extreme environments, what in particular was it about the evolutionary history of haloarchaea that drove the fundamental remodeling of protein translocation? Since the temporal and spatial management of protein folding largely dictates the pathway used to transport a protein to the membrane, an obvious answer centers on a shift in the protein folding dynamics of haloarchaea compared to other organisms. Many haloarchaeal proteins fold in a manner resulting in a highly negative charged surface, preventing the formation of protein aggregation in high salt environments (Frolow et al. 1996; Kennedy et al. 2001). Moreover, to avoid the formation of these kinds of aggregates, protein folding is also likely to be an extremely efficient process in haloarchaea, whether it occurs in the cytoplasm or upon transport into the extracytoplasmic environment. Efficient protein folding in the cytoplasm can be facilitated by ATP-dependent chaperones, however, similar chaperone activity outside the cytoplasm must be ATP-independent. Alternatively, rather than rely on extracytoplasmic chaperones, a relatively large portion of secreted proteins could fold in the cytoplasm prior to transport via the Tat pathway, perhaps explaining the extensive use of this pathway in haloarchaea. If this hypothesis is correct, one might anticipate that other halophilic organisms would follow a similar strategy. However, in silico predictions indicate that halophilic bacteria use the Tat pathway to transport only a limited number of proteins (Mongodin et al. 2005; Fig. 11.4). Given that halophilic bacterial species rarely use the Tat pathway and halophilic archaea use it extensively, either (a) high salt environments have not significantly impacted the mechanisms underlying protein translocation or (b) fundamental differences existed between bacteria and archaea prior to the acquisition of distinct adaptations in environments containing high concentrations of extracellular salt. Natural selection can only act upon existing characteristics. Therefore, if haloarchaea and halophilic bacteria have descended from species that already displayed significant differences in characteristics that affect protein transport, then the selective pressures caused by high salt concentrations would have resulted in the acquisition of divergent protein transport strategies. For example, as noted previously, although the Sec pathway is universally conserved, archaea appear to lack a translocation ATPase that supplies the energy required for post-translational protein transport in bacteria (Fig. 11.1d). Perhaps in bacteria improvements in the functionality of the accessory components involved in co-translational and post-translational transport via the Sec pathway was a more favorable means of adapting protein transport to high salt environments than rerouting secreted proteins to the Tat pathway. Conversely, if archaea lack an efficient post-translational transport system and folding proteins upon transport creates a significant bottleneck, then rerouting substrates to the Tat pathway may be the most favorable adaptation to high salt conditions in archaea. Co-translational transport is generally more efficient than post-translational transport via the Tat pathway, perhaps explaining why the Tat pathway is not used extensively in non-halophilic archaea.

Why most haloarchaeal Tat substrates are anchored to membranes via lipid tethers also remains an open question. Generally, lipid membrane-anchoring is less stable than other mechanisms that anchor proteins to membranes, and therefore, lipid-anchored proteins can be shed relatively quickly, perhaps an important characteristic of membrane-associated proteins in haloarchaea that produce large quantities of bacteriorhodopsin. However, lipid-anchored proteins are also common on the membranes of haloarchaeal species, such as *Hfx. volcanii*, that lack bacteriorhodopsin but not on non-halophilic archaeal membranes. Alternatively, perhaps the Tat pore obstructs lateral insertion of transmembrane segments into the cytoplasmic membrane in which case C-terminal hydrophobic anchors must be transported across the membrane in an unfolded conformation prior to insertion. However, maintaining the solubility of long hydrophobic regions of haloarchaeal proteins might be difficult in high salt conditions since enhanced hydrophobic interactions might result in chaperones becoming affixed to substrates.

Acknowledgements Support was provided to M.P. by a grant from the National Science Foundation (reference no. MCB02-39215) and the National Aeronautics and Space Administration grant (ref. no. NNX10AR84G).

References

Alam M, Oesterhelt D (1984) Morphology, function and isolation of halobacterial flagella. J Mol Biol 176:459–475

Albers SV, Pohlschröder M (2009) Diversity of archaeal type IV pilin-like structures. Extremophiles 13:403–410

Anisimova M, Gascuel O (2006) Approximate likelihood-ratio test for branches: a fast, accurate, and powerful alternative. Syst Biol 55:539–552

Bagos PG, Tsirigos KD, Liakopoulos TD, Hamodrakas SJ (2008) Prediction of lipoprotein signal peptides in Gram-positive bacteria with a hidden Markov Model. J Proteome Res 7:5082–5093

Bagos PG, Tsirigos KD, Plessas SK, Liakopoulos TD, Hamodrakas SJ (2009) Prediction of signal peptides in archaea. Protein Eng Des Sel 22:27–35

Bagos PG, Nikolaou EP, Liakopoulos TD, Tsirigos KD (2010) Combined prediction of Tat and Sec signal peptides with hidden Markov models. Bioinformatics 26:2811–2817

Bendtsen JD, Nielsen H, Widdick D, Palmer T, Brunak S (2005) Prediction of twin-arginine signal peptides. BMC Bioinform 6:167

Calo D, Eichler J (2010) Crossing the membrane in Archaea, the third domain of life. Biochim Biophys Acta 1808(3):885–891

Cao TB, Saier MH Jr (2003) The general protein secretory pathway: phylogenetic analyses leading to evolutionary conclusions. Biochim Biophys Acta 1609:115–125

Cline K, Mori H (2001) Thylakoid ΔpH-dependent precursor proteins bind to a cpTatC-Hcf106 complex before Tha4-dependent transport. J Cell Biol 154:719–729

Craig L, Li J (2008) Type IV pili: paradoxes in form and function. Curr Opin Struct Biol 18:267–277

De Castro RE, Ruiz DM, Gimenez MI, Silveyra MX, Paggi RA, Maupin-Furlow JA (2008) Gene cloning and heterologous synthesis of a haloalkaliphilic extracellular protease of *Natrialba magadii* (Nep). Extremophiles 12:677–687

DeLisa MP, Samuelson P, Palmer T, Georgiou G (2002) Genetic analysis of the twin arginine translocator secretion pathway in bacteria. J Biol Chem 277:29825–29831

Dilks K, Rose RW, Hartmann E, Pohlschröder M (2003) Prokaryotic utilization of the twin-arginine translocation pathway: a genomic survey. J Bacteriol 185:1478–1483

Dilks K, Gimenez MI, Pohlschröder M (2005) Genetic and biochemical analysis of the twin-arginine translocation pathway in halophilic archaea. J Bacteriol 187:8104–8113

Driessen AJ, Nouwen N (2008) Protein translocation across the bacterial cytoplasmic membrane. Annu Rev Biochem 77:643–667

Duong F, Wickner W (1997) The SecDFyajC domain of preprotein translocase controls preprotein movement by regulating SecA membrane cycling. EMBO J 16:4871–4879

Edgar RC (2004) MUSCLE: multiple sequence alignment with high accuracy and high throughput. Nucleic Acids Res 32:1792–1797

Egea PF, Stroud RM (2010) Lateral opening of a translocon upon entry of protein suggests the mechanism of insertion into membranes. Proc Natl Acad Sci USA 107:17182–18177

Eichler J (2002) Archaeal signal peptidases from the genus *Thermoplasma*: structural and mechanistic hybrids of the bacterial and eukaryal enzymes. J Mol Evol 54:411–415

Eichler J, Adams MW (2005) Posttranslational protein modification in Archaea. Microbiol Mol Biol Rev 69:393–425

Fine A, Irihimovitch V, Dahan I, Konrad Z, Eichler J (2006) Cloning, expression, and purification of functional Sec11a and Sec11b, type I signal peptidases of the archaeon *Haloferax volcanii*. J Bacteriol 188:1911–1919

Fink-Lavi E, Eichler J (2008) Identification of residues essential for the catalytic activity of Sec11b, one of the two type I signal peptidases of *Haloferax volcanii*. FEMS Microbiol Lett 278:257–260

Francetic O, Buddelmeijer N, Lewenza S, Kumamoto CA, Pugsley AP (2007) Signal recognition particle-dependent inner membrane targeting of the PulG pseudopilin component of a type II secretion system. J Bacteriol 189:1783–1793

Frolow F, Harel M, Sussman JL, Mevarech M, Shoham M (1996) Insights into protein adaptation to a saturated salt environment from the crystal structure of a halophilic 2Fe-2S ferredoxin. Nat Struct Biol 3:452–458

Gerard F, Cline K (2006) Efficient twin arginine translocation (Tat) pathway transport of a precursor protein covalently anchored to its initial cpTatC binding site. J Biol Chem 281:6130–6135

Gimenez MI, Dilks K, Pohlschröder M (2007) *Haloferax volcanii* twin-arginine translocation substrates include secreted soluble, C-terminally anchored and lipoproteins. Mol Microbiol 66:1597–1606

Gohlke U, Pullan L, McDevitt CA, Porcelli I, de Leeuw E, Palmer T, Saibil HR, Berks BC (2005) The TatA component of the twin-arginine protein transport system forms channel complexes of variable diameter. Proc Natl Acad Sci USA 102:10482–10486

Grudnik P, Bange G, Sinning I (2009) Protein targeting by the signal recognition particle. Biol Chem 390:775–782

Haddad A, Rose RW, Pohlschröder M (2005) The *Haloferax volcanii* FtsY homolog is critical for haloarchaeal growth but does not require the A domain. J Bacteriol 187:4015–4022

Hand NJ, Klein R, Laskewitz A, Pohlschröder M (2006) Archaeal and bacterial SecD and SecF homologs exhibit striking structural and functional conservation. J Bacteriol 188:1251–1259

Hartmann E, Sommer T, Prehn S, Gorlich D, Jentsch S, Rapoport TA (1994) Evolutionary conservation of components of the protein translocation complex. Nature 367:654–657

Hutchings MI, Palmer T, Harrington DJ, Sutcliffe IC (2009) Lipoprotein biogenesis in Gram-positive bacteria: knowing when to hold 'em, knowing when to fold 'em. Trends Microbiol 17:13–21

Jarrell KF, McBride MJ (2008) The surprisingly diverse ways that prokaryotes move. Nat Rev Microbiol 6:466–476

Juncker AS, Willenbrock H, Von Heijne G, Brunak S, Nielsen H, Krogh A (2003) Prediction of lipoprotein signal peptides in Gram-negative bacteria. Protein Sci 12:1652–1662

Kall L, Krogh A, Sonnhammer EL (2004) A combined transmembrane topology and signal peptide prediction method. J Mol Biol 338:1027–1036

Kennedy SP, Ng WV, Salzberg SL, Hood L, DasSarma S (2001) Understanding the adaptation of *Halobacterium species* NRC-1 to its extreme environment through computational analysis of its genome sequence. Genome Res 11:1641–1650

Kobayashi T, Kanai H, Aono R, Horikoshi K, Kudo T (1994) Cloning, expression, and nucleotide sequence of the α-amylase gene from the haloalkaliphilic archaeon *Natronococcus sp.* strain Ah-36. J Bacteriol 176:5131–5134

Kwan D, Bolhuis A (2010) Analysis of the twin-arginine motif of a haloarchaeal Tat substrate. FEMS Microbiol Lett 308:138–143

Kwan DC, Thomas JR, Bolhuis A (2008) Bioenergetic requirements of a Tat-dependent substrate in the halophilic archaeon *Haloarcula hispanica*. FEBS J 275:6159–6167

Matsuyama S, Fujita Y, Mizushima S (1993) SecD is involved in the release of translocated secretory proteins from the cytoplasmic membrane of *Escherichia coli*. EMBO J 12:265–270

McDonough JA, McCann JR, Tekippe EM, Silverman JS, Rigel NW, Braunstein M (2008) Identification of functional Tat signal sequences in *Mycobacterium tuberculosis* proteins. J Bacteriol 190:6428–6438

Mongodin EF, Nelson KE, Daugherty S, Deboy RT, Wister J, Khouri H, Weidman J, Walsh DA, Papke RT, Sanchez Perez G, Sharma AK, Nesbø CL, MacLeod D, Bapteste E, Doolittle WF, Charlebois RL, Legault B, Rodríguez-Valera F (2005) The genome of *Salinibacter ruber*: convergence and gene exchange among hyperhalophilic bacteria and archaea. Proc Natl Acad Sci USA 102:18147–18152

Natale P, Bruser T, Driessen AJ (2008) Sec- and Tat-mediated protein secretion across the bacterial cytoplasmic membrane – distinct translocases and mechanisms. Biochim Biophys Acta 1778:1735–1756

Ng SY, Chaban B, VanDyke DJ, Jarrell KF (2007) Archaeal signal peptidases. Microbiology 153:305–314

O'Connor EM, Shand RF (2002) Halocins and sulfolobicins: the emerging story of archaeal protein and peptide antibiotics. J Ind Microbiol Biotechnol 28:23–31

Ortenberg R, Mevarech M (2000) Evidence for post-translational membrane insertion of the integral membrane protein bacterioopsin expressed in the heterologous halophilic archaeon *Haloferax volcanii*. J Biol Chem 275:22839–22846

Osborne AR, Rapoport TA, van den Berg B (2005) Protein translocation by the Sec61/SecY channel. Annu Rev Cell Dev Biol 21:529–550

Paetzel M, Karla A, Strynadka NCJ, Dalbey RE (2002) Signal peptidases. Chem Rev 102:4549–4580

Pelicic V (2008) Type IV pili: e pluribus unum? Mol Microbiol 68:827–837

Pohlschröder M, Dilks K, Hand NJ, Wesley Rose R (2004) Translocation of proteins across archaeal cytoplasmic membranes. FEMS Microbiol Rev 28:3–24

Pohlschröder M, Gimenez MI, Jarrell KF (2005a) Protein transport in Archaea: Sec and twin arginine translocation pathways. Curr Opin Microbiol 8:713–719

Pohlschröder M, Hartmann E, Hand NJ, Dilks K, Haddad A (2005b) Diversity and evolution of protein translocation. Annu Rev Microbiol 59:91–111

Rose RW, Pohlschröder M (2002) In vivo analysis of an essential archaeal signal recognition particle in its native host. J Bacteriol 184:3260–3267

Rose RW, Bruser T, Kissinger JC, Pohlschröder M (2002) Adaptation of protein secretion to extremely high-salt conditions by extensive use of the twin-arginine translocation pathway. Mol Microbiol 45:943–950

Sankaran K, Wu HC (1994) Lipid modification of bacterial prolipoprotein. Transfer of diacylglyceryl moiety from phosphatidylglycerol. J Biol Chem 269:19701–19706

Sumper M, Berg E, Mengele R, Strobel I (1990) Primary structure and glycosylation of the S-layer protein of *Haloferax volcanii*. J Bacteriol 172:7111–7118

Szabo Z, Stahl AO, Albers SV, Kissinger JC, Driessen AJ, Pohlschröder M (2007) Identification of diverse archaeal proteins with class III signal peptides cleaved by distinct archaeal prepilin peptidases. J Bacteriol 189:772–778

Tripepi M, Imam S, Pohlschröder M (2010) *Haloferax volcanii* flagella are required for motility but are not involved in PibD-dependent surface adhesion. J Bacteriol 192:3093–3102

Tschumi A, Nai C, Auchli Y, Hunziker P, Gehrig P, Keller P, Grau T, Sander P (2009) Identification of apolipoprotein N-acyltransferase (Lnt) in mycobacteria. J Biol Chem 284:27146–27156

Tuteja R (2005) Type I signal peptidase: an overview. Arch Biochem Biophys 441:107–111

Uthandi S, Saad B, Humbard MA, Maupin-Furlow JA (2010) LccA, an archaeal laccase secreted as a highly stable glycoprotein into the extracellular medium by *Haloferax volcanii*. Appl Environ Microbiol 76:733–743

van den Berg B, Clemons WM, Collinson I, Modis Y, Hartmann E, Harrison SC, Rapoport TA (2004) X-ray structure of a protein-conducting channel. Nature 427:36–44

Widdick DA, Dilks K, Chandra G, Bottrill A, Naldrett M, Pohlschröder M, Palmer T (2006) The twin-arginine translocation pathway is a major route of protein export in *Streptomyces coelicolor*. Proc Natl Acad Sci USA 103:17927–17932

Yuan J, Zweers JC, van Dijl JM, Dalbey RE (2010) Protein transport across and into cell membranes in bacteria and archaea. Cell Mol Life Sci 67:179–199

Zimmermann R, Eyrisch S, Ahmad M, Helms V (2011) Protein translocation across the ER membrane. Biochim Biophys Acta 1808(3):912–924

Zolghadr B, Weber S, Szabo Z, Driessen AJ, Albers SV (2007) Identification of a system required for the functional surface localization of sugar binding proteins with class III signal peptides in *Sulfolobus solfataricus*. Mol Microbiol 64:795–806

Zwieb C, Bhuiyan S (2010) Archaea signal recognition particle shows the way. Archaea 2010:485051

Chapter 12
Salty and Sweet: Protein Glycosylation in *Haloferax volcanii*

Jerry Eichler, Doron Calo, Lina Kaminski, Lina Kandiba, Zvia Konrad, Hilla Magidovich, Shai Naparstek, and Sophie Yurist-Doutsch

12.1 Introduction

It would be easy to understand why many researchers would shy away from using the halophiles as a model system when addressing molecular questions with regard to Archaea. Yet, despite the apparent hurdle to experimentation presented by their requirement for hypersaline surroundings, halophilic archaea have been the source of many scientific breakthroughs. For many years, halophilic archaea represented the sole members of the third domain of life for which transformation was possible (Allers and Mevarech 2005). Accordingly, *Halobacterium* sp. NRC-1 was one of the first Archaea to have its genome sequenced (Ng et al. 2000). The *Haloarcula marismortui* ribosome structure, solved in 2000 (Ban et al. 2000), has provided understanding of the process of protein translation. Several years earlier, Henderson and Unwin (1975) had reported on the structure of the haloarchaeal light-driven proton pump, bacteriorhodopsin, thus providing the first structural description of a membrane protein. With the description of small archaeal modifier proteins in *Haloferax volcanii*, a novel version of proteasome-mediated protein degradation has been proposed (Humbard et al. 2010). Likewise, the study of protein glycosylation in halophilic archaea has provided novel insight into this post-translational modification both in Archaea and across evolution.

J. Eichler (✉) • D. Calo • L. Kaminski • L. Kandiba • Z. Konrad • H. Magidovich • S. Naparstek • S. Yurist-Doutsch
Department of Life Sciences, Ben Gurion University, Beersheva 84105, Israel
e-mail: jeichler@bgu.ac.il

A. Ventosa et al. (eds.), *Halophiles and Hypersaline Environments*,
DOI 10.1007/978-3-642-20198-1_12,

12.2 Halophilic Archaea Provided the First Glimpses into Glycoproteins and Protein Glycosylation in the Third Domain

Ever since Neuberger first reported that a carbohydrate group was an integral part of ovalbumin in 1938 (Neuberger 1938), it was accepted that N-glycosylation was a unique trait of Eukarya. This belief was, however, challenged in 1976 with the demonstration that the surface (S)-layer glycoprotein of *Halobacterium salinarum* experiences a similar modification (Mescher and Strominger 1976a). Subsequent detailed analysis of the bound glycan fraction showed the S-layer glycoprotein to be modified by two different Asn-linked oligosaccharides, with Asn-2 being decorated with a repeating sulfated pentasaccharide, bound via an N-glycosylamine bond, and ten other Asn residues being modified with a different sulfated glycan, linked through a glucose residue (Lechner et al. 1985a; Lechner and Wieland 1989; Wieland et al. 1980, 1983) (Fig. 12.1). After the *Hbt. salinarum* S-layer glycoprotein was identified as the first non-eukaryal N-modified glycoprotein, haloarchaea provided additional examples of glycoproteins, including the *Hbt. salinarum* flagellin, shown to bear the same glycan-linked sulfated polysaccharide as the S-layer glycoprotein from this species (Wieland et al. 1985), and the S-layer glycoprotein of *Hfx. volcanii* (Sumper et al. 1990).

In addition to providing the first examples of archaeal proteins bearing N-linked glycans, such haloarchaea-based efforts also revealed that in Archaea, a variety of saccharides can serve as linking sugar, namely that sugar serving to attach an oligosaccharide to a modified Asn residue, including glucose and N-acetylglucosamine (cf. Eichler and Adams, 2005). By contrast, in bacterial N-glycoproteins characterized thus far, bacillosamine (2,4-diamino-2,4,6-trideoxyglucopyranose) serves as the linking sugar (Young et al. 2002), while in the vast majority of eukaryal N-glycoproteins, N-acetylglucosamine (GlcNAc) fulfills this role (Spiro 1973). In the particular case of the *Hbt. salinarum* S-layer glycoprotein, two different linking sugars are employed within the same glycoprotein (Lechner and Wieland 1989), despite the fact that *Hbt. salinarum* seemingly only encodes a single oligosaccharyltransferase, the enzyme responsible for adding a glycan to its target Asn residue (Magidovich and Eichler 2009).

As well as providing the first examples of archaeal glycoproteins and descriptions of their N-linked glycan composition, initial efforts aimed at defining the pathway responsible for the archaeal version of this post-translational modification also largely relied on haloarchaea. In Eukarya, Bacteria and Archaea, N-linked glycans are first assembled on a lipid carrier. As in Eukarya (Burda and Aebi 1999), the archaeal N-linked oligosaccharide is assembled on a dolichol carrier, rather than the undecaprenol carrier used in bacterial N-glycosylation (Szymanski and Wren 2005). Accordingly, Archaea contain both dolichol phosphate and dolichol pyrophosphate bearing either mono- or polysaccharides. Specifically, *Hbt. salinarum* was reported as containing phosphorylated C_{60} dolichol bearing glucose, mannose and GlcNAc units, as well as a sulfated tetrasaccharide

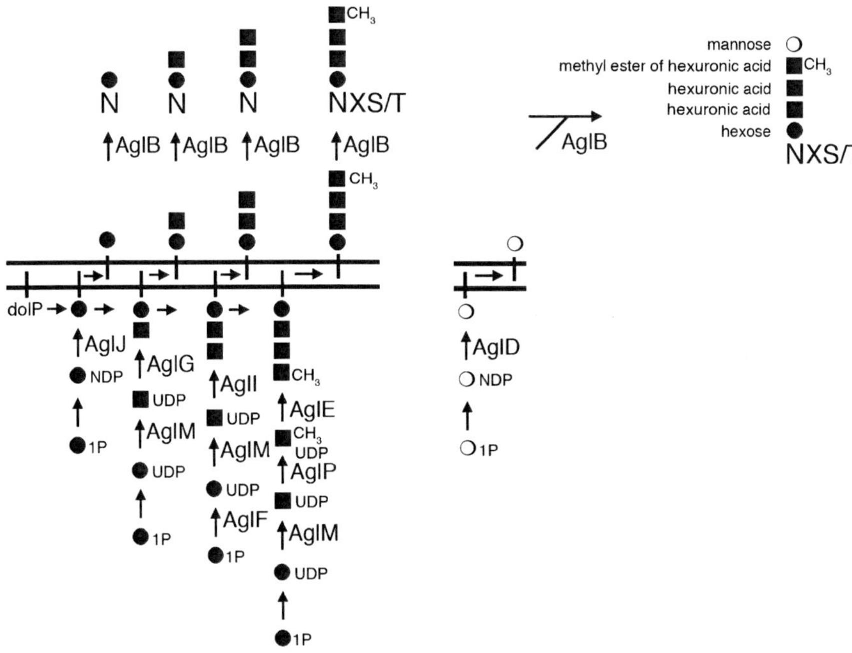

Fig. 12.1 Working model of the *Hfx. volcanii* N-glycosylation pathway. In *Hfx. volcanii*, a reporter glycoprotein, the S-layer glycoprotein, experiences N-glycosylation on at least two Asn residues, Asn-13 and Asn-83. N-glycosylation involves modification by a pentasaccharide comprising a hexose, two hexuronic acids, a methyl ester of hexuronic acid and mannose. The assembly and attachment of the pentasaccharide requires the actions of a series of Agl proteins. AglJ, AglG, AglI, and AglE sequentially add the first four pentasaccharide subunits onto a common dolichol phosphate carrier. AglD adds the mannose that corresponds to the final pentasaccharide subunit onto a distinct dolichol phosphate carrier. AglF, AglM, and AglP are involved in pentasaccharide subunit biogenesis. AglF is a glucose-1-phosphate uridyltransferase involved in the biosynthesis of the hexuronic acid found at pentasaccharide position three. AglM, a UDP-glucose dehydrogenase, was shown to participate in the biogenesis of the hexuronic acid found at pentasaccharide position two and likely of the hexuronic acids found at positions three and four, as well. AglP is a S-adenosyl-L-methionine-dependent methyltransferase responsible for the formation of the methyl ester of hexuronic acid found at position four of the pentasaccharide. Once the first four pentasaccharides have been assembled on the common dolichol phosphate carrier, the glycan-charged lipid is flipped to face the cell exterior, at which point the oligosaccharyltransferase, AglB, transfers the tetrasaccharide to select S-layer glycoprotein Asn residues. The mannose-charged dolichol phosphate carrier is also flipped, after which the mannose residue is transferred to the N-linked tetrasaccharide. In addition, dolichol phosphates bearing precursor tri-, di-, and monosaccharides are flipped and the sugar moieties transferred to the S-layer glycoprotein. The responsible flippase(s) have yet to be identified. For further details, see the text

(Mescher et al. 1976; Lechner et al. 1985a). Treatment with bacitracin, a reagent that interferes with recycling of glycan-charged pyrophosphodolichol carriers following release of their bound oligosaccharides (Stone and Strominger 1971), was able to prevent modification of the *Hbt. salinarum* S-layer glycoprotein by the glucose-linked sulfated glycan otherwise added to ten sequons of the protein

(Mescher and Strominger 1978). By contrast, such bacitracin treatment failed to prevent addition of the GlcNAc-linked repeating sulfated pentasaccharide normally found at *Hbt. salinarum* S-layer glycoprotein Asn-2, suggesting the use of dolichol phosphate rather than dolichol pyrophosphate as the carrier for this glycan (Wieland et al. 1980). In *Hbt. salinarum*, the glycan moiety of the pyrophospho-dolichol-bound sulfated polysaccharide is also detected on the S-layer glycoprotein and flagellins in this species (Lechner et al. 1985a; Wieland et al. 1985). Moreover, the sulfated polysaccharide is methylated in the pyrophosphodolichol-linked form but not when protein-bound (Lechner et al. 1985b). In *Hfx. volcanii*, C_{55} and/or C_{60} dolichols containing mannosyl-(β1-4)-galactosyl moieties, lesser quantities of a dihexosyl group and a tetrasaccharide comprising mannose, galactose, and rhamnose subunits were reported (Kuntz et al. 1997). Moreover, the transfer of radiolabeled glucose from UDP-[^{3}H]glucose to *Hfx. volcanii* glycoproteins was shown to proceed through a glucose-containing phosphopolyisoprenol intermediate (Zhu et al. 1995). More recently, it was reported that *Hfx. volcanii* contains dolichol phosphate species modified by the first four subunits of the pentasaccharide shown to decorate at least two Asn residues of the S-layer glycoprotein in this species, as well as a distinct dolichol phosphate species modified with mannose, the fifth pentasaccharide subunit (Guan et al. 2010; see below). Finally, the saccharide-charged dolichol species in *Hfx. volcanii* are unusual in that both the α- and ω-terminal isoprene units are saturated and since only monophosphorylated dolichol is observed (Kuntz et al. 1997; Guan et al. 2010).

When considering the topology of N-glycosylation, evidence obtained from haloarchaea points to this processing event as occurring on the outer surface of the cell, the topological equivalent of the lumenal-facing leaflet of the endoplasmic reticulum (ER) membrane bilayer, the site of N-glycosylation in Eukarya. Localization of archaeal N-glycosylation to the external cell surface is supported by studies showing the ability of *Hbt. salinarum* cells to modify cell-impermeable, sequon-bearing hexapeptides with sulfated oligosaccharides (Lechner et al. 1985a). Indeed, although unable to cross the haloarchaeal membrane (Mescher and Strominger 1978), bacitracin is nonetheless able to interfere with N-glycosylation in *Hbt. salinarum*, blocking the transfer of sulfated oligosaccharides to the S-layer glycoprotein (Mescher and Strominger 1978; Wieland et al. 1980). Finally, studies supporting a co-translational mode of membrane protein insertion in *Hbt. salinarum* and *Hfx. volcanii* also lend support to protein glycosylation as taking place on the exterior surface of the cell (Gropp et al. 1992; Dale and Krebs 1999; Ring and Eichler 2004). On the other hand, glycan charging of the dolichol phosphates used as carriers in the *Hfx. volcanii* N-glycosylation pathway transpires on the cytoplasmic face of the plasma membrane (Plavner and Eichler 2008).

Thus, while earlier studies on haloarchaeal protein N-glycosylation provided insight into the initial and final steps of the process, little was known of the intervening events. However, over the last 5 years or so, with the availability of a growing number of archaeal genome sequences, detailed understanding of the archaeal version of protein glycosylation became possible. In this manner,

Abu-Qarn and Eichler (2006) identified a series of *Hfx. volcanii* gene products experimentally verified as participating in this post-translational modification.

12.3 Sweet and Salty: N-glycosylation in *Haloferax volcanii*

As the major polypeptide in the species, the *Hfx. volcanii* S-layer glycoprotein represents an excellent model for addressing archaeal N-glycosylation. At the time of its description (Sumper et al. 1990; Mengele and Sumper 1992), it was reported that of the seven putative S-layer glycoprotein N-glycosylation sites, Asn-13 and Asn-505 were each modified by a linear chain of β-1-4 linked glucose residues, while Asn-274 and/or Asn-279 were described as bearing a polysaccharide containing glucose, idose and galactose groups. Later efforts relying on more sophisticated mass spectrometry tools revealed the *Hfx. volcanii* S-layer glycoprotein to instead be modified at Asn-13 and Asn-83 by a pentasaccharide comprising two hexoses, two hexuronic acids and a methyl ester of hexuronic acid (Abu-Qarn et al. 2007; Magidovich et al. 2010). Such efforts also revealed that the sequon at Asn-370 is not modified (Abu-Qarn et al. 2007). These studies not only redefined the N-linked glycan profile of the S-layer glycoprotein but also served to define components of the *Hfx. volcanii* N-glycosylation pathway.

Bioinformatics tools were first recruited to identity *Hfx. volcanii* homologues of eukaryal and bacterial N-glycosylation genes (Abu-Qarn and Eichler 2006). This led to the identification of open reading frames, subsequently remained *agl* (*a*rchaeal *gl*ycosylation) genes following the nomenclature of Chaban et al. (2006), shown to encode for glycosyltransferases (i.e., AglD, AglE, AglI, and AglJ), an oligosaccharyltransferase (i.e., AglB) and other sugar-processing enzymes (i.e., AglF) involved in the assembly and attachment of the *Hfx. volcanii* S-layer glycoprotein N-linked pentasaccharide (Abu-Qarn et al. 2007; Abu-Qarn et al. 2008; Yurist-Doutsch et al. 2008). The putative glycosyltransferase-encoding *aglG* gene was next identified on the basis of it being positioned between the *aglB* and *aglI* sequences (Yurist-Doutsch et al. 2008). Indeed, mapping the various *agl* sequences to the *Hfx. volcanii* genome revealed that, apart from *aglD*, all were sequestered to a gene island stretching from *aglJ* to *aglB* (Yurist-Doutsch and Eichler 2009). However, as the current annotation of the *Hfx. volcanii* genome did not recognize *aglE* as an open reading frame (Abu-Qarn et al. 2008), the entire *agl* gene island was subjected to manual re-annotation. As a result, additional sequences, i.e., *aglP*, *aglQ* and *aglR*, were identified (Yurist-Doutsch and Eichler 2009), as was *algM*, shown to lie beyond the original gene island borders (Yurist-Doutsch et al. 2010).

Through a combination of gene deletion and mass spectrometry approaches, each of the *Hfx. volcanii* Agl proteins has been assigned a specific role in the N-glycosylation process. AglJ, AglG, AglI and AglE sequentially add pentasaccharide subunits one through four onto a common dolichol phosphate carrier (Abu-Qarn et al. 2008; Yurist-Doutsch et al. 2008; Kaminski et al. 2010; Guan et al. 2010), while AglD adds a mannose residue, the fifth pentasaccharide subunit, onto a distinct dolichol phosphate (Abu-Qarn et al. 2007; Guan et al. 2010). AglB, the oligosaccharyltransferase, is responsible for adding the glycan to the target Asn residue (Abu-Qarn et al. 2007) Still, only for AglF, AglM and AglP has biochemical characterization been performed. AglF, was shown to be a glucose-1-phosphate uridyltransferase involved in the biosynthesis of the hexuronic acid found at position three of the S-layer glycoprotein-linked pentasaccharide (Yurist-Doutsch et al. 2010). AglM, a UDP-glucose dehydrogenase, was shown to participate in the biogenesis of the hexuronic acid found at pentasaccharide position two and likely of the hexuronic acids found at positions three and four, as well (Yurist-Doutsch et al. 2010). In a combined in vitro reconstitution experiment, AglF and AglM were shown to work in a coordinated manner to generate UDP-glucuronic acid from glucose-1-phosphate and UTP in a NAD^+-dependent manner (Yurist-Doutsch et al. 2010). Finally, AglP was confirmed to be a *S*-adenosyl-L-methionine-dependent methyltransferase responsible for the formation of the methyl ester of hexuronic acid found at position four of the pentasaccharide (Magidovich et al. 2010). The current working model of the *Hfx. volcanii* N-glycosylation pathway is presented in Fig. 12.1.

12.4 Haloarchaeal N-glycosylation: An Aid to Life in Hypersaline Conditions?

As *aglB* can be deleted from *Hfx. volcanii* (Abu-Qarn and Eichler 2006), it would appear that N-glycosylation is not essential for survival of this species, at least under the conditions tested. Nonetheless, N-glycosylation may contribute to the ability of haloarchaea and their proteins to survive or adapt to the hypersaline environments in which these organisms exist. In comparing the N-linked glycan profiles of two haloarchaeal S-layer glycoproteins (Mengele and Sumper 1992), it was noted that in *Hbt. salinarum*, the S-layer glycoprotein not only experiences a higher degree of N-glycosylation than does its *Hfx. volcanii* counterpart but that the glycans of former are highly sulfated, as opposed to the neutral sugars found in the latter. In this manner, the *Hbt. salinarum* S-layer glycoprotein presents a drastically increased surface charge density, possibly contributing to protein stability in the face of molar salt concentrations (Madern et al. 2000). Moreover, the *Hbt. salinarum* S-layer glycoprotein also contains 20% more acidic amino acid residues than does the *Hfx. volcanii* S-layer glycoprotein (Lechner and Sumper 1987; Sumper et al. 1990). These considerations could help explain how *Hbt. salinarum* is able to grow at higher salinity than does *Hfx. volcanii*. In *Hfx. volcanii*, however, absent or perturbed N-glycosylation greatly compromised growth at increasing salt concentrations (Abu-Qarn et al. 2007). At the same time, the transcription of *agl* genes involved in *Hfx. volcanii* N-glycosylation occurs in a coordinated (yet differential) manner in the face of different growth paradigms, pointing to this post-translational modification as serving an adaptive role (Yurist-Doutsch et al. 2008, 2010).

In addition to possibility affording advantages in the face of environmental challenges, archaeal protein glycosylation is thought to provide structural support to the cell. This is best exemplified in *Hbt. salinarum*, where bacitracin treatment transformed rod-shaped cells into spheres (Mescher and Strominger 1976b). Defective *Hfx. volcanii* N-glycosylation resulted in an unstructured surface layer, while the absence of N-glycosylation compromised S-layer stability (Abu-Qarn et al. 2007). Finally, N-glycosylation appears to play a role in stabilization against proteolysis, as reflected by the loss of *Hfx. volcanii* S-layer glycoprotein resistance to added protease in strains lacking Agl proteins involved in assembling and attaching the pentasaccharide N-linked to the protein (Yurist-Doutsch et al. 2008, 2010; Kaminski et al. 2010).

12.5 Conclusion

Although haloarchaea offered the first insight into archaeal protein glycosylation over 35 years ago, even today haloarchaea remain central to understanding the archaeal version of this protein-processing event. Indeed, what has been learned to date suggests that continued examination of haloarchaeal versions of this universal post-translational modification will provide additional insight into the process not obtained from analysis of either the eukaryal or bacterial pathways. Indeed, the ongoing development of improved molecular tools for working with haloarchaea will continue to expand our understanding of protein glycosylation. For instance, in addition to N-glycosylation, archaeal proteins can also experience O-glycosylation. Specifically, in both the *Hbt. salinarum* and the *Hfx. volcanii* S-layer glycoproteins, Thr-rich regions adjacent to the predicted membrane-spanning domain of the proteins are modified with galactose-glucose disaccharides (Mescher and Strominger 1976a; Sumper et al. 1990). At present, virtually nothing is known of the archaeal O-glycosylation pathway.

Acknowledgments Research in the Eichler laboratory is supported by the Israel Science Foundation (grant 30/07). S.Y.-D. is the recipient of a Negev-Faran Associates Scholarship, while L.K. is the recipient of a Negev-Zin Associates Scholarship.

References

Abu-Qarn M, Eichler J (2006) Protein N-glycosylation in Archaea: defining *Haloferax volcanii* genes involved in S-layer glycoprotein glycosylation. Mol Microbiol 61:511–525

Abu-Qarn M, Yurist-Doutsch S, Giordano A, Trauner A, Morris HR, Hitchen P, Medalia O, Dell A, Eichler J (2007) *Haloferax volcanii* AglB and AglD are involved in N-glycosylation of the S-layer glycoprotein and proper assembly of the surface layer. J Mol Biol 374:1224–1236

Abu-Qarn M, Giordano A, Battaglia F, Trauner A, Hitchen P, Morris HR, Dell A, Eichler J (2008) Identification of AglE, a second glycosyltransferase involved in N-glycosylation of the *Haloferax volcanii* S-layer glycoprotein. J Bacteriol 190:3140–3146

Allers T, Mevarech M (2005) Archaeal genetics – the third way. Nat Rev Genet 6:58–73

Ban N, Nissen P, Hansen J, Moore PB, Steitz TA (2000) The complete atomic structure of the large ribosomal subunit at 2.4 Åresolution. Science 289:905–920

Burda P, Aebi M (1999) The dolichol pathway of N-linked glycosylation. Biochim Biophys Acta 1426:239–257

Chaban B, Voisin S, Kelly J, Logan SM, Jarrell KF (2006) Identification of genes involved in the biosynthesis and attachment of *Methanococcus voltae* N-linked glycans: insight into N-linked glycosylation pathways in Archaea. Mol Microbiol 61:259–268

Dale H, Krebs MP (1999) Membrane insertion kinetics of a protein domain in vivo. The bacterioopsin N terminus inserts co-translationally. J Biol Chem 274:22693–22698

Eichler J, Adams MWW (2005) Post-translational protein modification in Archaea. Microbiol Mol Biol Rev 69:393–425

Gropp R, Gropp F, Betlach MC (1992) Association of the halobacterial 7S RNA to the polysome correlates with expression of the membrane protein bacterioopsin. Proc Natl Acad Sci USA 89:1204–1208

Guan Z, Naparstek S, Kaminski L, Konrad Z, Eichler J (2010) Distinct glycan-charged phosphodolichol carriers are required for the assembly of the pentasaccharide N-linked to the *Haloferax volcanii* S-layer glycoprotein. Mol Microbiol 78:1294–1303

Henderson R, Unwin PN (1975) Three-dimensional model of purple membrane obtained by electron microscopy. Nature 257:28–32

Humbard MA, Miranda HV, Lim JM, Krause DJ, Pritz JR, Zhou G, Chen S, Wells L, Maupin-Furlow JA (2010) Ubiquitin-like small archaeal modifier proteins (SAMPs) in *Haloferax volcanii*. Nature 463:54–60

Kaminski L, Abu-Qarn M, Guan Z, Naparstek S, Ventura VV, Raetz CRH, Hitchen PG, Dell A, Eichler J (2010) AglJ participates in adding the linking saccharide in the *Haloferax volcanii* N-glycosylation pathway. J Bacteriol 192:5572–5579

Kuntz C, Sonnenbichler J, Sonnenbichler I, Sumper M, Zeitler R (1997) Isolation and characterization of dolichol-linked oligosaccharides from *Haloferax volcanii*. Glycobiology 7:897–904

Lechner J, Sumper M (1987) The primary structure of a procaryotic glycoprotein. Cloning and sequencing of the cell surface glycoprotein gene of halobacteria. J Biol Chem 262:9724–9729

Lechner J, Wieland F (1989) Structure and biosynthesis of prokaryotic glycoproteins. Annu Rev Biochem 58:173–194

Lechner J, Wieland F, Sumper M (1985a) Biosynthesis of sulfated saccharides N-glycosidically linked to the protein via glucose. Purification and identification of sulfated dolichyl monophosphoryl tetrasaccharides from halobacteria. J Biol Chem 260:860–866

Lechner J, Wieland F, Sumper M (1985b) Transient methylation of dolichyl oligosaccharides is an obligatory step in halobacterial sulfated glycoprotein biosynthesis. J Biol Chem 260: 8984–8989

Madern D, Ebel C, Zaccai G (2000) Halophilic adaptation of enzymes. *Extremophiles* 4:91–98

Magidovich H, Eichler J (2009) Glycosyltransferases and oligosaccharyltransferases in Archaea: putative components of the N-glycosylation pathway in the third domain of life. FEMS Microbiol Lett 300:122–130

Magidovich H, Yurist-Doutsch S, Konrad Z, Ventura VV, Hitchen PG, Dell A, Eichler J (2010) AglP is a cytoplasmic S-adenosyl-L-methionine-dependent methyltransferase that participates in the N-glycosylation pathway of *Haloferax volcanii*. Mol Microbiol 76:190–199

Mengele R, Sumper M (1992) Drastic differences in glycosylation of related S-layer glycoproteins from moderate and extreme halophiles. J Biol Chem 267:8182–8185

Mescher MF, Strominger JL (1976a) Purification and characterization of a prokaryotic glucoprotein from the cell envelope of *Halobacterium salinarium*. J Biol Chem 251: 2005–2014

Mescher MF, Strominger JL (1976b) Structural (shape-maintaining) role of the cell surface glycoprotein of *Halobacterium salinarium*. Proc Natl Acad Sci USA 73:2687–2691

Mescher MF, Hansen U, Strominger JL (1976) Formation of lipid-linked sugar compounds in *Halobacterium salinarium*. Presumed intermediates in glycoprotein synthesis. J Biol Chem 251:7289–7294

Mescher MF, Strominger JL (1978) Glycosylation of the surface glycoprotein of *Halobacterium salinarium* via a cyclic pathway of lipid-linked intermediates. FEBS Lett 89:37–41

Neuberger A (1938) Carbohydrates in proteins. The carbohydrate component of crystalline egg albumin. Biochem J 32:1435–1451

Ng WV, Kennedy SP, Mahairas GG, Berquist B, Pan M, Shukla HD, Lasky SR, Baliga NS, Thorsson V, Sbrogna J, Swartzell S, Weir D, Hall J, Dahl TA, Welti R, Goo YA, Leithauser B, Keller K, Cruz R, Danson MJ, Hough DW, Maddocks DG, Jablonski PE, Krebs MP, Angevine CM, Dale H, Isenbarger TA, Peck RF, Pohlschroder M, Spudich JL, Jung KW, Alam M, Freitas T, Hou S, Daniels CJ, Dennis PP, Omer AD, Ebhardt H, Lowe TM, Liang P, Riley M, Hood L, DasSarma S (2000) Genome sequence of *Halobacterium* species NRC-1. Proc Natl Acad Sci USA 97:12176–12181

Plavner N, Eichler J (2008) Defining the topology of the N-glycosylation pathway in the halophilic archaea *Haloferax volcanii*. J Bacteriol 190:8045–8052

Ring G, Eichler J (2004) In the Archaea *Haloferax volcanii*, membrane protein biogenesis and protein synthesis rates are affected by decreased ribosomal-binding to the translocon. J Biol Chem 279:53160–53166

Spiro RG (1973) Glycoproteins. Adv Protein Chem 27:349–467

Stone KJ, Strominger JL (1971) Mechanism of action of bacitracin: complexation with metal ion and C 55-isoprenyl pyrophosphate. Proc Natl Acad Sci USA 68:3223–3227

Sumper M, Berg E, Mengele R, Strobel I (1990) Primary structure and glycosylation of the S-layer protein of *Haloferax volcanii*. J Bacteriol 172:7111–7118

Szymanski CM, Wren BW (2005) Protein glycosylation in bacterial mucosal pathogens. Nat Rev Microbiol 3:225–237

Wieland F, Dompert W, Bernhardt G, Sumper M (1980) Halobacterial glycoprotein saccharides contain covalently linked sulphate. FEBS Lett 120:110–114

Wieland F, Heitzer R, Schaefer W (1983) Asparaginylglucose: novel type of carbohydrate linkage. Proc Natl Acad Sci USA 80:5470–5474

Wieland F, Paul G, Sumper M (1985) Halobacterial flagellins are sulfated glycoproteins. J Biol Chem 260:15180–15185

Young NM, Brisson JR, Kelly J, Watson DC, Tessier L, Lanthier PH, Jarrell HC, Cadotte N, St Michael F, Aberg E, Szymanski CM (2002) Structure of the N-linked glycan present on multiple glycoproteins in the Gram-negative bacterium, *Campylobacter jejuni*. J Biol Chem 277:42530–42539

Yurist-Doutsch S, Eichler J (2009) Manual annotation, transcriptional analysis and protein expression studies reveal novel genes in the agl cluster responsible for N-glycosylation in the halophilic archaeon *Haloferax volcanii*. J Bacteriol 191:3068–3075

Yurist-Doutsch S, Abu-Qarn M, Battaglia F, Morris HR, Hitchen PG, Dell A, Eichler J (2008) *aglF*, *aglG* and *aglI*, novel members of a gene cluster involved in the N-glycosylation of the *Haloferax volcanii* S-layer glycoprotein. Mol Microbiol 69:1234–1245

Yurist-Doutsch S, Magidovich H, Ventura VV, Hitchen PG, Dell A, Eichler J (2010) N-glycosylation in Archaea: on the coordinated actions of *Haloferax volcanii* AglF and AglM. Mol Microbiol 75:1047–1058

Zhu BC, Drake RR, Schweingruber H, Laine RA (1995) Inhibition of glycosylation by amphomycin and sugar nucleotide analogs PP36 and PP55 indicates that *Haloferax volcanii* β-glucosylates both glycoproteins and glycolipids through lipid-linked sugar intermediates: evidence for three novel glycoproteins and a novel sulfated dihexosyl-archaeol glycolipid. Arch Biochem Biophys 319:355–364

Chapter 13
Effect of Anoxic Conditions and Temperature on Gas Vesicle Formation in *Halobacterium salinarum*

Felicitas Pfeifer, Regina Frommherz, Karin Faist, Torsten Hechler, Katharina Teufel, and Larissa Marschaus

13.1 Introduction

Many cyanobacteria and haloarchaea form gas vesicles. These gas-filled structures contain neither lipids nor carbohydrates and the wall is mainly constituted of a single protein, the 8-kDa GvpA (*g*as *v*esicle *p*rotein A). This protein forms the gas vesicle wall that is hydrophobic at the inner surface and hydrophilic at the outside. GvpC is attached on the outer surface and stabilizes the structure. The GvpA monomers are not covalently bound; rather hydrophobic and ionic interactions are the basis for the strong interactions between GvpA (Walsby 1994). Due to the hydrophobic inner surface water is excluded whereas gases dissolved in the cytoplasm freely diffuse. When present in large amounts and of appropriate size, gas vesicles reduce the buoyancy of the cells and enable them to float to the surface where light and oxygen conditions might be more favorable for growth (Walsby 1994). Photosynthetic haloarchaea and cyanobacteria are thus able to reach positions in the water column that are optimal for light absorption and energy conversion (Fig. 13.1a). However, also some anaerobic bacteria and methanogenic archaea produce gas vesicles, implying that these structures must have additional functions besides enabling microorganisms to float. One of the possibilities is a reduction of the cytoplasmatic volume without changing the size of the cell surface. Cells tightly filled with gas vesicles contain a much smaller cytoplasm that keeps diffusion pathways short. This might be beneficial, for instance, in the cold. Up to 91% of the bacteria found in sea ice of Antarctica possess gas vesicles, such as the filamentous *Rhodoferax antarctica* found in the near-freezing water column (Karr et al. 2003; Jung et al. 2004).

The haloarchaea *Halobacterium salinarum*, *Haloferax mediterranei*, *Haloquadratum walsbyi* and the haloalkaliphilic *Halorubrum vacuolatum* all

F. Pfeifer (✉) • R. Frommherz • K. Faist • T. Hechler • K. Teufel • L. Marschaus
Mikrobiologie & Genetik, Technische Universität Darmstadt, Schnittspahnstrasse 10, 64287 Darmstadt, Germany
e-mail: pfeifer@bio.tu-darmstadt.de

A. Ventosa et al. (eds.), *Halophiles and Hypersaline Environments*,
DOI 10.1007/978-3-642-20198-1_13,

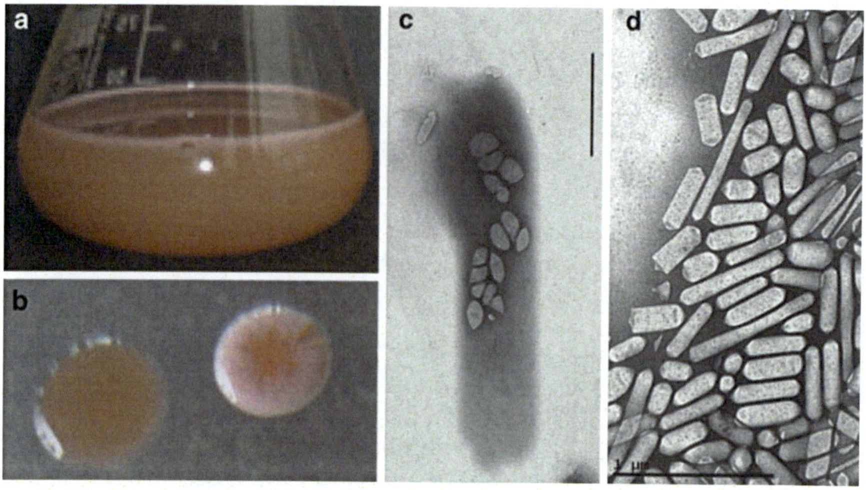

Fig. 13.1 Gas vesicles of *Hbt. salinarum*. (**a**) Liquid culture of *Hbt. salinarum* PHH1 containing cells floating at the surface. (**b**) Colonies grown on solid media. *Red* transparent colonies lack gas vesicles whereas *pink* and turbid colonies contain these structures. (**c**) PHH4 inspected by TEM, gas vesicles are seen as *light grey* bodies; (**d**) isolated gas vesicles inspected by TEM. The *bars* depict 1 μm

produce gas vesicles. These archaea grow well between temperatures of 20 and 40°C under oxic and often also anoxic conditions. The presence of gas vesicles is recognized by the white and turbid appearance of liquid cultures or colonies grown on solid media, whereas cells lacking gas vesicles are red and transparent (Fig. 13.1b). Light refractile bodies seen by phase contrast (light) microscopy occur during growth and are seen as spindle- and cylinder-shaped gas vesicles by transmission electron microscopy (TEM) (Fig. 13.1c). The mutant strain *Hbt. salinarum* PHH4 and also *Haloferax mediterranei* produce gas vesicles in the stationary growth phase only. Despite the early discovery of gas vesicles in *Halobacterium salinarum* not very much is known about the process of their formation.

Gas vesicles are easily isolated by disruption of the salt-adapted haloarchaeal cells with water, centrifugation of the lysate at low *g*-values and collection of the floating particles from the surface. Such a gas vesicle suspension can by analyzed by TEM (Fig. 13.1d) or biochemical procedures. When SDS-polyacrylamide gels electrophoresis (SDS-PAGE) is performed gas vesicles remain in the loading well, since the gas vesicle wall consisting of GvpA is not disrupted in SDS-containing buffers. Western analysis using an antiserum detecting GvpA results in an empty blot (Fig. 13.2a, left). However, the lysate of *Hbt. salinarum* contains GvpA monomers and oligomers as observed by Western analysis (Fig. 13.2a, right), suggesting that GvpA itself is able to enter the SDS-polyacrylamide gel, but once aggregated and incorporated to form the gas vesicle wall neither monomers nor oligomers are released. The monomer and oligomers observed in the lysate of gas vesicle producing cells are presumably intermediates in the process of gas vesicle formation (or degradation).

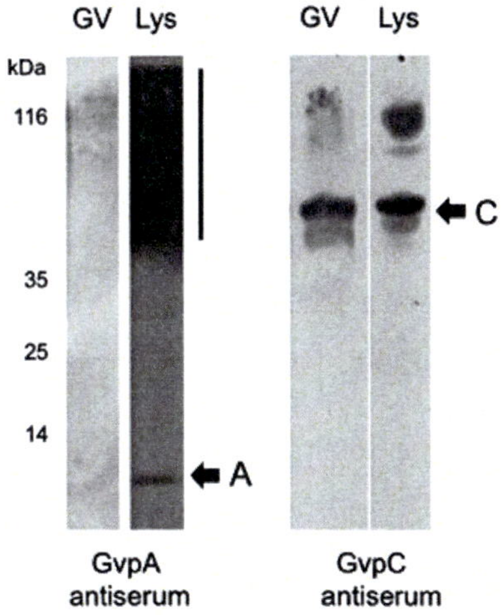

Fig. 13.2 Western analysis to determine GvpA and GvpC in a gas vesicle preparation (GV) or lysate of *Hbt. salinarum* PHH1 (Lys). Gas vesicles were isolated by centrifugation-enhanced flotation as described (Shukla and DasSarma 2004). Two microgram of total protein was applied in case of the lysate and 20 μg in case of GV. GV or lysate were separated by SDS-PAGE, transferred to a nitrocellulose membrane and reacted with a gas vesicle antiserum detecting GvpA (Englert et al. 1990) or GvpC (Englert and Pfeifer 1993). The reacting gas vesicle aggregates (*bar*) and the 8-kDa GvpA monomer (A), and the reacting 42-kDa GvpC (C) are marked

Gas vesicles purified by repeated centrifugation-enhanced flotation in Tris-buffered water almost exclusively contain GvpA (Englert et al. 1990). However, gas vesicles purified by repeated flotation in 5% NaCl contain in addition to GvpA also the second structural protein, GvpC that is attached to the outer gas vesicle surface (Fig. 13.2b, left) (Englert and Pfeifer 1993; Shukla and DasSarma 2004). The 42-kDa GvpC stabilizes the gas vesicle structure (Halladay et al. 1993; Englert and Pfeifer 1993; Offner et al. 1996). Very minor amounts of a few other Gvp proteins are found in gas vesicle preparations purified in 5% NaCl, namely GvpF, GvpG, GvpJ, GvpL, and GvpM (Shukla and DasSarma 2004). These accessory proteins can only be detected by immunological methods using 150 μg of gas vesicles per lane for Western analysis. Since gas vesicle preparations purified in water lack these additional Gvp proteins, these are either very minor constituents of the gas vesicle envelope and/or washed off during the purification procedure. Besides forming a putative tiny initial protein aggregate that is further enlarged by the addition of GvpA to form the air-filled gas vesicle, some of these additional Gvp proteins might support the formation of the gas vesicle structure as chaperone, scaffolding, or stabilizing protein. Others, like GvpD und GvpE are involved in the regulation of *gvp* gene expression. A total of 14 *gvp* genes are involved in gas vesicle formation, and 8 of these are required as demonstrated by genetic analyses (Offner et al. 2000). Two of the additional proteins encoded here, GvpD and GvpN, contain a nucleotide-binding motif (p-loop).

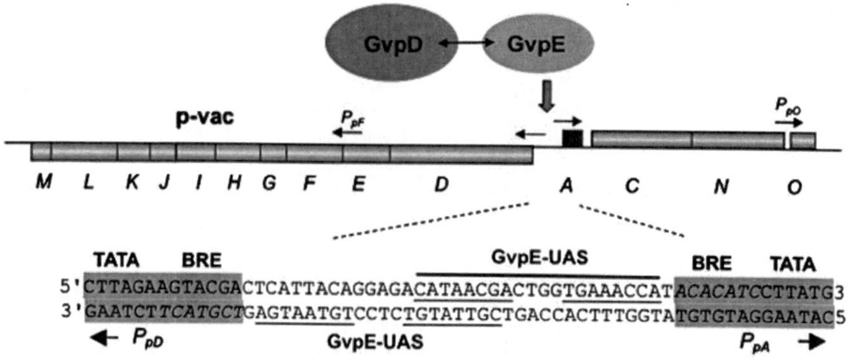

Fig. 13.3 The p-vac region of *Hbt. salinarum* PHH1 and nucleotide sequence of P_{pD}-P_{pA}. The 14 *gvp* genes are shown as boxes labeled ACNO and D through M. *Arrows* above the map depict the four promoters and the direction of transcription. The sequence of P_{pD}–P_{pA} is given below. The TATA-box and BRE are shaded in *grey* and the two 20-nt GvpE-responsive elements (GvpE-UAS) are *underlined*

The 14 *gvp* genes are located in two oppositely oriented gene clusters, *gvpACNO* and *gvpDEFGHIJKLM* and constitute the vac-region (Fig. 13.3). The wild-type *Hbt. salinarum* PHH1 contains the p-vac region in the 150-kb plasmid pHH1 (Englert et al. 1992a). An almost identical *gvp* gene cluster (*gvp1*) is found in pNRC100 of *Hbt. salinarum* NRC-1 (Ng et al. 2000). The deletion mutant *Hbt. salinarum* PHH4 lacks the p-vac region and surrounding sequences and produces gas vesicles due to the expression of c-vac (Krüger and Pfeifer 1996). The *gvp2* gene cluster of NRC-1 is almost identical to c-vac but not expressed in this strain. *Hbt. salinarum* is able to grow by arginine fermentation under anoxic conditions (Hartmann et al. 1980). In this report, we summarize our data on the regulation of *gvp* gene expression and the effect of anoxic conditions, heat shock and cold temperature on the formation of gas vesicles in *Hbt. salinarum* PHH1 and PHH4.

13.2 Transcription of *gvp* Is Regulated by GvpD and GvpE

Gas vesicles formed in *Hbt. salinarum* PHH1 are based on the expression of p-vac, whereas the c-*gvpACNO* genes are not transcribed in this strain. Transcription of p-vac starts at four promoters, P_{pA}, P_{pD}, P_{pF} and P_{pO} (Fig. 13.3) (Englert et al. 1992a; Hofacker et al. 2004). In contrast, the c-vac region of *Hbt. salinarum* PHH4 is transcribed from two promoters, P_{cA} and P_{cD} (Krüger and Pfeifer 1996). Two endogenous regulator proteins, the activator GvpE and the repressing protein GvpD are involved in the regulation of *gvp* transcription.

GvpE is able to dimerize and presumably binds to a 20-nt sequence (GvpE-UAS) upstream and adjacent to the BRE of the oppositely oriented P_{pA} and P_{pD} (Fig. 13.3) (Gregor and Pfeifer 2005; Bauer et al. 2008). Since both BRE are separated by 35 nt only, the two GvpE-UAS overlap significantly in their distal portion in the center of this region. The P_{pA} promoter shows a stronger activation by GvpE compared to P_{pD}, and the exchange of the respective GvpE-UAS elements leads to an increased activation of the P_{pD}. Quantification of the amount of transcripts derived from P_{pA} and comparison of the basal and GvpE-induced amounts of p-*gvpA* mRNA demonstrates a 50-fold increase in the presence of pGvpE and 75-fold increase in the presence of cGvpE suggesting that cGvpE is the stronger activator (Gregor and Pfeifer 2005). Neither P_{pF} nor P_{pO} is activated by pGvpE.

The binding sites of GvpE are adjacent to BRE and thus to the basal transcription apparatus consisting of RNA polymerase, the transcription factor TFB and the TATA-box binding protein TBP. The question arises whether GvpE is able to contact TFB and/or TBP. Five distinct *tbp* genes are present in *Hbt. salinarum* PHH1 and all five TBP proteins are able to interact with GvpE as demonstrated by protein–protein affinity chromatography (Teufel et al. 2008; Teufel and Pfeifer 2010). Similar analyses performed with the five different transcription factors TFB present of PHH1 are still in progress, but the results obtained so far suggest a similar behavior (Bleiholder A, Teufel K, Pfeifer F, unpublished results). It is possible that GvpE enhances the *gvp* transcription by facilitating the binding of the basal transcription machinery. *Hbt. salinarum* PHH4 only contains TbpE as single TBP protein (Teufel et al. 2008). This strain grows well under oxic and anoxic conditions suggesting that TbpE is able to recruit the RNA polymerase to all promoters of house-keeping genes of *Hbt. salinarum.*

GvpD is involved in the repression of gas vesicle formation as indicated by a strong overproduction of gas vesicles in the ΔD transformants carrying the related mc-vac region of *Hfx. mediterranei* (Englert et al. 1992b). Construct ΔD contains all mc-*gvp* genes except for mc-*gvpD*. Mutations in the nucleotide binding motif (p-loop motif) near the N terminus of GvpD result in GvpD proteins that are unable to repress *gvp* transcription, demonstrating that the p-loop motif is required for the repressing function. Also, variants of GvpD that incurred amino acid substitutions in two arginine rich regions lack the repressing function (Pfeifer et al. 2001; Scheuch and Pfeifer 2007). GvpD and GvpE are able to interact, and this interaction leads to the lack or breakdown of GvpE that finally results in the repression of *gvp* transcription and gas vesicle formation (Zimmermann and Pfeifer 2003; Scheuch and Pfeifer 2007). Thus, the presence or absence of the transcriptional activator GvpE is controlled by GvpD. Once the activator GvpE is present, the promoters are induced. However, in the presence of GvpD, GvpE is lacking. From these results, the level of transcription appears to be the important step in the regulation of *gvp* gene expression.

13.3 Environmental Factors Influencing Gas Vesicle Formation

Environmental factors such as salt concentration, carbon source, growth stage, oxygen availability, light and temperature influence the amount of gas vesicles in haloarchaea (Englert et al. 1990; Hechler et al. 2008; Hechler and Pfeifer 2009).

Whether and how these factors are signaled to influence the presence or absence and/or the activities of the two regulators GvpD and GvpE is not yet known. The environmental factors could affect the interaction of GvpD and GvpE, or the production or degradation of both proteins. To gain further insights into the regulatory processes involved and to uncover the signal transduction pathways affecting *gvp* gene expression we started to investigate the effect of such conditions on the gas vesicle formation of *Hbt. salinarum* PHH1 and PHH4.

13.3.1 Effect of Anoxic Growth

Cells of *Hbt. salinarum* PHH1 and PHH4 grown under oxic and anoxic conditions at 37°C were inspected for the presence of gas vesicles by phase contrast microscopy and TEM (Fig. 13.4a). Growth under oxic conditions leads to a continuous presence of gas vesicles in *Hbt. salinarum* PHH1, whereas PHH4 starts to produce gas vesicles in stationary growth phase. Gas vesicles are often confined to both cell poles of PHH4 (Fig. 13.4a). Growth under anoxic conditions reduces the amount of gas vesicles significantly. Groups of small gas vesicles are found in PHH1, whereas PHH4 completely lacks gas vesicles (Fig. 13.4a). The amount of p-*gvpA* mRNA was determined by Northern analysis (Fig. 13.4b). Total RNA derived from the

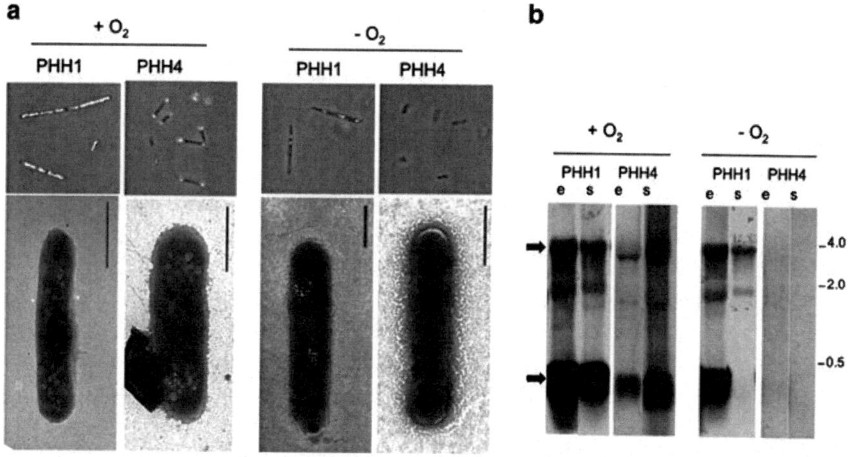

Fig. 13.4 *Hbt. salinarum* PHH1 and PHH4 grown under oxic (+O_2) and anoxic (−O_2) conditions (**a**) and Northern analysis of RNA samples derived from these cultures (**b**). (**a**) Phase-contrast microscopy (*top*) and TEM (*bottom*) of cells derived from the cultures indicated on *top*. Anoxic growth was supported by arginine fermentation. Gas vesicles appear as light refractile bodies. The bar in the TE micrographs depicts 1 μm. (**b**) Northern analyses to study *gvpA*-specific transcripts. Samples were taken in exponential (e) and stationary (s) growth phase from cultures grown under oxic or anoxic conditions at 37°C (Hechler and Pfeifer 2009). The *arrows* on the *left* depict the *gvpA* (0.3 kb) and *gvpACNO* (3 kb) transcripts. RNA marker sizes are given in kilobasepairs on the *right*

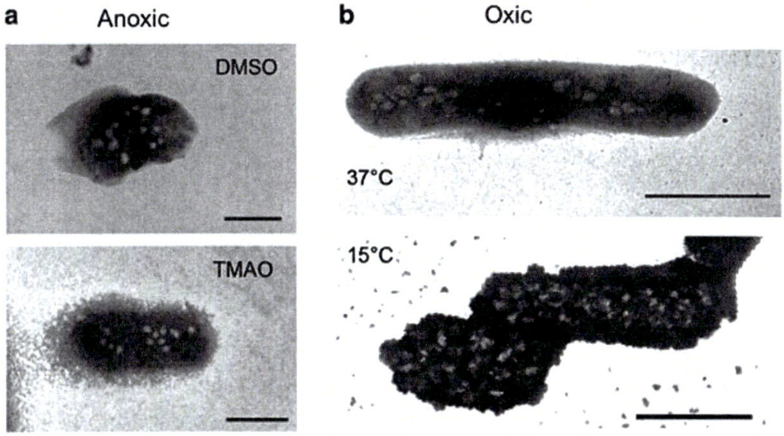

Fig. 13.5 Transmission electron micrographs of *Hbt. salinarum* PHH1 grown at 37°C under anoxic conditions plus DMSO or TMAO (**a**) and under oxic conditions at 37°C and 15°C (**b**)

exponential and stationary growth phase probed with a *gvpA*-specific probe indicate the presence of the 0.3-kb *gvpA* and 3.0-kb *gvpACNO* mRNAs under oxic conditions (Fig. 13.4b). Large amounts of these mRNAs are observed in PHH1, whereas PHH4 contains larger amounts of c-*gvpA* in the stationary growth phase only. Anoxic growth conditions result in a strong reduction in p-*gvpA* in case of PHH1, especially in the stationary growth phase, whereas the expression of c-*gvpACNO* in PHH4 is completely inhibited (Fig. 13.4b) (Hechler and Pfeifer 2009). The expression of *gvpDEFGHIJKLM* is also reduced under anoxic conditions (Hechler and Pfeifer 2009). These results demonstrate that a majority of the repression occurs at the level of p-vac or c-vac transcription.

However, anoxic growth of *Hbt. salinarum* PHH1 using DMSO or TMAO as terminal electron acceptor results in small, but gas vesicle containing cells (Fig. 13.5a) (Müller and DasSarma 2005; Hechler and Pfeifer 2009). The gas vesicles observed are of normal size, suggesting that the absence of oxygen does not always result in an inhibition of their synthesis. It is possible that the amount of energy gained by arginine fermentation is less and not sufficient for the formation of gas vesicles of normal size.

13.3.2 Effect of Temperature

The effect of temperature on gas vesicle formation also was investigated with *Hbt. salinarum* PHH1 and PHH4. Both strains were grown at 15°C or 37°C, and in addition a heat shock was applied to a culture grown at 37°C. The six growth curves obtained are shown in Fig. 13.6. The 15°C cultures grow much slower compared to

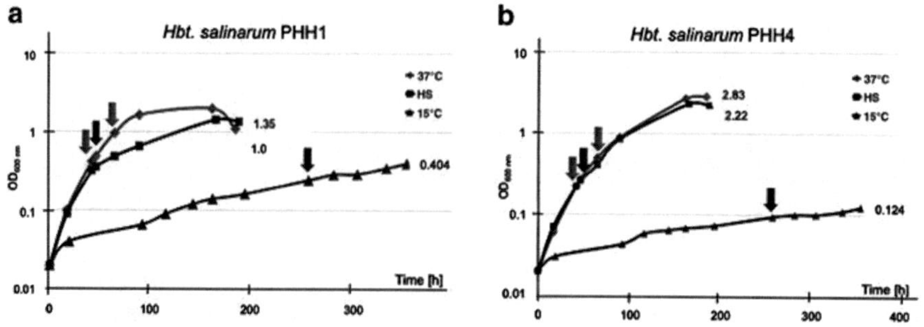

Fig. 13.6 Growth curves of *Hbt. salinarum* PHH1 (**a**) and PHH4 (**b**) obtained at different temperatures. The cells were grown at 15°C, 37°C and under heat shock conditions (HS). The HS culture was grown at 37°C to mid-exponential phase, treated for 1 h at 49°C followed by 3 h at 56°C. The *arrows* depict the time points when samples were taken for inspection of cells and RNA isolation

the two 37°C cultures and reach an optical density of 0.4 (PHH1) and 0.12 (PHH4) after 400 h of growth (Fig. 13.6). Growth of PHH4 is even more retarded compared to PHH1. However, this strain reaches a higher cell density as PHH1 when grown at 37°C and also turns later into stationary growth (Fig. 13.6). The heat shock (HS) cultures were grown at 37°C to mid-exponential growth, treated for 1 h at 49°C followed by 3 h at 56°C and then grown at 37°C again (Coker and DasSarma 2007). This treatment slows down the growth of *Hbt. salinarum* PHH1, whereas PHH4 continues to grow as fast as the untreated 37°C culture (Fig. 13.6). These differences in growth in response to temperature might be due to the deletions in the genome that strain PHH4 incurred compared to PHH1. Among the genes lacking are all of the *tbp* genes except for *tbpE* (Teufel et al. 2008).

Samples were taken in exponential and stationary growth of the 37°C and the HS culture and the cells were inspected by PCM (Fig. 13.7a). In addition, total RNA was isolated to determine the amounts of p-*gvpA* (in case of PHH1) or c-*gvpA* transcripts (PHH4). The amount of *gvpA* mRNA was quantified by dot-blot analysis using a serial dilution of 1 μg of total RNA (1:1 each) and probing the dot-blot with the respective *gvpA* gene probe (Fig. 13.7b).

In the case of *Hbt. salinarum* PHH1, the number of light refractile bodies increased during growth at 37°C whereas the amount of gas vesicles remained at the same level after heat shock treatment (Fig. 13.7a, left). The amount of p-*gvpA* mRNA was similar in the samples derived from exponential and stationary growth of the 37°C culture, but an eightfold reduction in the amount of p-*gvpA* mRNA was observed after heat shock (Fig. 13.7b). The p-*gvpA* mRNA remained at this low level also later in growth. These results suggest that the expression of p-vac is blocked at high temperature applied during heat shock and cannot be induced in later growth stages when the culture grows at 37°C again. It is possible that the expression of p-vac and the start of gas vesicle formation is confined to exponential

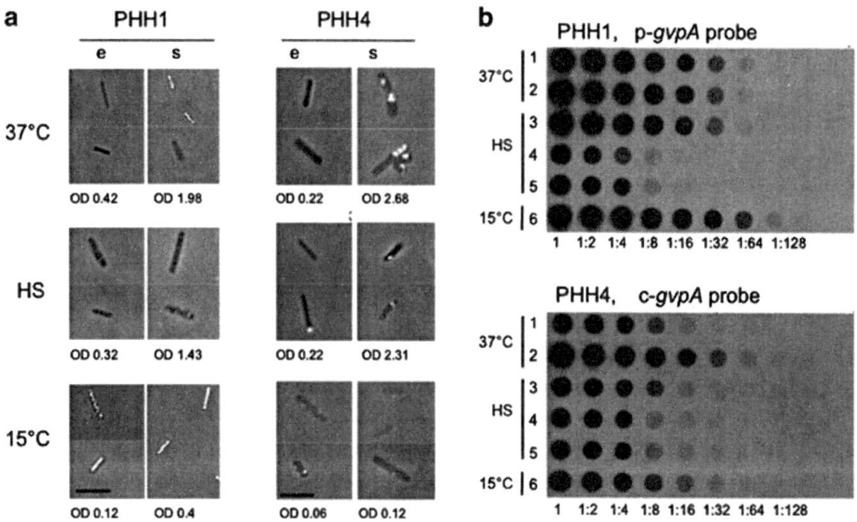

Fig. 13.7 Phase-contrast micrographs of cells grown at different temperatures (**a**) and dot-blot analysis to determine the amount of *gvpA* transcripts (**b**). (**a**) Samples were taken in exponential (e) and stationary (s) growth phase and analyzed by PCM. The optical densities (oD) are indicated at the bottom of each micrograph. Groups of gas vesicles are seen as light refractive bodies. The bar depicts 1 μm. *HS* heat shock culture. (**b**) Dot-blot analysis using a serial 1:1 dilution of total RNA (starting with 1 μg RNA on the left) to semi-quantify the amount of *gvpA* mRNAs present in the three different cultures of PHH1 (*top*) and PHH4 (*bottom*). The probe used for hybridization is indicated on top. RNA samples are: 1, 37°C, exponential growth; 2, 37°C, late exp. growth; 3, HS, prior to heat shock; 4, after HS; 5, HS after further growth at 37°C; 6, 15°C, 250 h of growth (see Fig. 13.6)

growth and that the conditions in the stationary growth phase might not support a further induction of *gvp* gene transcription.

Hbt. salinarum PHH4 grown at 37°C lacks gas vesicles in exponential growth but contains several light refractile bodies in the stationary growth phase (Fig. 13.7a, right). A fourfold increase in the amount of c-*gvpA* mRNA was seen by dot-blot analysis with RNA samples derived from the stationary growth phase (Fig. 13.7b). The HS culture did not increase the amount of gas vesicles after heat shock treatment, and this was accompanied by a twofold reduction in the amount of c-*gvpA* mRNA (Fig. 13.7b). It is interesting to note that both p- and c-*gvpA* transcripts are reduced after heat shock and that neither the p-*gvpA* nor the c-*gvpA* expression recovers after the application of the higher temperatures.

The influence of cold temperatures was analyzed in cultures grown at 15°C. Growth of *Hbt. salinarum* is almost not detectable below this temperature. Both PHH1 and PHH4 grow very slowly at 15°C (Fig. 13.6). Cells of PHH1 were tightly filled with gas vesicles throughout growth (Figs. 13.5b and 13.7a), similar as described for *Hbt. salinarum* NRC-1 (Coker and DasSarma 2007). Quantification of the p-*gvpA* mRNA by dot-blot analysis indicated a twofold increase compared to the 37°C culture (Fig. 13.7b). In contrast, strain PHH4 did not increase the amount of gas vesicles when grown at 15°C, and also the amount of c-*gvpA* mRNA was not increased (Fig. 13.7). It is possible that PHH4 did not yet reach the late exponential growth phase when the c-*gvpACNO* expression usually is induced. Another explanation could be that the c-*gvpA* promoter is not activated in the cold because strain PHH4 lacks the respective TBP protein required under these conditions.

All these results demonstrate that temperature has a major effect on the formation of gas vesicles in *Hbt. salinarum* PHH1 and PHH4. Cells exposed to heat shock inhibit the gas vesicle formation via a strong reduction of the amount of *gvp* mRNA. PHH1 increases the production of gas vesicles when grown at 15°C, whereas the expression of c-vac in PHH4 is inhibited at 15°C.

The generally low growth rate at 15°C suggests that the activity of the basal transcription machinery is low. Nevertheless, strain PHH1 overproduces gas vesicles under these conditions and increased the p-vac expression. Since the amount of p-*gvpACNO* mRNA is raised either the transcription is activated or the degradation of this mRNA is inhibited. If transcription activation occurs, the activity and/or the interaction of the two regulator proteins GvpE and GvpD might be affected by low temperatures. The large amounts of p-*gvpA* mRNA could be due to larger amounts of pGvpE or a more active pGvpE, resulting from the lack of pGvpD repressing activity. Furthermore, the interaction of pGvpE-pGvpD (leading to the breakdown of GvpE) could be disturbed at 15°C, or proteases involved in the degradation of pGvpE could be affected at cold temperatures. Further experiments are required to determine the reason for the overproduction of gas vesicles in PHH1.

13.4 Conclusion

Gas vesicle formation of *Halobacterium salinarum* is strongly affected by temperature and oxygen concentration. These environmental factors influence the transcription of the *gvp* gene clusters. However, different effects are observed with the expression of p-vac in wild-type strain PHH1 and c-vac in PHH4. PHH1 overproduces gas vesicles in the cold, but PHH4 does not. In contrast, heat shock conditions negatively affect gas vesicle formation in both cases. The *gvp* gene transcription is the major level of control, and it is possible that one or both of the regulator proteins GvpE and GvpD are the target of the signal transduction. The amounts of the *gvp* transcripts depend on the presence and/or activity of the activator protein GvpE. How these signals are sensed and signaled to the DNA is still under investigation.

Acknowledgments We thank Arnulf Kletzin for continuous discussions. This work was financially supported by DFG grant Pf165/10-1.

References

Bauer M, Marschaus L, Reuff M, Besche V, Sartorius-Neef S, Pfeifer F (2008) Overlapping activator sequences determined for two oppositely oriented promoters in halophilic Archaea. Nucleic Acids Res 36:598–606

Coker J, DasSarma S (2007) Genetic transcriptomic analysis of transcription factor genes in the model halophilic Archaeon: coordinate action of TbpD and TfbA. BMC Genet 8:61

Englert C, Pfeifer F (1993) Analysis of gas vesicle gene expression in *Haloferax mediterranei* reveals that GvpA and GvpC are both gas vesicle structural proteins. J Biol Chem 268:9329–9336

Englert C, Horne M, Pfeifer F (1990) Expression of the major gas vesicle protein in the halophilic archaebacterium *Haloferax mediterranei* is modulated by salt. Mol Gen Genet 222:225–232

Englert C, Krüger K, Offner S, Pfeifer F (1992a) Three different but related gene clusters encoding gas vesicles in halophilic archaea. J Mol Biol 227:586–592

Englert C, Wanner G, Pfeifer F (1992b) Functional analysis of the gas vesicle gene cluster of the halophilic archaeon *Haloferax mediterranei* defines the vac-region boundary and suggests a regulatory role for the *gvpD* gene or its product. Mol Microbiol 6:3543–3550

Gregor D, Pfeifer F (2005) *In vivo* analyses of constitutive and regulated promoters in halophilic archaea. Microbiology 151:25–33

Halladay J, Jones J, Lin F, MacDonald A, DasSarma S (1993) The rightward gas vesicle operon in *Halobacterium* plasmid pNRC100: identification of the *gvpA* and *gvpC* gene products by use of antibody probes and genetic analysis of the region downstream of *gvpC*. J Bacteriol 175:684–692

Hartmann R, Sickinger H, Oesterhelt D (1980) Anaerobic growth of halobacteria. Proc Natl Acad Sci USA 77:3821–3825

Hechler M, Pfeifer F (2009) Anaerobiosis inhibits gas vesicle formation in halophilic Archaea. Mol Microbiol 71:132–145

Hechler T, Frech M, Pfeifer F (2008) Glucose inhibits the formation of gas vesicles in *Haloferax volcanii* transformants. Environ Microbiol 10:20–30

Hofacker A, Schmitz KM, Cichonczyk A, Sartorius-Neef S, Pfeifer F (2004) GvpE- and GvpD-mediated transcription regulation of the p-*gvp* genes encoding gas vesicles in *Halobacterium salinarum*. Microbiology 150:1829–1838

Jung D, Achenbach L, Karr E, Takaichi S, Madigan M (2004) A gas vesiculate planktonic strain of the purple non-sulfur bacterium *Rhodoferax antarcticus* isolated from Lake Fryxell, Dry Valleys, Antarctica. Arch Microbiol 182:236–243

Karr E, Sattley W, Jung D, Madigan M, Achenbach L (2003) Remarkable diversity of phototropic purple bacteria in a permanently frozen Antarctica lake. Appl Environ Microbiol 69:4910–4914

Krüger K, Pfeifer F (1996) Transcript analysis of the c-vac region, and differential synthesis of the two regulatory gas-vesicle proteins GvpD and GvpE in *Halobacterium salinarium* PHH4. J Bacteriol 178:4012–4019

Müller J, DasSarma S (2005) Genomic analysis of anerobic respiration in the archaeon *Halobacterium* sp. strain NRC-1: dimethyl sulfoxide and trimethylamine N-oxide as terminal electron acceptors. J Bacteriol 187:1659–1667

Ng WV, Kennedy SP, Mahairas GG, Berquist B, Pan M, Shukla HD, Lasky SR, Baliga NS, Thorsson V, Sbrogna J, Swartzell S, Weir D, Hall J, Dahl TA, Welti R, Goo YA, Leithausen B, Keller K, Cruz R, Danson MJ, Hough DW, Maddocks DG, Jablonski PE, Krebs MP, Angevine GM, Dale H, Isenbarger TA, Peck RF, Pohlschröder M, Spudich JL, Jung K-H, Alam M, Freitas T, Hou S, Daniels CJ, Dennis PP, Omer AD, Ebhardt H, Lowe TM, Liang P, Riley M, Hood L, DasSarma S (2000) Genome sequence of *Halobacterium* species NRC-1. Proc Natl Acad Sci USA 97:12176–12181

Offner S, Wanner G, Pfeifer F (1996) Functional studies of the *gvpACNO* operon of *Halobacterium salinarium* reveal that the GvpC protein shapes gas vesicles. J Bacteriol 178:2071–2078

Offner S, Hofacker A, Wanner G, Pfeifer F (2000) Eight of fourteen *gvp* genes are sufficient for formation of gas vesicles in halophilic archaea. J Bacteriol 182:4328–4336

Pfeifer F, Zotzel J, Kurenbach B, Röder R, Zimmermann P (2001) A p-loop motif and two basic regions in the regulatory protein GvpD are important for the repression of gas vesicle formation in the archaeon *Haloferax mediterranei*. Microbiology 147:63–73

Scheuch S, Pfeifer F (2007) GvpD-induced breakdown of the transcriptional activator GvpE of halophilic archaea requires a functional p-loop and an arginine-rich region of GvpD. Microbiology 153:947–958

Shukla HD, DasSarma S (2004) Complexity of gas vesicle biogenesis in *Halobacterium* sp. strain NRC-1: identification of five new proteins. J Bacteriol 186:3182–3186

Teufel K, Pfeifer F (2010) Interaction of transcription activator GvpE with TATA-box-binding proteins of *Halobacterium salinarum*. Arch Microbiol 192:143–149

Teufel K, Bleiholder A, Griesbach T, Pfeifer F (2008) Variations in the multiple *tbp* genes in different *Halobacterium salinarum* strains and their expression during growth. Arch Microbiol 190:309–318

Walsby AE (1994) Gas vesicles. Microbiol Rev 190:309–318

Zimmermann P, Pfeifer F (2003) Regulation of the expression of gas vesicle genes in *Haloferax mediterranei:* the two regulatory proteins GvpD and GvpE interact. Mol Microbiol 49:783–794

Chapter 14
Halophiles Exposed Concomitantly to Multiple Stressors: Adaptive Mechanisms of Halophilic Alkalithermophiles

Noha M. Mesbah and Juergen Wiegel

14.1 Introduction

The world of halophilic microorganisms is highly diverse. Microorganisms adapted to life at high salt concentrations are found in all three domains of life: Archaea, *Bacteria* and *Eucarya*. In many ecosystems, halophiles are present with such high numbers that their presence is immediately recognized; the bright red color of marine saltern and crystallizer ponds found world-wide is due to populations of halophilic Archaea (order *Halobacteriales*), *Bacteria* (*Salinibacter*) and *Eucarya* (*Dunaliella salina*).

Most environments explored for the presence of halophiles are thalassohaline environments, i.e., that originated from the evaporation of seawater. Their chemical and ionic compositions reflect that of seawater and have a neutral or slightly alkaline pH. There also exist deep-sea brines, found at the bottom of the Red Sea, the Mediterranean Sea and the Gulf of Mexico. These brines are often anoxic and have very high salinity (approaching saturation) accompanied by elevated temperature and metal ion concentration. A number of interesting extremophiles have been isolated from deep-sea brines, including *Flexistipes sinusarabici* (an anaerobic heterotroph growing optimally at 2.7 M NaCl, pH 7 and 45–50°C (Fiala et al. 1990)), *Salinisphaera shabaensis* (an anaerobic heterotroph growing optimally at 1.5 M NaCl, 37°C and pH 7.5 (Antunes et al. 2003)), and *Halorhabdus tiamatea* (an anaerobic archaeon with optimal growth at 4 M NaCl, pH 7.0 and 45°C (Antunes et al. 2008)).

In many athalassohaline environments, i.e., hypersaline environments not originating from sea water, life at extreme salinity is combined with the need to

N.M. Mesbah (✉)
Department of Biochemistry, Suez Canal University, Ismailia 41522, Egypt
e-mail: noha_mesbah@pharm.suez.edu.eg

J. Wiegel
Department of Microbiology, University of Georgia, Athens, GA 30602, USA
e-mail: jwiegel@uga.edu

A. Ventosa et al. (eds.), *Halophiles and Hypersaline Environments*,
DOI 10.1007/978-3-642-20198-1_14, © Springer-Verlag Berlin Heidelberg 2011

survive at alkaline pH and in some cases, extreme temperature. These environments are fed by surface or river water and hypersalinity arises as a result of evaporative concentration driven by sunlight. Most of these environments are solar-heated, though a few also contain hot springs, such as saline Mono Lake, California or some of the South African Rift lakes. The ionic composition of athalassohaline environments is different from that of seawater, being dominated by potassium, magnesium, sodium and carbonate ions. Examples of athalassohaline environments include the alkaline soda lakes of Egypt (Wadi An Natrun), the Dead Sea, the East African soda lakes of the Kenyan–Tanzanian Rift Valley, the soda lakes of northern China and Inner Mongolia, salt flats of Nevada and Oregon, and Mono Lake, California.

As many athalassohaline environments are characterized by alkaline pH and/or elevated temperature in addition to high salinity and intense solar radiation, microorganisms inhabiting these environments must be adapted for survival and growth under this combination of extreme conditions. These microorganisms, termed the halophilic alkalithermophiles, not only survive, but grow optimally under conditions of high salt concentrations, alkaline pH and elevated temperature (Mesbah and Wiegel 2008; Bowers et al. 2009). The mechanisms enabling this tripartite lifestyle are essential for understanding how microorganisms grow under inhospitable conditions and will be explored in this review article.

14.2 Halophilic Alkalithermophiles

14.2.1 Growth Conditions of Halophilic Alkalithermophiles

Halophilic alkalithermophiles are a unique group of extremophiles which grow optimally at the combined extreme conditions of elevated salt concentration (>1.5 M), alkaline pH (greater than 8.5 measured at optimal temperature for growth, see Mesbah and Wiegel (2006) and Wiegel (1998) for details) and elevated temperature (>50°C). Table 14.1 contains definitions for different categories of halophiles, alkaliphiles and thermophiles.

In describing the optimal and marginal growth conditions for halophilic alkalithermophiles, care must be taken as the value of one extreme growth condition is affected by the other. For example, the measured pH value of a medium is dependent on temperature because of changing p*K*a values of different medium components at different temperatures (Wiegel 1998). Thus, the pH of the medium when measured at room temperature will be different from its pH when it is measured at the elevated growth temperature using temperature-calibrated electrodes and pH meters. For neutral pH, the difference in pH is small, usually less than 0.3 pH units. However, at acidic or alkaline pH values, the difference can be larger than 1 pH unit. Thus, to facilitate comparison of published data, it is important to know the conditions under which the pH was determined. It has been

Table 14.1 Definitions of different extremophiles

Growth characteristic	Minimum	Optimum	Maximum
Thermotolerant	T_{min} –	$T_{opt} < 50°C$	$T_{max} < 60°C$
Thermophile	T_{min} –	$T_{opt} \geq 50°C$	$T_{max} \geq 60°C$
Extreme thermophile	$T_{min} \geq 35°C$	$T_{opt} \geq 65°C$	$T_{max} < 85°C$
Hyperthermophile	$T_{min} \geq 60°C$	$T_{opt} \geq 80°C$	$T_{max} \geq 85°C$
Alkalitolerant	$pH_{min} \geq 6.0$	$pH_{opt} < 8.5$	$pH_{max} > 9.0$
Alkaliphile[a]			
Facultative	$pH_{min} < 7.5$	$pH_{opt} \geq 8.5$	$pH_{max} \geq 10.0$
Obligate	$pH_{min} \geq 7.5$	$pH_{opt} \geq 8.5$	$pH_{max} \geq 10.0$
Halotolerant	$NaCl_{min}$ –	$NaCl_{opt}$ 0.25–1.5 M[b]	$NaCl_{max} \leq 2.5$ M
Halophile	$NaCl_{min}$ 1 M	$NaCl_{opt} \geq 1.5$ M	$NaCl_{max}$ –
Extreme halophile	$NaCl_{min} \geq 1.5$ M	$NaCl_{opt} \geq 2.5$ M	$NaCl_{max}$ –

[a]For thermophiles and psychrophiles, the pH must be measured at the growth temperature for the microorganism
[b]1 M NaCl = 5.84% NaCl wt/vol

proposed (Mesbah and Wiegel 2008; Wiegel 1998) that the temperature at which the pH is measured and the pH meter calibrated be indicated with a superscript, e.g., $pH^{55°C}$.

Another interaction observed is the effect elevated salt concentration has on pH measurements. In solutions with high Na^+ concentrations, Na^+ can be read as H^+ by the pH electrode, and this is particularly pronounced at high pH values (>10). This phenomenon, known as the "Na^+ error," can be overcome to some extent with the use of a specific glass combination electrode. The glass membrane in this electrode is specifically constructed to reduce the Na^+ error at high-pH values, and it may also allow use of the electrode over a wider range of temperatures.

The first anaerobic, moderately halophilic alkalithermophilic eubacterium to be described was *Halonatronum saccharophilum*. It is a spore forming bacterium that belongs to the order *Halanaerobiales* (Zhilina et al. 2004). The optima for growth are 1.1–1.5 M NaCl, pH^{RT} 8.5 and 36–55°C, which represents a very broad temperature optimum. *H. saccharophilum* has a fermentative metabolism and grows on mono- and disaccharides, starch and glycogen. *H. saccharophilum* is the first alkaliphilic representative of the order *Halanaerobiales*, which is dominated by extremely halophilic anaerobes.

With a larger Na^+ ion requirement and more alkaline pH optimum, *Natranaerobius thermophilus* represents the first true anaerobic halophilic alkalithermophilic bacterium described (Mesbah et al. 2007b), and is the first representative of the novel order *Natranaerobiales* within the phylum *Firmicutes*. *N. thermophilus* grows (at $pH^{55°C}$ 9.5) between 35 and 56°C, with an optimum at 53°C. The $pH^{55°C}$ range for growth is 8.5–10.6, with an optimum at $pH^{55°C}$ 9.5 and no detectable growth at $\leq pH^{55°C}$ 8.2 or $\geq pH^{55°C}$ 10.8 (Fig. 14.1a). At the optimum pH and temperature, *N. thermophilus* grows in the Na^+ range of 3.1–4.9 M (1.5–3.3 M of added NaCl, the remainder from added Na_2CO_3 and $NaHCO_3$) and optimally between 3.3 and 3.9 M Na^+ (1.7–2.3 M added NaCl).

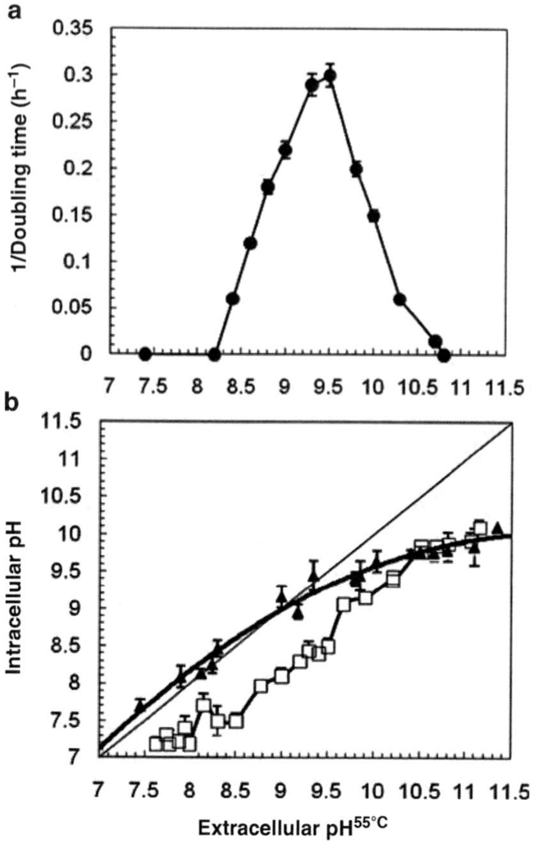

Fig. 14.1 Bioenergetic parameters in *Natranaerobius thermophilus*. (**a**) Effect of external pH$^{55°C}$ on growth of *N. thermophilus* in batch culture. (**b**) Effect of external pH on intracellular pH in sucrose-energized cell suspensions (*open square*) and non-energized cell suspensions (*filled triangle*). The *diagonal line* represents absence of a ΔpH

Two other anaerobic halophilic alkalithermophiles have been described recently. *Natranaerobius trueperi* grows (at pH$^{55°C}$ 9.5) between 26 and 55°C, with an optimum at 52°C. The pH$^{55°C}$ range for growth is 8.0–10.8, with an optimum at pH$^{55°C}$ 9.5 and no detectable growth at ≤ pH$^{55°C}$ 7.8 or ≥ pH$^{55°C}$ 11.0. At the optimum pH and temperature, *N. trueperi* grows in the Na^+ range of 3.1–5.4 M (1.5–3.8 M of added NaCl, the remainder from added Na_2CO_3 and $NaHCO_3$), and optimally at 3.8 M Na^+ (2.2 M added NaCl) (Mesbah and Wiegel 2009).

Natronovirga wadinatrunensis (the type species for the second genus described within the novel order *Natranaerobiales*) grows (at pH$^{55°C}$ 9.9) between 26 and 56°C, with an optimum at 51°C. The pH$^{55°C}$ range for growth is 8.5–11.5, with an optimum at pH$^{55°C}$ 9.9 and no detectable growth at ≤ pH$^{55°C}$ 8.3 or ≥ pH$^{55°C}$ 11.7. At the optimum pH and temperature, *Nv. wadinatrunensis* grows in the Na^+ range of 3.3–5.3 M (1.7–3.7 M of added NaCl, the remainder from added Na_2CO_3 and $NaHCO_3$), and optimally at 3.9 M Na^+ (2.3 M added NaCl) (Mesbah and Wiegel 2009).

A number of other aerobic and facultatively anaerobic halotolerant alkalithermophiles have been isolated, but their names have not been validly published and in some instances they were not characterized beyond the genus level. "Caloramator halophilus" is an obligately proteolytic facultatively aerobic member of the *Firmicutes* isolated in our laboratory from salt flats located in northern Nevada. It is thermophilic, growing optimally at 64°C (temperature range 42–75°C), and alkaliphilic, growing optimally at $pH^{60°C}$ 9.2, and not growing below $pH^{60°C}$ 8.1 or above 10.8. It is unusual in having a large NaCl range for growth: it grows between 0 and 3 M NaCl, with an optimum at 1.5 M (NM Mesbah, P Maurizio and J Wiegel, unpublished).

"Bacillus thermoalcaliphilus" is a chemoorganotrophic, facultatively anaerobic bacterium isolated from mound soil infested with the termite *Odontotermes obesus* (Sarkar 1991). The optima for growth are 1.5 M NaCl, pH^{RT} 8.5–9.0 and 60°C. *Bacillus* STS1, isolated from the same soil termite mounds as "B. thermoalcaliphilus," grows optimally at 60°C, pH 9.0 and 1 M NaCl (Sarkar 1991). Another example is *Bacillus* sp. BG-11, isolated from an alkaline thermal environment in India. It is a facultatively aerobic, non-motile, spore-forming rod that is capable of growth at temperatures greater than 55°C, NaCl concentrations greater than 1.5 M and pH values 7.5–9.0 (Kevbrin et al. 2004).

Among the Archaea, *Natronolimnobius aegyptiacus* grows (at $pH^{55°C}$ 9.5) between 29 and 65°C, with an optimum at 54–57°C. The $pH^{55°C}$ range for growth is 7.3–10.8, with an optimum at $pH^{55°C}$ 9.5 and no detectable growth at $\leq pH^{55°C}$ 7.1 or $\geq pH^{55°C}$ 10.9. At the optimum pH and temperature, *Nl. aegyptiacus* grows in the Na^+ range of 2.9–5.5 M (2.8–5.3 M of added NaCl, the remainder from added Na_2CO_3 and $NaHCO_3$), and optimally at 4.5 M Na^+ (4.3 M added NaCl) (Bowers et al. in press). To our knowledge, *Nl. aegyptiacus* represents most extreme halophilic alkalithermophile among the Archaea.

14.2.2 Habitats of Halophilic Alkalithermophiles

The halophilic alkalithermophiles isolated and characterized thus far, *N. thermophilus, N. trueperi, Nv wadinatrunensis, Nl aegyptiacus* and *H. saccharophilum*, as well as other undescribed species "*N. grantii*" and "*N. jonesii*" (Bowers et al. 2009; KJ Bowers, NM Mesbah and J Wiegel, unpublished) were isolated from sediments of soda lakes in the Wadi An Natrun, Egypt and Lake Magadi in the Kenyan–Tanzanian Rift Valley (Zhilina et al. 2004; Mesbah et al. 2007b; Mesbah and Wiegel 2009; Bowers et al. in press). As will become apparent below, the ability of the halophilic alkalithermophiles to survive and grow optimally in the presence of multiple stressors is an adaptation to life in their natural environments.

14.2.2.1 Wadi An Natrun, Egypt

The Wadi An Natrun is a depression in the Sahara desert located 90 km northwest of Cairo. The bottom of the valley is 23 m below sea level and 38 m below the Rosette branch of the river Nile. Along the valley stretches a chain of seven large alkaline, hypersaline lakes in addition to a number of ephemeral pools. Water is supplied by underground seepage from the river Nile and occasional winter precipitation. The depth of the lakes ranges between 0.5 and 2 m, and is regulated by seasonal changes in influx seepage and evaporation. High evaporation rates and arid climatic conditions during the summer months cause the salinity to rise above 5 M (Mesbah et al. 2007a).

The Wadi An Natrun lakes are extreme in several aspects; total dissolved salt concentrations are between 91.0 and 394 g L^{-1} in all seven of the lakes and the water in all lakes has pH values between 9 and 11. The lakes are solar heated. Salinity and temperature remain the same throughout the water column. Temperatures between ambient and 60°C were measured in the water below the salt crusts formed on the surface (NM Mesbah and J Wiegel, unpublished). Salinity in most of the lakes approaches saturation, with measured values between 2.5 and 5 M. The water in the most hypersaline lakes is anoxic, with dissolved oxygen concentrations below 0.2 ppm.

The Wadi An Natrun lakes are populated by dense communities of halophilic alkaliphilic microorganisms. The water displays different shades of red, purple and green according to their content of halophilic Archaea, photosynthetic purple bacteria and *Cyanobacteria*. Microbial mats occur along the lake floors and margins. The mats grow under a thin (1–2 mm) layer of sand. They consist of thin pink layers of purple photosynthetic bacteria and red halophilic Archaea followed by thick black layers containing decayed organic matter and sulfide minerals. Molecular analysis of the prokaryotic community of three large lakes of the Wadi An Natrun revealed a diverse range of prokaryotes, a high proportion of which had less than 90% 16S rRNA sequence identity to any sequences deposited in GenBank. The majority of the sequences retrieved were affiliated with the *Firmicutes*, *Alphaproteobacteria*, *Bacteroidetes*, and *Halobacteriales* (Mesbah et al. 2007a).

14.2.2.2 The Kenyan–Tanzanian Rift Valley

The Kenyan–Tanzanian rift valley harbors a series of six highly alkaline lakes. The salinities of these lakes vary from 1 M in the northerly lakes to 5 M in the lakes in the south, with roughly equal proportions of sodium carbonate and chloride as the major salts (Grant et al. 1999). The Rift valley is a volcanically active region and many of the soda lakes are fed by hot springs at temperatures of 45–96°C. The microbial community of these soda lakes contains alkaliphilic representatives of all major trophic groups of *Bacteria* and Archaea. It is hypothesized that there is active cycling of carbon, sulfur and nitrogen between these groups under both aerobic and anaerobic conditions. The aerobic environment of these soda lakes is dominated by *Cyanobacteria* and proteobacteria of the genera *Ectothiorhodospira* and *Halorhodospira*, contributing to photosynthetic primary production in the lakes (Jones et al. 1998; Rees et al. 2004) and haloalkaliphilic archaea (Grant et al. 1999). The soda lake anaerobic environment is dominated by chemoorganotrophic and sulfate-reducing members of the *Firmicutes* (Jones et al. 1998). In addition to *Halonatronum saccharophilum*, a fermentative Gram-negative microorganism belonging to the *Thermotogales* was isolated. It is capable of growth at temperatures up to 78°C and pH values above 10.5, but cannot tolerate large sodium chloride concentrations (Jones et al. 1998). It has not been characterized beyond the genus stage.

14.3 Adaptive Mechanisms of Anaerobic Halophilic Alkalithermophiles

Life at high salt concentrations, alkaline pH values, high temperatures in addition to intense UV radiation undoubtedly requires special adaptive physiological mechanisms. Each extreme growth condition, whether high salt concentration, alkaline pH or high temperature, poses a number of physiological and bioenergetic problems.

14.3.1 Adaptation of Anaerobic Halophilic Alkalithermophiles to High Salinity

Halophilic microorganisms must maintain their cytoplasm at least isoosmotic with their surroundings in order to prevent loss of water to the environment. Maintenance of a turgor pressure requires a hyperosmotic cytoplasm. All halophilic microorganisms tested thus far, with the possible exception of the halophilic Archaea of the family *Halobacteriaceae*, maintain a turgor pressure (Oren 1999).

There are two mechanisms that enable halophilic microorganisms to live in high salt concentrations, the "high-salt-in" strategy involves accumulation of equimolar concentrations of inorganic ions (predominantly potassium) in the cytoplasm. This method necessitates that all intracellular proteins be stable and active in the presence of molar concentrations of potassium and other salts (Oren 2002). This method has been observed in *Salinibacter* spp., *Halanaerobiales* and Archaea of the *Halobacteriaceae*. The "low-salt-organic-solutes-in" strategy is based on the accumulation and/or de novo synthesis of water soluble organic solutes which do not interfere with the activity or stability of normal enzymes (Oren 2002). This mode has been observed in moderate aerobic halophiles tolerating Na^+ concentrations up to 1.5 M (Oren 2008; Ventosa et al. 1998). In rare cases, both strategies can be combined (see below).

Analysis of intracellular ions in energized cell suspensions of the anaerobic, halophilic alkalithermophilic *N. thermophilus* showed that at its optimal growth conditions of 53°C, $pH^{55°C}$ 9.5 and 3.5 M Na^+, intracellular K^+ concentration was 250 mM. The intracellular K^+ concentration did not vary with changes in the

K^+ concentration of the culture medium. This concentration remained constant below extracellular $pH^{55°C}$ 9.5, but increased sharply at more alkaline extracellular pH values reaching 540 mM at $pH^{55°C}$ 10.6, the maximum growth pH for *N. thermophilus*. These values were consistent with those measured in exponentially growing cells under the same growth conditions. The intracellular Na^+ concentration in both exponentially growing and energized cell suspensions was 8 mM at extracellular $pH^{55°C}$ 9.5, and increased to 33 mM at $pH^{55°C}$ 10.5 (Mesbah et al. 2009). Concentrations of other ions (Mg^{2+}, Mn^{2+}, Li^+) were in the nanomolar range. These data indicate that *N. thermophilus* does not rely solely on the "high-salt-in" strategy for osmotic adaptation.

Analysis of the complete genome sequence of *N. thermophilus* showed that it contains genes for the biosynthesis and uptake of the osmotic solute glycine betaine. Physiological assays showed that, when grown at 52°C and $pH^{55°C}$ 9.5, intracellular concentration of glycine betaine increased more than twofold, from 410 mM at 3.3 Na^+ to 1.07 M at 4.5 M Na^+ (Table 14.2; B Zhao and J Wiegel, unpublished). Intracellular concentration of the amino acid glutamate (the only amino acid present in millimolar amounts in the cells) also increased greatly, from 19 mM in the presence of 3.3 M Na^+ to 207 mM at 4.5 M Na^+. The intracellular concentration of glycine betaine was only affected by changes in the extracellular Na^+ concentration but was not significantly affected by alterations in either temperature or extracellular pH (Table 14.2). On the other hand, intracellular concentration of glutamate increased almost threefold in response to an increase in temperature (Table 14.2). These data indicate that the accumulation of the amino acid plays roles in the adaptation to both high salinity and temperature.

The data above imply that the anaerobic, halophilic, alkalithermophilic *N. thermophilus* utilizes a combination of both the "high salt-in" and the "low-salt-organic-solutes-in" strategies for osmotic adaptation. The finding of glycine

Table 14.2 Effect of salinity, $pH^{55°C}$ and temperature on accumulation of glycine betaine and glutamate in cells of *Natranaerobius thermophilus* (data from B Zhao and J Wiegel)

Growth condition	Glycine betaine Intracellular concentration (mM)	Glutamate Intracellular concentration (mM)
Effect of salinity		
$pH^{55°C}$ 9.5, 52°C		
3.3 M Na^+	410	19
3.9 M Na^+	570	150
4.5 M Na^+	1,070	207
Effect of temperature		
$pH^{55°C}$ 9.5, 3.9 M Na^+		
37°C	500	61
54°C	630	178
Effect of $pH^{55°C}$		
52°C, 3.9 M Na^+		
$pH^{55°C}$ 8.5	630	148
$pH^{55°C}$ 10.5	500	121

betaine in molar concentration inside cells of *N. thermophilus* is significant, as it has been previously thought that anaerobic halophiles usually adapt to high salinity using the "high-salt-in" strategy alone, and do not contain osmotic solutes (Oren 2008). The majority of anaerobic halophiles isolated thus far belong to the order *Halanaerobiales*, and, with the exception of *Halonatronum saccharophilum*, are mesophilic/thermotolerant and tolerate neutral to slightly alkaline pH. The question therefore arises that the "high-salt-in" strategy could pose bioenergetic problems during growth at high salinity in combination with alkaline pH and high temperature. Indeed, as described below, *N. thermophilus* uses a dual mechanism for cytoplasm acidification, one involving cytoplasmic buffering mediated by an acidic proteome (p*I* between 4 and 5 based on amino acid analysis) and possibly by accumulation of acidic amino acids (glutamate) and unusual polyamines. The presence of molar concentrations of cations in the cytoplasm, as would occur during the "high-salt-in" strategy, would prematurely neutralize the interior of the cell, thereby compromising cytoplasm acidification towards the alkaline end of the pH growth profile for *Natranaerobius thermophilus*. It is also plausible that a high intracellular cation concentration will destabilize intracellular enzymes already adapted to be structurally stable and function at an intracellular pH that is alkaline compared to non-halophilic alkaliphiles and alkalithermophiles (Cook et al. 1996; Olsson et al. 2003).

N. thermophilus has also been shown to have a requirement of chloride in its growth medium (KJ Bowers and J Wiegel, unpublished results). *N. thermophilus* required 1.2 M of chloride ions for growth in the presence of 3.3 M Na^+, and showed no growth at or below 1 M chloride. A requirement for chloride has been previously described for several microorganisms, such as *Halobacillus halophilus*, *Salinibacter ruber* and *Halanaerobium praevalens* (Müller and Oren 2003; Roeßler and Müller 1998). Microorganisms of the *Halanaerobiales* such as *H. praevalens* use chloride and other inorganic ions to maintain the cytoplasm iso-osmotic with their environments i.e., "high-salt-in" strategy. *H. praevalens* possesses an intracellular chloride concentration of 2.2 M, roughly equivalent to that of its extracellular environment (Oren 2002). Similar results were reported for *S. ruber* (Oren et al. 2002). In *Halobacillus halophilus*, flagellin biosynthesis is strongly dependent upon the presence of chloride, as is organic compatible solute transport and synthesis (Roeßler and Müller 2001, 2002). Chloride has also been identified as a signal molecule for organic compatible solute modification in environments with constantly changing salt concentrations (Saum and Müller 2008; Müller and Saum 2005).

Since *N. thermophilus* uses a combination of both the "high-salt-in" and "low-salt-organic-solutes-in" strategies for osmotic adaptation, it is difficult to predict the role that intracellular chloride ions play during growth at high salinity, alkaline pH and elevated salt concentration. It is possible that intracellular chloride ions play a role in stabilization of intracellular proteins, as has been reported for other negatively charged compatible solutes (Martins et al. 1997).

14.3.2 Adaptation of Halophilic Alkalithermophiles to Alkaline pH

Halophilic alkalithermophiles not only cope with high salinity but also with alkaline pH. They are faced with all the bioenergetic problems faced by alkaliphiles, primarily an inverted ΔpH, suboptimal proton motive force and the need to constantly acidify the cytoplasm whilst growing in a dearth of protons. The extent to which the two stressors i.e., high salinity and alkaline pH interact and produce syntropic actions is not yet understood. Therefore, the adaptations of halophilic alkalithermophiles to alkaline pH are discussed from the viewpoint of alkaliphiles.

Generally, microorganisms must maintain a cytoplasmic pH that is compatible with optimal functional and structural integrity of the cytoplasmic proteins that support growth. Many non-extremophilic microorganisms grow over a broad range of external pH values, from 5.5 to 9.0, and maintain a cytoplasmic pH that lies within the narrow range of pH 7.4–7.8 (Padan et al. 2005). The consequences of not being able to do so are profound. The anaerobic *Caloramator fervidus* (basonym *Clostridium fervidum*), (Collins et al. 1994) has bioenergetic processes that are entirely Na^+-coupled and lacks active H^+ extrusion or uptake systems that can support pH homeostasis. *C. fervidus* is unable to generate a pH gradient or regulate cytoplasmic pH. As a result, it grows only within the narrow pH range of 6.3–7.7 (Speelmans et al. 1993).

The key bioenergetic difficulty faced by all alkaliphiles is that, in contrast to neutralophiles and acidophiles, they have a reversed ΔpH ($pH_{in}-pH_{out}$), i.e., their cytoplasmic pH is more acidic than the extracellular pH. Thus, alkaliphiles must have mechanisms for cytoplasmic acidification and/or homeostasis. A number of adaptive strategies are used for intracellular pH homeostasis. These strategies include (a) increased expression and activity of monovalent cation/proton antiporters, (b) increased ATP synthase activity that couples H^+ entry to ATP generation, (c) changes in cell surface properties, and (d) increased metabolic acid production through amino acid deaminases and sugar fermentation. Among these strategies, monovalent cation/proton antiporters play an essential and dominant role in cytoplasmic pH regulation and also have a role in Na^+ and volume homeostasis (Krulwich et al. 1998; Padan et al. 2001, 2005; Slonczewski et al. 2009).

14.3.2.1 Cytoplasm Acidification and ΔpH in *Natranaerobius thermophilus*

Bioenergetic analyses on energized cell suspensions of the anaerobic, halophilic, alkalithermophilic *N. thermophilus* showed that it is capable of cytoplasm acidification. *N. thermophilus* maintains a ΔpH of approximately 1 unit ($acid_{in}$) over an extracellular $pH^{55°C}$ range of 7.5–11.5 (Fig. 14.1b). This ΔpH of ~1 unit was due to active cytoplasm acidification as indicated by the collapse of the gradient (measured at optimal extracellular conditions) in the presence of inhibitors that dissipate the ΔpH (e.g., 3,3′,4′,5-tetrachlorosalicylanilide, nigericin). Furthermore, the ΔpH was absent in non-energized suspensions of *N. thermophilus* (Fig. 14.1b) (Mesbah et al. 2009). As the external $pH^{55°C}$ was increased from 8.0 to 11.0, the intracellular $pH^{55°C}$ increased from 7.2 to 9.9 (Fig. 14.1b). The ΔpH across the cell membrane (1 pH unit, $acid_{in}$) did not collapse, even at the upper and lower $pH^{55°C}$ boundaries for growth. This indicates that cessation of growth at more alkaline pH values is not due to a decrease in ΔpH as has been reported for non-halophilic alkaliphiles and alkalithermophiles, but rather due to excessive alkalinization of the cytoplasm (Sturr et al. 1994; Cook et al. 1996; Olsson et al. 2003). The intracellular pH of *N. thermophilus* increases to 9.9 at the maximum extracellular pH allowing growth ($pH^{55°C}$ 10.6). To our knowledge, this is one of the highest intracellular pH values measured in an alkaliphile.

Analysis of the intracellular pH in non-energized cell suspensions of *N. thermophilus* (under aerobic conditions) showed a ΔpH close to zero at external $pH^{55°C}$ less than 9.5, but appearance of a ΔpH of approximately 1 unit at external $pH^{55°C}$ values greater than 9.5. This ΔpH was partially collapsed by protonophores and ionophores. These data suggest that at the alkaline range of the pH growth profile of *N. thermophilus*, cytoplasm acidification is achieved by an energy-independent mechanism (cytoplasmic buffering, discussed below).

14.3.2.2 Active Cytoplasm Acidification and Cation/Proton Antiporters in *Natranaerobius thermophilus*

Cation-proton antiporters catalyze the active efflux of Na^+ and/or K^+ in exchange for H^+ from outside the cell. $Na^+(K^+)/H^+$ antiporters are secondary active transporters and use the energy of the electrochemical gradient of ions across the cell membrane to drive antiport. Cation-proton antiporters support cytoplasmic pH homeostasis and allow tolerance to alkali (Slonczewski et al. 2009).

Bacterial genomes typically contain 11–14 gene loci predicted to encode $Na^+(K^+)/H^+$ antiporters. The completed genome of *N. thermophilus* was found to contain 17 predicted genes for monovalent cation-proton antiporters (Mesbah et al. 2009; Krulwich et al. 2009). Cloning and heterologous expression of 12 of these genes identified in the draft genome of *N. thermophilus* showed that 8 of them were capable of complementing Na^+ and alkali-sensitivities of triple-antiporter deficient *Escherichia coli* KNabc. The remaining four complemented the K^+ uptake deficiency in K^+-uptake deficient *E. coli* TK2420 (Mesbah et al. 2009). The predicted antiporters belonged to different antiporter families, including the cation-proton antiporter (CPA)-1, CPA-2 and NhaC-families. In addition, there was one antiporter, Nt-Nha that according to sequence analyses was a stand-alone Mrp-type protein (Mesbah et al. 2009). All these protein families contain transporters that play roles in cytoplasmic pH regulation, extrusion of intracellular Na^+ and cell volume regulation (Radchenko et al. 2006; Wei et al. 2007; Slonczewski et al. 2009).

All eight of the putative antiporter proteins displayed electrogenic antiport when heterologously expressed in membranes of triple-antiporter deficient *E. coli* KNabc, indicating that they catalyze antiport at the expense of the electrochemical membrane potential (−174 mV, relatively negative inside the cell) present across the cell membrane of *N. thermophilus*. Electrogenicity is an important property of cation/proton antiporters that support alkali resistance (Padan et al. 2005; Slonczewski et al. 2009). All eight putative antiporters showed Na^+ and K^+-dependent consumption of the membrane potential.

The kinetic and biochemical properties of the eight antiporter proteins were well suited to the intracellular conditions of *N. thermophilus*. Seven antiporter proteins exhibited strong $Na^+(K^+)/H^+$ antiport activity, and one showed only K^+/H^+ antiport activity. Together, the eight antiporters functioned over a range of Na^+ and K^+ concentrations, consistent with the ability of *N. thermophilus* to grow over a range of salinities. Antiporters of *N. thermophilus* have alkaline pH optima for activity, ranging between $pH^{37°C}$ 8.5 and 8.8, consistent with the intracellular pH measured in cells of *N. thermophilus* (Fig. 14.2). One antiporter had a more alkaline profile (pH 8.5–10) with an optimum at pH 9.5. This could potentially allow the bacterium to survive and grow at the alkaline intracellular pH reached in *N. thermophilus* when exposed to alkaline stress. The pH profiles for the different antiporters were overlapping, indicating that they play concomitant roles in intracellular pH and/or salt tolerance of *N. thermophilus*. Concomitant roles were further confirmed when it

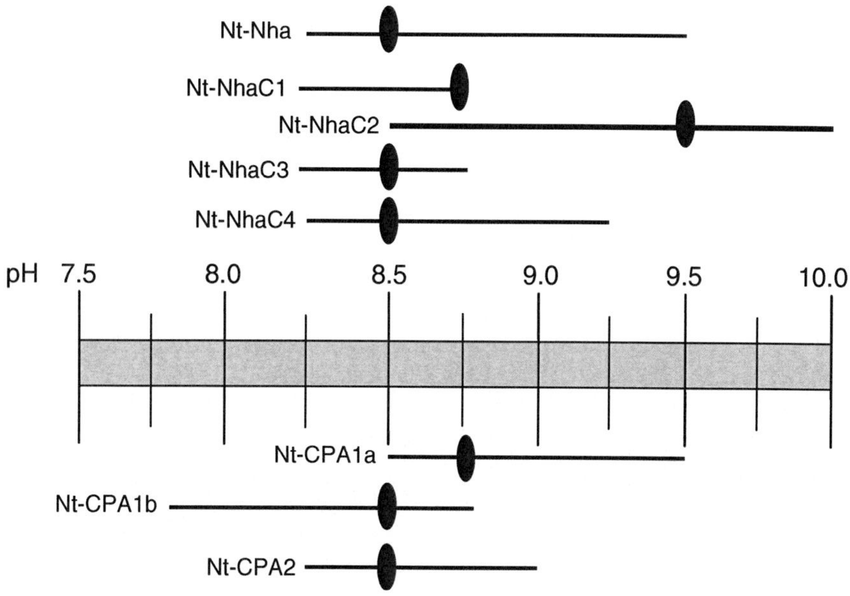

Fig. 14.2 pH activity profiles of cation/proton antiporters of *Natranaerobius thermophilus*. Optimal pH is indicated with a *filled circle*

was shown that they are constitutively expressed when *N. thermophilus* was exposed to both acid- and alkaline-stress (Mesbah et al. 2009).

A number of notable features exist about antiporter-mediated cytoplasm acidification of *N. thermophilus*. The unusually large number of predicted antiporter gene loci in the genome (the most observed in an alkaliphile thus far (Krulwich et al. 2009)) suggests that the total number of antiporters in the antiporter-complement of a bacterium is influenced by the number of environmental challenges faced by the bacterium and not simply the intensity of a single environmental challenge. *N. thermophilus* has 17 predicted antiporters; the extremely alkaliphilic, non-halophilic and mesophilic *Bacillus halodurans* C-125 has only five predicted antiporters (Takami et al. 2000). It is possible that extremophiles facing only one extreme require only a limited number of antiporters which are specifically adapted to face the bioenergetic difficulties posed by that particular extreme. On the other hand, poly-extremophiles such as *N. thermophilus* need a large complement of antiporters, their overlapping roles and properties will serve as defense mechanisms against the large array of bioenergetic problems posed by multiple extreme conditions. This assumption is further supported by the constitutive expression of antiporter genes under both acid and alkaline stress, indicating that they are important for growth under different environmental conditions. Genetic studies (deletion/mutation) are needed to determine whether individual antiporters play specific roles under certain growth conditions. However, no genetic systems exist for halophilic, alkalithermophilic bacteria.

Another notable feature of *N. thermophilus* antiporters is their ability to transport both Na^+ and K^+ during cytoplasm acidification, and the presence of a specific K^+/H^+ antiporter. This is in contrast to the aerobic alkaliphiles studied thus far, which use Na^+ exclusively as a substrate for cytoplasm acidification (Padan et al. 2005). The ability to use K^+/H^+ antiport in addition to Na^+/H^+ antiport for cytoplasm acidification is beneficial for an anaerobic halophilic alkalithermophile. Aerobic alkaliphiles typically use H^+-coupled bioenergetics; there is no competition for the intracellular Na^+ substrate. On the other hand, anaerobic, non-halophilic alkalithermophiles have Na^+-coupled mechanisms for pH homeostasis and solute transport (Speelmans et al. 1993; Prowe et al. 1996; Ferguson et al. 2006). *N. thermophilus* has been shown to have a Na^+-coupled F_1F_0-type ATPase which, when tested in vitro, functions primarily in the ATP hydrolysis direction, i.e., it expels Na^+ from the cell at the expense of ATP (NM Mesbah, J Wiegel, manuscript in review). This Na^+-coupled ATPase will compete with the Na^+/H^+ antiporters for the intracellular Na^+ substrate. In this case, K^+/H^+ antiporters can continue to acidify the cytoplasm after the Na^+/H^+ antiporters and the Na^+-coupled ATPase sufficiently reduce the intracellular Na^+ concentration. *N. thermophilus* accumulates approximately 250 mM of K^+ inside its cytoplasm when grown at optimal conditions (extracellular K^+ was 8.4–500 mM), this concentration more than doubles to 540 mM when grown under alkaline stress (extracellular $pH^{55°C}$ 10.2). Accumulation of intracellular K^+ can fuel K^+-dependent antiport. Four of the 12 predicted antiporter genes did not show cation/proton activity but did support K^+ uptake in a K^+-uptake deficient strain of *E. coli*. It is possible that these transporters couple K^+ uptake with exchange of organic molecules, as has been reported for the *Bacillus subtilis* YqkI malic/Na^+-lactate antiporter (Wei et al. 2000).

Another notable feature of *N. thermophilus* antiporters is the significant contribution of NhaC antiporters to cytoplasm acidification and alkali resistance. Four of the eight putative *N. thermophilus* antiporters showing cation/proton antiport activity when heterologously expressed in antiporter-deficient *E. coli* were homologous to other antiporters of the NhaC family (Mesbah et al. 2009). This is in contrast to what has been reported in extremely alkaliphilic aerobes. In these microorganisms, NhaC antiporters play minor roles in alkali-resistance; deletion of the NhaC antiporter in extremely alkaliphilic, aerobic *Bacillus pseudofirmus* OF4 did not significantly affect growth and cytoplasm acidification at pH 10.5 (Ito et al. 1997). Its major role was Na^+ extrusion from the cell at lower pH values. *Bacillus subtilis* NhaC does not have a prominent role in pH homeostasis (Padan et al. 2005). Among the cyanobacteria, the only cation/proton antiporter reported to contribute to alkaline pH resistance of the aerobic, alkaliphilic cyanobacterium *Synechococcus elongatus* belonged to the CPA-2 family (Billini et al. 2008). In the aerobic, alkaliphilic and moderately halophilic *Alkalimonas amylolytica*, the three antiporters shown to participate in alkali resistance belonged to the calcium/cation transporter, CPA-2 and CPA-1 families (Wei et al. 2007). It follows that NhaC antiporters play larger roles in alkali resistance in anaerobes than in aerobes. Indeed, the gene encoding the Na^+/H^+ antiporter NhaC-2 in the anaerobic, non-halophilic *Desulfovibrio vulgaris* was up-regulated during alkaline stress. An *nhaC-2* deletion mutant showed increased susceptibility to alkaline pH compared to the wild-type, showing that this NhaC antiporter plays an important role in adaptation to alkaline stress (Stolyar et al. 2007).

14.3.2.3 Cytoplasmic Buffering in *Natranaerobius thermophilus*

As discussed above, comparison of the intracellular pH in energized and non-energized cells of *N. thermophilus* showed that it utilizes a dual-mechanism for cytoplasm acidification; at extracellular pH at and below the optimal growth $pH^{55°C}$ of 9.5, the cytoplasm is acidified via the action of at least eight electrogenic cation/proton antiporters. As the extracellular pH increases beyond the optimum, antiport-dependent cytoplasm acidification stops, but the cytoplasm remains approximately 1 pH unit more acidic than the pH outside the cell. Under these conditions, cytoplasmic buffering capacity contributes to cytoplasm ΔpH homeostasis. Analysis of the proteome of *N. thermophilus* showed that, consistent with being a halophile, the isoelectric point of proteins is predominantly acidic, ranging between 4 and 5. Thus, at the alkaline intracellular pH of *N. thermophilus*, the majority of proteins will be negatively charged. As the internal pH increases, the charge on the proteome will be become more negative. In order to reach overall neutrality inside the cell, cations, such as H^+, K^+ and Na^+ will enter the cell, leading to a more acidic interior (Mesbah et al. 2009). Thereby, it is hypothesized that cessation of growth of

N. thermophilus at the alkaline pH range is due to cytoplasm alkalinization due to saturation of the cytoplasmic buffering capacity, and not due to ΔpH collapse as was reported for the non-halophilic, anaerobic alkalithermophile, *Clostridium paradoxum* (Cook et al. 1996).

Cytoplasmic buffering has been reported in aerobic alkaliphiles where it works concomitantly with antiporters to acidify the cytoplasm (Slonczewski et al. 2009). Cytoplasmic buffering does not replace the role of antiporters in cytoplasm acidification. The fact that antiporter-based homeostasis comes to a stop when approaching the pH_{max} of *N. thermophilus* and is replaced by cytoplasmic buffering suggests that the buffering capacity of cytoplasm in anaerobic halophilic alkalithermophiles is stronger than that of aerobic, non-halophilic, mesophilic aerobes. This could be due to differences in the isoelectric point and ionization of the cells proteome, or due to differences in the intracellular content of the cells.

14.3.3 Adaptation of Halophilic Alkalithermophiles to Elevated Temperature

The third stressor faced by halophilic alkalithermophiles is that of elevated temperature. Generally, thermophiles are faced with the challenge of controlling cytoplasmic membrane permeability at high temperatures. Increased motion of lipid molecules at elevated temperatures results in increased permeability to protons. Due to this motion, water molecules become trapped in the lipid core of the membranes allowing protons to hop from one molecule to the other. Other ions, unlike protons, can diffuse through the membrane. Diffusion is a temperature dependent process hence membrane permeability to ions will increase as well (Konings et al. 2002).

Bacterial cell membranes contain lipids composed of two fatty acyl chains esterified to glycerol. The third hydroxyl group of glycerol is linked to hydrophilic phosphor- or glyco-containing headgroups. These lipids are organized in a lipid bilayer such that the polar headgroups are exposed to the water phases and the acyl chains are directed towards the hydrophobic interior of the membrane. To keep the cytoplasmic membrane in a liquid crystalline state, thermophilic bacteria have been shown to increase the chain length of the lipid acyl chain, the ratio of iso/anteiso branching and/or the degree of saturation of the acyl chain (Albers et al. 2006).

Analysis of the phospholipid fatty acid (PLFA) profile for *N. thermophilus* showed that the polar and neutral fatty acid compositions were similar, with a predominance of iso- and, to a lesser extent, anteiso-branched 15:0 fatty acids (Table 14.3). Unexpectedly, the PLFA profile did not change significantly in response to either acid and alkaline stress or hyposalinity and hypersalinity (Table 14.3). The PLFA profile of *N. thermophilus* does not show the hallmarks of a thermophilic bacterium, there were no fatty acids longer than 18 carbons and some branched and unsaturated fatty acids (Table 14.3). This could possibly explain

Table 14.3 Effect of pH$^{55°C}$ and salinity on the membrane lipid composition of *Natranaerobius thermophilus*

Growth condition[a]	Na^+ 3.9 M pH$^{55°C}$ 9.5	Na^+ 3.1 M pH$^{55°C}$ 8.5	Na^+ 4.9 M pH$^{55°C}$ 8.5	Na^+ 3.1 M pH$^{55°C}$ 10.5	Na^+ 4.9 M pH$^{55°C}$ 10.5
Polar lipids					
Terminally branched saturated fatty acids					
i15:0	77.2[b]	65.1	74.0	78.4	74.9
a15:0	11.3	7.6	10.8	10.2	12.2
i17:0	1.8	2.8	2.5	2.6	2.6
a17:0	0.9	0.8	0.8	0.8	1.0
Monoenoic fatty acids					
16:1 ω 7c	0.7	4.4	1.1	0.9	0.8
Branched monoenoic fatty acids					
i17:1 ω 7c	1.2	0.9	1.5	1.3	1.2
Normal saturated fatty acids					
14:0	2.4	3.6	3.2	1.9	2.9
16:0	2.1	7.0	3.3	1.9	2.4
Polyenoic fatty acids					
18:2 ω 6	0.0	1.4	0.0	0.0	0.0
Neutral lipids					
Terminally branched saturated fatty acids					
i15:0	54.4	42.4	57.8	61.5	53.1
a15:0	11.5	6.4	10.3	9.2	10.8
i17:0	5.9	11.3	9.6	10.3	13.9
a17:0	2.4	1.7	2.2	2.3	3.0
Monoenoic fatty acids					
16:1 ω 7c	2.7	3.9	2.6	1.7	0.0
18:1 ω 9c	3.0	8.5	0.0	0.0	0.0
Branched monoenoic fatty acids					
i17:1 ω 7c	2.4	2.0	0.0	3.2	3.9
Normal saturated fatty acids					
14:0	2.2	2.3	3.0	1.7	2.3
16:0	7.3	12.3	9.9	5.6	9.1
18:0	4.8	4.0	3.2	1.9	2.5
Polyenoic fatty acids					
18:2w6	0.0	1.9	0.0	0.0	0.0

[a]All cultures were grown at 52°C
[b]Numbers denote the percentage of total polar and neutral lipids, respectively

why *N. thermophilus* is a moderate thermophile, showing no growth at temperatures greater than 56°C. While it was not possible to examine the effect of temperature on the PLFA profile of *N. thermophilus* due to poor biomass yield at the extremes of the temperature profile (NM Mesbah and J Wiegel, unpublished), it is not expected that there will be variation in the PLFA profile. *N. thermophilus* grows optimally at a temperature of 52°C, at temperatures greater than 54°C the growth rate plummets and growth ceases at 56°C (Mesbah et al. 2007b). Such a small difference between the optimal growth temperature and the maximum growth temperature is not sufficient to induce massive changes in the PLFA profile. The maintenance of a ΔpH even at alkaline extracellular pH values indicates that the cell membrane of *N. thermophilus* remains relatively impermeable to protons, i.e., protons are not lost from the cytoplasm during alkaline stress.

Polyamines are low-molecular weight, aliphatic polycations found in the cells of all living organisms. The most common polyamines in microorganisms are putrescine (1,4-diaminobutane), spermine and spermidine. Due to their negative charge, polyamines bind to macromolecules such as DNA, RNA and proteins. Polyamines are involved in diverse processes, and have been reported to play critical roles in the stabilization of proteins and nucleic acids during exposure to extremes of temperature (either hot or cold) (Tabor and Tabor 1985). Polyamines have been shown to stabilize DNA and RNA in thermophilic cells (Terui et al. 2005). The presence of the polyamines tetrakis(3-aminopropyl)ammonium and spermidine were shown to be critical for thermophile protein biosynthesis in *Thermus thermophilus* near the optimal growth temperature of 65°C (Uzawa et al. 1993).

Analysis of the polyamine content of *N. thermophilus* showed the presence of spermine, spermidine and putrescine, in addition to two unidentified polyamines (Table 14.4). One of the unidentified polyamines had an elution time that coincides with that of tetrakis(3-aminopropyl)ammonium, a branched polyamine found in hyperthermophiles (Terui et al. 2005). These polyamines could play a role in stabilization of nucleic acids and proteins in the thermophilic *N. thermophilus*. Analysis of the changes in intracellular concentrations of polyamines in response to changing temperature and genetic analyses are needed to determine the precise role each polyamine plays in the adaption of *N. thermophilus* to high temperature in the presence of high salinity and alkaline pH.

It is interesting that, in addition to increasing in response to high salinity, the intracellular concentration of the amino acid glutamate increases from 61 to 178 mM in response to an increase in growth temperature from 37 to 54°C (Table 14.2). Amino acids typically serve as organic compatible solutes in halophilic bacteria (Roberts 2005). A number of studies have revealed that the role of organic compatible solutes extends beyond protection against osmotic stress to protection against high temperature, freezing and desiccation (Santos and da Costa 2002; Welsh 2000). In many thermophiles studied, the level of compatible solutes increases when cells are grown at supraoptimal temperatures, indicating that compatible solutes play a role in protection of cellular components from the effects of heat (Silva et al. 1999; Gonçalves et al. 2003). Compatible solutes of thermophiles

Table 14.4 Polyamines of *Natranaerobius thermophilus*

Polyamine	Intracellular concentration (nmol/mg dry weight)
Putrescine	22.8
Spermidine	45.5
Spermine	59.0
N^1-Aminopropylagmatine[a]	22.8
Unknown	7.9

[a]Further structural analyses needed for confirmation

are generally negatively charged, where as mesophiles accumulate neutral solutes (Martins et al. 1997). The increase in intracellular content of the negatively charged amino acid glutamate therefore indicates that it plays a role in adaption of the anaerobic halophilic alkalithermophile *Natranaerobius thermophilus* to heat.

14.3.4 Resistance to UV Radiation and Other Adaptive Mechanisms of Anaerobic Halophilic Alkalithermophiles

Since anaerobic halophilic alkalithermophiles are exposed to intense solar radiation and periods of desiccation in their natural environments, it is expected that they have the ability to survive desiccation and γ-radiation. However, cells of *Natranaerobius thermophilus* did not show desiccation resistance for periods greater than 12 h (KJ Bowers, M.Sc. thesis). *N. thermophilus* also did not show resistance to ionizing radiation; cells exposed to 1 kGy of γ-radiation had a survival rate of 0.01%, cells exposed to 3 kGy of γ-radiation lost viability. The resistance of *N. thermophilus* to γ-radiation was only slightly higher than that of *E. coli* (KJ Bowers, M.Sc. thesis).

Both type species of the genera of the family *Natranaerobiaceae*, *N. thermophilus* and *Natronovirga wadinatrunensis* showed remarkable resistance when exposed to the three wavelengths of UV A, B, and C radiation (385, 312 and 254 nm, respectively). *N. thermophilus* displayed greater resistance, as evidenced by a greater number of viable cells (KJ Bowers, M.Sc. thesis) (Table 14.5). Such high adaptation to UV exposure is most probably an adaptation to the intense solar irradiation commonly encountered in hypersaline habitats.

Microorganisms have developed multiple strategies for surviving UV radiation. These include cell cycle arrest and activation of various pathways for repair of UV-damaged DNA. Tolerance mechanisms also exist, such as DNA recombination and lesion by-pass which allow cells to survive unrepaired lesions in their DNA (Boubriak et al. 2008). Bacterial nucleotide excision repair has also been shown to be required for repair of UV damage in the extremely halophilic archaeon *Halobacterium* NRC-1 (Crowley et al. 2006). Analysis of the genome sequence of *Natranaerobius thermophilus* showed that it possesses homologs for the *recFOR* genes, the *radA* and *radC* genes, in addition to homologs for DNA mismatch repair proteins MutS and MutN. The RecFOR pathway is a pathway of homologous recombination that repairs DNA; it can repair single- and double-stranded gaps

Table 14.5 Survival of *Natranaerobius thermophilus* and *Natronovirga wadinatrunensis* after UV radiation exposure (J Blamey, unpublished results)

	N. thermophilus	*N. wadinatrunensis*	*E. coli*
385 nm (34 J cm^{-2})	75% at 30 h	60% at 30 h	0% at 2 h
312 nm (46 J cm^{-2})	80% at 28 h	70% at 28 h	0% at 2 h
254 nm (60 J cm^{-2})	50% at 28 h	45% at 28 h	0% at 2 h

(Handa et al. 2009; Hiom 2009). RecO plays a vital role in radioresistance of *Deinococcus radiodurans* (Xu et al. 2008). A *recO* null mutant of *Deinococcus radiodurans* was extremely sensitive to irradiation with gamma rays and UV light. MutS and MutN play roles in DNA mismatch repair, which recognizes and repairs erroneous incorporation of bases that can arise during DNA damage (Kunkel and Erie 2005). RadA and RadC are archaeal proteins and play roles in DNA repair in the extremely halophilic archaea of the *Halobacteriales* (Boubriak et al. 2008). The presence of several putative DNA repair systems in the genome of *N. thermophilus* could explain its resistance to UV radiation. Functional genetic analyses are needed to confirm the role of each gene in the UV resistance of *N. thermophilus*.

Susceptibility to desiccation is interesting. It is possible that desiccation or a decrease in water activity results in irreversible destabilization of and denaturation of intracellular proteins. The proteome of *N. thermophilus* is acidic, with an isoelectric point between 4 and 5. It follows that intracellular proteins are stabilized primarily by polar and ionic interactions. Absence of water can compromise the polar bonds necessary for protein stabilization.

14.4 Conclusions

Anaerobic halophilic alkalithermophiles are an unusual group of extremophiles capable of robust growth under the combined extremes of high salinity, alkaline pH and elevated temperature. *Natranaerobius thermophilus* has been used as a prototype for investigation of the adaptive mechanisms allowing this unique lifestyle. *N. thermophilus* grows well when confronted with the combined challenges of high salinity (3.3 M Na^+), $pH^{55°C}$ and a temperature of 53°C. Several mechanisms appear to contribute to this extraordinary feat. First, *N. thermophilus* appears to use a combination of both the "high-salt-in" and "low-salt-organic-solutes-in" strategies for osmotic adaptation. *N. thermophilus* accumulates approximately 250 mM of K^+ (extracellular concentration 8.4 mM) and 1.1 M of the compatible solute glycine betaine when grown at 3.3 M Na^+, making it the first anaerobe known to accumulate compatible solutes. It has been assumed till now that anaerobic halophiles rely solely on the "high-salt-in" strategy for osmotic adaptation.

N. thermophilus displays an unusual pattern of cytoplasm pH homeostasis. It maintained a ΔpH of 1 unit throughout the entire pH range of growth, and the ΔpH did not collapse, even beyond the upper and lower ends of the pH growth profile. *N. thermophilus* tolerates a high cytoplasmic pH (when compared to anaerobic alkaliphiles and alkalithermophiles), and the growth rate drops significantly as the intracellular pH rises, indicating that growth cessation is due to excessive cytoplasm acidification and not due to collapse of the ΔpH as described for anaerobic, non-halophilic alkalithermophiles.

N. thermophilus possesses an unusually large number of cation/proton antiporters, and all of them contribute to cytoplasm acidification and pH homeostasis at the optimal growth pH for *N. thermophilus*. Kinetic properties of these antiporters showed that they have similar and overlapping roles, and expression analyses indicated genes encoding these antiporters are constitutively expressed under acid- and alkaline-stress. It appears that extremophiles growing under multiple extremes need a large complement of antiporters, their combined and overlapping roles necessary to cope with the multiple bioenergetic problems posed by multiple extreme conditions. Cation/proton antiporters of *N. thermophilus* are capable of using both Na^+ and K^+ for cytoplasm acidification. *N. thermophilus* couples different bioenergetic processes such as solute transport and ATP hydrolysis with Na^+-pumping. Therefore, the ability to use multiple cations for antiport allows them to continue to acidify the cytoplasm in the event of a decrease in cytoplasmic Na^+ content. Kinetic analyses of the individual antiporter genes also showed a new role for NhaC antiporters in cytoplasm acidification. Previously, antiporters of the NhaC family have been thought to function mainly in Na^+ extrusion from the cell and do not play major roles in alkali resistance in aerobic alkaliphiles. It appears that they play different roles in anaerobic halophilic alkalithermophiles than they do in aerobic alkaliphiles.

N. thermophilus utilizes a dual-mechanism for cytoplasm acidification. At extracellular pH values at and below the optimum, cytoplasm acidification is achieved by a large number of cation/proton antiporters. When the extracellular pH rises beyond the optimum however, antiport-mediated cytoplasm acidification comes to a stop, and the high cytoplasmic buffering capacity (mediated by an acidic proteome) contributes to cytoplasmic acidification and maintains the ΔpH at more alkaline pH values. Aerobic alkaliphiles have also been shown to have high cytoplasmic buffering capacity, but it works concomitantly with antiporters for cytoplasm acidification and does not replace antiporters in alkali resistance.

N. thermophilus does not appear to alter its cell membrane lipid composition in response to salt- and pH-stress. It also does not show an unusual phospholipid fatty acid profile, indicating that the combined extremes of high salt, alkaline pH and high temperature do not influence the membrane lipid composition. *N. thermophilus* is a moderate thermophile, it grows optimally at 53°C and the growth rate plummets at temperatures greater than 55°C. It is possible that the membrane lipid composition plays a role in limiting the upper temperature for growth of this poly-extremophile.

Similar to other thermophiles, *N. thermophilus* accumulates polyamines and the negatively charged amino acid glutamate in its cells, with intracellular glutamate concentrations more than doubling in response to elevated temperature. It follows that in anaerobic halophilic alkalithermophiles, polyamines and compatible solutes not only protect against osmotic stress, but also contribute to protection against temperature stress.

A model showing the bioenergetic processes of *N. thermophilus* is shown in Fig. 14.3. *N. thermophilus* is a fermentative bacterium and generates the bulk of its ATP via substrate level phosphorylation. The membrane bound F-type ATPase of *N. thermophilus* is Na^+-coupled and is geared primarily in the hydrolysis direction, fuelled by cytoplasmic ATP and expelling Na^+ from the cytoplasm. Na^+-pumping

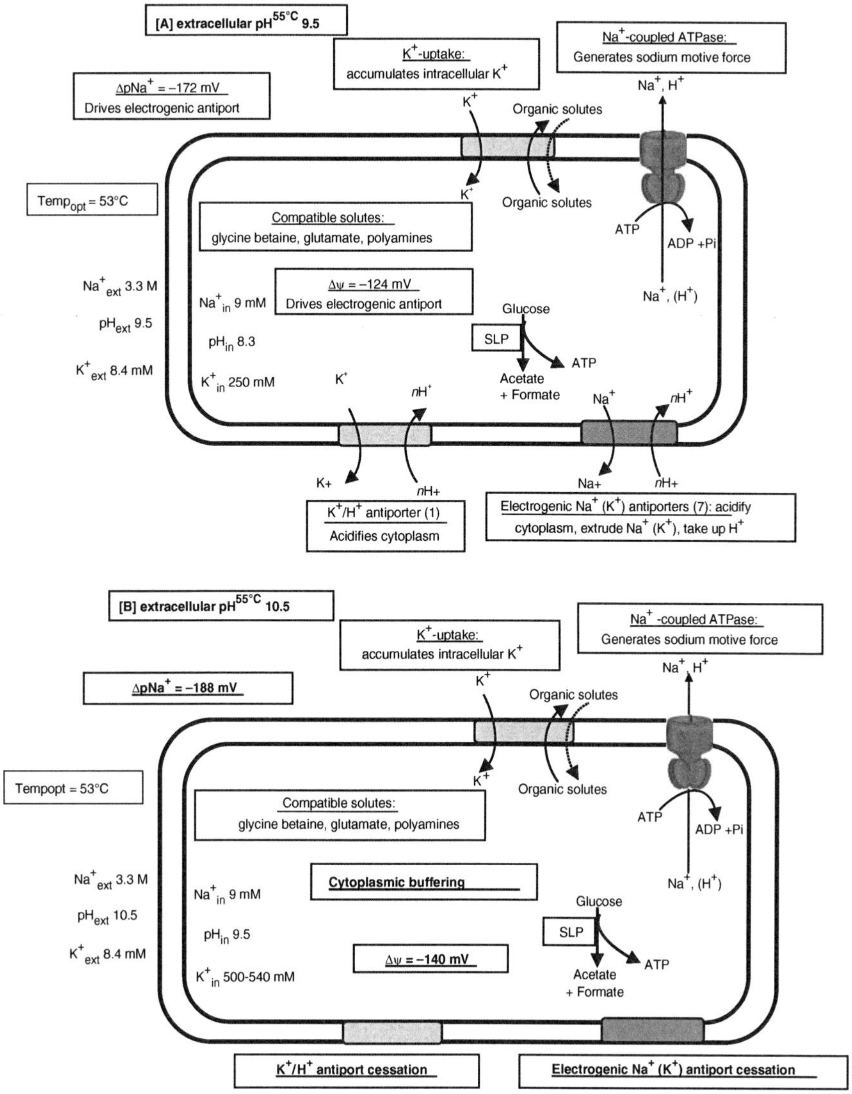

Fig. 14.3 Schematic diagram of bioenergetic processes of *Natranaerobius thermophilus* under optimal conditions (**a**) and alkaline stress (**b**)

contributes to the generation of an electrochemical gradient across the cell membrane, which is critical for driving electrogenic cation/proton antiport. *N. thermophilus* utilizes two distinct methods for cytoplasm acidification under conditions of high salt concentration, alkaline pH and elevated temperature. At extracellular $pH^{55^\circ C}$ values at and below the optimum, acidification of the cytoplasm is achieved via a large cohort of electrogenic cation/proton antiporters that are able to translocate Na^+ and K^+ ions in exchange for protons. As the extracellular $pH^{55^\circ C}$ increases, energy-dependent antiport activity stops, and acidification is probably achieved by physiochemical forces such as cytoplasmic buffering. *N. thermophilus* also uses two mechanisms for adaption to high salinity, accumulating both the inorganic cation K^+ and the organic compatible solute glycine betaine in its cytoplasm. The PLFA profile of *N. thermophilus* is not influenced by its combined growth extremes; however it appears that during growth at high temperature in the presence of alkaline pH and high salinity, both organic compatible solutes and polyamines play roles in adaptation to high temperature. All the above strategies allow *N. thermophilus* to adapt to combined extreme growth conditions, and also enable it to adapt to fluctuations in extracellular pH, Na^+ or temperature.

Acknowledgments Several people contributed to the work described in this article. We would like to thank Dr. Jenny Blamey for assistance with UV resistance experiments, Dr. Tairo Oshima for performing polyamines analysis and Dr. Baisuo Zhao for providing data on compatible solute accumulation in *N. thermophilus*. Karen Bowers performed experiments on the desiccation and gamma radiation resistance of *N. thermophilus*. The presented research herein was supported by grants MCB-060224 from the National Science Foundation and AFOSR 033835–01 from the Air Force Office of Scientific Research to J Wiegel.

References

Albers SV, Konings WN, Driessen AJM (2006) Membranes of thermophiles and other extremophiles. In: Rainey FA, Oren A (eds) Methods in microbiology: extremophiles, vol 35. Elsevier, London, pp 161–171

Antunes A, Eder W, Fareleira P, Santos H, Huber R (2003) *Salinisphaera shabanensis* gen. nov., sp. nov., a novel, moderately halophilic bacterium from the brine-seawater interface of the Shaban Deep, Red Sea. Extremophiles 7:29–34

Antunes A, Taborda M, Huber R, Moissl C, Nobre MF, da Costa MS (2008) *Halorhabdus tiamatea* sp. nov., a non-pigmented, extremely halophilic archaeon from a deep-sea, hypersaline anoxic basin of the Red Sea, and emended description of the genus *Halorhabdus*. Int J Syst Evol Microbiol 58:215–220

Billini M, Stamatakis K, Sophianopoulou V (2008) Two members of a network of putative Na^+/H^+ antiporters are involved in salt and pH tolerance of the freshwater cyanobacterium *Synechococcus elongatus*. J Bacteriol 190:6318–6329

Boubriak I, Ng WL, DasSarma P, DasSarma S, Crowley DJ, McCready SJ (2008) Transcriptional responses to biologically relevant doses of UV-B radiation in the model archaeon, *Halobacterium* sp. NRC-1. Saline Systems 4:13

Bowers KJ, Mesbah NM, Wiegel J (2009) Biodiversity of poly-extremophilic bacteria: does combining the extremes of high salt, alkaline pH and elevated temperature approach a physico-chemical boundary for life? Saline Systems 5:9

Bowers KJ, Sarmiento F, Mesbah NM, Wiegel J *Natronolimnobius aegyptiacus* sp. nov., an aerobic, extremely halophilic alkalithermophilic archaeon isolated from the athalassohaline Wadi An Natrun, Egypt. Extremophiles (in press)

Collins MD, Lawson PA, Willems A, Cordoba JJ, Fernandez-Garayzabel J, Garcia P, Cai J, Hipper H, Farrow JA (1994) The phylogeny of the genus *Clostridium*: proposal of five new genera and eleven new species combinations. Int J Syst Bacteriol 44:812–826

Cook G, Russell J, Reichert A, Wiegel J (1996) The intracellular pH of *Clostridium paradoxum*, an anaerobic, alkaliphilic, and thermophilic bacterium. Appl Environ Microbiol 62:4576–4579

Crowley DJ, Boubriak I, Berquist BR, Clark M, Richard E, Sullivan L, DasSarma S, McCready SJ (2006) The *uvrA, uvrB* and *uvrC* genes are required for repair of ultraviolet light induced DNA photoproducts in *Halobacterium* sp. NRC-1. Saline Systems 2:11

Ferguson SA, Keis S, Cook GM (2006) Biochemical and molecular characterization of a Na^+-translocating F_1F_o-ATPase from the thermoalkaliphilic bacterium *Clostridium paradoxum*. J Bacteriol 188:5045–5054

Fiala G, Woese CR, Langworthy TA, Stetter KO (1990) *Flexistipes sinusarabici*, a novel genus and species of eubacteria occurring in the Atlantis II Deep brines of the Red Sea. Arch Microbiol 154:120–126

Gonçalves LG, Huber R, da Costa M, Santos H (2003) A variant of the hyperthermophile *Archaeoglobus fulgidus* adapted to grow at high salinity. FEMS Microbiol Lett 218:239–244

Grant S, Grant WD, Jones BE, Kato C, Li L (1999) Novel archaeal phylotypes from an East African alkaline saltern. Extremophiles 3:139–145

Handa N, Morimatsu K, Lovett ST, Kowalczykowski SC (2009) Reconstitution of initial steps of dsDNA break repair by the RecF pathway of *E. coli*. Genes Dev 23:1234–1245

Hiom K (2009) DNA repair: common approaches to fixing double-strand breaks. Curr Biol 19: R523–R525

Ito M, Guffanti AA, Zemsky J, Ivey DM, Krulwich TA (1997) Role of the *nhaC*-encoded Na^+/H^+ antiporter of alkaliphiic *Bacillus firmus* OF4. J Bacteriol 179:3851–3857

Jones BE, Grant WD, Duckworth AW, Owenson GG (1998) Microbial diversity of soda lakes. Extremophiles 2:191–200

Kevbrin VV, Romanek CS, Wiegel J (2004) Alkalithermophiles: a double challenge from extreme environments. In: Seckbach J (ed) Origins: genesis, evolution and diversity of life. Kluwer Academic Publishers, Dordrecht, pp 395–412

Konings WN, Albers S-V, Koning S, Driessen AJM (2002) The cell membrane plays a crucial role in survival of bacteria and archaea in extreme environments. Antonie Leeuwenhoek 81:61–72

Krulwich TA, Ito M, Gilmour R, Hicks DB, Guffanti AA (1998) Energetics of alkaliphilic *Bacillus* species: physiology and molecules. Adv Microb Physiol 40:401–438

Krulwich TA, Hicks DB, Ito M (2009) Cation/proton antiporter complements of bacteria: why so large and diverse? Mol Microbiol 74:257–260

Kunkel T, Erie DA (2005) DNA mismatch repair. Annu Rev Biochem 74:681–710

Martins LO, Huber R, Huber H, Stetter KO, da Costa M, Santos H (1997) Organic solutes in hyperthermophilic *Archaea*. Appl Environ Microbiol 63:896–902

Mesbah NM, Wiegel J (2006) Isolation, cultivation and characterization of alkalithermophiles. In: Rainey FA, Oren A (eds) Methods in microbiology, vol 35, Extremophiles. Academic Press/ Elsevier, London, pp 451–468

Mesbah NM, Wiegel J (2008) Life at extreme limits: the anaerobic halophilic alkalithermophiles. In: Wiegel J, Adams MWW, Maier R (eds) The incredible anaerobes: from physiology to genomics to fuels. Annals of the New York Academy of Sciences, New York

Mesbah NM, Wiegel J (2009) *Natronovirga wadinatrunensis* gen. nov. sp. nov. and *Natranaerobius trueperi* sp. nov., two halophilic alkalithermophilic microorganisms from soda lakes of the Wadi An Natrun, Egypt. Int J Syst Evol Microbiol 59:2042–2048

Mesbah NM, Abou-El-Ela SH, Wiegel J (2007a) Novel and unexpected prokaryotic diversity in water and sediments of the alkaline, hypersaline lakes of the Wadi An Natrun, Egypt. Microb Ecol 54:598–617

Mesbah NM, Hedrick DB, Peacock AD, Rohde M, Wiegel J (2007b) *Natranaerobius thermophilus* gen. nov. sp. nov., a halophilic, alkalithermophilic bacterium from soda lakes of the Wadi An Natrun, Egypt, and proposal of *Natranaerobiaceae* fam. nov. and *Natranaerobiales* ord. nov. Int J Syst Evol Microbiol 57:2507–2512

Mesbah NM, Cook GM, Wiegel J (2009) The halophilic alkalithermophile *Natranaerobius thermophilus* adapts to multiple environmental extremes using a large repertoire of $Na^+(K^+)/H^+$ antiporters. Mol Microbiol 74:270–281

Müller V, Oren A (2003) Metabolism of chloride in halophilic prokaryotes. Extremophiles 7:261–266

Müller V, Saum SH (2005) The chloride regulon of *Halobacillus halophilus*: a novel regulatory network for salt perception and signal transduction in bacteria. In: Gunde-Cimerman N, Oren A, Plemenitaš A (eds) Adaptation to life at high salt concentrations in Archaea, Bacteria, and Eukarya. Springer, Dordrecht, pp 303–310

Olsson K, Keis S, Morgan HW, Dimroth P, Cook GM (2003) Bioenergetic properties of the thermoalkaliphilic *Bacillus* sp. strain TA2.A1. J Bacteriol 185:461–465

Oren A (1999) Bioenergetic aspects of halophilism. Microbiol Mol Biol Rev 63:334–348

Oren A (2002) Halophilic microorganisms and their environments. Kluwer Academic Publishers, Dordrecht

Oren A (2008) Microbial life at high salt concentrations: phylogenetic and metabolic diversity. Saline Systems 4:2

Oren A, Heldal M, Norland S, Galinski EA (2002) Intracellular ion and organic solute concentrations of the extremely halophilic *Salinibacter ruber*. Extremophiles 6:491–498

Padan E, Venturi M, Gerchman Y, Dover N (2001) Na^+/H^+ antiporters. Biochim Biophys Acta 1505:144–157

Padan E, Bibi E, Masahiro I, Krulwich TA (2005) Alkaline pH homeostasis in bacteria: new insights. Biochim Biophys Acta 1717:67–88

Prowe S, van de Vossenberg J, Driessen A, Antranikian G, Konings W (1996) Sodium-coupled energy transduction in the newly isolated thermoalkaliphilic strain LBS3. J Bacteriol 178:4099–4104

Radchenko MV, Waditee R, Oshimi S, Fukuhara M, Takabe T, Nakamura T (2006) Cloning, functional expression and primary characterization of *Vibrio parahaemolyticus* K^+/H^+ antiporter genes in *Escherichia coli*. Mol Microbiol 59:651–663

Rees HC, Grant WD, Jones BE, Heaphy S (2004) Diversity of Kenyan soda lake alkaliphiles assessed by molecular methods. Extremophiles 8:63–71

Roberts M (2005) Organic compatible solutes of halotolerant and halophilic microorganisms. Saline Systems 1:5

Roeßler M, Müller V (1998) Quantitative and physiological analysis of chloride dependence of growth in *Halobacillus halophilus*. Appl Environ Microbiol 64:3813–3817

Roeßler M, Müller V (2001) Chloride, a new environmental signal molecule involved in gene regulation in a moderately halophilic bacterium, *Halobacillus halophilus*. J Bacteriol 184:6207–6215

Roeßler M, Müller V (2002) Chloride dependence of glycine betaine transport in *Halobacillus halophilus*. FEBS Lett 489:125–128

Santos H, da Costa M (2002) Compatible solutes of organisms that live in hot saline environments. Environ Microbiol 4:501–509

Sarkar A (1991) Isolation and characterization of thermophilic, alkaliphilic, cellulose-degrading *Bacillus thermoalcaliphilus* sp. nov. from termite (*Odontotermes obesus*) mound soil of a semiarid area. Geomicrobiol J 9:225–232

Saum SH, Müller V (2008) Regulation of osmoadaptation in the moderate halophile *Halobacillus halophilus*: chloride, glutamate and switching osmolyte strategies. Saline Systems 4:4

Silva Z, Borges N, Martins LO, Wait R, da Costa M, Santos H (1999) Combined effect of the growth temperature and salinity of the medium on the accumulation of compatible solutes by *Rhodothermus marinus* and *Rhodothermus obamensis*. Extremophiles 3:163–172

Slonczewski JL, Fujisawa M, Dopson M, Krulwich TA (2009) Cytoplasmic pH measurement and homeostasis in bacteria and archaea. Adv Microb Physiol 55(1–79):317

Speelmans G, Poolman B, Abee T, Konings W (1993) Energy transduction in the thermophilic anaerobic bacterium *Clostridium fervidus* is exclusively coupled to sodium ions. Proc Natl Acad Sci USA 90:7975–7979

Stolyar S, He Q, Joachimiak MP, He Z, Yang ZK, Borglin SE, Joyner DC, Huang K, Alm E, Hazen TC, Zhou J, Wall JD, Arkin AP, Stahl DA (2007) Response of *Desulfovibrio vulgaris* to alkaline stress. J Bacteriol 189:8944–8952

Sturr MG, Guffanti AA, Krulwich TA (1994) Growth and bioenergetics of alkaliphilic *Bacillus firmus* OF4 in continuous culture at high pH. J Bacteriol 176:3111–3116

Tabor CW, Tabor H (1985) Polyamines in microorganisms. Microbiol Rev 49:81–99

Takami H, Nakasone K, Takaki Y, Maeno G, Sasaki R, Masui N, Fuji F, Hirama C, Nakamura Y, Ogasawara N, Kuhara S, Horikoshi K (2000) Complete genome sequence of the alkaliphilic bacterium *Bacillus halodurans* and genomic sequence comparison with *Bacillus subtilis*. Nucleic Acids Res 28:4317–4331

Terui Y, Otnuma M, Hiraga K, Kawashima E, Oshima T (2005) Stabilization of nucleic acids by unusual polyamines produced by an extreme thermophile, *Thermus thermophilus*. Biochem J 388:427–433

Uzawa T, Hamasaki N, Oshima T (1993) Effects of novel polyamines on cell-free polypeptide synthesis catalyzed by *Thermus thermophilus* HB8 extract. J Biochem 114:478–486

Ventosa A, Nieto JJ, Oren A (1998) Biology of moderately halophilic aerobic bacteria. Microbiol Mol Biol Rev 62:504–544

Wei Y, Guffanti AA, Ito M, Krulwich TA (2000) *Bacillus subtilis* YqkI is a novel malic/Na^+-lactate antiporter that enhances growth on malate at low protonmotive force. J Biol Chem 275:30287–30292

Wei Y, Liu J, Ma Y, Krulwich TA (2007) Three putative cation/proton antiporters from the soda lake alkaliphile *Alkalimonas amylolytica* N10 complement and alkali-sensitive *Escherichia coli* mutant. Microbiology 153:2168–2179

Welsh DT (2000) Ecological significance of compatible solute accumulation by microorganisms: from single cells to global climate. FEMS Microbiol Rev 24:263–290

Wiegel J (1998) Anaerobic alkalithermophiles, a novel group of extremophiles. Extremophiles 2:257–267

Xu G, Wang L, Chen H, Lu H, Ying N, Tian B, Hua Y (2008) RecO is essential for DNA damage repair in *Deinococcus radiodurans*. J Bacteriol 190:2624–2628

Zhilina TN, Garnova ES, Tourova TP, Kostrikina NA, Zavarzin GA (2004) *Halonatronum saccharophilum* gen. nov. sp. nov.: a new haloalkaliphilic bacterium of the order *Haloanaeobiales* from Lake Magadi. Microbiology 70:64–72

Chapter 15
Cellular Adjustments of *Bacillus subtilis* and Other Bacilli to Fluctuating Salinities

Marco Pittelkow and Erhard Bremer

15.1 The Osmotic Challenge

Bacilli are exposed, either suddenly or on a sustained basis, to changes in the osmotic conditions of the varied ecological niches that they can colonize. This can be vividly visualized by considering the main habitat of *Bacillus subtilis*, the upper layers of the soil (Earl et al. 2008). Rainfall and drying of the soil cause fluctuations in water availability and in the osmotic potential of this ecological niche. Such changes pose a considerable challenge to the microbial cell since they elicit water fluxes across the cytoplasmic membrane. These water fluxes are driven by the differential in the osmotic potential between the cell's interior and that of the surrounding microenvironment (Bremer and Krämer 2000). Water entry at low osmolarity can increase turgor to such an extent that the elastic and stress-bearing peptidoglycan sacculus can no longer cope with it and the cell will rupture. Water efflux at high osmolarity will trigger dehydration of the cytoplasm and the ensuing reduction or even collapse of turgor will cause growth arrest or even cell death. Turgor, an intracellular hydrostatic pressure considered to be essential for cell expansion and growth, is difficult to determine experimentally but has been estimated at 1.9 MPa (19.37 atm) for *B. subtilis* (Whatmore and Reed 1990), a pressure that is close to ten times the pressure present in a standard car tire.

To maintain turgor within physiologically acceptable boundaries, the *B. subtilis* cell needs to take active countermeasures to redirect the flow of water in or out of the cell when it faces decreases or increases in the external osmolarity. A considerable number of bacteria, including several Bacilli (but not *B. subtilis*), possess AqpZ-type aquaporins, water-selective channels embedded in the cytoplasmic membrane that can mediate accelerated water fluxes along osmotic gradients. However, the potential role of AqpZ-type water channels in the osmotic adjustment

M. Pittelkow • E. Bremer (✉)
Department of Biology, Laboratory for Microbiology, Philipps-University Marburg, Karl-von-Frisch-Strasse 8, 35032 Marburg, Germany
e-mail: bremer@biologie.uni-marburg.de

A. Ventosa et al. (eds.), *Halophiles and Hypersaline Environments*,
DOI 10.1007/978-3-642-20198-1_15,

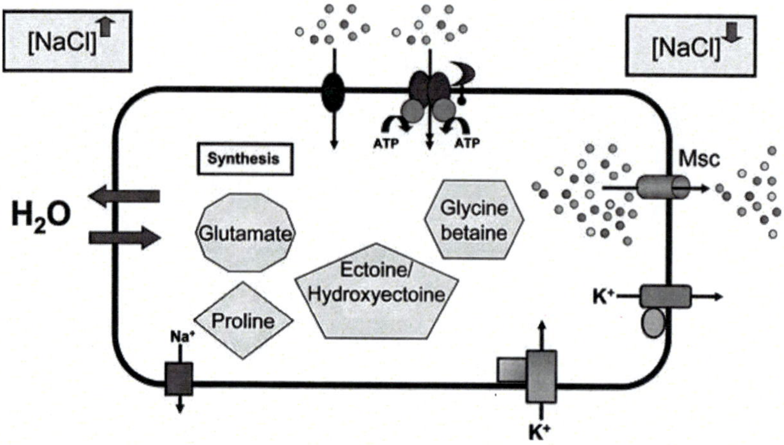

Fig. 15.1 The core of the osmostress response of Bacilli. Schematic overview of the initial and sustained cellular stress responses to high salinity through the uptake of K^+, synthesis and import of various compatible solutes and the active export of K^+ and Na^+ ions. Non-selective expulsion of ions and organic solutes occur in response to sudden osmotic downshifts via mechanosensitive channels (Msc)

processes of microorganisms has not been rigorously worked out (Tanghe et al. 2006). It is important to recall that no microorganism can actively move water across the cytoplasmic membrane in a directed fashion; in other words: through an energy-consuming transport process. Hence, microorganisms cannot counteract the environmentally imposed water fluxes by actively pumping water in or out of the cell. However, by carefully setting the differential in the osmotic potential of the cytoplasm and that of the environment, bacteria can actively determine the direction and magnitude of the water fluxes across their cytoplasmic membrane (Fig. 15.1) (Bremer and Krämer 2000).

15.2 Exposure to Hypo-osmolarity: Opening the Floodgates

A *Bacillus* cell living in the upper layers of dried-out soil will experience a sudden osmotic downshift when rain sets in. This creates an emergency situation, which the cell must deal with very rapidly. The stress-bearing peptidoglycan sacculus is elastic (Vollmer and Seligman 2010), but only to a certain degree. The rapid inflow of water following an osmotic downshift can drive up turgor almost instantaneously and to such a degree that the cell will burst. A calculation by Booth and Louis (1999) suggests that the turgor of *Escherichia coli*, which is estimated to lie in the range between 2 and 6 atm (Booth et al. 2007), will be driven up in a few milliseconds by about 10 atm when cells that were pre-adapted to high osmolarity media are suddenly transferred to low-osmolarity media (Booth and Louis 1999). Detailed genetic, electrophysiological, and structural studies have demonstrated that microorganisms transiently open mechanosensitive channels embedded in their cytoplasmic membrane to rapidly jettison water-attracting solutes when they are suddenly exposed to hypo-osmotic surroundings (Booth et al. 2007; Kung et al. 2010). This reduces the osmotic potential of the cytoplasm quickly and thereby curbs water influx; consequently, the undue rise in turgor is counteracted and cell lysis is prevented.

Mechanosensitive channels remain in a closed state when turgor is properly balanced, but they can sense, via their protein–lipid interface, tension in the cytoplasmic membrane that is caused by increased turgor. Consequently, they are stretched opened to form huge holes through which both solvents and solutes can rapidly pass. Microorganisms typically possess several types of these channels (e.g., MscM, MscS, MscK, MscL) and their gating at different threshold values of membrane tension provides the cell with the tools for a graded response to osmotic down-shifts (Booth et al. 2007; Booth and Louis 1999; Kung et al. 2010). The opening of the channel of large conductance, MscL, is typically the measure of last resort, since MscL opens just below the turgor pressure threshold value at which the cells are about to lyse (Levina et al. 1999). The diameters of the *E. coli* MscS and MscL channels have been estimated to about 16 Å and 30 Å, respectively (Kung et al. 2010), and thereby exceed the diameters (about 12 Å) of the permanently open channels of porins (e.g., OmpC and OmpF) present in the outer membrane of Gram-negative bacteria (Nikaido 2003). Hence, the transient opening of mechanosensitive channels embedded in the cytoplasmic membrane will not only reduce turgor at a time when this is desperately needed, but also result in the loss of valuable metabolites and ions. However, the cell has to endure this negative side effect in order to survive. For obvious reasons, the timing and duration of the opening of such floodgates in the cytoplasmic membrane needs to be very carefully controlled by the microbial cell. The critical role of mechanosensitive channels for managing the transition from high to low salinities is manifest by the fact that mutants lacking these channels typically do not survive sudden osmotic downshifts (Levina et al. 1999; Hoffmann et al. 2008).

Electrophysiological studies with whole cells have shown that *B. subtilis* possesses mechanosensitive channels that can gate at different pressure set points (Szabo et al. 1992). Inspection of the *B. subtilis* genome sequence revealed one MscL-type and three MscS-type channel-forming proteins (YhdY, YfkC, YkuT) (Hoffmann et al. 2008; Wahome and Setlow 2008). Mutational analysis demonstrated that the simultaneous disruption of the structural genes for the MscS-type YhdY, YfkC, YkuT proteins does not cause a noticeable effect on cell survival when the corresponding *B. subtilis* triple mutant strain is subjected to a severe osmotic downshift. In contrast, 70% of the cells carrying an *mscL* gene disruption mutation do not survive such a downshift (Hoffmann et al. 2008), indicating that the channel-activity of MscL is key for the management of a rapid transition of *B. subtilis* from hyper- to hypo-osmotic surroundings (Hoffmann et al. 2008; Wahome and Setlow 2008). The inactivation of the MscS-type YkuT protein has no phenotype by itself, but the combination of a *ykuT* mutation with that of an *mscL* gene disruption potentiates the phenotype of the *mscL* mutant to such an extent that almost all cells are killed upon a severe osmotic down-shock (Hoffmann et al. 2008). Indeed, Wahome and Setlow (2008) have shown that the overproduction of the YkuT protein in an *mscL* mutant background can rescue the osmotically down-shock-sensitive phenotype of a *B. subtilis mscL* mutant strain, thereby providing further evidence for a channel function of the MscS-related YkuT protein. It is currently unclear whether the MscS-related YhdY and YfkC proteins are actually channel-forming proteins and what the functions of such channels might be in the physiological context of the *B. subtilis* cell.

Interestingly, the *ykuT* gene is part of the SigB-controlled general stress regulon of *B. subtilis*, an emergency stress response system that provides environmentally or nutritionally challenged *B. subtilis* cells with a multifaceted and pre-emptive stress resistance (Hecker et al. 2007). This includes cellular resistance against the imposition of a severe and growth-restricting salt shock; such a salt shock elicits the transient induction of almost the entire SigB regulon (Höper et al. 2005, 2006). Consistent with the SigB control of the *ykuT* gene, genome-wide transcriptional profiling studies of *B. subtilis* cells subjected to an osmotic up-shift with 6% (wt/vol) NaCl revealed that *ykuT* expression is transiently up-regulated upon the hyperosmotic challenge. In contrast, the expression of the SigB-independent *mscL* gene was suppressed under these conditions (Hahne et al. 2010). Using GFP- and YFP-tagged Msc-fusion proteins from *B. subtilis*, Wahome and Setlow (2008) found that the cellular levels of the MscL, YhdY, YfkC, YkuT proteins varies with growth phase, suggesting that some of these mechanosensitive channels might play not yet understood roles during certain periods of the growth cycle of individual cells.

The central role of mechanosensitive channels for the challenging transition from high to low osmolarity surroundings is reflected in the speed in which osmotically down-shifted *B. subtilis* cells can jettison radiolabeled glycine betaine that was pre-accumulated under high salinity (0.8 M NaCl) growth conditions. Both the osmotic properties of the *B. subtilis* cell (Whatmore and Reed 1990) and the elastic properties of the peptidoglycan sacculus (Vollmer and Seligman 2010) appear to contribute to the timing and the speed by which glycine betaine is released from the cells through the transient opening of mechanosensitive channels. Moderate osmotic downshifts (equivalent to the withdrawal of about 0.285 mM NaCl) elicit little release of glycine betaine. This finding indicates that the elastic properties of the peptidoglycan sacculus allow the *B. subtilis* cell to restrain moderate increases in turgor and thereby obviate the need for the opening of mechanosensitive channels. Osmotic downshifts that exceed this threshold value lead to a progressive increase in the amount of the released glycine betaine and the extent of the release correlates with the severity of the imposed osmotic downshift. Upon an osmotic downshift equivalent to the withdrawal of about 0.57 mM NaCl, almost the entire glycine betaine content of the cells was released. Expulsion of glycine betaine under these conditions was very rapid (in less than 30 s) and was definitely not caused by cell lysis. A wild-type *B. subtilis* culture survives such a severe osmotic down-shift without any reduction in viability, whereas the cells of a MscL-YkuT double mutant strain were almost all killed under such a regimen (Hoffmann et al. 2008).

Recently reported data for the soil bacterium *Corynebacterium glutamicum* indicate that an MscS-type mechanosensitive channel (MscCG) also contributes to the fine-tuning of the cellular solute pool and turgor of high-osmolarity-grown cells that are not subjected to rapid osmotic downshifts. The data obtained by Börngen et al. (2010) indicate that MscCG may act in the osmoadaptation process of *C. glutamicum* by fine-tuning the steady state level of the glycine betaine solute pool resulting from import of this osmoprotectant from environmental sources. Hence, *C. glutamicum* cells apparently respond to temporary imbalances in turgor through a "pump and leak" mechanism – import of the compatible solute glycine betaine through the high-affinity BetP carrier and its release via the MscCG channel – to attain a level of turgor that is "just right" for the prevailing osmotic and cellular conditions (Börngen et al. 2010). *B. subtilis* might employ similar mechanisms for the fine-tuning of its cellular compatible solute pool.

15.3 Exposure to Hyperosmolarity: Salt-In or Salt-Out

To cope physiologically with high osmolarity surroundings, prokaryotes have developed two principally different approaches. One is the so-called salt-in-cytoplasm strategy in which the internal salt concentration is actively maintained through transport processes at a higher level than that present in the environment. Primarily K^+ and Cl^- ions are pumped into the cell and cytotoxic Na^+ ions are pumped out. The salt-in cytoplasm strategy is energetically favorable (Oren 2010), but it requires far reaching adaptations of the entire intracellular enzyme machinery and cellular physiology to the permanently high ion content of the cell. The evolutionary adjustment to a high ion content of the cytoplasm has left an "acid signature" on the entire proteome of those Archaea and Bacteria that have adopted the salt-in-cytoplasm strategy in order to maintain proper protein solubility, stability and catalytic activity in a cytoplasm of high ionic strength (Coquelle et al. 2010; Rhodes et al. 2010). Although energetically favorable (Oren 2010), the salt-in cytoplasm strategy is not prevalent among the different phylogenetic and physiological groups of halophilic Archaea and Bacteria (Oren 2008; Ventosa et al. 1998). This is probably because considerable amendments in protein composition and cellular physiology had to be made by cells in the course of evolution to cope on a sustained basis with molar concentrations of KCl in their cytoplasm.

No such evolutionary adjustments in protein composition or cellular physiology are imposed on those microorganisms that use a salt-out cytoplasm strategy and that employ a selected class of organic osmolytes, the compatible solutes, to balance turgor (Kempf and Bremer 1998; Empadinhas and da Costa 2008; Bremer and Krämer 2000; Wood et al. 2001). The amassing of compatible solutes through synthesis is energetically more costly then the import of ions (Oren 2010), but it provides great flexibility to the osmotic stress response of those microorganisms that frequently encounter fluctuating osmotic conditions. It also does not tie their wellbeing and cellular integrity to permanently high salinity habitats, as the salt-in cytoplasm strategy often does (Ventosa et al. 1998; Oren 2008). Furthermore, almost all microorganisms can take up pre-formed compatible solutes, or their biosynthetic precursors, from environmental sources (Kempf and Bremer 1998; Bremer and Krämer 2000; Wood et al. 2001; Ziegler et al. 2010). This allows the bacterial cell to take advantage of precious environmental resources (Welsh 2000). Import of compatible often partially suppresses the expression of biosynthetic genes for these types of stress protectants, indicating that the uptake of compatible solutes is physiologically (or energetically) more advantageous for osmotically stressed cells than their biosynthesis.

It should, however, be stressed that there are microorganisms that can use a combination of both the salt-in and the salt-out cytoplasm strategies to cope with high salinity habitats on a sustained basis. One example is the poly-extremophile *Natranaerobius thermophilus*. It uses both high intracellular K^+ concentrations and the accumulation of compatible solutes to balance the osmotic potential of its cytoplasm and employs Na^+/H^+ antiporters to keep the intracellular sodium concentration very low (about 8 mM) in an environment that contains 3–4 M NaCl (Mesbah et al. 2009).

15.4 Properties of Compatible Solutes

Compatible solutes are operationally defined as organic osmolytes that can be amassed by cells to exceedingly high concentrations without interfering with cellular physiology and growth. They are widely employed as osmostress protectants not only by microorganisms but also by plant, animal and even human cells, attesting to the effectiveness of this strategy for coping with high osmolarity in different environmental and cellular settings (Kempf and Bremer 1998; Yancey 2005; Burg and Ferraris 2008). Compatible solutes used by members of the *Bacteria* are usually non-charged highly water soluble compounds and comprise a rather restricted number of organic osmolytes. Important representatives are the sugar trehalose, the heteroside glycosylglycerol, the imino acid proline, the trimethylammonium compound glycine betaine and the tetrahydropyrimidines ectoine and 5-hydroxyectoine (Kempf and Bremer 1998; Bremer and Krämer 2000; Empadinhas and da Costa 2008; Hagemann 2011; Ziegler et al. 2010). *B. subtilis* synthesizes proline as its primary compatible solute (Whatmore et al. 1990), whereas many other Bacilli synthesize ectoines (Kuhlmann and Bremer 2002; Zhao et al. 2006; Bursy et al. 2007; Kuhlmann et al. 2008; Rajan et al. 2008; Saum and Müller 2008a). Bacilli can also use exogenously provided proline, glycine betaine and ectoine as osmoprotectants using various types of osmotically controlled high affinity transport systems for their retrieval (Fig. 15.1) (Bremer and Krämer 2000; Bremer 2002).

A hallmark of compatible solutes is their preferential exclusion from the immediate hydration shell of proteins that is due to unfavorable interactions between the osmolytes and the protein backbone. This uneven distribution of the compatible solutes in the cell water generates a thermodynamic driving force that promotes the proper folding and conformation of proteins and enhances their stability (Street et al. 2006). Hence, not only do compatible solutes contribute to osmotic balance by serving as water-attracting osmolytes, they also benefit for the cell by helping maintain proper protein function (Bourot et al. 2000; Diamant et al. 2003; Fisher 2006; Ignatova and Gierasch 2006). These solutes are therefore sometimes referred to as chemical chaperones.

In addition to their osmoprotective role, compatible solutes also have notable effects on the growth of microorganisms at the very upper and lower temperature boundaries. For instance, glycine betaine acts both as a thermo-stress and cold-stress protectant for *B. subtilis*, regardless of whether it is synthesized from the precursor choline or taken up from the environment (Brigulla et al. 2003; Holtmann and Bremer 2004; Hoffman and Bremer 2011). The biochemical and biophysical mechanisms that underlie the temperature-stress protective functions of compatible solutes are far from being understood and might not necessarily be connected with their well-studied function as water-attracting osmolytes. However, the combined osmotic-stress and temperature-stress protective properties of compatible solutes can have undesired consequences for human health. It allows food-pathogens such as *Bacillus cereus* and *Bacillus weihenstephanensis* to counteract commonly used food-preservation measures such as salting and refrigeration since plant and animal materials present in foodstuff contain osmoprotectants such as choline, glycine betaine and carnitine. The same compatible solute transporters (Opu) are used under osmotic and temperature stress conditions by *B. subtilis* to acquire these compounds from environmental sources (see below).

15.5 *B. subtilis*: A Genetically Tractable Model System to Study Cellular Adjustment to High Salinity

Although *B. subtilis* is certainly not a salt-loving bacterium, it can grow, albeit slowly, in minimal and rich media containing a substantial amount of salt (e.g., 1.2 M NaCl) (Boch et al. 1994). The development of such a considerable degree of salt stress resistance probably stems from the exposure of *B. subtilis* cells to prolonged periods of desiccation in natural habitats; e.g., the well-oxygenated upper layers of the soil. *B. subtilis* does not respond to high salinity conditions by initiating the sporulation process (Ruzal et al. 1998) that under nutrient-limited growth conditions leads to the formation of highly desiccation resistant endospores. Hence, *B. subtilis* needs to engage in defense mechanisms that allow the vegetative cell to confront high salinity environments effectively.

One facet in the osmostress response system of *B. subtilis* and other Bacilli is the salt-induction of the SigB-controlled general stress regulon (Hecker et al. 2007). However, induction of the SigB-regulon of *B. subtilis* by a salt-shock is only transient and a *sigB*-mutant grows like a wild-type strain in minimal medium with 1.2 M NaCl (Spiegelhalter and Bremer 1998). Therefore, the general stress regulon is certainly not the central osmotic stress response system for vegetative *B. subtilis* cells, although the disruption of the structural gene for *sigB* itself and of many members of the SigB regulon frequently causes a salt-sensitive phenotype (Höper et al. 2005).

Proteome and genome-wide transcriptional profiling studies of high salinity-challenged cells have highlighted the complexity and multifaceted nature of the osmotic stress response systems of *B. subtilis* (Steil et al. 2003; Höper et al. 2006; Hahne et al. 2010). However, detailed physiological and molecular studies have shown beyond doubt that the effective water-management is the cornerstone of the cell's acclimatization to either sudden or sustained rises in the environmental osmolarity and salinity (Bremer and Krämer 2000; Bremer 2002). The accumulation of compatible solutes, either via synthesis or uptake, plays a key role in this process. This is evident by the salt-sensitive growth phenotypes of *B. subtilis* mutants that cannot synthesize the compatible solute proline (Brill J. and Bremer E.; unpublished results) or by the strongly impaired ability of mutants with defects in compatible solute uptake system, to grow efficiently in high-salt minimal media (Kappes et al. 1996, 1999).

As in other bacteria, the cellular reaction of *B. subtilis* to a sudden rise in the external osmolarity is multiphasic (Kempf and Bremer 1998; Bremer and Krämer 2000; Wood et al. 2001; Booth et al. 2007). A rapid rise in the cellular K^+ content through transport processes is typically the initial cellular response that follows the loss of cell water subsequent to an osmotic up-shift (Whatmore et al. 1990). In Gram-negative bacteria, glutamate is used as the counter-ion for K^+ to maintain electro-neutrality. However, the nature of the counter-ion used by *B. subtilis* to balance the positive electric charge of the accumulated K^+ ion has not been rigorously established. The size of the glutamate-pool increases only moderately, whereas the K^+ content of osmotically challenged cells increases strongly from a basal level of about 350 mM to about 650 mM within 1 h subsequent to the osmotic up-shock (Whatmore et al. 1990). The Ktr-type potassium uptake systems KtrAB and KtrCD play a central role for K^+ uptake, both in osmotically non-stressed and in osmotically stressed *B. subtilis* cells (Holtmann et al. 2003). This type of transport system consists of a dimer of the K^+-translocating membrane-embedded subunit (KtrB), probably evolutionarily derived from a subunit of an ancestral K^+ channel, and a soluble regulatory subunit that can bind both ATP and NAD^+/NADH via its RCK/KTN domain; an octameric ring structure has been proposed for the RCK/KTN domain-containing KtrA subunit (Albright et al. 2006). Ktr-systems are Na^+-coupled K^+ uptake systems (Corratge-Faillie et al. 2010) and hence, the massive import of K^+ that follows an osmotic up-shift (Whatmore et al. 1990) will simultaneously lead to a large-scale import of cytotoxic Na^+ ions into the *B. subtilis* cell.

Concomitant with the rise in the K^+ pool in osmotically up-shocked cells, *B. subtilis* starts to synthesize the compatible solute proline on a very large scale (Whatmore et al. 1990). This leads eventually to a reduction in the K^+ pool, and hence, of the ionic strength of the cytoplasm. Such a reduction in K^+ content can also be detected when the osmotically stressed cells are allowed to take up glycine betaine from exogenous sources (Whatmore et al. 1990). It is obvious that K^+ extrusion systems must function in the second acclimatization stage of *B. subtilis* to high-salinity surroundings, a time when primarily compatible solutes are amassed to raise the osmotic potential of the cytoplasm (Fig. 15.1). Indeed, one such K^+ export system (YhaTU) has already been identified and characterized biochemically (Fujisawa et al. 2007). The expression of the structural genes for the YhaTU system is strongly induced in response to an osmotic up-shock (Hahne et al. 2010). Hence, during the initial acclimatization reactions of *B. subtilis* to a sudden rise in the external osmolarity, fluxes of K^+ ions in and out of the cell apparently play an important role. This must be certainly also true for the Na^+ ions that enter the cell in co-transport with K^+ via the Ktr transporters (Corratge-Faillie et al. 2010) and that need to be effectively exported from the cell due to the cytotoxicity of Na^+ ions. Na^+ ions will also enter the *B. subtilis* cell as a by-product of the import of compatible solutes through secondary transport systems. For instance, the BCCT-type glycine betaine uptake system OpuD and the SSSF-type proline uptake system OpuE (see below) of *B. subtilis* are Na^+-coupled transporters (von Blohn et al. 1997; Kappes et al. 1996; Ziegler et al. 2010). Interestingly, the expression of the operon encoding the multi-subunit Mrp Na^+ extrusion system and the genes that encoding the single-component NhaK and NhaC Na^+ exporter are all up-regulated in salt-shocked *B. subtilis* cells (Hahne et al. 2010). However, *B. subtilis* cannot rely exclusively on the cycling of ions in and out of the cell to adjust to truly high salinity environments. This is deducible from the phenotype of a mutant strain (*proHJ*) that cannot synthesize the compatible solute proline. Such a strain exhibits a salt-sensitive growth defect in a minimal medium containing 1.2 M NaCl; however, this phenotype can be effectively rescued by the addition of a compatible solute (e.g., 1 mM glycine betaine) to the growth medium (J. Brill and E. Bremer; unpublished results).

15.5.1 Synthesis and Uptake of Compatible Solutes by **B. subtilis**

The above reported findings highlight the importance of compatible solute accumulation for the sustained salt-stress response of *B. subtilis*. *B. subtilis* can synthesize the compatible solute proline *de novo* from its considerable glutamate pool (Whatmore et al. 1990), and it can synthesize glycine betaine by first importing the precursor choline and then oxidizing it to glycine betaine (Boch et al. 1994, 1996). Furthermore, *B. subtilis* can take up a large number of pre-formed compatible solutes from environmental sources as effective osmo-stress protectants (Bremer 2002).

15.5.1.1 Synthesis of Proline

Since the pioneering studies of Measures in 1975, it has been known that *B. subtilis* synthesizes large quantities of the compatible solute proline when it is subjected to high salinity environments (Measures 1975). Subsequent detailed physiological studies conducted by Reed and co-workers (Whatmore et al. 1990) revealed that the cellular proline pool build-up by de novo synthesis rises from a basal level of 16 mM to about 500–700 mM within 7 h of growth subsequent to a moderate osmotic up-shift with 0.4 M NaCl. Since *B. subtilis* can grow in a minimal medium with 1.2 M NaCl (Boch et al. 1994), one can expect that these severely osmotically stressed cells will amass proline in excess of 1 M through de novo synthesis. We have elucidated the genetic basis for the production of the vast quantities of proline produced by *B. subtilis* as an osmostress protectant and found that the anabolic proline biosynthetic route and that of the osmoadaptive route are interconnected via the γ-glutamyl phosphate reductase (ProA) (Brill J. and Bremer E.; unpublished results).

In many microorganisms, proline biosynthesis proceeds from the precursor glutamate and involves three enzyme-catalyzed steps. It begins with the ATP-dependent phosphorylation of glutamate by the γ-glutamyl kinase (ProB). The resulting γ-glutamyl phosphate is then reduced to γ-glutamic semialdehyde by the γ-glutamyl phosphate reductase (ProA). The formed γ-glutamic semialdehyde spontaneously cyclizes to Δ^1-pyrroline-5-carboxylate, which is then further reduced by the Δ^1-pyrroline-5-carboxylate reductase (ProC) to the end product proline. This pathway for the production of proline as a building block for protein biosynthesis is also present in *B. subtilis* (Belitsky et al. 2001) and comprises the ProB-ProA-ProI enzymes (several ProC-type enzymes are present; hence, the designation ProI) (Belitsky et al. 2001). The anabolic proline biosynthesis route from the precursor glutamate is frequently regulated in microorganisms though allosteric feedback inhibition of the biochemical activity of first proline-biosynthetic enzyme (ProB) by the end product proline. This is also the case in *B. subtilis* (Chen et al. 2007). Furthermore, the expression of the anabolic *proBA* and *proI* genes is genetically controlled via a proline-responsive T-box system, a regulatory device (Green et al. 2010) that allows the induction of the *proBA* and *proI* genes only when the *B. subtilis* cells starve for proline (Brill et al. 2011). Hence, the anabolic ProB-ProA-ProI proline biosynthetic route in *B. subtilis* is genetically and biochemically wired in such a fashion that a wasteful overproduction of proline is strictly prevented when it is synthesized as a building block for protein biosynthesis. The content of free proline in the cytoplasm of osmotically non-stressed *B. subtilis* cells is about 16 mM (Whatmore et al. 1990). It is immediately apparent that the above described biochemical and genetic control mechanisms make the anabolic ProB-ProA-ProI proline biosynthetic route unsuitable to provide the *B. subtilis* cell with the very large amounts (0.5–1 M) of proline it needs as an osmostress protectant (Bremer 2002).

A second proline biosynthetic route formed by the ProJ-ProA-ProH enzymes, accomplishes this task (Brill J. and Bremer E.; unpublished results). ProJ and ProH are iso-enzymes of ProB and ProI, respectively, and catalyze the first and last steps of the osmo-adaptive proline biosynthetic route. The implication of this finding is that the activity of ProJ enzyme, in contrast to that of ProB, is not (or at least not very strongly) subjected to feedback inhibition by proline, although this prediction still needs to be confirmed biochemically. The transcription of the *proHJ* structural genes is strongly induced both subsequent to an osmotic up-shock and during continued growth of the *B. subtilis* cells in high salinity media (Steil et al. 2003; Hahne et al. 2010). As a matter of fact, the *proHJ* operon has the highest induction ratio of all of the approximately 100 high-salinity-induced genes of *B. subtilis* cells cultivated under sustained high osmolarity conditions (Steil et al. 2003). Attesting to the central role of the ProJ and ProH enzymes for the adjustment of the cell to high salinity habitats is the finding that the disruption of the *proHJ* operon causes osmotic sensitivity (Brill J. and Bremer E.; unpublished data).

Since no paralogous enzyme for the γ-glutamyl phosphate reductase (ProA) is present in *B. subtilis*, the anabolic (ProB-ProA-ProI) and the osmostress adaptive (ProJ-ProA-ProH) proline biosynthetic routes are interconnected. It is not immediately obvious why this interconnection has occurred in the course of evolution since the *B. subtilis* cell has to rely under osmotic stress conditions on the basal level of the T-box-controlled expression of the *proBA* operon for the supply of the ProA enzyme. Such an interconnection of the anabolic and osmoadaptive proline biosynthetic routes does not always occur in those Bacilli that produce proline as a compatible solute. For instance, in the industrial workhorse *Bacillus licheniformis*, two complete proline biosynthetic routes are present whose regulation is precisely tailed to two different tasks: proline production for protein biosynthesis and proline production for osmostress protection. As in *B. subtilis*, the expression of the genes (*proBA* and *proI*) for the anabolic route (ProB-ProA-ProI) is controlled via T-box systems (Bleisteiner M., Putzer H. and Bremer E.; unpublished results) and consequently these genes are only strongly induced under proline-starvation conditions. However, in contrast to *B. subtilis*, a full set of osmoadaptive proline biosynthetic enzymes (ProJ-ProAA-ProH) is present in *B. licheniformis*. The expression of the corresponding gene cluster (*proH-proJ-proAA*) is strongly induced at high salinity (M. Bleisteiner, T. Hoffmann, and E. Bremer; unpublished results). Likewise, in the proline-producing moderate halophile *Bacillus halophilus*, an osmotically inducible gene cluster (*proJ-proH-proA*) is present that encodes an entire set of proline biosynthetic enzymes (Saum and Müller 2008b). Interestingly, the solute glutamate serves as a messenger to turn on the expression of *proJ-proH-proA* gene cluster and thereby switch the osmolytes strategy of *B. halophilus* from glutamate production to proline production during the transition from moderate to high salinity environments (Saum and Müller 2008b).

15.5.1.2 Synthesis of Glycine Betaine

Microorganisms can synthesize glycine betaine by two different routes: (a) through a step-wise methylation of the amino acid glycine or (b) through the oxidation of choline using a variety of enzymes. The de novo synthesis of glycine betaine from glycine is catalyzed by several methylases with overlapping substrate specificities that use *S*-adenosylmethionine (AdoMet) as the methyl donor. This is an energetically very expensive way to produce glycine betaine since the reduction of a single methyl group and its activation via AdoMet requires 12 ATP equivalents (Nyyssölä et al. 2000). Consequently, the de novo synthesis pathway for glycine betaine production is not prevalent in the microbial world. Instead, most microorganisms that are capable of glycine betaine synthesis produce it via oxidation of choline that is acquired through transport processes from external sources. This is also the way though which *B. subtilis* synthesizes glycine betaine as an osmostress protectant (Boch et al. 1994; Kappes et al. 1999). Choline is a component of the lipids of plant cells and it can be liberated from phosphatidylcholine by the action of lipases. Hence, the precursor for glycine betaine is introduced, albeit at low and variable concentrations, into the soil habitat of *B. subtilis*. Choline is taken up by *B. subtilis* via two osmotically inducible ABC-type importers: the OpuC and OpuB systems (Kappes et al. 1999). Both transporters exhibit high-affinity for choline with K_m values in the low μM range, but only OpuB is highly substrate specific; in contrast, the OpuC possesses a broad substrate specificity for compatible solutes (Bremer 2002). Once imported via OpuB and OpuC, choline is oxidized by the type III alcohol dehydrogenase (GbsB) to glycine betaine aldehyde, and this chemically highly reactive and toxic intermediate is then further oxidized to glycine betaine via the glycine betaine aldehyde dehydrogenase (GbsA) (Boch et al. 1996). In this way, *B. subtilis* can achieve a considerable degree of osmotic stress tolerance, but the synthesis of glycine betaine from choline is somewhat less effective in terms of stress protection than the uptake of pre-formed glycine betaine (Boch et al. 1994). It should be noted in this context, that choline has no osmoprotective properties per se in *B. subtilis*.

15.5.1.3 Uptake of Compatible Solutes

The cellular content of compatible solutes is sensitively determined by the degree of the osmotic stress perceived by the microbial cell (Bremer and Krämer 2000; Ziegler et al. 2010). These compounds can reach molar concentrations in severely stressed cells. Sources of compatible solutes in natural settings are osmotically down-shocked or decaying microbial and eukaryotic cells and excretion products of animals and plants (Welsh 2000). Since these compounds are typically found in very low concentrations (nM or μM) in the environment, microbial cells must possess effective transport systems to scavenge them (Bremer and Krämer 2000; Wood et al. 2001; Ziegler et al. 2010). *B. subtilis* possesses five osmotically inducible transport systems for compatible solutes (Bremer 2002). We have christened these transporters Opu: *o*smo*p*rotectant *u*ptake (Kempf and Bremer 1995; Kappes et al. 1996, 1999; von Blohn et al. 1997). These systems have homologues in many microbial species and our genetic, physiological and structural characterizations of these transporters have served as a blueprint for the functional annotation of compatible solute uptake systems in many microbial genome projects and in databases.

Glycine betaine is a particularly important osmostress protectant for *B. subtilis*. Not only can *B. subtilis* synthesize it from an exogenous supply of the precursor choline (see above), but it also can take it up via three osmotically inducible transport systems: the ABC transporters OpuA and OpuC and the BCCT-type carrier OpuD (Kempf and Bremer 1995; Kappes et al. 1996, 1999). Each of these systems exhibits a high affinity for its substrate with K_m values in the low μM range, and each of them possess considerable transport capacity. Furthermore, each of these transporters is sufficient to mediate effective osmostress protection for *B. subtilis* cells in the presence of glycine betaine but the OpuA transporter is the dominant uptake system due to its high V_{max} (Kappes et al. 1996). The OpuA system also has a considerable transport activity for glycine betaine in cells that are not osmotically stressed, resulting in a substantial cellular glycine betaine pool of 130–170 mM when *B. subtilis* is grown in standard laboratory minimal media (Whatmore et al. 1990; Holtmann and Bremer 2004). Considerable glycine betaine pools are also expected to be present in *B. subtilis* cells grown in rich media containing yeast extract (e.g., LB medium), since yeast extract is a rich source for glycine betaine (Dulaney et al. 1968) and LB-medium is routinely prepared with a considerable amount of NaCl (5 g/l).

The OpuA, OpuC, and OpuD transporters from *B. subtilis* not only function in the uptake of glycine betaine but also for the transport of other compatible solutes, most of which are structurally related to glycine betaine (Bremer 2002). Some of these compatible solutes are substrates for more than one of these uptake systems, whereas others are taken up only through a single Opu transporter. Examples for this latter type of uptake profile are the osmoprotectants L-carnitine, crotonobetaine, γ-aminobutyrobetaine, choline-*O*-sulfate, and ectoine which are all exclusively acquired by the osmotically stressed *B. subtilis* cell via the OpuC ABC transporter (Jebbar et al. 1997; Kappes and Bremer 1998; Nau-Wagner et al. 1999). The OpuC system is noteworthy because it can transport with high affinity (K_m values in the low μM range) in total 11 osmoprotectants that are all structurally related to glycine betaine, and with very low-affinity ectoine (K_i of approximately 1.6 mM), a compatible solute structurally unrelated to glycine betaine (Jebbar et al. 1997).

It should be noted, that from the 12 currently known osmoprotectants acquired by *B. subtilis* through transport processes, 11 are metabolically inert (Bremer 2002). Hence, they are amassed for a sole purpose: stress protection! The only exception is the amino acid proline that can be efficiently used both as a sole carbon and nitrogen source (S. Moses and E. Bremer; unpublished results).

15.5.1.4 Structural Analysis of the Glycine Betaine-Binding Protein OpuAC

High-affinity interactions between compatible solutes and transport proteins must take place in order to achieve effective import. However, as mentioned above, the preferential exclusion from protein surfaces is a hallmark of these types of solutes (Street et al. 2006). Since transport systems for compatible solutes have been detected in practically every microorganism studied in the context of adaptation to high osmolarity environments, a fundamental problem arises: how can a compound be bound by a protein with high affinity and specificity when this compound is typically excluded from the immediate hydration shell of the very same protein? What are the structural determinants for the selective binding of compatible solutes by components of transport systems?

The answer, at least for the sub-group of compatible solutes chemically related to glycine betaine, has recently come through crystallographic and mutational studies of soluble ligand-binding proteins from several microbial ABC transporters. One of the studied proteins is OpuAC, the ligand-binding protein of the OpuA transporter from *B. subtilis* (Kempf and Bremer 1995; Horn et al. 2006). This protein is tethered via a lipid anchor attached to an N-terminal Cys-residue on the outer surface of the cytoplasmic membrane of *B. subtilis* so that it can capture exogenously provided glycine betaine with high affinity (K_D of about 20 μM). The crystal structure of the OpuAC protein in complex with glycine betaine revealed a bilobal organization of the protein (Fig. 15.2; left panel), a topology that is typical for ligand-binding proteins of ABC transporters. The two lobes of the OpuAC protein are connected through a flexible hinge region that allows a rigid-body movement of the lobes between the open non-liganded form of the protein and a closed liganded form; the ligand shifts the equilibrium towards the substrate-loaded conformation. As a consequence of this movement, the ligand glycine betaine is captured and buried deep in a cleft formed by the two domains of the OpuAC protein (Fig. 15.2; left panel) (Horn et al. 2006). The open and closed crystal structures of the OpuAC protein from *Lactococcus lactis* have recently been

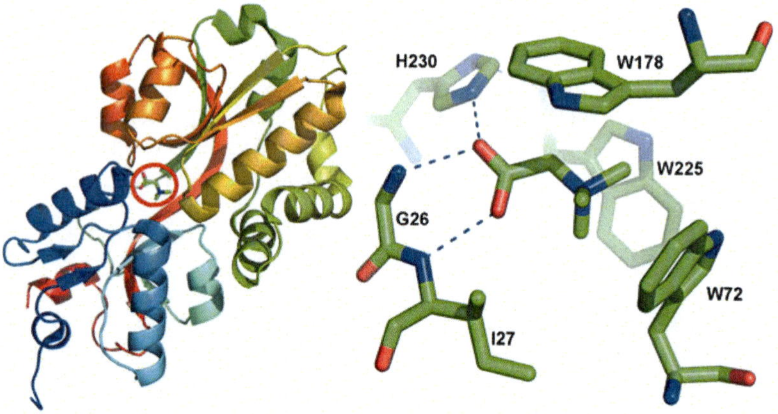

Fig. 15.2 Crystal structure of the glycine betaine binding protein OpuAC from *B. subtilis*. (*Left panel*) Overall structure of the OpuAC protein in complex with its ligand glycine betaine; the ligand is circled in *red*. (*Right panel*) Architecture of the glycine betaine-binding site present in OpuAC. The pictures were prepared with PyMol (http://www.pymol.org/) using the atomic coordinates deposited in the PDB file 2B4L (http://www.pdb.org/) (Horn et al. 2006)

reported and can thus be used to trace the movements of the two domains toward each other and the capturing of the glycine betaine ligand (Wolters et al. 2010).

The crystallographic analysis of the *B. subtilis* OpuAC protein revealed a remarkably structured ligand-binding site that consists of three Trp residues (the aromatic cage) into which the positively charged trimethylammonium headgroup of glycine betaine is wedged and is coordinated via cation–pi interactions (Fig. 15.2; right panel). The carboxylic group of the glycine betaine ligand protrudes out of the aromatic cage and forms directed hydrogen bonds with either the backbone or side chains of specific amino acid residues (Horn et al. 2006). Mutational analysis established the key contributions of cation–pi interactions for the effective binding of the glycine betaine ligand by OpuAC within the aromatic cage (Fig. 15.2; right panel) (Smits et al. 2008). The OpuAC protein from *B. subtilis* can also bind the compatibe solutes proline betaine and dimethylsulfonioacetate (DMSA), a sulfur analog of glycine betaine, with reasonable affinities (Horn et al. 2006; Smits et al. 2008).

Crystallographic studies of the glycine betaine-binding proteins from *E. coli* (ProX), the hyperthermophilic archaeon *Archaeoglobus fulgidus* (ProX), the OpuAC proteins from *L. lactis*, and the choline/acetylcholine-binding protein ChoX from the root-associated soil bacterium *Sinorhizobium meliloti* all revealed similarly structured ligand-binding sites. Although the precise architecture of the aromatic cage in the ligand-binding sites varies between these proteins, common denominators for substrate binding have emerged. In each substrate-binding protein, the bulky and positively charged head groups of glycine betaine, choline, acetylcholine, proline betaine, and DMSA are similarly accommodated within the aromatic ligand-binding site via cation–pi interactions. The tails of the various substrates protrude from the aromatic cage and are coordinated through H-bonds, salt-bridges, and water-networks by residues stemming from the two domains of the ligand-binding proteins. The ProX proteins from *E. coli* and *A. fulgidus*, the OpuAC proteins from *B. subtilis* and *L. lactis*, and the ChoX protein from *S. meliloti* have homologues in many microbial species, and in almost all of these sequence-related proteins those residues forming the aromatic cage are conserved.

The aromatic cages discussed so far are all present in soluble ligand-binding proteins of ABC transport systems. What about the ligand-binding sites in membrane-embedded carriers for compatible solutes? The crystal structure of the glycine betaine transporter BetP from *C. glutamicum*, a member of the ubiquitously found BCCT carriers (Ziegler et al. 2010), revealed a glycine betaine-binding site (Ressl et al. 2009) that is virtually super-imposable onto that present in the ProX protein from *E. coli* (Schiefner et al. 2004). Since the soluble periplasmic binding protein ProX and the integral membrane protein BetP are certainly not closely evolutionarily related, nature has apparently adopted common design principles to construct a high-affinity ligand-binding site for a solute (glycine betaine) that typically is preferentially excluded from the surface of proteins.

15.6 Osmotically Controlled Synthesis of Compatible Solutes by Bacilli Other Than *B. subtilis*

Having focused on the synthesis of proline in *B. subtilis*, we wondered whether most other Bacilli would also produce proline as their primary compatible solute. We therefore embarked on a study that evaluated the synthesis of compatible solutes of a large group of Bacilli (26 species altogether) via natural abundance ^{13}C-NMR spectroscopy (Kuhlmann and Bremer 2002; Bursy et al. 2007). Inspection of the recorded data revealed that the studied 26 Bacilli could be grouped into three main classes (a) those that used only L-glutamate as their primary compatible solute [6 species], (b) those that produced proline in response to osmotic stress [5 species], and (c) those that synthesized ectoine [15 species] when they were osmotically challenged. This later class can be further sub-grouped into microorganisms that produce ectoine alone, or in combination with hydroxyectoine, or in combination with proline. Our studies therefore revealed that the most widely synthesized compatible solute within the studied Bacilli is the tetrahydropyrimidine ectoine (Fig. 15.3). The ectoine biosynthetic gene cluster has been studied at the molecular level in several Bacilli (Kuhlmann and Bremer 2002; Zhao et al. 2006; Bursy et al. 2007; Kuhlmann et al. 2008; Rajan et al. 2008; Saum and Müller 2008a).

Although this was not systematically studied by us, those species that rely on L-glutamate production alone tend to be rather salt-sensitive (e.g., *B. cereus*), those that synthesize proline exhibit an intermediate osmotic stress resistance (e.g., *B. subtilis* and *Bacillus licheniformis*), and those species that produce ectoines (e.g., *Virgibacillus salexigens*) or a combination of ectoine and proline (e.g., *Halobacillus halophilus*) typically are rather salt tolerant. With the exception of *Paenibacillus polymyxa*, the sugar trehalose was not detected in the compatible solute pools of the studied Bacilli; furthermore, trehalose was only produced in stationary-phase cultures of *P. polymyxa*, indicating that it might be produced as a stress protectant for purposes other than osmotic balance. Hence, our findings indicate that Bacilli do not usually produce trehalose as an osmostress protectant, whereas this non-reducing sugar is the dominant compatible solute synthesized by *E. coli* (Wood et al. 2001).

Inspection of the spectrum of compatible solutes synthesized by the Bacilli studied by us experimentally and of additional data compiled from the literature or derived from genome sequencing projects revealed that the type of compatible solute(s) produced does not always closely follows phylogenetic constraints (Fig. 15.3). For instance, *B. licheniformis*, *B. subtilis*, *Bacillus amyloliquefaciens*, *Bacillus vallismortis* and *Bacillus mojavensis* are phylogenetically closely related and all synthesize proline as a compatible solute. However, *B. mojavensis* is capable to synthesize ectoine as well (Fig. 15.3). The type(s) of compatible solute produced is therefore not a very informative taxonomic marker and the inspection of the data compiled by us in Fig. 15.3 suggest that lateral gene transfer events might have significantly shaped the compatible solute profile of Bacilli. This seems

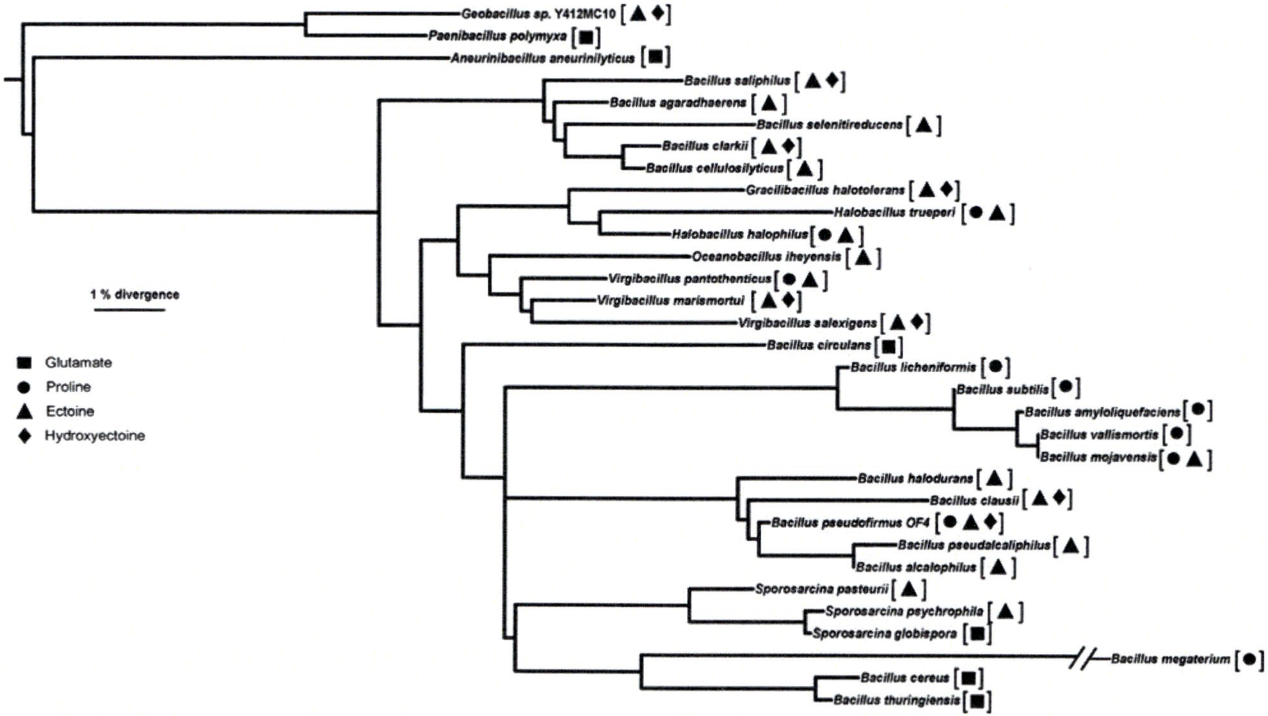

Fig. 15.3 Pattern of the synthesis of the compatible solutes glutamate, proline, ectoine, and hydroxyectoine in various Bacilli. The data on the synthesis of compatible solutes of the 32 listed Bacilli were compiled either from reports in the literature (Kuhlmann and Bremer 2002; Kuhlmann et al. 2008; Bursy et al. 2007; Saum and Müller 2008a; Romano et al. 2005; Rajan et al. 2008) or gleaned from the genome sequences of *Geobacillus* sp. Y412MC10, *Bacillus clausii* KSM K16, *Oceanobacillus iheyensis*, *Bacillus cellulosilyticus* and *Bacillus selenitireducens*. The phylogenetic tree was constructed using the ARB software package and is derived from a distance matrix of 16S rRNA sequences using the neighbor-joining method

in particular to be the case when the pattern of ectoine and hydroxyectoine production is considered (Fig. 15.3). One should take note that none of the 26 Bacilli studied by us for the production of compatible solutes with ^{13}C-NMR-spectroscopy actually synthesized glycine betaine de novo (Kuhlmann and Bremer 2002; Bursy et al. 2007). Romano et al. (2005) evaluated the compatible solute pool of the highly salt tolerant haloalkaliphic *Bacillus saliphilus* and reported that this *Bacillus* species accumulates glycine betaine as its dominant osmoprotectant (81% of the compatible solute pool) and only minor amounts of ectoine (12%) and hydroxyectoine (3%) (Romano et al. 2005). This finding can in all likelihood be explained by the fact that the cells were grown in a minimal medium that contained yeast extract, a well-known rich source of glycine betaine (Dulaney et al. 1968). We therefore stress the need to use truly chemically defined media when the spectrum of compatible solutes produced by a given microorganism is evaluated to clearly distinguish de novo synthesized compatible solutes from those that are imported. In this way, one also can avoid the pitfall that exogenously provided compatible solutes suppress the expression of compatible solute biosynthetic genes and thereby skew the profile of the composition of the cellular compatible solute pool.

15.6.1 Synthesis of Ectoine and Hydroxyectoine

Although our evaluation of the synthesis of compatible solutes by Bacilli (Kuhlmann and Bremer 2002; Bursy et al. 2007) is certainly not representative for all members of this very large order, it is apparent from our study of 26 type strains that ectoine and its derivative 5-hydroxyectoine are widely synthesized by Bacilli as osmostress protectants. Ectoine and 5-hydroxyectoine were initially regarded as rather uncommon compatibe solutes when they first were discovered in *Ectothiorhodospira halochloris* and *Streptomyces parvulus*, respectively. However, improved ^{13}C-NMR- and HPLC-based screening procedures and finally the discovery of the ectoine/hydroxyectoine biosynthetic genes (Louis and Galinski 1997) revealed their widespread occurrence in the microbial world. Our recent database searches of finished genome sequences of *Bacteria* (1,391 genome sequences) and Archaea (80 genome sequences) identified 241 putative ectoine producers (based on searches with the amino acid sequence of the ectoine synthase EctC, a search criterion that might slightly overestimate the number of actual ectoine producers). With the single exception of the archaeon *Nitrosopumilus maritimus* SCM1, all these putative ectoine producers are members of the *Bacteria*. There are currently no *Eukarya* that are known to synthesize ectoines naturally.

Three enzymes mediate ectoine biosynthesis from L-aspartate-β-semialdehyde, a central intermediate in amino acid metabolism, through the sequential enzymatic reactions of the L-2,4-diaminobutyrate transaminase (EctB), the L-2,4, diaminobutyrate acetyltransferase (EctA), and the ectoine synthase (EctC), an enzyme of the cupin protein family that mediates the ring-closure of the ectoine biosynthetic intermediate N-acetyl-L-2,4-diaminobutyrate. 5-hydroxyectoine is formed from ectoine in a subset of the ectoine producers though the catalytic activity of the ectoine hydroxylase (EctD). In all ectoine- and hydroxyectoine-producing bacteria analyzed so far, biosynthesis of these compatible solutes is strongly enhanced under high osmolarity growth conditions. This is largely due to the osmotic induction of the expression of the ectoine biosynthetic genes, although stimulation of the activity of the ectoine biosynthetic enzymes by high salinity or high ionic strength might also contribute to ectoine production in osmotically stressed cells. The genetic disruption of the ectoine biosynthetic gene cluster typically causes an osmosensitive growth phenotype.

The genes (*ectABC*) for the ectoine biosynthetic enzymes are typically organized as an operon (Louis and Galinski 1997; Kuhlmann and Bremer 2002; Bursy et al. 2007; Kuhlmann et al. 2008; Schwibbert et al. 2010) that also can contain in the hydroxyectoine producers the gene (*ectD*) for the ectoine hydroxylase (Prabhu et al. 2004). However, the *ectD* gene is not always part of the *ectABC* gene cluster and can be frequently found at a separate position on the genome (Garcia-Estepa et al. 2006; Bursy et al. 2007). Some of the *ectABC* or *ectABCD* gene clusters also contain a gene for an aspartokinase (*ask_ect*) (Reshetnikov et al. 2006), a genetic configuration that might serve to boost the production of the ectoine precursor L-aspartate-β-semialdehyde under osmotic stress conditions. The aspartokinase (Ask) enzyme is often subjected to complex feed-back regulatory mechanisms which might generate a bottle-neck for an adequate supply of L-aspartate-β-semialdehyde to meet the considerable demands of this precursor for ectoine biosynthesis (Bestvater et al. 2008). Hence, the Ask encoded by the *ask_ect* gene might encode an enzyme with special biochemical properties to aid ectoine production under osmotic stress conditions.

Transcription of the *ect* gene cluster is typically strongly up-regulated in cells challenged by high salinity. Osmotically controlled promoters for *ect* gene clusters have been identified in several microorganisms (Kuhlmann and Bremer 2002; Calderon et al. 2004; Bursy et al. 2007; Kuhlmann et al. 2008; Schwibbert et al. 2010). It is worth noting that a subgroup of ectoine producers possesses a regulatory protein, termed EctR, which acts as a repressor of the ectoine biosynthetic genes. This regulatory protein was first discovered in the halotolerant methanotroph *Methylomicrobium acidiphilum* 20Z and is a member of the MarR family of transcriptional regulators (Mustakhimov et al. 2010). EctR contains a predicted winged helix-turn-helix motive for DNA binding, and DNA footprinting analysis has shown that EctR binds to a region overlapping the promoter for the *ect* gene cluster. Disruption of the *ectR* gene in *M. acidiphilum* 20Z substantially derepresses the transcriptional activity of the promoter for the ectoine biosynthetic genes, but this promoter remained osmotically controlled in the *ectR* mutant (Mustakhimov et al. 2010). Hence, the EctR regulator functions as a repressor, but it is apparently not a sensor for high salinity or high ionic strength of the cytoplasm. Its potential effector ligand (or covalent modification) that would trigger the displacement of EctR from its operator sequence at the *ect* promoter remains to be discovered.

There are many speculations in the literature on the genetic regulation of the ectoine biosynthetic genes in response to osmotic stress. However, in no case has the signal transduction cascade been worked out experimentally at a level that would give a firm understanding of those molecular mechanisms that enable the cell to detect increases in the environmental osmolarity. Furthermore, it is unclear how such an osmotic stimulus might be processed and how this physical signal might be converted into a genetic signal that would allow the cell to set the activity of the promoter(s) for ectoine biosynthetic gene clusters. It is worth noting that signals other than increases in the environmental osmolarity can contribute to the level of *ect* expression (Calderon et al. 2004). For instance, in *V. pantothenticus* expression of the *ectABC* genes can be triggered not only by high salinity but also by decreasing growth temperature, suggesting that the observed ectoine production in the cold-stressed cells serves protective function against chill-stress (Kuhlmann et al. 2008).

15.6.2 The Ectoine Hydroxylase EctD: Biochemistry and Structural Analysis

Despite the fact that ectoine and hydroxyectoine are chemically very closely related, the properties and physiological functions of these solutes differ. This is evident from differences in vitro assays addressing the protein stabilization function of both ectoines and from physiological studies with *Chromohalobacter salexigens*, where the EctD-mediated formation of hydroxyectoine was found to be critical for the development of full heat stress resistance (Garcia-Estepa et al. 2006). Likewise, exogenously provided ectoine and hydroxyectoine confer heat stress protection to *Streptomyces coelicolor*, with a mixture of ectoine and hydroxyectoine being the superior thermoprotectant (Bursy et al. 2008). One therefore wonders why not all ectoine-producing bacteria also synthesize hydroxyectoine. But this is clearly not the case: our genome-based database searches indicate that from the 241 potential ectoine producers, only 93 species are also potential hydroxyectoine producers as judged by the presence of the structural gene for the ectoine hydroxylase (EctD). Again, *N. maritimus* SCM1 is the only archaeon that is a potential hydroxyectoine producer, indicating that this marine microorganism has acquired the *ectABCD* gene cluster via a lateral gene transfer event from a member of the *Bacteria* that shares its aquatic habitat.

The structural gene (*ectD*) for the ectoine hydroxylase was discovered through genetic and physiological approaches in *Streptomyces chrysomallus*, *Chromohalobacter salexigens* and *Virgibacillus* (*Salibacillus*) *salexigens* (Prabhu et al. 2004; Garcia-Estepa et al. 2006; Bursy et al. 2007). Biochemical studies with the purified EctD protein from *V. salexigens* and *S. coelicolor* revealed that the ectoine hydroxylase is a member of the non-heme iron(II)-containing and 2-oxoglutarate-dependent dioxygenases (Bursy et al. 2007, 2008).

Recently, the high-resolution crystal structure of the EctD protein from *V. salexigens* in complex with a Fe^{3+} ligand was solved (Reuter et al. 2010). It has not yet been possible to obtain an EctD crystal structure containing the substrate ectoine, or the co-substrate 2-oxoglutarate or the reaction product, 5-hydroxyectoine. The EctD protein has a β-barrel fold commonly found in cupin-type proteins ("*cupa*" is the Latin form for small barrel). As typically observed for members of the non-heme iron(II)-containing and 2-oxoglutarate-dependent dioxygenase super-family, EctD consists of a double-stranded β-helix core decorated with and stabilized by a number of α-helices (Fig. 15.4; left panel). The enzymatic function of the dioxygenases depends on a highly reactive iron species. The iron ligand (Fe^{2+}) is most-often coordinated by side chains of three amino acid residues, the so-called 2-His-1-carboxylate facial triad. This iron-ligand-binding motive is also present in the ectoine hydroxylase and is formed in the *V. salexigens* EctD enzyme by His-146, Asp-148, and His-248 (Fig. 15.4; left panel) and the iron-ligand is well resolved in the high-resolution (1.85 Å) EctD crystal structure (Reuter et al. 2010). As mentioned above, the EctD crystal structure does not contain 2-oxoglutarate but educated guesses for the binding of this co-factor can be made by inspecting crystal structures of other members of dioxygenase superfamily that are structurally closely related to EctD, in particular the human phytanoyl-CoA 2-hydroxylase PhyH and the halogenase SyrB2 from *Pseudomonas syringae*. In this way, Phe-143, Ser-250, and Arg-259 were implicated in 2-oxoglutarate binding (Reuter et al.

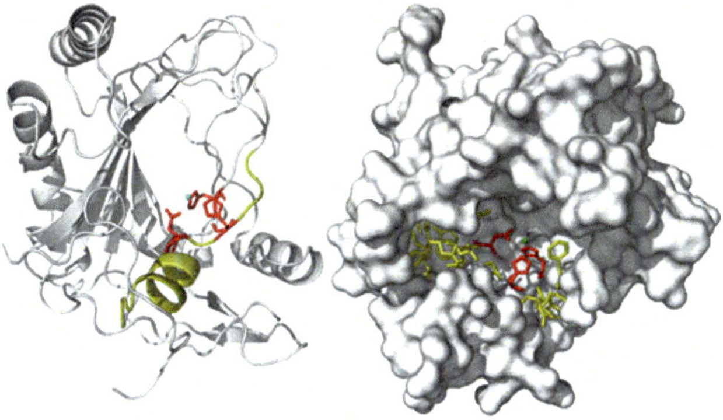

Fig. 15.4 Crystal structure of the ectoine hydroxylase EctD from the moderate halophile *Virgibacillus salexigens*. The EctD protein is a member of the super-family of the non-heme-containing ferrous iron and 2-oxoglutarate dependent oxygenases that typically contain a double-stranded β-helix (DSBH) fold at its core. (*Left panel*) Ribbon diagram of the EctD protein highlighting the side-chains of those three residues (in *red*) that coordinate the iron co-factor (in *green*). The position of the 17 amino acid long segment that serves as the signature-sequence motive for EctD-type ectoine hydroxylases is highlighted in *yellow*. (*Right panel*) Surface representation of the EctD protein. The position of the iron-binding residues, the coordinated iron ligand and the spacial orientation of the ectoine hydroxylase signature-sequence motive are highlighted. The pictures of the EctD crystal structure were prepared with PyMol (http://www.pymol.org/) using the atomic coordinates deposited in the PDB file 3EMR (http://www.pdb.org/) (Reuter et al. 2010)

2010). The crystal structure of EctD does not immediately reveal firm clues for the identity of those residues that position the substrate ectoine within the ligand-binding cavity.

Using the *V. salexigens* EctD enzyme as a search template, more than 90 EctD-type proteins can be currently identified in the databases of finished and unfinished microbial genome sequences. An alignment of the amino acid sequences of these proteins revealed a closely related protein sub-family with amino acid sequence identities that range from 72% for the EctD protein from *B. pseudofirmus* to 44% for the EctD protein from *Kytococcus sedentarius*. The most conserved region of the compiled and aligned 93 EctD-type proteins is a 17-amino-acid segment (F^{143}-X-W-H-S-D-F-E-T-W-H-X-E-D-G-M/L-P^{159}) that we suggested as a signature sequence for bona fide ectoine hydroxylases (Reuter et al. 2010). In the meantime, we have purified and biochemically characterized additional representatives of EctD-type proteins, and each of these enzymes exhibited ectoine hydroxylase activity (our unpublished data) with kinetic parameters resembling those of the *V. salexigens* and *S. coelicolor* enzymes (Bursy et al. 2007, 2008). The EctD signature sequence motive contains two (His-146 and Asp-148) of the three residues involved in iron binding (Fig. 15.4) and one of the residues (Phe-143) implicated in the binding of the 2-oxoglutarate co-factor. The two other residues (Ser-250 and Arg-259) considered in the binding of 2-oxoglutarate as well, are completely conserved in our compilation of 93 EctD-type proteins, lending credence to our suggestion for a functional role of these residues in the functioning of the EctD enzyme (Reuter et al. 2010). When one considers the topological organization of those 17 residues forming the ectoine hydroxylase signature sequence motive within the framework of the available EctD crystal structure from *V. salexigens* (Fig. 15.4; left panel), one finds that this segments lines one side of the ligand-binding cavity of the EctD enzyme (Fig. 15.4; right panel). It thus appears that this segment of the EctD protein is of both structural and functional relevance because it not only contributes to the formation of the overall architecture of the ligand-binding cavity but it also positions the side chains of functionally important residues within the active site of the enzyme (Fig. 15.4; right panel). The strict conservation of the 17-residue signature sequence can thus be rationally understood within the framework of the overall fold of the EctD protein and the architecture of its catalytic core.

15.7 Conclusions, Perspectives and Challenges

The complexity and multifaceted nature of the osmostress response systems of *B. subtilis* have been highlighted by proteome and genome-wide transcriptional profiling studies (Steil et al. 2003; Höper et al. 2006; Hahne et al. 2010). As exemplified by a comprehensive gene disruption study of members of the salt-inducible SigB general stress regulon of *B. subtilis*, physiological function can often not be gleaned with certainty from bioinformatic approaches and data base searches, despite the fact that many of the disrupted SigB-controlled genes exhibited a salt-sensitive phenotype (Höper et al. 2005). Further complications arise when one considers that osmotic stress caused by salt shock or continued cultivation of cells in high-salinity media apparently requires different, but partially overlapping, cellular stress responses (Steil et al. 2003; Höper et al. 2006; Hahne et al. 2010). Hence, detailed physiological studies are required to uncover the function of individual osmotic inducible genes within the wider osmostress response network of the *B. subtilis* cell. Genome-wide assessments of the transcriptional and protein-biosynthetic profile of osmotically stressed cells continue to provide clues and inspiration for such gene-by-gene functional studies.

As outlined in this chapter with a focus on *Bacillus subtilis* and other Bacilli, effective water management by the microbial cell is without doubt the cornerstone of its acclimatization to either sudden or sustained rises in the environmental osmolarity and the osmotic downshift that inevitably will follow hyperosmotic growth conditions. Although much has already been learned about the genetics, biochemistry, and physiology of salt-stressed *B. subtilis* cells (Bremer and Krämer 2000; Bremer 2002), one issue has been a very hard nut to crack: how do microorganisms sense fluctuations in the external salinity and osmolarity? Do microorganisms actually sense differences in the external osmolarity or do they deduce changes in the environmental conditions by monitoring changes in intracellular parameters such ionic strength or the size and composition of their solute pool? How do bacteria transform the information gleaned about osmotic changes in their environment into a genetic signal that allows the cell to sensitively set gene expression of osmoadaptive systems? Molecular studies of the regulation of the osmotic control of the activity of transport systems for osmoprotectants (Mahmood et al. 2009; Keates et al. 2010; Ziegler et al. 2010) and studies focusing on the osmotic control of the transcription of genes encoding synthesis and uptake systems for compatible solutes (Krämer 2010; Hagemann 2011) will certainly continue to enlighten and inform us about the ins-and outs of microbial salt-stress responses in the coming years.

Acknowledgments We would like to extend our special thanks to Tamara Hoffmann for intensive discussions on the "ins and outs" of osmotic stress responses of Bacilli. We thank our colleagues Lutz Schmitt and Sander H. J. Smits (University of Düsseldorf; Germany) and Michael Hecker and Uwe Völker (University of Greifswald; Germany) for very fruitful collaborations on the structural analysis of microbial binding proteins for compatible solutes and on the proteomic and transcriptomic assessment of osmotically stressed *B. subtilis* cells, respectively. We are very grateful to Vickie Koogle for the carefully editing of the manuscript and thank Tamara Hoffmann and Nadine Stöveken for critically reading of drafts of this book chapter.

Financial support for our studies on osmoregulation in microorganisms and the synthesis and uptake of compatible solutes were generously provided by the Deutsche Forschungsgemeinschaft, the Fonds der Chemischen Industrie, the LOEWE program of the State of Hessen via the Centre for Synthetic Microbiology [SynMicro; Marburg] and by a grant from the BMBF via the Bacell-SysMo2 consortium.

References

Albright RA, Ibar JL, Kim CU, Gruner SM, Morais-Cabral JH (2006) The RCK domain of the KtrAB K^+ transporter: multiple conformations of an octameric ring. Cell 126:1147–1159

Belitsky BR, Brill J, Bremer E, Sonenshein AL (2001) Multiple genes for the last step of proline biosynthesis in *Bacillus subtilis*. J Bacteriol 183:4389–4392

Bestvater T, Louis P, Galinski EA (2008) Heterologous ectoine production in *Escherichia coli*: by-passing the metabolic bottle-neck. Saline Systems 4:12

Boch J, Kempf B, Bremer E (1994) Osmoregulation in *Bacillus subtilis*: synthesis of the osmoprotectant glycine betaine from exogenously provided choline. J Bacteriol 176: 5364–5371

Boch J, Kempf B, Schmid R, Bremer E (1996) Synthesis of the osmoprotectant glycine betaine in *Bacillus subtilis*: characterization of the *gbsAB* genes. J Bacteriol 178:5121–5129

Booth IR, Louis P (1999) Managing hypoosmotic stress: aquaporins and mechanosensitive channels in *Escherichia coli*. Curr Opin Microbiol 2:166–169

Booth IR, Edwards MD, Black S, Schumann U, Miller S (2007) Mechanosensitive channels in bacteria: signs of closure? Nat Rev Microbiol 5:431–440

Börngen K, Battle AR, Moker N, Morbach S, Marin K, Martinac B, Kramer R (2010) The properties and contribution of the *Corynebacterium glutamicum* MscS variant to fine-tuning of osmotic adaptation. Biochim Biophys Acta 1798:2141–2149

Bourot S, Sire O, Trautwetter A, Touze T, Wu LF, Blanco C, Bernard T (2000) Glycine betaine-assisted protein folding in a *lysA* mutant of *Escherichia coli*. J Biol Chem 275:1050–1056

Bremer E (2002) Adaptation to changing osmolarity. In: Sonenshein AL, Hoch JA, Losick R (eds) *Bacillus subtilis* and its closest relatives. ASM Press, Washington, DC, pp 385–391

Bremer E, Krämer R (2000) Coping with osmotic challenges: osmoregulation through accumulation and release of compatible solutes. In: Storz G, Hengge-Aronis R (eds) Bacterial stress responses. ASM Press, Washington, DC, pp 79–97

Brigulla M, Hoffmann T, Krisp A, Völker A, Bremer E, Völker U (2003) Chill induction of the SigB-dependent general stress response in *Bacillus subtilis* and its contribution to low-temperature adaptation. J Bacteriol 185:4305–4314

Brill J, Hoffman T, Putzer H, Bremer E (2011) T-bor-mediated control of the anabolic proline biosynthetic genes of *Bacillus subtilis*. Microbiology 157:977–987

Burg MB, Ferraris JD (2008) Intracellular organic osmolytes: function and regulation. J Biol Chem 283:7309–7313

Bursy J, Pierik AJ, Pica N, Bremer E (2007) Osmotically induced synthesis of the compatible solute hydroxyectoine is mediated by an evolutionarily conserved ectoine hydroxylase. J Biol Chem 282:31147–31155

Bursy J, Kuhlmann AU, Pittelkow M, Hartmann H, Jebbar M, Pierik AJ, Bremer E (2008) Synthesis and uptake of the compatible solutes ectoine and 5-hydroxyectoine by *Streptomyces coelicolor* A3(2) in response to salt and heat stresses. Appl Environ Microbiol 74:7286–7296

Calderon MI, Vargas C, Rojo F, Iglesias-Guerra F, Csonka LN, Ventosa A, Nieto JJ (2004) Complex regulation of the synthesis of the compatible solute ectoine in the halophilic bacterium *Chromohalobacter salexigens* DSM 3043^T. Microbiology 150:3051–3063

Chen M, Wei H, Cao J, Liu R, Wang Y, Zheng C (2007) Expression of *Bacillus subtilis proBA* genes and reduction of feedback inhibition of proline synthesis increases proline production and confers osmotolerance in transgenic *Arabidopsis*. J Biochem Mol Biol 40:396–403

Coquelle N, Talon R, Juers DH, Girard E, Kahn R, Madern D (2010) Gradual adaptive changes of a protein facing high salt concentrations. J Mol Biol 404:493–505

Corratge-Faillie C, Jabnoune M, Zimmermann S, Very AA, Fizames C, Sentenac H (2010) Potassium and sodium transport in non-animal cells: the Trk/Ktr/HKT transporter family. Cell Mol Life Sci 67:2511–2532

Diamant S, Rosenthal D, Azem A, Eliahu N, Ben-Zvi AP, Goloubinoff P (2003) Dicarboxylic amino acids and glycine-betaine regulate chaperone-mediated protein-disaggregation under stress. Mol Microbiol 49:401–410

Dulaney EL, Dulaney DD, Rickes EL (1968) Factors in yeast extract which relieve growth inhibition of bacteria in defined medium of high osmolarity. Dev Ind Microbiol 9:260–269

Earl AM, Losick R, Kolter R (2008) Ecology and genomics of *Bacillus subtilis*. Trends Microbiol 16:269–275

Empadinhas N, da Costa MS (2008) Osmoadaptation mechanisms in prokaryotes: distribution of compatible solutes. Int Microbiol 11:151–161

Fisher MT (2006) Proline to the rescue. Proc Natl Acad Sci USA 103:13265–13266

Fujisawa M, Ito M, Krulwich TA (2007) Three two-component transporters with channel-like properties have monovalent cation/proton antiport activity. Proc Natl Acad Sci USA 104:13289–13294

Garcia-Estepa R, Argandona M, Reina-Bueno M, Capote N, Iglesias-Guerra F, Nieto JJ, Vargas C (2006) The *ectD* gene, which is involved in the synthesis of the compatible solute hydroxyectoine, is essential for thermoprotection of the halophilic bacterium *Chromohalobacter salexigens*. J Bacteriol 188:3774–3784

Green NJ, Grundy FJ, Henkin TM (2010) The T box mechanism: tRNA as a regulatory molecule. FEBS Lett 584:318–324

Hagemann M (2011) Molecular biology of cyanobacterial salt acclimation. FEMS Microbiol Rev 35:87–123

Hahne H, Mäder U, Otto A, Bonn F, Steil L, Bremer E, Hecker M, Becher D (2010) A comprehensive proteomics and transcriptomics analysis of *Bacillus subtilis* salt stress adaptation. J Bacteriol 192:870–882

Hecker M, Pane-Farre J, Völker U (2007) SigB-dependent general stress response in *Bacillus subtilis* and related gram-positive bacteria. Annu Rev Microbiol 61:215–236

Hoffmann T, Boiangiu C, Moses S, Bremer E (2008) Responses of *Bacillus subtilis* to hypotonic challenges: physiological contributions of mechanosensitive channels to cellular survival. Appl Environ Microbiol 74:2454–2460

Hoffmann T, Bremer E (2011) Protection of *Bacillus subtilis* against cold stress via compatible-solute aquisition. J Bacteriol 193:1552–1562

Holtmann G, Bremer E (2004) Thermoprotection of *Bacillus subtilis* by exogenously provided glycine betaine and structurally related compatible solutes: involvement of Opu transporters. J Bacteriol 186:1683–1693

Holtmann G, Bakker EP, Uozumi N, Bremer E (2003) KtrAB and KtrCD: two K^+ uptake systems in *Bacillus subtilis* and their role in adaptation to hypertonicity. J Bacteriol 185:1289–1298

Höper D, Volker U, Hecker M (2005) Comprehensive characterization of the contribution of individual SigB-dependent general stress genes to stress resistance of *Bacillus subtilis*. J Bacteriol 187:2810–2826

Höper D, Bernhardt J, Hecker M (2006) Salt stress adaptation of *Bacillus subtilis*: a physiological proteomics approach. Proteomics 6:1550–1562

Horn C, Sohn-Bosser L, Breed J, Welte W, Schmitt L, Bremer E (2006) Molecular determinants for substrate specificity of the ligand-binding protein OpuAC from *Bacillus subtilis* for the compatible solutes glycine betaine and proline betaine. J Mol Biol 357:592–606

Ignatova Z, Gierasch LM (2006) Inhibition of protein aggregation in vitro and in vivo by a natural osmoprotectant. Proc Natl Acad Sci USA 103:13357–13361

Jebbar M, von Blohn C, Bremer E (1997) Ectoine functions as an osmoprotectant in *Bacillus subtilis* and is accumulated via the ABC-transport system OpuC. FEMS Microbiol Lett 154:325–330

Kappes R, Bremer E (1998) Response of *Bacillus subtilis* to high osmolarity: uptake of carnitine, crotonobetaine and γ-butyrobetaine via the ABC transport system OpuC. Microbiology 144:83–90

Kappes RM, Kempf B, Bremer E (1996) Three transport systems for the osmoprotectant glycine betaine operate in *Bacillus subtilis*: characterization of OpuD. J Bacteriol 178:5071–5079

Kappes RM, Kempf B, Kneip S, Boch J, Gade J, Meier-Wagner J, Bremer E (1999) Two evolutionarily closely related ABC-transporters mediate the uptake of choline for synthesis of the osmoprotectant glycine betaine in *Bacillus subtilis*. Mol Microbiol 32:203–216

Keates RA, Culham DE, Vernikovska YI, Zuiani AJ, Boggs JM, Wood JM (2010) Transmembrane helix I and periplasmic loop 1 of *Escherichia coli* ProP are involved in osmosensing and osmoprotectant transport. Biochemistry 49:8847–8856

Kempf B, Bremer E (1995) OpuA, an osmotically regulated binding protein-dependent transport system for the osmoprotectant glycine betaine in *Bacillus subtilis*. J Biol Chem 270:16701–16713

Kempf B, Bremer E (1998) Uptake and synthesis of compatible solutes as microbial stress responses to high osmolality environments. Arch Microbiol 170:319–330

Krämer R (2010) Bacterial stimulus perception and signal transduction: response to osmotic stress. Chem Rec 10:217–229

Kuhlmann AU, Bremer E (2002) Osmotically regulated synthesis of the compatible solute ectoine in *Bacillus pasteurii* and related *Bacillus* spp. Appl Environ Microbiol 68:772–783

Kuhlmann AU, Bursy J, Gimpel S, Hoffmann T, Bremer E (2008) Synthesis of the compatible solute ectoine in *Virgibacillus pantothenticus* is triggered by high salinity and low growth temperature. Appl Environ Microbiol 74:4560–4563

Kung C, Martinac B, Sukharev S (2010) Mechanosensitive channels in microbes. Annu Rev Microbiol 64:313–329

Levina N, Totemeyer S, Stokes NR, Louis P, Jones MA, Booth IR (1999) Protection of *Escherichia coli* cells against extreme turgor by activation of MscS and MscL mechanosensitive channels: identification of genes required for MscS activity. EMBO J 18:1730–1737

Louis P, Galinski EA (1997) Characterization of genes for the biosynthesis of the compatible solute ectoine from *Marinococcus halophilus* and osmoregulated expression in *Escherichia coli*. Microbiology 143:1141–1149

Mahmood NA, Biemans-Oldehinkel E, Poolman B (2009) Engineering of ion sensing by the cystathionine beta-synthase module of the ABC transporter OpuA. J Biol Chem 284:14368–14376

Measures JC (1975) Role of amino acids in osmoregulation of non-halophilic bacteria. Nature 257:398–400

Mesbah NM, Cook GM, Wiegel J (2009) The halophilic alkalithermophile *Natranaerobius thermophilus* adapts to multiple environmental extremes using a large repertoire of $Na^+(K^+)/H^+$ antiporters. Mol Microbiol 74:270–281

Mustakhimov II, Reshetnikov AS, Glukhov AS, Khmelenina VN, Kalyuzhnaya MG, Trotsenko YA (2010) Identification and characterization of EctR1, a new transcriptional regulator of the ectoine biosynthesis genes in the halotolerant methanotroph *Methylomicrobium alcaliphilum* 20Z. J Bacteriol 192:410–417

Nau-Wagner G, Boch J, Le Good JA, Bremer E (1999) High-affinity transport of choline-*O*-sulfate and its use as a compatible solute in *Bacillus subtilis*. Appl Environ Microbiol 65:560–568

Nikaido H (2003) Molecular basis of bacterial outer membrane permeability revisited. Microbiol Mol Biol Rev 67:593–656

Nyyssölä A, Kerovuo J, Kaukinen P, von Weymarn N, Reinikainen T (2000) Extreme halophiles synthesize betaine from glycine by methylation. J Biol Chem 275:22196–22201

Oren A (2008) Microbial life at high salt concentrations: phylogenetic and metabolic diversity. Saline Systems 4:2

Oren A (2010) Thermodynamic limits to life at high salt concentrations. Environ Microbiol (in press). (doi: 10.1111/j.1462-2920.2010.02365.x)

Prabhu J, Schauwecker F, Grammel N, Keller U, Bernhard M (2004) Functional expression of the ectoine hydroxylase gene (*thpD*) from *Streptomyces chrysomallus* in *Halomonas elongata*. Appl Environ Microbiol 70:3130–3132

Rajan LA, Joseph TC, Thampuran N, James R, Chinnusamy V, Bansal KC (2008) Characterization and phylogenetic analysis of ectoine biosynthesis genes from *Bacillus halodurans*. Arch Microbiol 190:481–487

Reshetnikov AS, Khmelenina VN, Trotsenko YA (2006) Characterization of the ectoine biosynthesis genes of haloalkalotolerant obligate methanotroph "*Methylomicrobium alcaliphilum* 20Z". Arch Microbiol 184:286–297

Ressl S, Terwisscha van Scheltinga AC, Vonrhein C, Ott V, Ziegler C (2009) Molecular basis of transport and regulation in the Na^{+}/betaine symporter BetP. Nature 458:47–52

Reuter K, Pittelkow M, Bursy J, Heine A, Craan T, Bremer E (2010) Synthesis of 5-hydroxyectoine from ectoine: crystal structure of the non-heme iron(II) and 2-oxoglutarate-dependent dioxygenase EctD. PLoS One 5:e10647

Rhodes ME, Fitz-Gibbon ST, Oren A, House CH (2010) Amino acid signatures of salinity on an environmental scale with a focus on the Dead Sea. Environ Microbiol 12:2613–2623

Romano I, Lama L, Nicolaus B, Gambacorta A, Giordano A (2005) *Bacillus saliphilus* sp. nov., isolated from a mineral pool in Campania, Italy. Int J Syst Evol Microbiol 55:159–163

Ruzal SM, Lopez C, Rivas E, Sanchez-Rivas C (1998) Osmotic strength blocks sporulation at stage II by impeding activation of early sigma factors in *Bacillus subtilis*. Curr Microbiol 36:75–79

Saum SH, Müller V (2008a) Growth phase-dependent switch in osmolyte strategy in a moderate halophile: ectoine is a minor osmolyte but major stationary phase solute in *Halobacillus halophilus*. Environ Microbiol 10:716–726

Saum SH, Müller V (2008b) Regulation of osmoadaptation in the moderate halophile *Halobacillus halophilus*: chloride, glutamate and switching osmolyte strategies. Saline Systems 4:4

Schiefner A, Breed J, Bosser L, Kneip S, Gade J, Holtmann G, Diederichs K, Welte W, Bremer E (2004) Cation–pi interactions as determinants for binding of the compatible solutes glycine betaine and proline betaine by the periplasmic ligand-binding protein ProX from *Escherichia coli*. J Biol Chem 279:5588–5596

Schwibbert K, Marin-Sanguino A, Bagyan I, Heidrich G, Lentzen G, Seitz H, Rampp M, Schuster SC, Klenk HP, Pfeiffer F, Oesterhelt D, Kunte HJ (2010) A blueprint of ectoine metabolism from the genome of the industrial producer *Halomonas elongata* DSM 2581^{T}. Environ Microbiol (in press). (doi: 10.1111/j.1462-2920.2010.02336.x)

Smits SH, Höing M, Lecher J, Jebbar M, Schmitt L, Bremer E (2008) The compatible-solute-binding protein OpuAC from *Bacillus subtilis*: ligand binding, site-directed mutagenesis, and crystallographic studies. J Bacteriol 190:5663–5671

Spiegelhalter F, Bremer E (1998) Osmoregulation of the *opuE* proline transport gene from *Bacillus subtilis* – contributions of the SigA- and SigB-dependent stress-responsive promoters. Mol Microbiol 29:285–296

Steil L, Hoffmann T, Budde I, Völker U, Bremer E (2003) Genome-wide transcriptional profiling analysis of adaptation of *Bacillus subtilis* to high salinity. J Bacteriol 185:6358–6370

Street TO, Bolen DW, Rose GD (2006) A molecular mechanism for osmolyte-induced protein stability. Proc Natl Acad Sci USA 103:13997–14002

Szabo I, Petronilli V, Zoratti M (1992) A patch-clamp study of *Bacillus subtilis*. Biochim Biophys Acta 1112:29–38

Tanghe A, Van Dijck P, Thevelein JM (2006) Why do microorganisms have aquaporins? Trends Microbiol 14:78–85

Ventosa A, Nieto JJ, Oren A (1998) Biology of moderately halophilic aerobic bacteria. Microbiol Mol Biol Rev 62:504–544

Vollmer W, Seligman SJ (2010) Architecture of peptidoglycan: more data and more models. Trends Microbiol 18:59–66 von Blohn C, Kempf B, Kappes RM, Bremer E (1997) Osmostress response in *Bacillus subtilis*: characterization of a proline uptake system (OpuE) regulated by high osmolarity and the alternative transcription factor sigma B. Mol Microbiol 25:175–187

Wahome PG, Setlow P (2008) Growth, osmotic downshock resistance and differentiation of *Bacillus subtilis* strains lacking mechanosensitive channels. Arch Microbiol 189:49–58

Welsh DT (2000) Ecological significance of compatible solute accumulation by micro-organisms: from single cells to global climate. FEMS Microbiol Rev 24:263–290

Whatmore AM, Reed RH (1990) Determination of turgor pressure in *Bacillus subtilis*: a possible role for K^+ in turgor regulation. J Gen Microbiol 136:2521–2526

Whatmore AM, Chudek JA, Reed RH (1990) The effects of osmotic upshock on the intracellular solute pools of *Bacillus subtilis*. J Gen Microbiol 136:2527–2535

Wolters JC, Berntsson RP, Gul N, Karasawa A, Thunnissen AM, Slotboom DJ, Poolman B (2010) Ligand binding and crystal structures of the substrate-binding domain of the ABC transporter OpuA. PLoS One 5:e10361

Wood JM, Bremer E, Csonka LN, Krämer R, Poolman B, van der Heide T, Smith LT (2001) Osmosensing and osmoregulatory compatible solute accumulation by bacteria. Comp Biochem Physiol A Mol Integr Physiol 130:437–460

Yancey PH (2005) Organic osmolytes as compatible, metabolic and counteracting cytoprotectants in high osmolarity and other stresses. J Exp Biol 208:2819–2830

Zhao B, Lu W, Yang L, Zhang B, Wang L, Yang SS (2006) Cloning and characterization of the genes for biosynthesis of the compatible solute ectoine in the moderately halophilic bacterium *Halobacillus dabanensis* D-8^T. Curr Microbiol 53:183–188

Ziegler C, Bremer E, Kramer R (2010) The BCCT family of carriers: from physiology to crystal structure. Mol Microbiol 78:13–34

Chapter 16
The Nature and Function of Carotenoids in the Moderately Halophilic Bacterium *Halobacillus halophilus*

Saskia Köcher and Volker Müller

16.1 Introduction

With more than 750 different molecular structures (Britton et al. 2004) the carotenoids are the most important group of pigments in nature. In contrast to phototrophic organisms for which the presence of carotenoids as photoprotectants is essential, formation of carotenoids is found only in a few heterotrophic microorganisms (Goodwin 1980). However, it is striking that carotenoids are widely distributed in extremophiles. Examples are the thermophilic bacterium *Thermus thermophilus,* which synthesizes zeaxanthin and β-cryptoxanthin glucoside fatty acid esters that help in membrane stabilization (Yokoyama et al. 1995, 1996a), and the psychrotrophic bacterium *Arthrobacter agilis,* which increases its content in C50 bacterioruberin glycosides in response to low temperature (Fong et al. 2001). Bacterioruberin is also the major carotenoid found in the radioresistant bacterium *Rubrobacter radiotolerans* (Saito et al. 1994) and in halophilic archaea of the family *Halobacteriaceae* (Oren 2002). It is striking that saltern ponds are frequently colored orange to deep red. This is due to the presence of mainly the β-carotene-rich alga *Dunaliella salina* (Ben-Amotz and Avron 1990), to C50 carotenoids including bacterioruberin (Straub 1987) produced by halophilic archaea, or the C40 acyl glycoside produced by the extremely halophilic bacterium *Salinibacter ruber* (Lutnæs et al. 2002). Interestingly, also many endospore-forming bacteria isolated from saline environments like salt marshes contain carotenoids, whereas their non-halophilic relatives do not (Turner and Jervis 1963), indicating that carotenoids may play a crucial role in salt adaptation in these organisms.

All these observations suggest an important role and ecological function of carotenoids in marine environments. In this review, we concentrate on the type

S. Köcher • V. Müller (✉)
Molecular Microbiology & Bioenergetics, Institute of Molecular Biosciences, Johann Wolfgang Goethe University Frankfurt, Max-von-Laue-Strasse 9, 60438 Frankfurt am Main, Germany
e-mail: vmueller@bio.uni-frankfurt.de

A. Ventosa et al. (eds.), *Halophiles and Hypersaline Environments*,

DOI 10.1007/978-3-642-20198-1_16,

and function of carotenoids produced by halophilic microorganisms with focus on *Halobacillus halophilus* (formerly *Sporosarcina halophila*). This Gram-positive bacterium is moderately halophilic (Claus et al. 1983; Spring et al. 1996) and can tolerate up to 3.0 M sodium chloride. This osmotic burden is compensated by the accumulation of compatible solutes (Roeßler and Müller 1998, 2001; Müller and Saum 2005; Saum et al. 2006; Saum and Müller 2007, 2008a, b). Like several other halophilic and halotolerant bacteria (Aasen et al. 1969; Duc et al. 2006), *H. halophilus* is pigmented.

16.2 Structure and Biosynthesis of Carotenoids

Carotenoids are isoprenoids containing a characteristic polyene chain of conjugated double bonds. The two general groups of pigments are the hydrocarbons (carotenes) and oxygenated derivatives (xanthophylls). Most of them consist of 40 carbon atoms, but in a few species carotenoids with a variation of the carbon atom number are formed. Especially among non-photosynthetic halophilic bacteria and archaea also carotenoids with a C30 or C50 backbone were identified (Kelly et al. 1970; Marshall and Wilmoth 1981; Takaichi et al. 1997; Krubasik et al. 2001; Köcher et al. 2009). The most prominent example of C50 carotenoids are the straight chain derivatives of α-bacterioruberin (Fig. 16.1 (I)) found in halophilic archaea (Kelly et al. 1970). Other carotenoids found in halophilic organisms, as already mentioned above, are β-carotene (Fig. 16.1 (III)) produced by the alga *Dunaliella salina* or the C40 acyl glycoside salinixanthin (Fig. 16.1 (II)) which is synthesized by the extremely halophilic bacterium *Salinibacter ruber*. But there are also new marine isolates which produce structurally novel and rare carotenoids. These pigments include 2-hydroxyastaxanthin (Fig. 16.1 (IV)) from *Brevundimonas* sp. SD212 (Yokoyama et al. 1996b), saproxanthin (Fig. 16.1 (V)) and myxol (Fig. 16.1 (VI)) from marine strains of the *Flavobacteriaceae* (Shindo et al. 2007), and deoxymyxol 1′-glucoside and 4-ketodeoxymyxol 1′glucoside from *Gordonia terrae* AIST-1 (Takaichi et al. 2008).

Despite the structural diversity of carotenoids, a similar biosynthesis pathway is known for all carotenogenic organisms. The starting point for assembly of the carbon backbone of all carotenoids is the isomerization of the five-carbon compound isopentenyl pyrophosphate (IPP) to its allelic isomer dimethyallyl pyrophosphate (DMAPP) to generate geranyl pyrophosphate (GPP) (Fig. 16.2). This compound undergoes further condensation with IPP to form farnesyl pyrophosphate (FPP) catalyzed by FPP synthase. FPP is the main precursor for all known C30 carotenoids, which is condensed with another FPP molecule to form diapophytoene. For the synthesis of C40, C45 and C50 carotenoids one further addition of IPP to FPP to generate geranylgeranylpyrophosphate (GGPP) is catalyzed by a GGPP synthase. GGPP is then condensed to a second GGPP

I

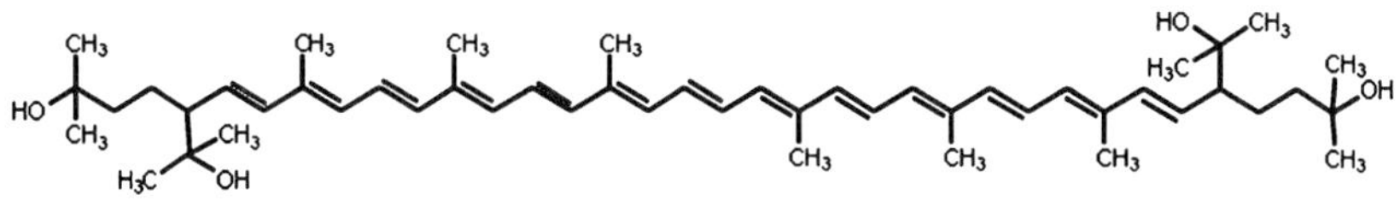

II

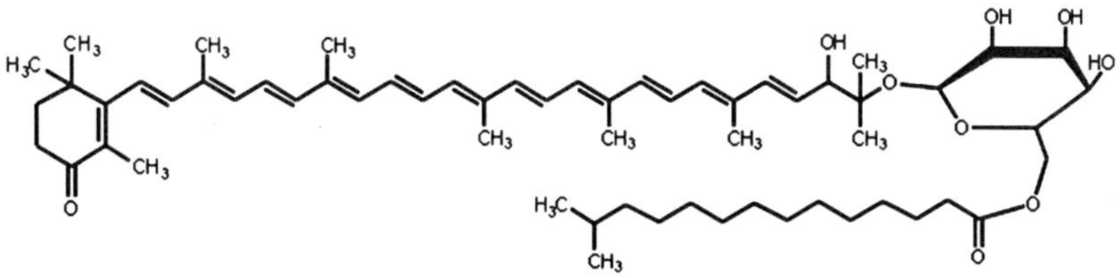

III

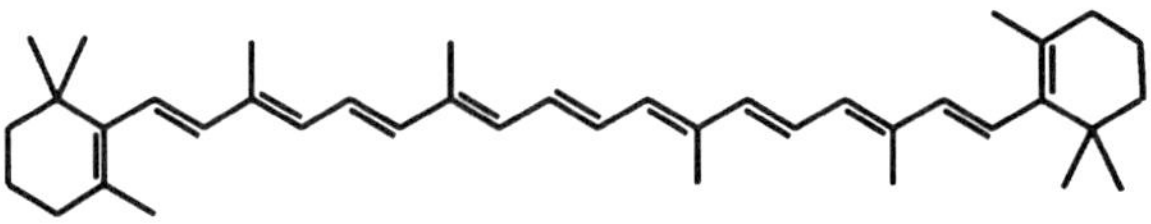

IV

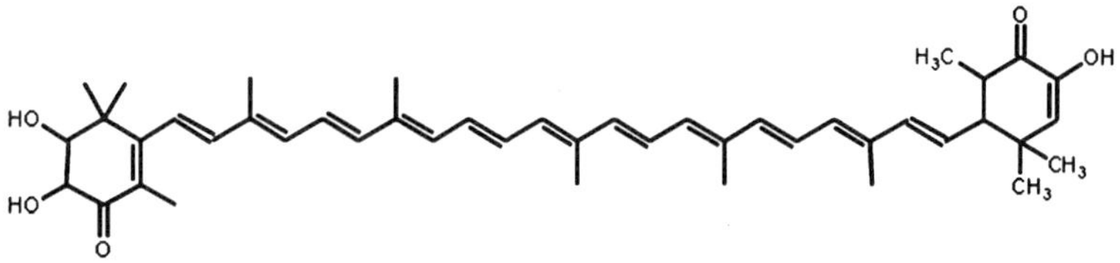

V

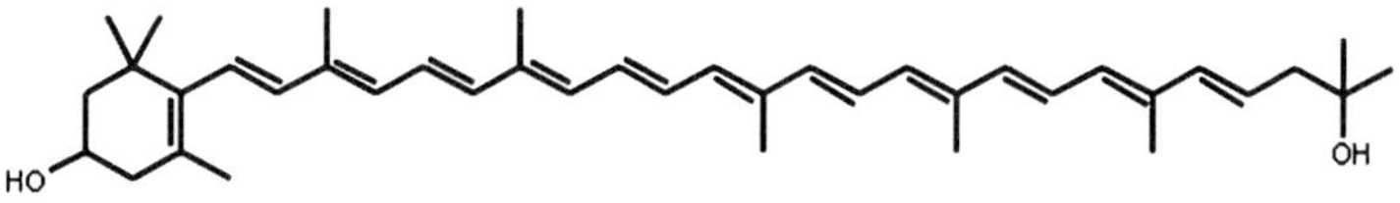

VI

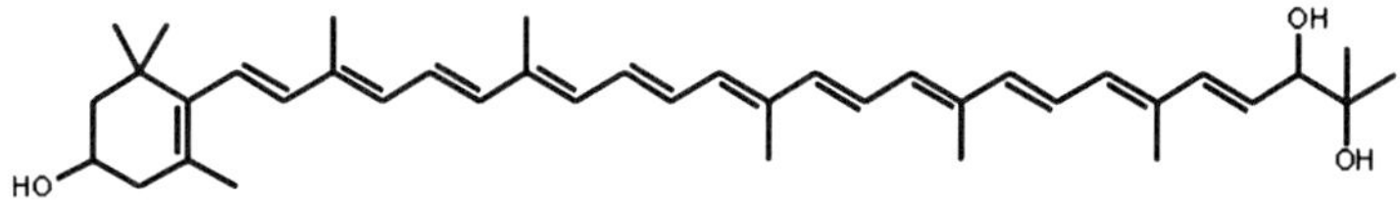

Fig. 16.1 Structure of carotenoids found in halophilic microorganisms. (I) α-bacterioruberin; (II) salinixanthin; (III) β-carotene; (IV) 2-hydroxyastaxanthin; (V) saproxanthin; (VI) myxol

molecule by the action of a phytoene synthase to form the C40 precursor phytoene. Phytoene as well diapophytoene are both colorless but they represent the first step of carotenoid biosynthesis. Next a diapophytoene/phytoene desaturase introduces double bonds into the molecules to yield diapo-/neurosporene, or diapo-/lycopene.

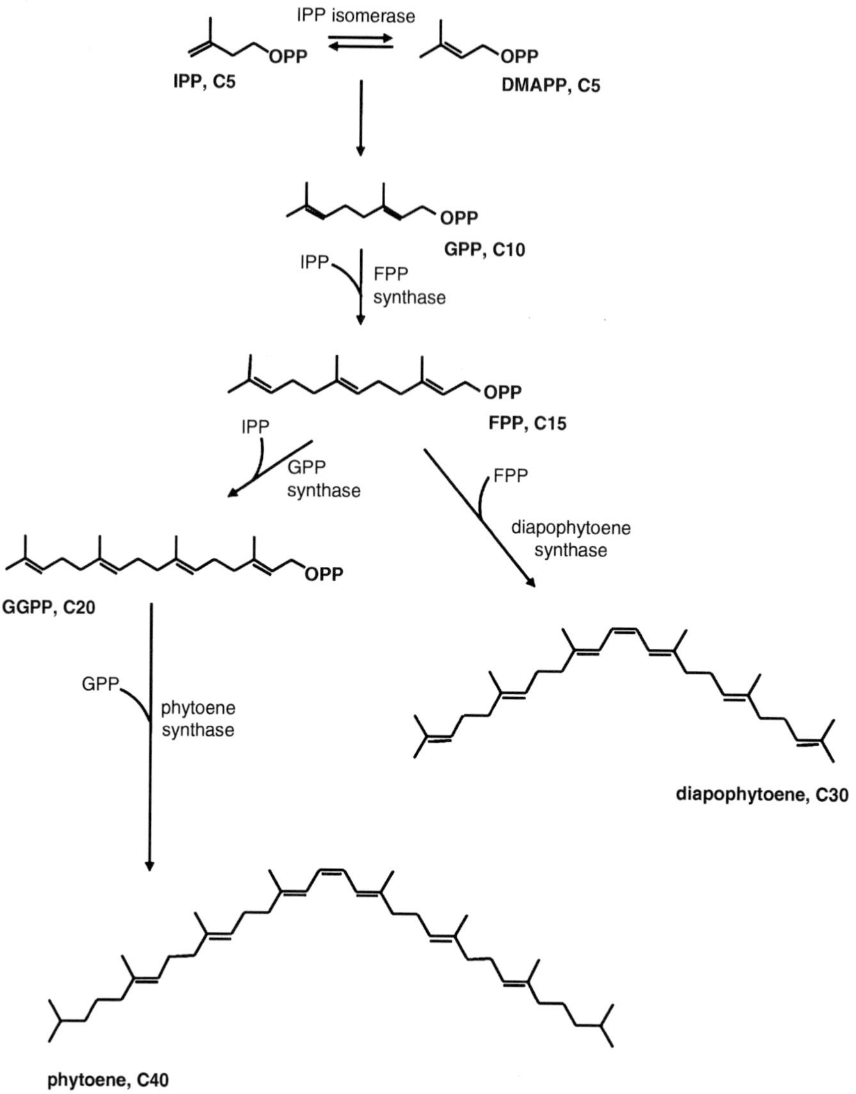

Fig. 16.2 Biosynthetic pathway of C30 and C40, C45 and C50 carotenoids. IPP, C5 (Isopentenyl pyrophosphate); DMAPP, C5 (dimethylallyl pyrophosphate); GPP, C10 (geranyl pyrophosphate); FPP, C15 (farnesyl pyrophosphate); GGPP, C20 (geranylgeranyl pyrophosphate); PP (pyrophosphate)

16.3 Structure of the Carotenoids Produced by *H. halophilus*

A first indication on the nature of the *H. halophilus* pigments was obtained by inhibition of carotene desaturation with diphenylamine (DPA). When cells were grown in the presence of this inhibitor the biosynthesis of colored carotenoids was

Fig. 16.3 Pigment mutants of *Halobacillus halophilus* generated by chemical mutagenesis with EMS. *WT* wild type; *OC* light orange pigmentation; *YC* yellow pigmentation; *WH* white pigmentation

blocked and they showed a white pigmentation. DPA prevents the insertion of double bonds into phytoene and diapophytoene (Sandmann and Fraser 1993). Biochemical analyses revealed that a C30 carotenoid was accumulated in *H. halophilus* and identified as 15-cis and all-trans 4,4′-diapophytoene (Köcher et al. 2009). The spectra showed the typical maxima at 275, 285 and 295 nm. In addition to inhibitors, pigment mutants are a useful tool to elucidate a carotenogenic pathway. Several mutants of *H. halophilus* were generated by chemical mutagenesis with EMS (ethyl methanesulfonate) (Köcher et al. 2009). They fell into three categories according to their pigmentation: light orange (mutant OC), bright yellow (mutant YC) and white (mutant WH) (Fig. 16.3). The carotenoids which accumulated were separated and identified. In OC, the prominent carotenoid had the same retention time and spectrum as diaponeurosporenoic acid, in WH the only detectable carotenoid was apo- or diapophytoene and in YC only apo-/diaponeurosporene was detectable.

The chemical nature of the carotenoid produced by *H. halophilus* was finally identified by HR-MS and NMR analyses. For this purpose the carotenoids were extracted three times with CH_2Cl_2:MeOH (1:1, [v/v]) and subjected to silica gel chromatography. The main compound was isolated and dissolved in CH_2Cl_2: MeOH (1:1, [v/v]). The carotenoid produced by *H. halophilus* could be identified as a C30 methyl glucosyl-3,4-dehydro-apo-8′-lycopenoate (Osawa et al. 2010) (Fig. 16.4a). This type of carotenoid is also produced by *Planococcus maritimus*, a marine Gram-positive bacterium (Shindo et al. 2008). To get a closer look on the biosynthesis of this new C30 carotenoid we analyzed the structure of the carotenoids of the OC mutant, which are most likely intermediates of the wild type carotenoid. And indeed two novel carotenoids, hydroxy-3,4-dehydro-apo-8′-lycopene and methyl hydroxy-3,4-dehydro-apo-8′-lycopenoate, could be identified (Fig. 16.4b).

16.4 Biosynthesis of a Unique C30 Carotenoid Produced by *H. halophilus*

Inspection of the genome of *H. halophilus* revealed two gene clusters probably involved in carotenoid biosynthesis (Fig. 16.5). Genes *crtNa*, *crtNc*, and *crtM* are arranged in the same orientation. *crtNa* and *crtNc* show an overlap of 3 bp,

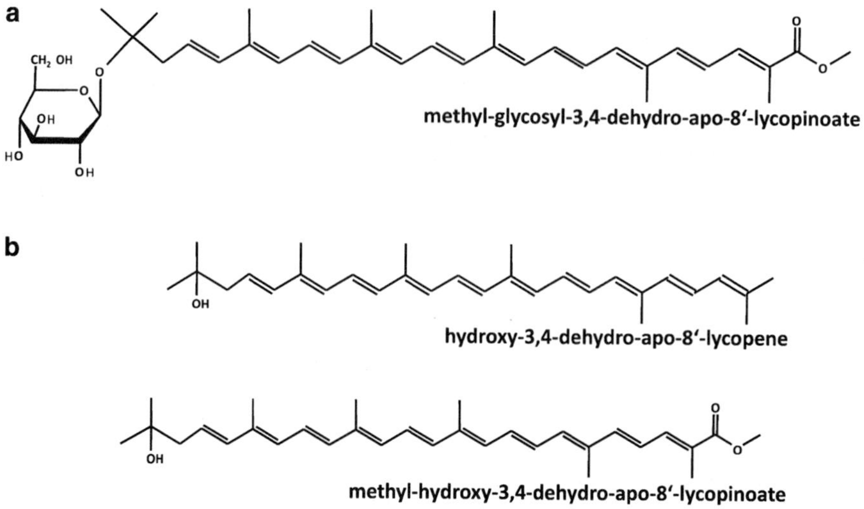

Fig. 16.4 The structure of carotenoids produced by the wild type and the mutant OC of *Halobacillus halophilus*

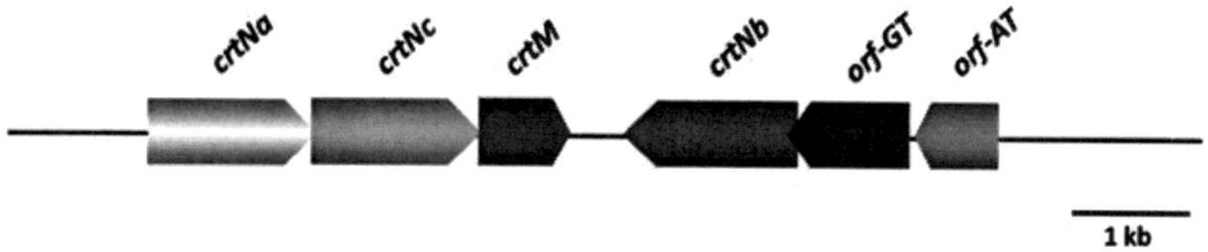

Fig. 16.5 Organization of carotenogenic genes in *Halobacillus halophilus*. *crtNa*: diapophytoene desaturase; *crtNc*: putative diapopyhtoene dehydrogenase; *crtM*: diapopyhtoene synthase; *crtNb*: putative diapolycopine oxidase; *orf-GT*: glucosyl transferase; and orf-AT: acyl transferase

and *crtNc* and *crtM* are separated by only 3 bp. 429 bp downstream of *crtM crtNb*, *orf-GT*, and *orf-AT* build a second cluster in the opposite orientation. They are separated by 21 bp between *orf-GT* and *orf-AT* and overlap 117 bp in the case of *crtNb* and *orf-GT*.

The arrangement of the putative carotenoid biosynthesis genes is indicative of their organization in transcriptional units. To prove the existence of such a polycistronic messenger, mRNA was isolated from *H. halophilus* cells and transcribed into cDNA. This cDNA was then used as a template in a PCR using primers that bridge the intergenic regions between *crtNa* and *crtNc*, between *crtNc* and *crtM*, between *crtNb* and *orf-GT* and between *orf-GT* and *orf-AT*. Products with the expected fragment sizes were obtained indicating that the cDNA originated from two different polycistronic mRNA. This result demonstrates that *crtNa*, *crtNc*, and *crtM* as well as *crtNb*, *orf-GT* and *orf-AT* are part of an operon, respectively.

Three of the genes products, designated CrtNa, CrtNb, and CrtNc, are similar to bacterial carotene desaturases. CrtNa has 48% amino acid identity (similarity 71%)

with that of a CrtN homologue in *Methylomonas* sp 16a. CrtNb had 31% amino acid identity (similarity 54%) with that of a 4,4′-diapolycopene oxidase in *Methylomonas* sp. 16a and 31% amino acid identity (similarity 54%) with the diapophytoene desaturase in *Psychroflexus torquis* ATCC 700755. We could also identify a third homologue of a diapophytoene desaturase. CrtNc is highly similar to the diapophytoene dehydrogenase of *Heliobacillus mobilis* (AAC84034) (55%). CrtM compares best to a putative phytoene/diapophytoene synthase that catalyses the condensation of two molecules of farnesyl-pyrophosphate. The protein of the latter has a length of 301 amino acids and is 55 and 44% identical (similarities, 71% and 61%) to CrtM from *Bacillus* sp. NRRL B-14911 and *Oceanobacillus iheyensis*, respectively. Besides these typical proteins for carotenoid biosynthesis, we could identify two additional proteins that appear to be involved in pigment biosynthesis in *H. halophilus*. The product of *orf-GT* with a deduced length of 369 amino acids is similar to a glycosyl transferase from *Chlorobaculum tepidum* TLS (57% similarity, 35% identity) involved in carotenoid synthesis. *orf-AT* is a putative acyltransferase with a deduced length of 236 amino acids. It is 57% similar (35% identity) to the protein found in *Exiguobacterium sibiricum* 255-15 (Köcher et al. 2009).

The biochemical and genetic data allowed us to postulate a biosynthesis pathway for the carotenoids in *H. halophilus*. It should be pointed out that the identified intermediates as well as methyl glucosyl-3,4-dehydro-apo-8′-lycopenoate are 8′-apo derivatives with asymmetrical arrangement of the methyl groups, unlike the 4,4′-diapo derivatives, which are typically synthesized with two molecules farnesyl diphosphate (Tao et al. 2005; Köcher et al. 2009). In addition, the identification of hydroxyl-3,4-dehydro-apo-8′lycopene without terminal end group oxidation excludes the possibility that apo-8′ products are derived by cleavage of C40 carotenoids at position 8′. Therefore, we propose that the biosynthesis pathway in *H. halophilus* initially starts with apo-8′-phytoene by the condensation of C20 geranylgeranyl pyrophosphate and C10 geranyl pyrophosphate, catalyzed by an apophytoene synthase, encoded by *crtM* (Fig. 16.6). In the next step, apo-8′-lycopene is produced by the action of CrtNa, an apophytoene desaturase. In the later biosynthesis steps towards methyl glucosyl-3,4-dehydro-apo-8′-lycopenoate hydroxyl-3,4-dehydro-apo-8′lycopene is glycosylated and oxidized to the corresponding aldehyde propably catalyzed by an CrtNb-like enzyme as it is described by Tao et al. (2005), and then oxidized with another oxidase, which may be encoded by crtNc.

16.5 Function of Carotenoids in Halophilic Microorganisms

Carotenoids play an important biological role not only in phototrophic but also in heterotrophic organisms. Especially for halophilic microorganisms the diversity of these functions which involve interactions with and without light is essential for surviving in such extreme habitats.

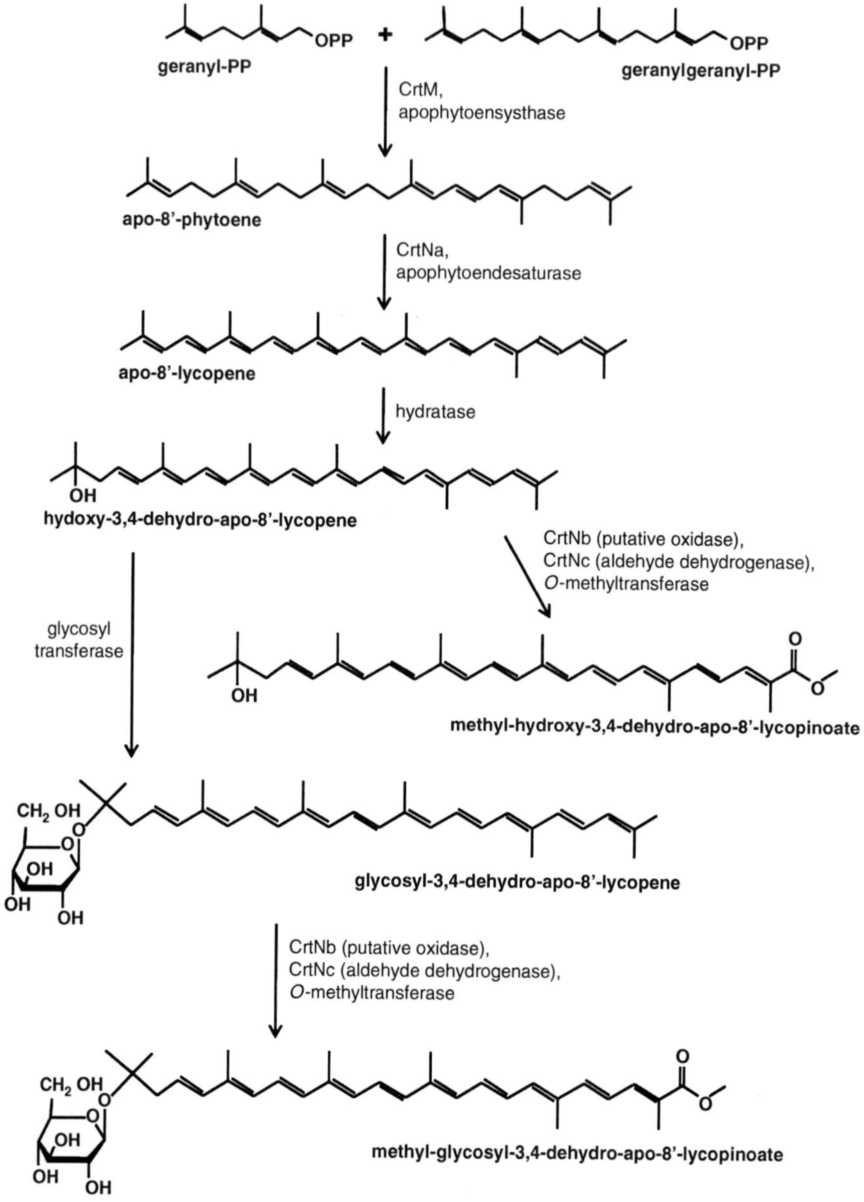

Fig. 16.6 Proposed carotenoid biosynthetic pathway in *Halobacillus halophilus*. Gene products catalyzing the individual reactions are indicated next to the *arrow*

16.5.1 Carotenoids and Light Harvesting

The most prominent function of carotenoids is their contribution to the harvesting of light energy by absorbing light and passing excitation energy on to (bacterio) chlorophyll, thereby extending the wavelength range of the light that can be harvested. Much simpler retinal-based energy transducers are found in halophilic archaea, bacteriorhodopsin and archaerhodopsin. Absorption of light and the proton transport take place within one protein molecule containing a single retinal chromophore. This protein undergoes a cycle of reactions, resulting in the translocation of a proton from inside the cell to the outside and the generation of a transmembrane potential that is usable for ATP synthesis, ion transport, and cell motility. The reaction is initiated by light-induced isomerization of the chromophore from all-*trans* to 13-*cis* (Mathies et al. 1991) and involves changes in the pK_a values of the buried carboxyl groups (Balashov 2000), and small scale (Lanyi and Schobert 2004) and large-scale (Subramaniam et al. 2002) conformational changes of the protein. The early events involve spectral evolution, which has been interpreted as relaxation through a series of excited states and photoproducts (denoted as H, I, and J) (Sharkov et al. 1985; Gai et al. 1998; Kobayashi et al. 2001; Herbst et al. 2002; Kahan et al. 2007) that leads to the formation of the 13-*cis* photoproduct (Kochendoerfer and Mathies 1995). Another retinal protein, xanthorhodopsin (Balashov et al. 2005), from the cell membrane of the extremely halophilic eubacterium *S. ruber*, showed an unexpected association with a C40 carotenoid, salinixanthin (Lutnæs et al. 2002). This carotenoid was found to undergo large reversible absorption changes when the retinal chromophore was removed and replaced in the protein (Balashov et al. 2005). The carotenoid is bound to the protein in a 1:1 ratio. The action spectrum for proton transport indicated that light absorbed not only by the retinal but also by the carotenoid is utilized for proton transport, with 40% efficiency (Balashov et al. 2005). Subsequent studies using steady-state fluorescence measurements demonstrated that there indeed exists an efficient energy-transfer channel between salinixanthin and retinal (Balashov et al. 2008), making xanthorhodopsin the simplest antenna protein known so far.

16.5.2 Carotenoids as Photoprotectants

High on the list of functions attributed to carotenoids is "photoprotection" against photoxidative damage by quenching singlet oxygen as well as other harmful radicals that are formed when cells are illuminated (Demmig-Adams and Adams 2002). Light energy especially in combination with oxygen, can be very harmful, causing damage to cells via singlet oxygen ($^1O_2$) and oxidizing free radicals. Living organisms have evolved defenses to prevent or minimize this damage. Carotenoids are a major part of this defense and can be effective in a number of ways. Not only photosynthetic organisms have to prevent the formation and damaging effects of $^1O_2$, also in non-photosynthetic organisms, exogenous or endogenous sensitizers (e.g., porphyrin such as protoporphyrin IX and heme) can be excited and cause photooxidative stress by damaging DNA, protein, lipids and other cell components (Sies and Mehlhorn 1986). When cells are illuminated photosensitizer can form the longer lived, lower energy triplet state. This in turn can undergo energy transfer to oxygen to form the highly reactive singlet oxygen. The triplet energy level of carotenoids is comparatively low, so that these pigments have the ability to accept energy from a triplet state sensitizer or from $^1O_2$, thereby preventing or minimizing damage. The triplet excited state of the carotenoid can then return to the ground state dissipating the energy excess as heat. Thus, the carotenoid acts as a catalyst in the deactivation of $^1O_2$ (Becker et al. 1991). It is known from many studies that light is the environmental factor with the most influence on carotenoid biosynthesis. The best understood examples for non-photosynthetic prokaryotes are the blue light induced carotenoid biosynthesis in *Myxococcus xanthus* (Burchard and Dworkin 1966) and the light induced pigment production in *Streptomyces coelicolor* (Takano et al. 2005). But also for halophilic organisms there is some evidence for the function of carotenoids to prevent photooxidative damage. For *Halobacterium salinarum* it could be shown that white mutants lacking bacterioruberins grown under high light intensities approaching those of full sunlight were outcompeted by the pigmented parent strain (Dundas and Larsen 1963). It has also been shown that bacterioruberin offers protection against ionizing radiation and UV both in vitro, reducing the occurrence of DNA strand breaks (Kottemann et al. 2005) and thymidine degradation (Saito et al. 1997), and in vivo with a decrease in survival of colorless *H. salinarum* mutants lacking bacterioruberin (Shahmohammadi et al. 1998). Also for *H. halophilus* the C30 carotenoid is essential for growth under oxidative stress (Köcher et al. 2009). Growth of non-stressed cells at conditions where the synthesis of colored carotenoids was inhibited was comparable to the non-inhibited culture. However, when oxidative stress was applied by the addition of duroquinone, the culture devoid of colored carotenoids did not grow. This result indicates that the carotenoids produced by *H. halophilus* cope with oxidative stress. This is in accordance with the fact that the main carotenoid produced by *H. halophilus* (methyl glucosyl-3,4-dehydro-apo-8′-lycopenoate) is a very potent antioxidant (Shindo et al. 2008). Also hydroxy-3,4-dehydro-apo-8′-lycopene and methyl hydroxy-3,4-dehydro-apo-8′-lycopenoate, the main carotenoids in the OC mutant, possess a potent antioxidative activity. But they were not as potent as the wild type carotenoid. This may be explained by their reduced hydrophobicity (Shimidzu et al. 1996). Further studies on the antioxidative activities of these carotenoids are in progress.

16.5.3 Carotenoids as Membrane Stabilizers

Membranes in extremophilic prokaryotes often contain polar carotenoids (Anwar et al. 1977; Yokoyama et al. 1995; Chattopadhyay et al. 1997; Jagannadham et al. 2000; Fong et al. 2001). All these organisms live under extreme conditions, and therefore, are dependent of solid membranes as an intact barrier to prevent an uncontrolled penetration of small molecules and ions.

Because of their lipophilic nature carotenoids are located within the hydrophobic inner membrane. The rod-like structure, the presence of polar end groups, and the molecular dimensions of a typical carotenoid, which match the thickness of the bilayer, are directly responsible for the localization and orientation of the molecule within the membrane and for effects on the membrane properties (Gruszecki 1999; Okulski et al. 2000; Sujak et al. 2000). Unpolar carotenoids like β-carotene are entirely lipophilic and remain within the hydrocarbon inner part of the bilayer. Resonance Raman spectroscopy or linear dichroism have indicated an orientation of these pigments with their long axis roughly parallel to the membrane surface (Johanson 1981). In contrast, polar dihydroxycarotenoids like zeaxanthin span across the bilayer with one polar end group associated with each polar surface (Gruszecki and Sielewiesiuk 1990, 1991). By EPR and NMR studies the effect of carotenoids on the structure and dynamics of the lipid membranes was determined (Gruszecki et al. 1999). For the polar carotenoids lutein and zeaxanthin it could be shown that they restrict the molecular motion of lipids and increase the rigidity of the membrane in its fluid state (Subczynski et al. 1992, 1993; Strzalka and Gruszecki 1994). It is concluded, therefore, that the carotenoid acts as a "rivet" mechanically reinforcing and strengthening the bilayer. This idea is supported by studies with *Acholeplasma laidlawii* whose membranes normally contain cholesterol, obtained from his host (Rottem and Markowitz 1979). When no cholesterol is available, the organism begins to synthesize polar carotenoids which take the place of cholesterol in the membranes. As mentioned above the carotenoid molecule is of suitable length to span across the membrane with polar substituents in the end groups associated with the polar outer faces of the bilayer and the hydrocarbon polyene chain in the hydrophobic lipid core. The length of carotenoids produced by the organisms is dependent in the thickness of the membrane of each organism, since it was shown that the incorporation ratio is higher when the molecular length of the carotenoid corresponds to the thickness of the phospholipid bilayer (Lazrak et al. 1987). Glycosylation of carotenoid end groups strengthens the association with the polar head groups of the bilayer. A good example is the acyclic xanthophylls bacterioruberin of *Halobacterium* species, which fits perfectly in the thick diphytanyl lipid membrane of these organisms. The incorporation of bacterioruberins in reconstituted *Halobacterium* lipid membranes was found to greatly increase their rigidity and to decrease their water permeability (Lazrak et al. 1987). For *Haloferax mediterranei*, which grows best in the presence of 1.5–2.5 M NaCl (Rodriguez-Valera et al. 1983), it was shown that with decreasing the NaCl concentrations to 1.0 M in the medium the carotenoid content increased ninefold compared to cells grown in the presence of 3.4 M NaCl (D'Souza et al. 1997). It was suggested, that carotenoids stabilize the membranes, and therefore, prevent the lysis of the cell at such low salt concentrations.

Acknowledgment Work from the authors' laboratory was supported by a fellowship from the Marianne and Dr. Fritz Walter Fisher-Stiftung to Saskia Köcher and by a grant from the Deutsche Forschungsgemeinschaft.

References

Aasen AJ, Francis GW, Liaaen-Jensen S (1969) Bacterial Carotenoids. XXIX. The carotenoids of two yellow halophilic cocci – including a new glycosidic methyl apo-lycopene. Acta Chem Scand 23:4–13

Anwar M, Khan TH, Prebble J, Zagalsky PF (1977) Membrane-bound carotenoid in *Micrococcus luteus* protects naphthoquinone from photodynamic action. Nature 270:538–540

Balashov SP (2000) Protonation reactions and their coupling in bacteriorhodopsin. Biochim Biophys Acta 1460:75–94

Balashov SP, Imasheva ES, Boichenko VA, Anton J, Wang JM, Lanyi JK (2005) Xanthorhodopsin: a proton pump with a light-harvesting carotenoid antenna. Science 309:2061–2064

Balashov SP, Imasheva ES, Wang JM, Lanyi JK (2008) Excitation energy-transfer and the relative orientation of retinal and carotenoid in xanthorhodopsin. Biophys J 95:2402–2414

Becker HGO, Böttcher H, Dietz F, Rehorek D, Roewer G, Schiller K, Timpe H-J (1991) Einführung in die Photochemie. Deutscher Verlag der Wissenschaften, Berlin

Ben-Amotz A, Avron M (1990) The biotechnology of cultivating the halotolerant alga *Dunaliella*. Trends Biotechnol 8:121–126

Britton G, Liaaen-Jensen S, Pfander H (2004) Handbook of carotenoids. Birkhäuser Verlag, Basel

Burchard RP, Dworkin M (1966) Light-induced lysis and carotenogenesis in *Myxococcus xanthus*. J Bacteriol 91:535–545

Chattopadhyay MK, Jagannadham MV, Vairamani M, Shivaji S (1997) Carotenoid pigments of an antarctic psychrotrophic bacterium *Micrococcus roseus*: temperature dependent biosynthesis, structure, and interaction with synthetic membranes. Biochem Biophys Res Commun 239: 85–90

Claus D, Fahmy F, Rolf HJ, Tosunoglu N (1983) *Sporosarcina halophila* sp. nov., an obligate, slightly halophilic bacterium from salt marsh soils. Syst Appl Microbiol 4:496–506

D'Souza SE, Altekar W, D'Souza SF (1997) Adaptive response of *Haloferax mediterranei* to low concentrations of NaCl (<20%) in the growth medium. Arch Microbiol 168:68–71

Demmig-Adams B, Adams WW 3rd (2002) Antioxidants in photosynthesis and human nutrition. Science 298:2149–2153

Duc LH, Fraser PD, Tam NK, Cutting SM (2006) Carotenoids present in halotolerant *Bacillus* spore formers. FEMS Microbiol Lett 255:215–224

Dundas ID, Larsen H (1963) A study on the killing by light of photosensitized cells of *Halobacterium salinarium*. Arch Mikrobiol 46:19–28

Fong NJ, Burgess ML, Barrow KD, Glenn DR (2001) Carotenoid accumulation in the psychrotrophic bacterium *Arthrobacter agilis* in response to thermal and salt stress. Appl Microbiol Biotechnol 56:750–756

Gai F, Hasson KC, McDonald JC, Anfinrud PA (1998) Chemical dynamics in proteins: the photoisomerization of retinal in bacteriorhodopsin. Science 279:1886–1891

Goodwin W (1980) The biochemnistry of carotenoids. Chapman and Hall, London

Gruszecki WI (1999) Carotenoids in membranes. In: Frank HA, Young AJ, Britton G, Cogdell R (eds) The Photochemistry of Carotenoids. Kluwer, Dordrecht, Netherlands, pp 363–379

Gruszecki WI, Sielewiesiuk J (1990) Orientation of xanthophylls in phosphatidylcholine multibilayers. Biochim Biophys Acta 1023:405–412

Gruszecki WI, Sielewiesiuk J (1991) Galactolipid multibilayers modified with xanthophylls: orientational and diffractometric studies. Biochim Biophys Acta 1069:21–26

Gruszecki WI, Sujak A, Strzalka K, Radunz A, Schmid GH (1999) Organisation of xanthophyll-lipid membranes studied by means of specific pigment antisera, spectrophotometry and monomolecular layer technique lutein versus zeaxanthin. Z Naturforsch C 54:517–525

Herbst J, Heyne K, Diller R (2002) Femtosecond infrared spectroscopy of bacteriorhodopsin chromophore isomerization. Science 297:822–825

Jagannadham MV, Chattopadhyay MK, Subbalakshmi C, Vairamani M, Narayanan K, Rao CM, Shivaji S (2000) Carotenoids of an antarctic psychrotolerant bacterium, *Sphingobacterium antarcticus*, and a mesophilic bacterium, *Sphingobacterium multivorum*. Arch Microbiol 173: 418–424

Johanson CE (1981) Commonalities and differences among reinforcers. NIDA Res Monogr 37: 235–237

Kahan A, Nahmias O, Friedman N, Sheves M, Ruhman S (2007) Following photoinduced dynamics in bacteriorhodopsin with 7-fs impulsive vibrational spectroscopy. J Am Chem Soc 129:537–546

Kelly M, Norgård S, Liaaen-Jensen S (1970) Bacterial carotenoids. 31. C50-carotenoids 5. Carotenoids of *Halobacterium salinarium*, especially bacterioruberin. Acta Chem Scand 24:2169–2182

Kobayashi T, Saito T, Ohtani H (2001) Real-time spectroscopy of transition states in bacteriorhodopsin during retinal isomerization. Nature 414:531–534

Kochendoerfer GG, Mathies RA (1995) Ultrafast spectroscopy of rhodopsins-photochemistry at its best. Isr J Chem 35:211–226

Kottemann M, Kish A, Iloanusi C, Bjork S, DiRuggiero J (2005) Physiological responses of the halophilic archaeon *Halobacterium sp.* strain NRC1 to desiccation and gamma irradiation. Extremophiles 9:219–227

Köcher S, Breitenbach J, Müller V, Sandmann G (2009) Structure, function and biosynthesis of carotenoids in the moderately halophilic bacterium *Halobacillus halophilus*. Arch Microbiol 191: 95–104

Krubasik P, Takaichi S, Maoka T, Kobayashi M, Masamoto K, Sandmann G (2001) Detailed biosynthetic pathway to decaprenoxanthin diglucoside in *Corynebacterium glutamicum* and identification of novel intermediates. Arch Microbiol 176:217–223

Lanyi JK, Schobert B (2004) Local-global conformational coupling in a heptahelical membrane protein: transport mechanism from crystal structures of the nine states in the bacteriorhodopsin photocycle. Biochemistry 43:3–8

Lazrak T, Milon A, Wolff G, Albrecht AM, Miehe M, Ourisson G, Nakatani Y (1987) Comparison of the effects of inserted C40- and C50-terminally dihydroxylated carotenoids on the mechanical properties of various phospholipid vesicles. Biochim Biophys Acta 903:132–141

Lutnæs BF, Oren A, Liaaen-Jensen S (2002) New C(40)-carotenoid acyl glycoside as principal carotenoid in *Salinibacter ruber*, an extremely halophilic eubacterium. J Nat Prod 65: 1340–1343

Marshall JH, Wilmoth GJ (1981) Pigments of *Staphylococcus aureus*, a series of triterpenoid carotenoids. J Bacteriol 147:900–913

Mathies RA, Lin SW, Ames JB, Pollard WT (1991) From femtoseconds to biology: mechanism of bacteriorhodopsin's light-driven proton pump. Annu Rev Biophys Biophys Chem 20:491–518

Müller V, Saum SH (2005) The chloride regulon of *Halobacillus halophilus*: a novel regulatory network for salt perception and signal transduction in bacteria. In: Gunde-Cimerman N, Oren A, Plemenitaš A (eds) Adaptation to life at high salt concentrations in Archaea, Bacteria, and Eukarya. Springer, Dordrecht, pp 303–310

Okulski W, Sujak A, Gruszecki WI (2000) Dipalmitoylphosphatidylcholine membranes modified with zeaxanthin: numeric study of membrane organisation. Biochim Biophys Acta 1509: 216–228

Oren A (2002) Diversity of halophilic microorganisms: environments, phylogeny, physiology, and applications. J Ind Microbiol Biotechnol 28:56–63

Osawa A, Ishii Y, Sasamura N, Morita M, Köcher S, Müller V, Sandmann G, Shindo K (2010) Hydroxy-3,4-dehydro-apo-8′-lycopene and methyl hydroxy-3,4-dehydro-apo-8′-lycopenoate, novel C(30) carotenoids produced by a mutant of marine bacterium *Halobacillus halophilus*. J Antibiot (Tokyo) 63:291–295

Rodriguez-Valera F, Nieto JJ, Ruiz-Berraquero F (1983) Light as an energy source in continuous cultures of bacteriorhodopsin-containing halobacteria. Appl Environ Microbiol 45:868–871

Roeßler M, Müller V (1998) Quantitative and physiological analyses of chloride dependence of growth of *Halobacillus halophilus*. Appl Environ Microbiol 64:3813–3817

Roeßler M, Müller V (2001) Chloride dependence of glycine betaine transport in *Halobacillus halophilus*. FEBS Lett 489:125–128

Rottem S, Markowitz O (1979) Carotenoids acts as reinforcers of the *Acholeplasma laidlawii* lipid bilayer. J Bacteriol 140:944–948

Saito T, Terato H, Yamamoto O (1994) Pigments of *Rubrobacter radiotolerans*. Arch Microbiol 162:414–421

Saito T, Miyabe Y, Ide H, Yamamoto O (1997) Hydroxyl radical scavenging ability of bacterioruberin. Radiat Phys Chem 50:267–269

Sandmann G, Fraser PD (1993) Differential inhibition of phytoene desaturases from diverse origins and analysis of resistant cyanobacterial mutants. Z Naturforsch C 48c:307–311

Saum SH, Müller V (2007) Salinity-dependent switching of osmolyte strategies in a moderately halophilic bacterium: glutamate induces proline biosynthesis in *Halobacillus halophilus*. J Bacteriol 189:6968–6975

Saum SH, Müller V (2008a) Growth phase-dependent switch in osmolyte strategy in a moderate halophile: ectoine is a minor osmolyte but major stationary phase solute in *Halobacillus halophilus*. Environ Microbiol 10:716–726

Saum SH, Müller V (2008b) Regulation of osmoadaptation in the moderate halophile *Halobacillus halophilus*: chloride, glutamate and switching osmolyte strategies. Saline Systems 4:4

Saum SH, Sydow JF, Palm P, Pfeiffer F, Oesterhelt D, Müller V (2006) Biochemical and molecular characterization of the biosynthesis of glutamine and glutamate, two major compatible solutes in the moderately halophilic bacterium *Halobacillus halophilus*. J Bacteriol 188: 6808–6815

Shahmohammadi HR, Asgarani E, Terato H, Saito T, Ohyama Y, Gekko K, Yamamoto O, Ide H (1998) Protective roles of bacterioruberin and intracellular KCl in the resistance of *Halobacterium salinarium* against DNA-damaging agents. J Radiat Res (Tokyo) 39:251–262

Sharkov AV, Matveets Iu A, Chekalin SV, Pakulev AV (1985) Subpicosecond spectroscopy of bacteriorhodopsin. Dokl Akad Nauk SSSR 281:466–470

Shimidzu N, Goto M, Miki W (1996) Carotenoids as singlet oxygen quenchers in marine organisms. Fish Sci 62:134–137

Shindo K, Kikuta K, Suzuki A, Katsuta A, Kasai H, Yasumoto-Hirose M, Matsuo Y, Misawa N, Takaichi S (2007) Rare carotenoids, (3R)-saproxanthin and (3R,2′S)-myxol, isolated from novel marine bacteria (*Flavobacteriaceae*) and their antioxidative activities. Appl Microbiol Biotechnol 74:1350–1357

Shindo K, Endo M, Miyake Y, Wakasugi K, Morritt D, Fraser PD, Kasai H, Misawa N (2008) Methyl glucosyl-3,4-dehydro-apo-8′-lycopenoate, a novel antioxidative glyco-C(30)-carotenoic acid produced by a marine bacterium *Planococcus maritimus*. J Antibiot (Tokyo) 61:729–735

Sies H, Mehlhorn R (1986) Mutagenicity of nitroxide-free radicals. Arch Biochem Biophys 251: 393–396

Spring S, Ludwig W, Marquez MC, Ventosa A, Schleifer K-H (1996) *Halobacillus* gen. nov., with descriptions of *Halobacillus litoralis* sp. nov. and *Halobacilus trueperi* sp. nov., and transfer of *Sporosarcina halophila* to *Halobacillus halophilus* comb. nov. Int J Syst Bacteriol 46:492–496

Straub O (1987) Key of carotenoids. Birkhäuser Verlag, Basel

Strzalka K, Gruszecki WI (1994) Effect of β-carotene on structural and dynamic properties of model phosphatidylcholine membranes. I. An EPR spin label study. Biochim Biophys Acta 1194:138–142

Subczynski WK, Markowska E, Gruszecki WI, Sielewiesiuk J (1992) Effects of polar carotenoids on dimyristoylphosphatidylcholine membranes: a spin-label study. Biochim Biophys Acta 1105:97–108

Subczynski WK, Markowska E, Sielewiesiuk J (1993) Spin-label studies on phosphatidylcholine-polar carotenoid membranes: effects of alkyl-chain length and unsaturation. Biochim Biophys Acta 1150:173–181

Subramaniam S, Hirai T, Henderson R (2002) From structure to mechanism: electron crystallographic studies of bacteriorhodopsin. Philos Transact A Math Phys Eng Sci 360:859–874

Sujak A, Okulski W, Gruszecki WI (2000) Organization of xanthophyll pigments lutein and zeaxanthin in lipid membranes formed with dipalmitoylphosphatidylcholine. Biochim Biophys Acta 1509:255–263

Takaichi S, Inoue K, Akaike M, Kobayashi M, Oh-oka H, Madigan MT (1997) The major carotenoid in all known species of *Heliobacteria* is the C30 carotenoid 4,4′-diaponeurosporene, not neurosporene. Arch Microbiol 168:277–281

Takaichi S, Maoka T, Akimoto N, Carmona ML, Yamaoka Y (2008) Carotenoids in a *Corynebacterineae*, *Gordonia terrae* AIST-1: carotenoid glucosyl mycoloyl esters. Biosci Biotechnol Biochem 72:2615–2622

Takano H, Obitsu S, Beppu T, Ueda K (2005) Light-induced carotenogenesis in *Streptomyces coelicolor* A3(2): identification of an extracytoplasmic function sigma factor that directs photodependent transcription of the carotenoid biosynthesis gene cluster. J Bacteriol 187:1825–1832

Tao L, Schenzle A, Odom JM, Cheng Q (2005) Novel carotenoid oxidase involved in biosynthesis of 4,4′-diapolycopene dialdehyde. Appl Environ Microbiol 71:3294–3301

Turner M, Jervis DI (1963) The distribution of pigmented *Bacillus* species in saltmarsh and other saline and non-saline soils. Nova Hedwig 16:293–298

Yokoyama A, Sandmann G, Hoshino T, Adachi K, Sakai M, Shizuri Y (1995) Thermozeaxanthin, new carotenoid-glycoside-esters from thermophilic eubacterium *Thermus thermophilus*. Tetrahedron Lett 36:4901–4904

Yokoyama A, Shizuri Y, Hoshino T, Sandmann G (1996a) Thermocryptoxanthins: novel intermediates in the carotenoid biosynthetic pathway of *Thermus thermophilus*. Arch Microbiol 165:342–345

Yokoyama A, Miki W, Izumida H, Shizuri Y (1996b) New trihydroxy-keto-carotenoids isolated from an astaxanthin-producing marine bacterium. Biosci Biotechnol Biochem 60:200–203

Chapter 17
Xanthorhodopsin

Janos K. Lanyi and Sergei P. Balashov

17.1 Introduction

The extremely halophilic Archaea have acquired a capability for phototrophy based on a single protein with an attached retinal chromophore. It captures light and utilizes the energy gained by performing transmembrane ion transfer to generate electrochemical gradients (Drachev et al. 1976; Michel and Oesterhelt 1980) that are used for ATP synthesis, motility, and transport. The classical examples are bacteriorhodopsin, the proton pump of *Halobacterium salinarum* (Oesterhelt and Stoeckenius 1973), and halorhodopsin, the light driven chloride pump (Schobert and Lanyi 1982). This type of simple energy transducer and light sensor was a very useful "invention" early in evolution, and it has spread far beyond the halophilic world in the form of the homologous proteorhodopsins found in the genomes of eubacteria, inhabiting sea waters (Béjà et al. 2000, 2001; Venter et al. 2004; Gomez-Consarnau et al. 2007) and fresh waters (Sharma et al. 2009). They are present in an ancient Siberian permafrost (Petrovskaya et al. 2010), and were found in many eukaryotes as well (Brown and Jung 2006). Xanthorhodopsin (Balashov et al. 2005) was discovered in and isolated from a cultured extremely halophilic eubacterium, *Salinibacter ruber*, which shares its environment with haloarchaea (Antón et al. 2002). It was soon recognized that, besides xanthorhodopsin, its genome contains three other retinal proteins, one homologous to halorhodopsin (Peña et al. 2005) and two sensory rhodopsins (Mongodin et al. 2005). Xanthorhodopsin is unique in that it has a light-harvesting antenna in addition to retinal.

Photosynthetic systems in biology depend on specialized colored molecules that are suitable both for collecting light and converting the absorbed electromagnetic energy into usable forms. Because of chemical constraints, these requirements are not always both optimally achieved, and the chromophores in reaction centers are often supplemented by auxiliary pigments, which harvest light and funnel energy to

J.K. Lanyi (✉) • S.P. Balashov
Department of Physiology & Biophysics, University of California, Irvine, CA 92697, USA
e-mail: jklanyi@uci.edu

A. Ventosa et al. (eds.), *Halophiles and Hypersaline Environments*,
DOI 10.1007/978-3-642-20198-1_17, © Springer-Verlag Berlin Heidelberg 2011

the center. Chlorophyll-based photosynthesis utilizes carotenoids and other pigments in large multi-chromophore protein complexes as antennae (McDermott et al. 1995; Polívka and Frank 2010). Bacteriorhodopsin of *Halobacterium salinarum*, a much simpler energy transducer, does not have such an antenna (Oesterhelt and Stoeckenius 1973; Litvin et al. 1977) whereas xanthorhodopsin contains a carotenoid, salinixanthin (Lutnaes et al. 2002), serving as antenna. Other than being a dual-chromophore system (with one retinal and one carotenoid molecule), this small heptahelical transmembrane protein (Balashov et al. 2005; Luecke et al. 2008) is similar to bacteriorhodopsin and especially its eubacterial counterparts, the proteorhodopsins (Béjà et al. 2000, 2001). Much is known already about this protein. Action spectra indicated that both chromophores participate in driving proton transport (Balashov et al. 2005; Boichenko et al. 2006), and the observed dependence of a tight and specific carotenoid binding on the presence of retinal (Balashov et al. 2005, 2006; Imasheva et al. 2008) suggested their close interaction. Recent steady-state and ultrafast spectroscopy confirmed the postulated energy transfer from the excited state of the carotenoid to the retinal, and described features of the excited states involved (Balashov et al. 2008; Polívka et al. 2009; Zhu et al. 2010). The deduced geometry of the two polyene chains from spectroscopic data (Balashov et al. 2008) is consistent with the recently solved crystal structure of the protein (Luecke et al. 2008), the first one for an eubacterial proton pump. The structure revealed not only the precise location of the carotenoid antenna but also entirely new features in architecture of the proton conducting pathways, especially on the extracellular side where proton release takes place. It is probable that these structural features are present in many eubacterial retinal proteins, proteorhodopsins and other proteins homologous to xanthorhodopsin.

The genome of *Salinibacter ruber* and the relationship of the xanthorhodopsin gene to other retinal proteins (Fig. 17.1a) were described (Mongodin et al. 2005). Alignment of the amino acid sequences indicates that xanthorhodopsin has slightly higher homology to proteorhodopsins than to bacteriorhodopsin (61 identities vs. 58, Fig. 17.1b). A xanthorhodopsin-like gene was found in an abundant coastal ocean methylotroph that utilizes methanol and formaldehyde as sources of carbon (Giovannoni et al. 2008). This gene forms a clade with the xanthorhodopsin of *Salinibacter ruber* and the rhodopsins of cyanobacterium *Gloeobacter violaceus* and the dinoflagellate *Pyrosystis lunula*, and homologous to several other divergent organisms (Sharma et al. 2008; Imasheva et al. 2009). It appears that xanthorhodopsin-like retinal proteins might be as widespread and numerous as the homologous proteorhodopsins (Béjà et al. 2001; Fuhrman et al. 2008).

This review examines current issues concerning xanthorhodopsin as a simple dual-chromophore system for absorption and utilization of light. The outstanding questions include the structure of eubacterial proton pump, the way the large carotenoid salinixanthin (Fig. 17.2) is accommodated by the relatively small protein, the mechanism of inter-chromophore excitation energy transfer, and the distinctive features of xanthorhodopsin in comparison with the earlier studied bacteriorhodopsin and archaerhodopsin of the halobacteria and the proteorhodopsins of marine bacteria. For background material, the reader is directed to

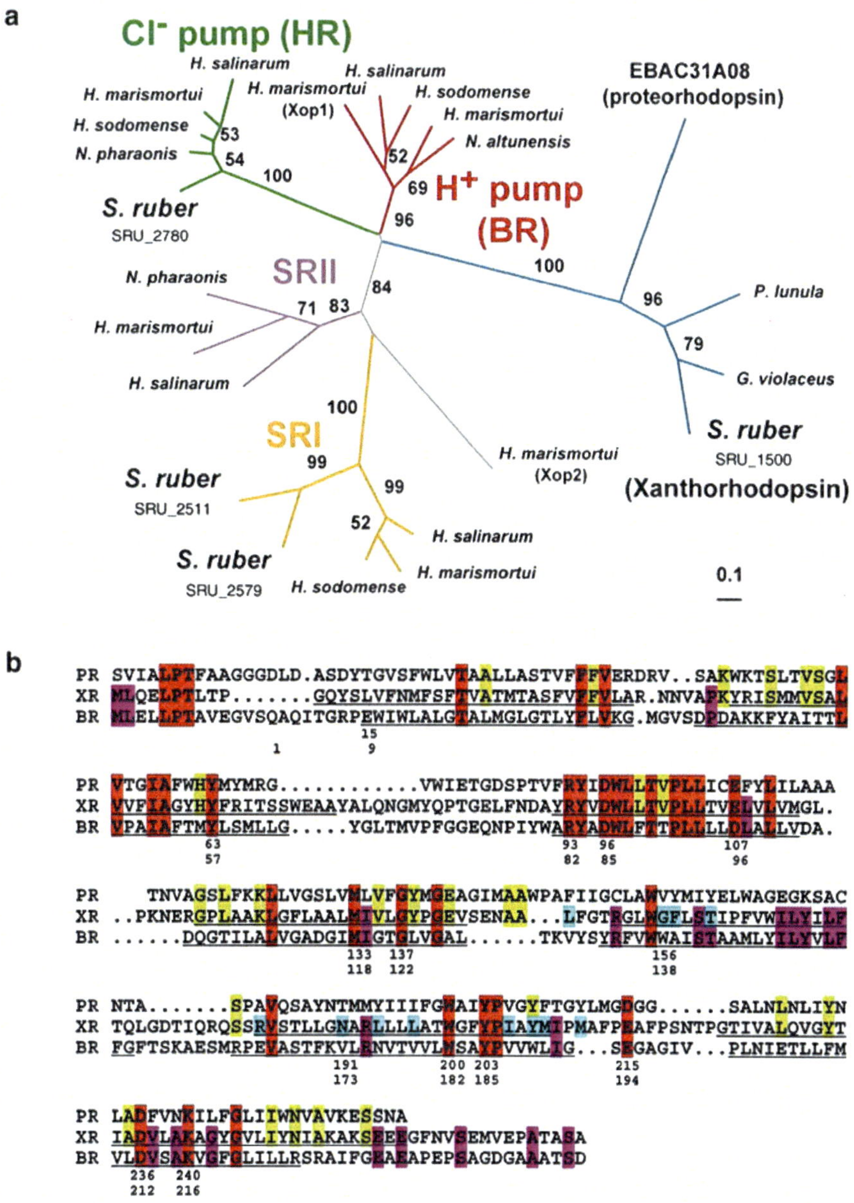

Fig. 17.1 (**a**) A likelihood phylogeny of retinal protein genes of *Salinibacter ruber*. Note the distant relationship of xanthorhodopsin to the archaeal proton pump bacteriorhodopsin, and the much greater homology to the rhodopsins of *Gloeobacter violaceus*. Three other genes in *S. ruber*, halorhodopsin (HR) and two sensory rhodopsin-I like proteins (SRI), are nearer their archaeal homologues and imply lateral gene transfer. From Mongodin et al. (2005). (**b**) Alignment of amino acid sequences of xanthorhodopsin (XR), proteorhodopsin (PR) and bacteriorhodopsin (BR). Red, residues conserved in all three proteins; yellow, conserved in PR and XR; purple, conserved in XR and BR; cyan, residues involved in carotenoid binding. *Top row* of numbers refer to the XR sequence; *bottom row*, to the BR sequence. From Luecke et al. (2008)

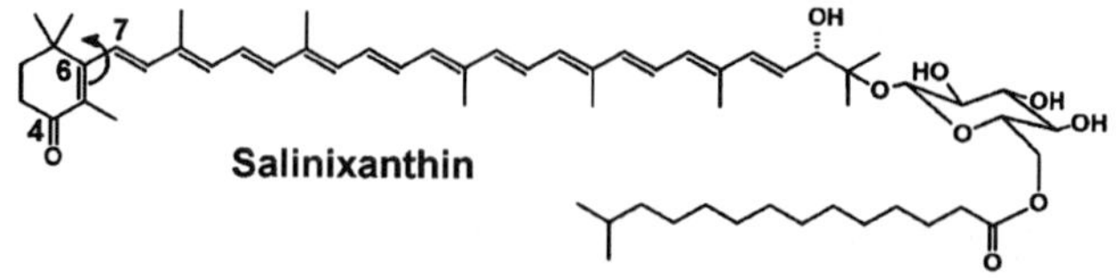

Fig. 17.2 Chemical structure of carotenoid salinixanthin from the cell membrane of *Salinibacter ruber*. From Lutnaes et al. (2002). An *arrow* around C6–C7 bond illustrates the turn of the ring in the xanthorhodopsin binding site as shown in (Balashov et al. 2006; Luecke et al. 2008)

earlier review articles on microbial retinal proteins (Oesterhelt 1998; Spudich and Jung 2005), bacteriorhodopsin (Balashov and Lanyi 2006; Hirai et al. 2009), proteorhodopsins (Fuhrman et al. 2008), xanthorhodopsin (Balashov and Lanyi 2007; Lanyi and Balashov 2008) and carotenoid antennae in different photosynthetic systems including xanthorhodopsin (Polívka and Frank 2010).

17.2 Action Spectrum of Xanthorhodopsin in Live *Salinibacter ruber* Cells

The first evidence for involvement of salinixanthin in proton transport was from measurements of action spectra for light-induced pH changes in vesicles produced by sonication of *Salinibacter ruber* cells (Balashov et al. 2005) and oxygen consumption in live cells (Balashov et al. 2005; Boichenko et al. 2006). The latter approach produced an especially detailed spectrum. Illumination of *Salinibacter ruber* cells causes decrease of respiration rate up to 50% (detected as increase in steady-state oxygen content of cell suspensions) from back-pressure of the light-induced proton electrochemical potential created by xanthorhodopsin on the electron transfer steps in the respiratory chain (Boichenko et al. 2006). Under excitation with single flashes the change in respiration rate occurred with a 0.2 s time constant, consistent with the turnover of the xanthorhodopsin photocycle (100–200 ms). The action spectrum for photoinhibition of respiration in *Salinibacter ruber* cells at high spectral resolution (4 nm) indicated participation of the carotenoid in proton pumping. The high accuracy of the spectrum made it possible to deconvolute it into two components: retinal and carotenoid (Fig. 17.3a). The derived spectrum of the carotenoid component is particularly useful for estimations of the efficiency of the antenna (ca. 40%), since the *Salinibacter* cell membrane preparations always contain a variable fraction of carotenoid unbound to xanthorhodopsin, which complicates determination of the exact spectrum of the bound component when using absorption spectroscopy.

Measurements of action spectra in archaeon *Halorubrum* sp. A1C cells containing another proton pump, archaerhodopsin (Mukohata et al. 1991), and the carotenoid bacterioruberin (Britton et al. 2004), showed that, unlike in xanthorhodopsin, in this system there is no energy transfer from bacterioruberin to retinal

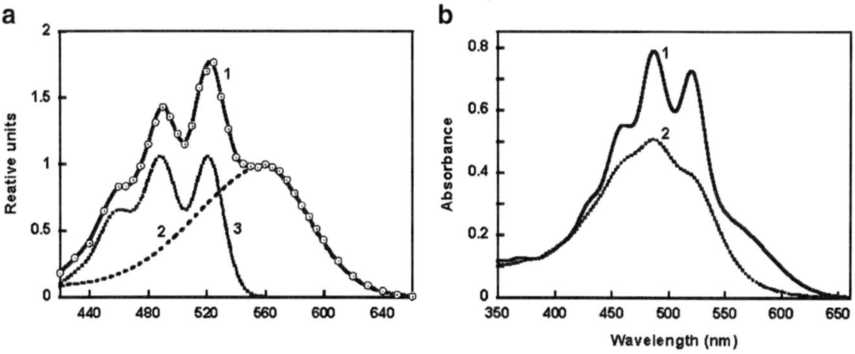

Fig. 17.3 (**a**) Action spectrum (1) of photoinhibition of respiration in *Salinibacter ruber* cells and its retinal (2) and carotenoid (3) components. Light absorbed by the carotenoid decreases the rate of cell respiration (as a result of back pressure from the light-induced electrochemical proton gradient), indicating its involvement in collecting energy for the transport process. From Boichenko et al. (2006). (**b**) Absorption spectra of fractions of *Salinibacter ruber* cell membrane containing: 1, xanthorhodopsin with bound salinixanthin; 2, salinixanthin not bound to xanthorhodopsin

(Boichenko et al. 2006). This finding emphasizes the uniqueness of proteins like xanthorhodopsin, and implies that antenna function utilizing excitation energy transfer did not emerge among the archaea but appeared later during evolution. The function of bacterioruberin in archaeal membranes must be primarily photoprotection and structural stability (Yoshimura and Kouyama 2008), whereas salinixanthin has a clear energetic role, perhaps in addition to the two others.

17.3 Spectroscopic Evidence for a Specific Binding Site for Salinixanthin and Its Interaction with Retinal

The cell membrane of *Salinibacter ruber* after dialysis was separated into two fractions (Balashov et al. 2005); the lighter fraction in the supernatant was characterized by an absorption spectrum with a maximum at 488 nm and poorly resolved shoulders on both sides. It contained salinixanthin which accounts for >95% of all carotenoids with a small amount of β-carotene also present (Lutnaes et al. 2002). The heavier fraction in the pellet exhibited sharp, well resolved bands at 522, 486 and 456 nm from the salinixanthin bound to xanthorhodopsin and a broad retinal band with maximum at ca. 560 nm (Fig. 17.3b), bands similar to those in the action spectrum.

Evidence for specific and rigid binding of salinixanthin in xanthorhodopsin is from experiments in which the retinal chromophore was removed and then reconstituted with all-*trans* retinal (Balashov et al. 2005, 2006). Incubation with hydroxylamine results in cleavage of the C=N bond that connects the retinal to the protein and elimination of the 560 nm retinal chromophore band, which shifts to

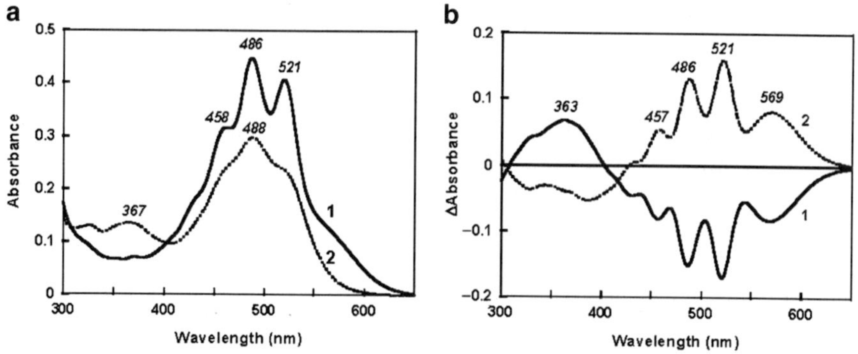

Fig. 17.4 (**a**) Absorption spectra of: 1, membranes containing xanthorhodopsin; 2, after incubation with 0.2 M hydroxylamine during illumination (to hydrolyze the C=N bond and remove retinal from the binding site). (**b**) Absorption changes accompanying: 1, hydrolysis of the Schiff base with hydroxylamine (curve 2 minus 1 in panel (**a**)); 2, reconstitution of xanthorhodopsin with all trans retinal. Adapted from Imasheva et al. (2008)

362 nm upon formation of the retinal oxime (Fig. 17.4). This is typical for retinal proteins. Surprisingly, the carotenoid bands are also strongly affected (spectra 1 and 2 in Fig. 17.4a and the difference spectrum 1 in Fig. 17.4b). They become less intense and broader than in untreated xanthorhodopsin. Addition of retinal after treatment with hydroxylamine results in the re-formation of the C=N bond and the reappearance of retinal protein band, accompanied by reversal of the changes in the absorption spectrum of the carotenoid (spectrum 2 in Fig. 17.4b): the vibronic bands become narrower and more intense as in the untreated protein. The structured spectrum was attributed to decreased conformational heterogeneity of the carotenoid because of restriction of the rotation of its ring around the C6–C7 bond by the retinal protein (Balashov et al. 2006); such rotations and conformational heterogeneity are typical for similar compounds with conjugated rings (Christensen and Kohler 1973; Buchecker and Noack 1995). The conjugated keto group is known to eliminate the vibronic structure of carotenoid spectra (Britton 1995), but it can be at least partially recovered in nonpolar solvents where interaction with carboxyl oxygen are minimized, and in frozen solvents where intramolecular motions are reduced (Ke et al. 1970).

The restoration of the well resolved structure of the carotenoid spectrum by insertion of the retinal into its binding site occurs before (and even in the absence of) formation of the protonated Schiff base, as experiments with retinal analogs indicate (Imasheva et al. 2008). In reconstitution experiments, the 13-desmethyl retinal analogue as well as the analogue with a "locked" C13=C14 bond, forms the protonated Schiff base linkage much more slowly than retinal (Imasheva et al. 2008). This provided the opportunity to observe carotenoid changes at various stages of retinal binding. When immobilization of the salinixanthin keto ring is followed by measuring the development of its highly structured bands, it is found to occur earlier than the appearance of the 560 nm band of the protonated Schiff base as retinal analogues with slower reconstitution rates are used. Thus, the carotenoid is "re-bound" (assumes its twisted and immobilized conformation) in its specific site even before the covalent bond to a lysine via a Schiff base is formed (Imasheva et al. 2008; Smolensky and Sheves 2009).This implicates the retinal chain and ionone ring in the interaction with the carotenoid, as was recently confirmed by X-ray crystal structure (Luecke et al. 2008), see below. When reconstitution is with retinol, which cannot form a Schiff base, the carotenoid exhibits spectral changes similar to those when retinal is added. This provides further evidence that restriction of conformational heterogeneity of the carotenoid is through steric interaction with the retinal, as the retinal enters the binding site.

Salinixanthin in an organic solvent does not exhibit optical activity in the visible (a Cotton effect), but when bound it becomes chiral, as detected by circular dichroism (CD) (Balashov et al. 2006). Thus, the native xanthorhodopsin complex has an asymmetric conformation. Like the structured absorption spectrum, the bands of the CD spectrum are controlled by the retinal. Hydrolysis of the retinal Schiff base with hydroxylamine and the ensuing removal of retinal from the binding site abolish optical activity (Balashov et al. 2006). This shows that in the native state the retinal is responsible for forcing the antenna into an asymmetric conformation. Analogies with other carotenoids (Buchecker and Noack 1995) suggest that this involves a turn of its ring moiety (Balashov et al. 2006). Part of the CD spectrum is likely to be from electronic interactions of the conjugated systems in the two chromophores, as in the light harvesting complexes of photosynthetic bacteria (Georgakopoulou et al. 2004).

Independent evidence for close interaction of the carotenoid antenna with the retinal chromophore was obtained in studies of spectral changes that accompany the photocycle reactions. Formation of the K intermediate causes a small (1–2 nm) blue shift of the carotenoid spectrum (Balashov et al. 2005) which might originate from changes in either electrostatic field or steric interaction with the ionone ring of retinal (see below). At a later time in the cycle, the carotenoid exhibits a different type of transient change, which can be described as a broadening of the spectrum from increased freedom for bond rotations, i.e., loosening of the binding site (Balashov et al. 2005).

The carotenoid antenna can be selectively oxidized with ammonium persulfate with relatively little effect on the retinal absorption spectrum and photocycle kinetics. The latter exhibits only a small decrease in the photocycle turn-on (Imasheva et al. 2011).

17.4 Xanthorhodopsin as a Proteorhodopsin Homologue: Similarities in the pK_a of the Counterion and the Photocycle

Proton pumps depend on the reversible protonation and deprotonation of groups inside the protein and at its surface, and thus their proton affinities are of interest, especially that of the counterion to the protonated retinal Schiff base. The pK_a of the counterion of xanthorhodopsin is lower value in detergent (Imasheva et al. 2006), more like that in proteorhodopsin, pK_a 7.5 (Dioumaev et al. 2003; Imasheva et al. 2004) than in bacteriorhodopsin, pK_a 2.6 (Balashov et al. 1996) or archaerhodopsin, pK_a 3.5 (Ming et al. 2006). In the latter microbial retinal proteins the pK_a of the counterion can be easily determined by a large (30–40 nm) red shift of retinal absorption band at low pH. Surprisingly, in xanthorhodopsin there is only a small (3 nm) red shift between pH 2 and 12 (Imasheva et al. 2006). A similarly small shift was observed for the highly homologous *Gloeobacter* rhodopsin (Miranda et al. 2009). The pH dependence of the yield of the M photointermediate, another indication of the counterion protonation (because one of the aspartate groups in the counterion, Asp85 in bacteriorhodopsin serves as a proton acceptor during the photocycle), yields the same pK_a. This provides independent evidence that the pK_a of the counterion in xanthorhodopsin is 6.0 in detergent (Imasheva et al. 2006). In bacteriorhodopsin, the counterion is a pair of aspartates (Asp85 and Asp212). The unusually small spectral shift in xanthorhodopsin indicates that the counterion in this protein might have a different structure. This was confirmed by the crystallographic structure. It showed rotation and close interaction of Asp96 (homologous to Asp85 in bacteriorhodopsin) with a histidine and presence of only single water at the active site, corresponding to Wat402 in bacteriorhodopsin. The high pK_a makes xanthorhodopsin a proteorhodopsin-like proton pump, different from bacteriorhodopsin. Mutagenesis of proteorhodopsin indicated that the aspartate serving as a counterion to the Schiff base interacts with histidine in this protein also (Bergo et al. 2009). Unlike the archaeal rhodopsins, the eubacterial rhodopsins do not function at neutral or acid pH, because the counterion will not function as proton acceptor below its pK_a (Dioumaev et al. 2003).

Studies of the pH dependence of the recovery of the initial state in the photocycle of xanthorhodopsin reveal two pK_a's, 6.0 and 9.3. The changes with pK_a 6 reflect the two kinds of photocycles, at acidic and alkaline pH. They originate from the protonation state of the counterion (presumably Asp96, or probably the Asp96–His62 pair, see below). The second pK_a, at 9.3, has been attributed to the pK_a of the acidic group (Glu107) that is the internal proton donor to the deprotonated retinal Schiff base in the M intermediate (Imasheva et al. 2006). It is 2 pH units higher than in bacteriorhodopsin (Balashov 2000).

Overall, the photocycle of xanthorhodopsin exhibits features more similar to those of proteorhodopsin than bacteriorhodopsin, with proton uptake occurring first, followed by release coincident with the last step at the end of the photocycle. The sequence of reactions includes formation of the K, L, M, N and O like intermediates. The kinetics at pH 8.8 can be fit with six exponentials (7.5 μs, 35 μs, 280 μs, 1.3 ms, 10 ms and 100 ms) (Balashov et al. 2005). The K to L reaction is slower, and occurs at a higher cryogenic temperature than in bacteriorhodopsin, >175 K versus 130 K (Litvin et al. 1975; Lozier et al. 1975; Balashov et al. 2005). That helped to establish the evolution of the primary photoproduct K as at least two subsequent states, XR $\leftrightarrow K_0 \rightarrow K_E$, where K_0 is observed almost as a pure species at 80 K and undergoes transition to K_E between 80 and 180 K (Dioumaev et al. 2010). The two K forms are analogous to those in bacteriorhodopsin

(Rothschild et al. 1985; Maeda et al. 2004). They differ substantially in their HOOPs bands and water bands and most likely represent untwisting and relaxation in the strained 13-cis retinal chromophore after its photoisomerization from the initial all-trans in the XR $\rightarrow$ K_0 photoreaction.

Illumination of xanthorhodopsin at low pH (pH 3) results in formation of long-lived photoproducts (Imasheva et al. 2006) similar to the 13-*cis*-states observed in proteorhodopsin (Imasheva et al. 2004, 2005). Remarkably, the spectral bands of the antenna carotenoid are greatly affected in these states (showing broadening and decrease in the maximum amplitude), indicating a connection between the isomeric state of the retinal chromophore and the carotenoid binding.

17.5 Excited States and Fluorescence of the Retinal and the Salinixanthin Chromophores of Xanthorhodopsin: Efficiency and Pathway of Energy Transfer

Direct evidence for excitation energy transfer from the carotenoid to the retinal was obtained by detecting fluorescence of the retinal chromophore induced by quanta absorbed by the carotenoid (Balashov et al. 2008). It eliminated alternative interpretations of the action spectra, e.g., that the carotenoid is involved in proton transfer through an unknown mechanism, or regulates in some manner the functioning of retinal but without supplying the energy for the pump. Fluorescence of the retinal can be detected from its strongly allowed $1B_u^+$ excited state (using the C_{2h} symmetry group notation), which in retinal proteins with protonated Schiff bases is the lowest excited state, the S_1 state (Birge 1990). The retinal fluorescence is weak (quantum yield ca. 2–5 $\times$ 10^{-4}, depending on pH) but the emission from the carotenoid is tenfold weaker (see below). The presence of the carotenoid bands in the excitation spectrum for the retinal emission at 720 nm (Fig. 17.5) provided the final evidence for excited-state energy transfer (Balashov et al. 2008; Lanyi and Balashov 2008).

The excitation spectrum for the retinal emission, sampled at 720 nm, is similar to the excitation spectrum obtained earlier for the physiological responses (Fig. 17.2), i.e., light-induced proton transport and photoinhibition of respiration (Balashov et al. 2005; Boichenko et al. 2006). From the comparison of the relative amplitude of the carotenoid bands in the excitation and absorption spectra (Fig. 17.5a, spectra 1 and 2), and taking into account a small contribution from the carotenoid fluorescence (see below), the efficiency of energy transfer from salinixanthin to the retinal was estimated to be 45 $\pm$ 5% (Balashov et al. 2008). This was determined without using extinction coefficients for the two chromophores. The earlier lower estimate at 33% (Boichenko et al. 2006) was obtained with the assumption that the extinction of the carotenoid is ca. threefold larger than that of the retinal chromophores. This appears to be an overestimate: a more accurate ratio is about 2.5, in the range of 130,000–150,000 L/(mol cm) (Balashov et al. 2008).

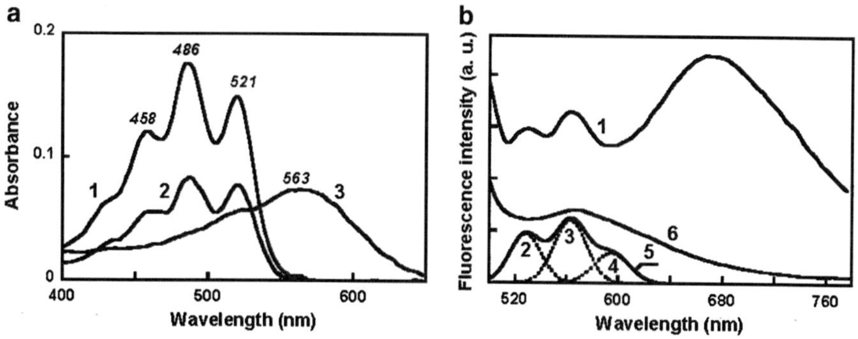

Fig. 17.5 (**a**) Components of the excitation and absorption spectra of xanthorhodopsin, pH 5.5: 1, carotenoid component in the absorption spectrum; 2, carotenoid component in the excitation spectrum; 3, retinal component in the absorption and excitation spectra. (**b**) 1, fluorescence spectrum of xanthorhodopsin upon excitation at 470 nm. The bands at 529 and 565 nm are from salinixanthin emission. A broader band with maximum at 690 nm belongs to retinal fluorescence; 2, 3 and 4, Gaussian fits (components) of the salinixanthin emission from the S_2 excited state; 5, their sum, approximating the short wavelength part of bound salinixanthin emission spectrum; 6, same as 1, but after hydrolysis of the Schiff base with hydroxylamine. Both retinal component and sharp bands of bound salinixanthin are missing from this spectrum. After Balashov et al. (2008)

The contribution of carotenoid emission to the total emission at 720 nm (where the retinal emission was sampled) is small, as indicated by the good agreement of the "physiological" action spectrum and the fluorescence excitation spectrum. Nevertheless, an estimate of the contribution from the fluorescence spectrum of carotenoid was necessary. Detection of this emission was crucial also for determining the mechanism and pathway of the energy transfer. Fluorescence of carotenoids with long conjugated chains has been detected before, but only from solutions of carotenoids in organic solvents. It occurs from a $^1B_u^+$ – like state (S_2), and is extremely weak (reviewed in (Frank and Cogdell 1993; Polívka and Sundström 2004)).

In the xanthorhodopsin fluorescence spectrum, three sharp bands (different from the retinal bands and the background signal) at 529, 565 and 595–605 nm were identified as the carotenoid emission (Balashov et al. 2008). These bands (shown in Fig. 17.5b) exhibit the features peculiar to the fluorescence from the S_2 excited state of carotenoids with long conjugated chains studied in organic solvents: very low quantum yield (4×10^{-5}), small Stokes shift (ca. 300 cm^{-1}), and an approximate mirror-image symmetry of the absorption and fluorescence spectra. The contribution of carotenoid emission to the total emission at 720 nm is less than 5%. The quantum yield corresponds to a very short (ca. 70 fs) lifetime for the excited state of bound carotenoid (Balashov et al. 2008), a value confirmed by direct time resolved experiment following the decay of the absorption band associated with the S_2 to S_n transition (Polívka et al. 2009).

The energy transfer occurs from the S_2 excited state of the carotenoid to the S_1 state of the retinal. The intense absorption bands of carotenoids are from electron transitions from the ground S_o to the S_2 excited state. Transition from S_o to the S_1 state is forbidden but the latter state is populated in internal conversion from S_2.

In the light-harvesting complexes of green plants and photosynthetic bacteria, the carotenoid to chlorophyll energy transfer occurs from both the S_2 and S_1 excited state levels of a carotenoid (Polívka and Sundström 2004). The S_1 level of salinixanthin is estimated to be in the near IR region, between 750 and 850 nm (Balashov et al. 2008). Thus, in xanthorhodopsin, the carotenoid S_1 level cannot serve as a donor because it is below the S_1 level of the retinal at 560 nm. Energy transfer must occur mostly from the S_2 excited state level of salinixanthin to the S_1 level of retinal (Balashov et al. 2005).

This prediction is confirmed in experiments with sodium borohydride (Balashov et al. 2008). Reduction of the double C=N bond of the retinal Schiff base to a single bond shifts the absorption of retinal from 560 to 360 nm, with no (or almost no) effect on the carotenoid absorption bands because retinal remains in the binding site. The large blue shift of the retinal band upon reduction eliminates any possibility of energy transfer since its energy level becomes much higher than that of the carotenoid. Under these conditions there is a twofold increase in the intensity of the carotenoid emission at 529 and 565 nm, characteristic of the S_2 level (Balashov et al. 2008). This provides evidence that the S_2 level is the only (or the major) source of energy for the S_1 state of retinal chromophores, the lowest resolved singlet excited state (Birge and Zhang 1990).

Measurements of the spectral changes in the excited state and their dynamics on the femtosecond time scale confirm and extend the conclusions from steady-state spectroscopy (Polívka et al. 2009). Here also, borohydride treatment, which reduces the retinal Schiff base $C{=}NH^+$ from double to single bond and shifts the absorption maximum far to the blue but unlike the hydroxylamine treatment retains the retinal in its binding site, affords suitable comparison of the system with and without energy transfer. The more rapid decay of the S_2 state of salinixanthin when the retinal chromophore is available as energy acceptor, 66 fs versus 110 fs, reveals the presence of the expected additional decay channel, and yields an efficiency of ca. 40% for the energy transfer (Polívka et al. 2009). No change in the decay of the carotenoid S_1 state is found (ca. 3 ps), consistent with the expectation that its level is below the S_1 state of the retinal (scheme in Fig. 17.6). These results are supported by a recent independent femtosecond study (Zhu et al. 2010).

The rate constant for the Förster resonance energy transfer is proportional to the spectral overlap integral of carotenoid emission and retinal absorption and the square of the coloumbic interaction (electronic coupling) of the two chromophores, V (Scholes 2003). From the spectral and kinetic data the latter was estimated to be in the range of 160–210 cm^{-1} (Polívka et al. 2009), a value comparable with those in carotenoid-bacteriochlorophyll antenna systems (Krueger et al. 1998). Theoretical calculations based on the crystal structure of xanthorhodopsin, that took into account the mutual geometry of the chromophores yielded a similar estimate (Fujimoto and Hayashi 2009).

The S_2 state of salinixanthin being very short-lived, the efficient energy transfer requires close proximity of the donor to the acceptor. The efficiency of the excitation energy transfer depends on the overlap integral of the donor fluorescence and acceptor absorption and their distance. The fluorescence from the S_2 level of

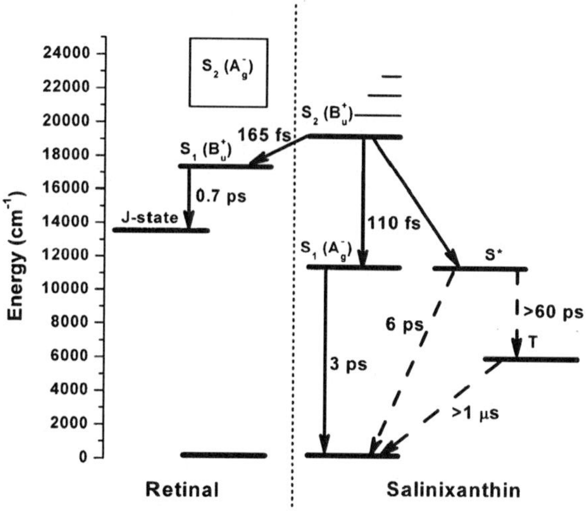

Fig. 17.6 Scheme of excited states of the retinal and the carotenoid chromophores of xanthorhodopsin, the energy transfer pathway from the S_2 of the carotenoid to S_1 of retinal and alternative relaxation processes as inferred from femtosecond time resolved absorption changes (Polívka et al. 2009). Numbers by the *arrows* are time constants for the pathways of energy conversion in xanthorhodopsin (pH 8). S* is a substrate of the S_1 state. T is a triplet state. The lifetime of the S_2 state is 66 fs (sum of the two processes with time constants 110 and 165 fs). *Dotted vertical line* separates the excited states of the retinal chromophore (to the *left*) and salinixanthin antennae (to the *right*). From Polívka et al. (2009)

carotenoid, with maxima at 529 nm and 565 nm, optimally overlaps with the retinal absorption with a maximum at 560 nm. Calculations using the Förster equation for the efficiency of energy transfer and the experimentally obtained values for overlap integrals and quantum yield of carotenoid S_2 fluorescence yielded the distance between the centers of the carotenoid polyene chain and the retinal as ca. 11 Å (Balashov et al. 2008). This is a very rough estimate only, because the dimensions of the two linear chromophores are comparable to this distance, a case that is not described accurately by the original Förster equation.

The contribution of the salinixanthin bands to the excitation spectrum of the retinal chromophore fluorescence strongly depends on the polarization of the excitation and emission beams (Balashov et al. 2008). This means that there is a substantial angle ($\gg 0°$) between the transition moments of the $S_0 \rightarrow S_2$ carotenoid absorption and $S_1 \rightarrow S_0$ retinal fluorescence. From measurement of the excitation anisotropy the angle was determined to be $56 \pm 3°$ (Balashov et al. 2008), near the 46° angle between the geometric axes of the two conjugated chains (Fig. 17.7a) determined later by X-ray diffraction (Luecke et al. 2008). The discrepancy is most likely because of an off-axis orientation of the transition moment, as in rhodopin (Georgakopoulou et al. 2003). The angle appears to be a compromise between the best efficiency for energy transfer (parallel) and the ability to collect incident light by the dual chromophore system at all angles of polarization (perpendicular). In bacteriorhodopsin the retinal is tilted 21° from the

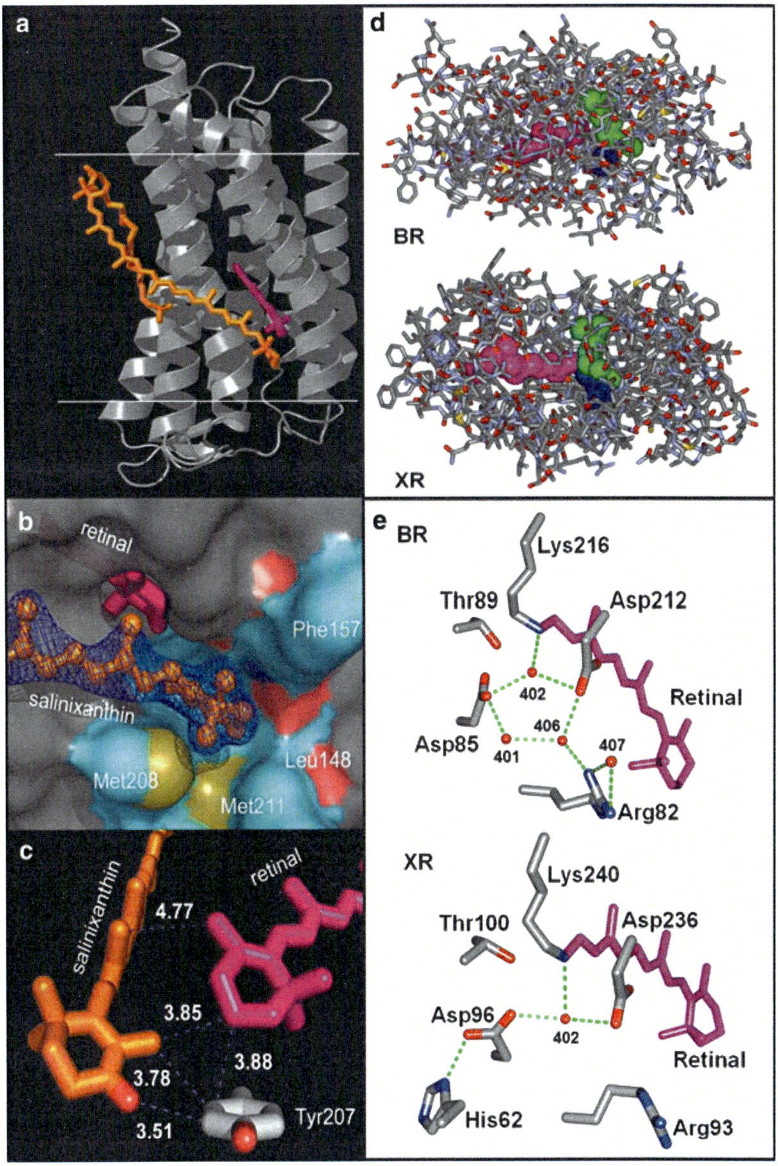

Fig. 17.7 *Left panel*. Location of salinixanthin (*orange*) and retinal (*magenta*) in the X-ray structure of xanthorhodopsin. (**a**) The extended carotenoid is tightly bound on the surface of xanthorhodopsin. The angle between the chromophore axes is 46°. *Horizontal lines* indicate the approximate boundaries of the lipid bilayer. (**b**) The binding pocket of the salinixanthin keto ring is formed by Leu148, Gly156, Phe157, Thr160, Met208, and Met211, as well as the retinal β-ionone ring. (**c**) The keto ring of the carotene is rotated 82° out of plane of the salinixanthin-conjugated system and is in van der Waals distance of the retinal β-ionone and the phenolic side chain of Tyr207. From Luecke et al. (2008). *Right panel*. Comparison of bacteriorhodopsin and xanthorhodopsin. (**d**) A view on bacteriorhodopsin and xanthorhodopsin from the extracellular side showing a cleft extended to the

membrane plane. The high degree of homology of residues in the retinal binding sites of the two proteins suggests that the retinal will be oriented similarly in xanthorhodopsin. In this case, the carotenoid will be tilted either 13° or 55° to the membrane normal, and the structure of the protein (Luecke et al. 2008) indicates that the latter is the case. The tilt ensures that all but the glycoside moiety of the long and roughly linear carotenoid is immersed in the hydrophobic region of the membrane.

17.6 Crystal Structure of Xanthorhodopsin on Carotenoid Binding Site and Novel Features of the Eubacterial Proton Pump

In the crystal structure (Luecke et al. 2008) the carotenoid is seen to be buried at the protein–lipid boundary, and lies transverse against the outer surface of helix F at a 56° angle to the membrane normal (Fig. 17.7a). Its keto ring contacts residues at the extracellular ends of helices E and F and the β-ionone ring of the retinal, and is rotated 82° from the plane of the methyl group of its polyene chain and therefore from the plane of the extended π-system (Fig. 17.7b). The keto group oxygen is not hydrogen-bonded. Immobilization of the ring and its acutely out-of-plane orientation minimizes participation of its two double bonds in the conjugated π-system, and explains the well-resolved vibronic bands of the bound carotenoid, the lack of a red-shift of the bands relative to the bands in organic solvents, and the strong CD bands in the visible region. The dependence of the carotenoid spectrum on the retinal is explained by the fact that the retinal β-ionone ring is part of the carotenoid binding site. The relatively rigid polyene chain is wedged in a slot on the outside of helix F, with one side formed by the Leu194 and Leu197 side-chains and the other by the Ile205 side-chain. The carotenoid glucoside is hydrogen-bonded to the C=O and the NH_2 of the amide side-chain of Asn191, as well as NH1 of Arg184.

The keto ring of the carotenoid is in the space occupied by Trp-138 in bacteriorhodopsin, whose bulky side-chain fixes the position of the retinal ionone ring in that protein. In xanthorhodopsin it is replaced by a glycine, whose smaller volume makes room for the carotenoid. Another difference is Glu141 of xanthorhodopsin, which is an alanine in bacteriorhodopsin but a conserved glutamate in proteorhodopsins involved in spectral tuning (Kralj et al. 2008).

The center-to-center distance of the two chromophores is 11.7 Å, i.e., about the same as the 11 Å estimate. The two polyenes may interact more intimately than this distance predicts, however, because the retinal β-ionone ring is within van der Waals distance of the carotenoid keto ring (Fig. 17.7c). Both rings are in contact

---

Fig. 17.7 (continued) retinal in xanthorhodopsin. Retinal is in pink, Arg82 and its homolog is in blue and Asp212 is in green. (**e**) Active site structure in bacteriorhodopsin and xanthorhodopsin based on high resolution crystal structures, 1.55 Å for bacteriorhodopsin, 1C3W, (Luecke et al. 1999b) and 1.9 Å for xanthorhodopsin, 3DLL (Luecke et al. 2008) with the aromatic ring of Tyr207 between them. This is unlike the crystal structure of archaerhodopsin (Yoshimura and Kouyama 2008), the other proton pump that contains a carotenoid, bacterioruberin, but without antenna function, where the corresponding inter-chromophore center-to-center distance is 17 Å, and the closest approach of bacterioruberin to the retinal at 12 Å. Energy transfer in that protein is obviously excluded by the large distance between the chromophores.

Solving the X-ray structure of xanthorhodopsin defined not only the geometry of carotenoid-retinal interaction, but provided also better understanding of the broader question of the transport mechanism in microbial light-driven proton pumps. Until now, a crystal structure was not available for any of the numerous proteorhodopsins, but at this stage, because they share many common features with xanthorhodopsin, the new structure may be considered as a model for eubacterial rhodopsins in general. The structure helped to improve the initial sequence alignment of xanthorhodopsin and proteorhodopsin (Balashov et al. 2005), and points to an even large homology with the latter (Fig. 17.1b).

Remarkably, in the 1.9-Å resolution structure of xanthorhodopsin there are great differences from the disposition of the main-chain of bacteriorhodopsin (Fig. 17.7a). Helices A and G are longer by four and nine residues, respectively, and their tilt and rotation, particularly of helix A, are considerably different. The 28 residues that comprise helix B are four residues shifted in the sequence toward the C terminus (i.e., toward the extracellular side). In bacteriorhodopsin, the interhelical B–C antiparallel β-sheet interacts with the D–E loop, while in xanthorhodopsin it reorients dramatically to interact with the Arg8 peptide C=O near the N terminus, where it forms a three-stranded β-sheet. As a result, the B–C loop is displaced, by as much as 30 Å, toward the periphery of the protein. This uncovers a large cleft at the extracellular surface that extends far into the interior and brings the aqueous interface near to functional residues that are buried in bacteriorhodopsin (Fig. 17.7d). The absence of a hydrogen-bonded network of polar groups and water in this region correlates with the observed lack of proton release to the extracellular surface upon deprotonation of the retinal Schiff base in the photocycle (Luecke et al. 2008; our unpublished results).

In bacteriorhodopsin, Wat402 receives a hydrogen-bond from the protonated retinal Schiff base and donates hydrogen-bonds to the two anionic residues, Asp85 and Asp212 (Luecke et al. 1998). This arrangement is conserved in xanthorhodopsin. However, the extracellular hydrogen-bonded aqueous network is missing entirely (Fig. 17.7e), and the single glutamate is far removed from Arg93, the homologue of Arg82 in bacteriorhodopsin (>18 Å vs. 7.3 Å in bacteriorhodopsin). Rearrangement of the Arg93 side-chain is unlikely to occur in the xanthorhodopsin photocycle, because its NH1 and NH2 are both hydrogen-bonded to the peptide carbonyl of Gln229 instead of water molecules. In bacteriorhodopsin, a pair of glutamate residues, Glu194 and Glu204 (Brown et al. 1995; Balashov et al. 1997), coordinates a hydrogen-bonded water network from which a proton is released to the extracellular surface (Garczarek et al. 2005) after the retinal Schiff base is deprotonated and the counterion is protonated in M intermediate and the side-chain of Arg-82 moves toward the glutamate pair (Luecke et al. 1999a). As

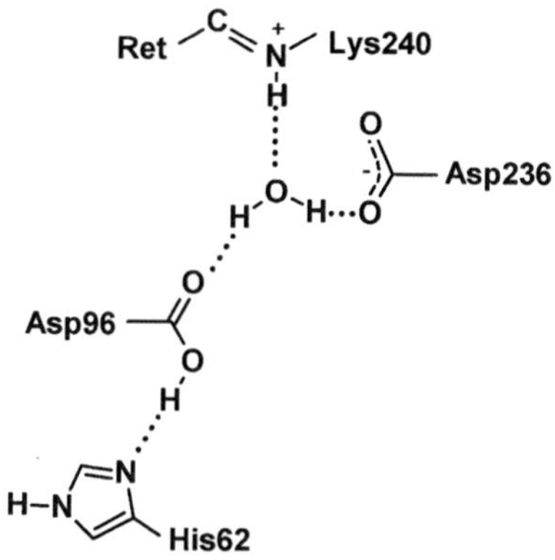

Fig. 17.8 Asp-His counterion to the Schiff base of xanthorhodopsin. The Schiff base interacts with the counterion through a water molecule (Wat402). After Luecke et al. (2008)

xanthorhodopsin, the other eubacterial pumps also contain only one of these carboxylic acid residues, homologous to Glu194 of bacteriorhodopsin.

One of the distinguishing features of eubacterial proton pumps is that the pK_a of the primary proton acceptor is not as low as 2.6 in bacteriorhodopsin, but near 6–7 (see above). The origin of the increased proton affinity had been an unsolved problem. In xanthorhodopsin, ND1 of His62 is hydrogen-bonded to OD1 of Asp96 (Figs. 17.7e and 17.8). With a length of 2.4–2.5 Å, this is a very short hydrogen-bond that suggests that the proton is strongly shared by the imidazole ring and the carboxylate. The aspartate–histidine complex, with an expected pK_a higher than the aspartate alone, must be thus regarded as the Schiff base counterion. The archaeal rhodopsins do not contain this histidine. If otherwise the analogy with bacteriorhodopsin holds, it is the anionic, rather than the neutral complex, that is the proton acceptor of the Schiff base in the photocycle.

A histidine at this position is highly conserved in the proteorhodopsins, making it likely that the aspartate–histidine complex is a general characteristic of eubacterial pumps. Once protonated in the photocycle, the His62–Asp96 complex would be a good candidate for the origin of the proton released to the medium upon deprotonation of the retinal Schiff base, but at neutral pH, at least, such early proton release does not occur (Luecke et al. 2008). Asp–His pairs connected with a strong short hydrogen bond are involved in catalysis of different reactions in many enzymes, such as α-chymotrypsin (Cleland et al. 1998).

In the cytoplasmic region of bacteriorhodopsin, the proton donor Asp96 is in an anhydrous environment that constitutes the hydrophobic barrier in the cytoplasmic half of the protein (Belrhali et al. 1999; Luecke et al. 1999b). This, and the fact that it donates a hydrogen-bond to OD1 of Thr46, raises its pK_a. The aspartic acid becomes a proton donor to the Schiff base during the photocycle only after hydration of this region that includes a hydrogen-bonded chain of four water molecules to connect the proton donor to its acceptor (Schobert et al. 2003). In xanthorhodopsin, as in the proteorhodopsins, these residues are replaced by a glutamic acid and a serine. The carboxyl is hydrogen-bonded to Wat502 that connects to the peptide carbonyl of Lys240. It appears therefore, that in xanthorhodopsin part of the cytoplasmic hydrogen-bonded chain of water molecules between the retinal and the proton donor is in position favorable for proton transport already in the initial state, which might explain accelerated reprotonation of the Schiff base in the photocycle.

17.7 Other Retinal Proteins with Light-Harvesting Antennae: Reconstitution of Gloeobacter Rhodopsin with Salinixanthin and Echinenone

The X-ray structure of xanthorhodopsin reveals that the 4-keto ring is in the space created by replacement of the bulky Trp138 of bacteriorhodopsin with a Gly in xanthorhodopsin. This gave the clue for a search of other retinal proteins that might contain an antenna similar to salinixanthin. Over a dozen of sequences of retinal proteins, presumably pumps and sensors, from various groups (*Alphaproteobacteria, Actinobacteria, Cyanobacteria, Flavobacteria* and others), especially in xanthorhodopsin clade, have Gly instead of Trp at this site (Table 17.1). To

Table 17.1 Representative list of organisms harboring genes homologous to xanthorhodopsin of *Salinibacter ruber*

Species carrying gene homologous to XR	Accession number[a]	Homology to XR (%)	Key homologous residues[b]
Salinibacter ruber DSM 13855	YP_445623	100	H D E G
Gloeobacter violaceus PCC 7421	NP_923144	53	H D E G
Thermus aquaticus Y51MC23 ctg62	ZP_03495873	53	H D E G
Roseiflexus sp. RS-1	YP_001277280	52	H D E G
Methylophilales bacterium HTCC2181	ZP_01551538	49	H D E W
Alphaproteobacterium BAL199	EDP63929	48	H D E G
Octadecabacter antarcticus 238	ZP_05063020	47	H D E G
Actinobacterium MWH-EgelM2-3.D6	ACN42852	46	H D E G
"*Candidatus* Aquiluna rubra"	ACN42850.1	45	H D E G
Fulvimarina pelagi HTCC2506	ZP_01440547	34	S N Q G
Geodermatophilus obscurus DSM 43160	ZP_03889903	32	A D E G
Kineococcus radiotolerans SRS30216	EAM73404	32	S D E G
Exiguobacterium sp. AT1b	YP_002885111	31	H D K G
Polaribacter sp. MED152	ZP_05108337	32	H D E G
Dokdonia donghaensis MED134	EAQ40507	31	H D E G
Polaribacter irgensii 23-P	ZP_01117885	31	H D E G
Flavobacteria bacterium BAL38	ZP_01734914	29	H D E G
Psychroflexus torquis ATCC 700755	ZP_01253360	29	H D E G

[a]Protein sequence accession numbers obtained from NCBI using BLASTP
[b]Four residues homologous to H62, D96 (proton acceptor), E108 (proton donor) and G178 of xanthorhodopsin. All sequences shown in the table except one contain glycine homologous to Gly178 of xanthorhodopsin, which makes these proteins potential candidates for binding of a carotenoid antenna with a keto ring similar to that in salinixanthin as shown for gloeobacter rhodopsin (Imasheva et al. 2009)

test that at least some of them are indeed capable of binding carotenoid antenna, we "implanted" (reconstituted) salinixanthin into gloeobacter rhodopsin expressed in *E. coli*, and showed that it transfers energy to the retinal chromophore (Imasheva et al. 2009). Replacing the Gly with a Trp abolishes carotenoid binding, thus confirming that accommodation of the ring is crucial for the antenna binding. Surprisingly, minor modification of salinixanthin that involves reduction of its 4-keto C=O group to C–OH also practically eliminates binding and energy transfer in both *Gloeobacter* rhodopsin (Balashov et al. 2010) and xanthorhodopsin (Imasheva et al. 2011). This suggests the importance of the 4-keto group in the carotenoid binding and energy transfer. This conclusion was supported by the experiments with β-carotene and echinenone (Balashov et al. 2010). The host organism, *Gloeobacter violaceus*, does not contain salinixanthin, but a simpler carotenoid, echinenone, also with a 4-keto ring but lacking glucoside and acyl tale, is present in addition to β-carotene and oscillol. We found that while β-carotene does not bind, its 4-keto derivative, echinenone, does, as follows from the characteristic changes in the absorption and CD spectra and fluorescence excitation spectrum which shows echinenone bands in addition to the retinal band (Balashov et al. 2010). This result again points to the 4-keto group as a key factor in the binding of carotenoids in these proteins. Reconstitution with echinenone is slower than with salinixanthin, apparently from lack of the 2′ hydroxy group and the acyl glycoside.

17.8 Conclusions

Despite the short history of xanthorhodopsin research, it has uncovered new features not seen before in light-driven retinal based pumps. This includes the existence of a light-harvesting carotenoid antenna, and a structure for the proton translocating pathways dramatically different from those in bacteriorhodopsin. Genetic homology with many members of xanthorhodopsin clade and proteorhodopsins suggest that these features might be common to other numerous eubacterial pumps and represent a considerable variation of a design of retinal-based proton pump as has been known from bacteriorhodopsin research.

Acknowledgments The authors would like to thank all collaborators involved in exploration of xanthorhodopsin, especially Prof. J Antón for bringing *S. ruber* to our attention and helpful advice, Profs. T Polívka and V Sundström and their colleagues for femtosecond experiments, Prof. H Luecke for refinement of xanthorhodopsin structure from X-ray diffraction data, Dr. B Schobert for crystallization of xanthorhodopsin, Dr. ES Imasheva for spectroscopic studies, purification, and help in preparation of the manuscript, JM Wang for isolation of *S. ruber* cell membranes and the late Dr. V Boichenko for action spectra in native cells. The research of xanthorhodopsin was supported in part by grants from the National Institutes of Health (GM29498), the Department of Energy (DEFG03-86ER13525) to JKL and the U.S. Army Research Office (W911NF-09-1-0243) to SPB.

References

Antón J, Oren A, Benlloch S, Rodríguez-Valera F, Amann R, Rosselló-Mora R (2002) *Salinibacter ruber* gen. nov., sp. nov., a novel, extremely halophilic member of the *Bacteria* from saltern crystallizer ponds. Int J Syst Evol Microbiol 52:485–491

Balashov SP (2000) Protonation reactions and their coupling in bacteriorhodopsin. Biochim Biophys Acta 1460:75–94

Balashov SP, Lanyi JK (2006) Bacteriorhodopsin. In: Rehn B (ed) Microbial biotechnology. Biological self-assembly systems and biopolymer-based nanostructures. Horizon Bioscience, Norfolk, pp 339–366

Balashov SP, Lanyi JK (2007) Xanthorhodopsin: proton pump with a carotenoid antenna. Cell Mol Life Sci 64:2323–2328

Balashov SP, Imasheva ES, Govindjee R, Ebrey TG (1996) Titration of aspartate-85 in bacteriorhodopsin: what it says about chromophore isomerization and proton release. Biophys J 70: 473–481

Balashov SP, Imasheva ES, Ebrey TG, Chen N, Menick DR, Crouch RK (1997) Glutamate-194 to cysteine mutation inhibits fast light-induced proton release in bacteriorhodopsin. Biochemistry 36:8671–8676

Balashov SP, Imasheva ES, Boichenko VA, Antón J, Wang JM, Lanyi JK (2005) Xanthorhodopsin: a proton pump with a light-harvesting carotenoid antenna. Science 309: 2061–2064

Balashov SP, Imasheva ES, Lanyi JK (2006) Induced chirality of the light-harvesting carotenoid salinixanthin and its interaction with the retinal of xanthorhodopsin. Biochemistry 45: 10998–11004

Balashov SP, Imasheva ES, Wang JM, Lanyi JK (2008) Excitation energy-transfer and the relative orientation of retinal and carotenoid in xanthorhodopsin. Biophys J 95:2402–2414

Balashov SP, Imasheva ES, Choi AR, Jung K-H, Liaaen-Jensen S, Lanyi JK (2010) Reconstitution of gloeobacter rhodopsin with echinenone: role of the 4-keto group. Biochemistry 49: 9792–9799

Béjà O, Aravind L, Koonin EV, Suzuki MT, Hadd A, Nguyen LP, Jovanovich SB, Gates CM, Feldman RA, Spudich JL, Spudich EN, DeLong EF (2000) Bacterial rhodopsin: evidence for a new type of phototrophy in the sea. Science 289:1902–1906

Béjà O, Spudich EN, Spudich JL, Leclerc M, DeLong EF (2001) Proteorhodopsin phototrophy in the ocean. Nature 411:786–789

Belrhali H, Nollert P, Royant A, Menzel C, Rosenbusch JP, Landau EM, Pebay-Peyroula E (1999) Protein, lipid and water organization in bacteriorhodopsin crystals: a molecular view of the purple membrane at 1.9 Å resolution. Structure 7:909–917

Bergo VB, Sineshchekov OA, Kralj JM, Partha R, Spudich EN, Rothschild KJ, Spudich JL (2009) His-75 in proteorhodopsin, a novel component in light-driven proton translocation by primary pumps. J Biol Chem 284:2836–2843

Birge RR (1990) Nature of the primary photochemical events in rhodopsin and bacteriorhodopsin. Biochim Biophys Acta 1016:293–327

Birge RR, Zhang C-F (1990) Two-proton double resonance spectroscopy of bacteriorhodopsin. Assignment of the electronic and dipolar properties of the low-lying $^1A_g^*$-like and $^1B_u^{*+}$-like π, π* states. J Chem Phys 92:7178–7195

Boichenko VA, Wang JM, Antón J, Lanyi JK, Balashov SP (2006) Functions of carotenoids in xanthorhodopsin and archaerhodopsin, from action spectra of photoinhibition of cell respiration. Biochim Biophys Acta 1757:1649–1656

Britton G (1995) UV/Visible Spectroscopy. In: Britton G, Liaaen-Jensen S, Pfander H (eds) Carotenoids, vol 1B, Spectroscopy. Birkhäuser Verlag, Basel, pp 13–62

Britton G, Liaaen-Jensen S, Pfander H (2004) Carotenoids handbook. Birkhauser Verlag, Basel, Boston, Berlin

Brown LS, Jung K-H (2006) Bacteriorhodopsin-like proteins of eubacteria and fungi: the extent of conservation of the haloarchaeal proton-pumping mechanism. Photochem Photobiol Sci 5: 538–546

Brown LS, Sasaki J, Kandori H, Maeda A, Needleman R, Lanyi JK (1995) Glutamic acid 204 is the terminal proton release group at the extracellular surface of bacteriorhodopsin. J Biol Chem 270:27122–27126

Buchecker R, Noack K (1995) Circular dichroism. In: Britton G, Liaaen-Jensen S, Pfander H (eds) Carotenoids, vol 1B, Spectroscopy. Birkhäuser Verlag, Basel, pp 63–116

Christensen RL, Kohler BE (1973) Low resolution optical spectroscopy of retinyl polyenes: low lying electronic levels and spectral broadness. Photochem Photobiol 1973:293–301

Cleland WW, Frey PA, Gerlt JA (1998) The low barrier hydrogen bond in enzymatic catalysis. J Biol Chem 273:25529–25532

Dioumaev AK, Wang JM, Bálint Z, Váró G, Lanyi JK (2003) Proton transport by proteorhodopsin requires that the retinal Schiff base counterion Asp-97 be anionic. Biochemistry 42:6582–6587

Dioumaev AK, Wang JM, Lanyi JK (2010) Low-temperature FTIR study of multiple K intermediates in the photocycles of bacteriorhodopsin and xanthorhodopsin. J Phys Chem B 114:2920–2931

Drachev LA, Frolov VN, Kaulen AD, Liberman EA, Ostroumov SA, Plakunova GV, Semenov AY, Skulachev VP (1976) Reconstitution of biological molecular generators of electric current. Bacteriorhodopsin. J Biol Chem 251:7059–7065

Frank HA, Cogdell RJ (1993) The photochemistry and function of carotenoids in photosynthesis. In: Young A, Britton G (eds) Carotenoids in photosynthesis. Chapman and Hall, London, pp 252–326

Fuhrman JA, Schwalbach MS, Stingl U (2008) Proteorhodopsins: an array of physiological roles? Nat Rev Microbiol 6:488–494

Fujimoto KJ, Hayashi S (2009) Electronic Coulombic coupling of excitation-energy transfer in xanthorhodopsin. J Am Chem Soc 131:14152–14153

Garczarek F, Brown LS, Lanyi JK, Gerwert K (2005) Proton binding within a membrane protein by a protonated water cluster. Proc Natl Acad Sci USA 102:3633–3638

Georgakopoulou S, Cogdell RJ, van Grondelle R, van Amerongen H (2003) Linear-dichroism measurements on the LH2 antenna complex of *Rhodopseudomonas acidophila* strain 10050 show that the transition dipole moment of the carotenoid rhodopin glucoside is not collinear with the long molecular axis. J Phys Chem B 107:655–658

Georgakopoulou S, van Grondelle R, van der Zwan G (2004) Circular dichroism of carotenoids in bacterial light-harvesting complexes: experiments and modeling. Biophys J 87:3010–3022

Giovannoni SJ, Hayakawa DH, Tripp HJ, Stingl U, Givan SA, Cho J-C, Oh H-M, Kitner JB, Vergin KL, Rappé MS (2008) The small genome of an abundant coastal ocean methylotroph. Environ Microbiol 10:1771–1782

Gomez-Consarnau L, Gonzalez JM, Coll-Llado M, Gourdon P, Pascher T, Neutze R, Pedros-Alio C, Pinhassi J (2007) Light stimulates growth of proteorhodopsin-containing marine Flavobacteria. Nature 445:210–213

Hirai T, Subramaniam S, Lanyi JK (2009) Structural snapshots of conformational changes in a seven-helix membrane protein: lessons from bacteriorhodopsin. Curr Opin Struct Biol 19: 433–439

Imasheva ES, Balashov SP, Wang JM, Dioumaev AK, Lanyi JK (2004) Selectivity of retinal photoisomerization in proteorhodopsin is controlled by aspartic acid 227. Biochemistry 43: 1648–1655

Imasheva ES, Shimono K, Balashov SP, Wang JM, Zadok U, Sheves M, Kamo N, Lanyi JK (2005) Formation of a long-lived photoproduct with a deprotonated Schiff base in proteorhodopsin, and its enhancement by mutation of Asp227. Biochemistry 44:10828–10838

Imasheva ES, Balashov SP, Wang JM, Lanyi JK (2006) pH-dependent transitions in xanthorhodopsin. Photochem Photobiol 82:1406–1413

Imasheva ES, Balashov SP, Wang JM, Smolensky E, Sheves M, Lanyi JK (2008) Chromophore interaction in xanthorhodopsin – retinal dependence of salinixanthin binding. Photochem Photobiol 84:977–984

Imasheva ES, Balashov SP, Choi AR, Jung K-H, Lanyi JK (2009) Reconstitution of *Gloeobacter violaceus* rhodopsin with a light-harvesting carotenoid antenna. Biochemistry 48:10948–10955

Imasheva ES, Balashov SP, Wang JM, Lanyi JK (2011) Removal and reconstitution of the carotenoid antenna of xanthorhodopsin. J Membr Biol 239:95–104

Ke B, Imsgard F, Kjosen H, Liaaen-Jensen S (1970) Electronic spectra of carotenoids at 77°K. Biochim Biophys Acta 210:139–152

Kralj JM, Bergo VB, Amsden JJ, Spudich EN, Spudich JL, Rothschild KJ (2008) Protonation state of Glu142 differs in the green- and blue-absorbing variants of proteorhodopsin. Biochemistry 47:3447–3453

Krueger BP, Scholes GD, Fleming GR (1998) Calculation of couplings and energy-transfer pathways between the pigments of LH2 by the ab initio transition density cube method. J Phys Chem B 102:5378–5386

Lanyi JK, Balashov SP (2008) Xanthorhodopsin: a bacteriorhodopsin-like proton pump with a carotenoid antenna. Biochim Biophys Acta 1777:684–688

Litvin FF, Balashov SP, Sineshchekov VA (1975) The investigation of the primary photochemical conversions of bacteriorhodopsin in purple membranes and cells of *Halobacterium halobium* by the low temperature spectrophotometry method. Bioorgan Khim 1:1767–1777

Litvin FF, Boichenko VA, Balashov SP, Dubrovskii VT (1977) Photoinduced inhibition and stimulation of respiration in cells of *Halobacterium halobium*: kinetics, action spectra, relation to photoinduction of Δ_pH. Biofizika 22:1062–1071

Lozier RH, Bogomolni RA, Stoeckenius W (1975) Bacteriorhodopsin: a light-driven proton pump in *Halobacterium halobium*. Biophys J 15:955–963

Luecke H, Richter H-T, Lanyi JK (1998) Proton transfer pathways in bacteriorhodopsin at 2.3 Angstrom resolution. Science 280:1934–1937

Luecke H, Schobert B, Richter H-T, Cartailler J-P, Lanyi JK (1999a) Structural changes in bacteriorhodopsin during ion transport at 2 Angstrom resolution. Science 286:255–260

Luecke H, Schobert B, Richter H-T, Cartailler J-P, Lanyi JK (1999b) Structure of bacteriorhodopsin at 1.55 Å resolution. J Mol Biol 291:899–911

Luecke H, Schobert B, Stagno J, Imasheva ES, Wang JM, Balashov SP, Lanyi JK (2008) Crystallographic structure of xanthorhodopsin, the light-driven proton pump with a dual chromophore. Proc Natl Acad Sci USA 105:16561–16565

Lutnaes BF, Oren A, Liaaen-Jensen S (2002) New C_{40}-carotenoid acyl glycoside as principal carotenoid in *Salinibacter ruber*, an extremely halophilic eubacterium. J Nat Prod 65: 1340–1343

Maeda A, Verhoeven MA, Lugtenburg J, Gennis RB, Balashov SP, Ebrey TG (2004) Water rearrangement around the Schiff base in the late K (KL) intermediate of the bacteriorhodopsin photocycle. J Phys Chem B 108:1096–1101

McDermott G, Prince SM, Freer AA, Hawthornthwaite-Lawless AM, Papiz MZ, Cogdell RJ, Isaacs NW (1995) Crystal structure of an integral membrane light-harvesting complex from photosynthetic bacteria. Nature 374:517–521

Michel H, Oesterhelt D (1980) Light-induced changes of the pH gradient and the membrane potential in *Halobacterium halobium*. FEBS Lett 65:175–178

Ming M, Lu M, Balashov SP, Ebrey TG, Li QG, Ding JD (2006) pH dependence of light-driven proton pumping by an archaerhodopsin from Tibet: comparison with bacteriorhodopsin. Biophys J 90:3322–3332

Miranda MRM, Choi AR, Shi L, Bezerra AG, Jung K-H, Brown LS (2009) The photocycle and proton translocation pathway in a cyanobacterial ion-pumping rhodopsin. Biophys J 96: 1471–1481

Mongodin EF, Nelson KE, Daugherty S, DeBoy RT, Wister J, Khouri H, Weidman J, Walsh DA, Papke RT, Sanchez Perez G, Sharma AK, Nesbø CL, MacLeod D, Bapteste E, Doolittle WF,

Charlebois RL, Legault B, Rodriguez-Valera F (2005) The genome of *Salinibacter ruber*: convergence and gene exchange among hyperhalophilic bacteria and archaea. Proc Natl Acad Sci USA 102:18147–18152

Mukohata Y, Ihara K, Uegaki K, Miyashita Y, Sugiyama Y (1991) Australian *Halobacteria* and their retinal-protein ion pumps. Photochem Photobiol 54:1039–1045

Oesterhelt D (1998) The structure and mechanism of the family of retinal proteins from halophilic archaea. Curr Opin Cell Biol 8:489–500

Oesterhelt D, Stoeckenius W (1973) Functions of a new photoreceptor membrane. Proc Natl Acad Sci USA 70:2853–2857

Peña A, Valens M, Santos F, Buczolits S, Antón J, Kämpfer P, Busse H-J, Amann R, Rosselló-Mora R (2005) Intraspecific comparative analysis of the species *Salinibacter ruber*. Extremophiles 9:151–161

Petrovskaya LE, Lukashev EP, Chupin VV, Sychev SV, Lyukmanova EN, Kryukova EA, Ziganshin RH, Spirina EV, Rivkina EM, Khatypov RA, Erokhina LG, Gilichinsky DA, Shuvalov VA, Kirpichnikov MP (2010) Predicted bacteriorhodopsin from *Exiguobacterium sibiricum* is a functional proton pump. FEBS Lett 584:4193–4196

Polívka T, Frank HA (2010) Molecular factors controlling photosynthetic light harvesting by carotenoids. Acc Chem Res 43:1125–1134

Polívka T, Sundström V (2004) Ultrafast dynamics of carotenoid excited states – from solution to natural and artificial systems. Chem Rev 104:2021–2071

Polívka T, Balashov SP, Chábera P, Imasheva ES, Yartsev A, Sundström V, Lanyi JK (2009) Femtosecond carotenoid to retinal energy transfer in xanthorhodopsin. Biophys J 96: 2268–2277

Rothschild KJ, Roepe P, Gillespie J (1985) Fourier transform infrared spectroscopic evidence for the existence of two conformations of the bacteriorhodopsin primary photoproduct at low temperature. Biochim Biophys Acta 808:140–148

Schobert B, Lanyi JK (1982) Halorhodopsin is a light-driven chloride pump. J Biol Chem 257: 10306–10313

Schobert B, Brown LS, Lanyi JK (2003) Crystallographic structures of the M and N intermediates of bacteriorhodopsin: assembly of a hydrogen-bonded chain of water molecules between Asp-96 and the retinal Schiff base. J Mol Biol 330:553–570

Scholes GD (2003) Long-range resonance energy transfer in molecular systems. Annu Rev Phys Chem 54:57–87

Sharma AK, Zhaxybayeva O, Papke RT, Doolitle WF (2008) Actinorhodopsins: proteorhodopsin-like gene sequences found predominantly in non-marine environments. Environ Microbiol 10:1039–1056

Sharma AK, Sommerfeld K, Bullerjahn GS, Matteson AR, Wilhelm SW, Jezbera J, Brandt U, Doolittle WF, Hahn MW (2009) Actinorhodopsin genes discovered in diverse freshwater habitats and among cultivated freshwater *Actinobacteria*. ISME J 3:726–737

Smolensky E, Sheves M (2009) Retinal-salinixanthin interactions in xanthorodopsin: a circular dichroism (CD) spectroscopy study with artificial pigments. Biochemistry 48:8179–8188

Spudich JL, Jung K-H (2005) Microbial rhodopsins: phylogenetic and functional diversity. In: Briggs WR, Spudich JL (eds) Handbook of photosensory receptors. Wiley-VCH, Darmstadt, pp 1–23

Venter JC, Remington K, Heidelberg JF, Halpern AL, Rusch D, Eisen JA, Wu D, Paulsen I, Nelson KE, Nelson W, Fouts DE, Levy S, Knap AH, Lomas MW, Nealson K, White O, Peterson J, Hoffman J, Parsons R, Baden-Tillson H, Pfannkoch C, Rogers Y-H, Smith HO (2004) Environmental genome shotgun sequencing of the Sargasso Sea. Science 304:66–74

Yoshimura K, Kouyama T (2008) Structural role of bacterioruberin in the trimeric structure of archaerhodopsin-2. J Mol Biol 375:1267–1281

Zhu JY, Gdor I, Smolensky E, Friedman N, Sheves M, Ruhman S (2010) Photoselective ultrafast investigation of xanthorhodopsnin and its carotenoid antenna salinixanthin. J Phys Chem B 114:3038–3045

Chapter 18
Potential Enhancement of Biofuel Production Through Enzymatic Biomass Degradation Activity and Biodiesel Production by Halophilic Microorganisms

Matthew B. Begemann, Melanie R. Mormile, Varun G. Paul, and Daniel J. Vidt

18.1 Introduction

Biofuels hold great promise as a replacement for some of our demand for fossil fuel. However, a number of environmental and economical costs have to be reduced before biofuels can be a suitable replacement for fossil fuels. These costs include the use of land, water and energy. Furthermore, these costs place the production of biofuels in direct competition with food production (Pimentel et al. 2009). To overcome some of these costs, solutions such as the efficient use of non-food biomass have to be developed and conservation of water via recycling.

Non-food plant biomass is an abundant, renewable source of stored energy. This material holds great potential for providing energy in the form of alcohols, hydrogen, and methane production, as well as electricity generation from microbial fuel cells. Typical forms of biomass include switch-grass, corn stover and wood chips. Biomass exists as cellulose and hemicellulose, encased by lignin. The purpose of the cellulose, hemicellulose, and lignin matrix is to provide structural support for the plant. Thus, this material is highly resistant to biodegradation. This also makes harsh pretreatments, high temperature and extreme pH conditions, necessary to

M.B. Begemann
Microbiology Doctoral Training Program, University of Wisconsin, Madison, WI 53706, USA

M.R. Mormile (✉)
Department of Biological Sciences, Missouri University of Science and Technology, Rolla, MO 65409, USA
e-mail: mmormile@mst.edu

V.G. Paul
Geology and Geophysics Program, Missouri University of Science and Technology, Rolla, MO 65409, USA

D.J. Vidt
Mining and Nuclear Engineering, Missouri University of Science and Technology, Rolla, MO 65409, USA

A. Ventosa et al. (eds.), *Halophiles and Hypersaline Environments*,
DOI 10.1007/978-3-642-20198-1_18,

prepare biomass for subsequent fermentation reactions (Mosier et al. 2005). Alkali pretreatment, such as lime, has been used to treat wheat straw, poplar wood, switch-grass, and cornstover. Lime can be substituted by alkaline salts during alkaline pretreatment. This can result in pH and salt concentrations similar to those found in alkaline saline lakes. Thus, bacteria that thrive under these conditions can possibly be used to further breakdown biomass after alkaline pretreatment and subsequent biofuel production.

The use of halophilic microorganisms to reduce demand for freshwater usage is not new (Woolard and Irvine 1995). To elevate the water costs of biofuel/biodiesel production, water recycle will have to be utilized. However, as water is recycled, the salts that are present will accumulate and become inhibitory to freshwater biodiesel-producing algae. Halotolerant/halophilic algae present a solution to this impediment.

18.1.1 Cellulose

Cellulose generally is a major component of biomass and exists in forms that are exceptionally stable in terms of both conformational structure and chemical bonds. This makes the molecules relatively recalcitrant due to limited solubility and high heterogeneity (Schwarz 2001). Additional hindrances to biodegradation result from the sheer size of cellulose molecules. The structure and size of cellulose preclude uptake by cells and enzymes must be secreted into the medium or bound to the outside of the cell (Beguin and Aubert 1994). However, the tremendous availability of cellulosic biomass, estimated at 30 Gt per year in global terrestrial production, has resulted in the development of very diverse, multi-faceted approaches of cellulose biodegradation in the biosphere (Cox et al. 2000). Despite the awareness of the array of cellulose biodegradative mechanisms, there have been very few biotechnologies developed from cellulolytic organisms. It is estimated that of all the microorganisms within the biosphere, only 1–2% have been transferred into an industrial or commercial applications (Gomes and Steiner 2004) and even this percentage appears to be a gross over-exageration. Given that context, there are obviously vast, untapped reserves of cellulolytic biotechnology still awaiting discovery.

18.1.2 Hemicellulose

Hemicelluloses are a broad group of heterogeneous branched and linear chain polysaccharides that account for 20–40% of lignocellulose. Within the lignocellulose complex, hemicellulose is covalently bound to lignin sheaths and interacts with cellulose through hydrogen bonds (Biely et al. 1985; Joseleau et al. 1992). These interactions aid in structural integrity and protect the cellulose from degradation by cellulases (Uffen 1997). Unlike cellulose and lignin, the composition of hemicellulose varies greatly between plant species in backbone structure, branching, and modifications. Hemicelluloses can be categorized based on the carbohydrate polymer composition. These include xylan (D-xylose), xyloglucan (D-xylose and D-glucose), glucomannan (D-glucose and D-mannose), galactoglucomannan (D-galactose, D-glucose, and D-mannose), and arabinogalactan (D-galactose and L-arabinose). A more comprehensive review of hemicellulose structure and biological function can be found in Puls and Schuseil (1993).

Due to the diversity of these polysaccharides, microorganisms involved in their degradation require the use of multiple enzymes with varying specificity and function. A genome analysis of ten representative hemicellulose-degrading bacteria revealed that each bacterium has between 4 and 14 annotated genes encoding hemicellulases (Shallom and Shoham 2003). The inherent nature of hemicellulose provides additional challenges including high molecular weight, insolubility, and rigid structure. In order to cope with these challenges, multiple hemicellulases are secreted to convert large polymers into smaller oligosaccharides, disaccharides, and monosaccharides, which can then be internalized and further metabolized (Wong et al. 1988). Total conversion of hemicellulose requires glycoside hydrolases to break the sugar backbone and branched sugar residues as well as esterases to remove acetyl and ferulic acid modifying groups (Shallom and Shoham 2003; Saha 2003).

18.1.3 Lignin

Lignin, a heterogeneous polymer made up of phenylpropanoid interunits linked by covalent bonds, is the second most abundant raw material on earth. It is a major component in the cell wall structure of many plants, providing strength and rigidity to the plant structure and helps in water transport. There are several isomers of lignin most of which are structurally complex and very resistant to microbial degradation (Li et al. 2009). Enzymes capable of degrading lignin under aerobic conditions are grouped into two main categories, peroxidases and phenol oxidases. Phenol oxidases can be further divided to include two groups of enzymes, laccases and polyphenol oxidases. On the other hand, anaerobic organisms primarily use enzymes involved in the degradation of phenols such as phenylphosphate synthase and phenylphosphate carboxylase. Many studies have focused on microbial breakdown of lignin and few have considered lignin degradation under extreme conditions of pH, temperature or high salt concentrations.

18.2 Biodegradation of Cellulose

The degradation of cellulose is generally accomplished by an inducible system of cellulolytic enzymes and proteins, working synergistically, comprising a distinct cellulase system (Wood 1975). Commonly, the term cellulase is used to describe an entire system that attacks the cellulose molecule in an extremely coordinated effort to synergistically hydrolyze cellulose within the environment and not the organism. This is important because cellulolytic microorganisms are also capable of degrading hemicellulose and coproduces xylanases along with cellulolytic enzymes (Lynd et al. 2002). The specific activity of cellulases is the hydrolysis of the β-1,4-glucosidic bonds between the glucosyl residues in cellulose (Beguin and Aubert 1994).

Cellulases can function in three different ways. Exoglucanases, cellodextrinases (1,4-β-D-glucanhydrolases) and cellobiohydrolases (1,4-β-D-glucan cellobiohydrolase) hydrolyze the glycosyl bonds at either end of the cellulose polymer chain while endoglucanases, such as 1,4-β-glucan-4-glucanhydrolases cleave the cellulose chain internally. The glycosyl bond of cellobiose can be cleaved by 1,4-β-D-glucan cellobiohydrolases and β-glucoside glycohydrolases (Voget et al. 2006). Along with the catalytic domains of the general cellulase complex, a fourth, non-catalytic component is central to cellulose degradation. In general, a carbohydrate binding module must be present in the cellulase complex to attach and position the enzymes onto the cellulose and facilitate the removal of single molecules resulting from enzymatic lysis (Schwarz 2001; Lynd et al. 2002).

18.2.1 Biodegradation of Cellulose by Halophilic Bacteria

Formal cellulase research began with the United States Army where investigations into cellulose degrading organisms were undertaken in response to alarming rates of "rot" noted in the natural materials (mostly cotton) used by combat units in the South Pacific campaigns of World War II (Reese 1976). By the time of the Vietnam War, the first research results quantifying three, functionally distinct cellulase components from crude filtrates of *Trichoderma viride* were published (Li et al. 1965). Since then, prospecting for relevant biotechnical resources has become more focused on extremophiles since many of the cellulases with industrial relevant characteristic have already been obtained from them (Voget et al. 2006). However, little work has been done thus far with halophiles specifically for cellulase. The current survey of biotechnical and enzymatic applications involving halophiles denotes interesting applications from the development of soy and fish sauce to holographic computer storage material to enchasing crude oil recovery (Oren 2002; Sanchez-Porro et al. 2003; Pikuta et al. 2007). However, there is no mention in the current literature of any cellulase biotechnologies developed from halophiles. This isn't surprising as limited research has been done to study the biodegradation of cellulose by halophilic bacteria even though other extremophilic organisms, specifically alkaliphiles and thermophiles, have long been employed for their cellulosic properties (Bhat and Bhat 1997; Lynd et al. 2002; Fujinami and Fujisawa 2010). However mixed cellulolytic cultures and a pure culture of *Haloarcula hispanica* isolated from the Waste Isolation Pilot Plant were shown to attach to and produce oxalacetic and pyruvic acids from cellulose (Vreeland et al. 1998).

Recently, a carboxymethyl cellulose-degrading *Marinimicrobium* was isolated from the Great Salt Lake, Utah (Møller et al. 2010). Furthermore, the reported incidence of enhanced enzymatic stability and activity with the addition of salt to thermophilic cellulases has been known for some time (Bronnenmeier et al. 1995). For reasons directly related to the structure and composition their proteins, halophilic enzymes systems have thus far demonstrated enhanced utility as they are not only stable in high salt environments, but the enzymes are also found to be thermotolerant and alkalphilic and thus can be considered "polyextremophilic" (Setati 2010).

18.2.2 Uniqueness of Halophilic Cellulosic Enzymes

Enzymes produced by halophilic microorganisms possess an excess of acidic amino acids providing extensive negative charges on the enzyme surfaces (Danson and Hough 1997). This unique feature of the halophilic proteome allows these proteins to remain flexible and resist aggregation that is common to non-halophilic proteins exposed to high salt concentrations (Mevarech et al. 2000). In general, halophilic enzymes actually require high salt concentrations (mainly Cl^{-}) to stimulate transcription and translation as well as transporter activity (Averhoff and Müller 2010). Furthermore, the solubility of halophilic enzymes at high salt concentrations has allowed for unique applications whereby aqueous/organic and non-aqueous media can be employed without substantially affecting the activity or the solubility of the enzyme (van den Burg 2003; Fukushima et al. 2005; Karbalaei-Heidari et al. 2007). While there has been a relatively long history of research and development of cellulose degrading enzymes, those specifically sourced from halophilic organisms have only come to the fore relatively recently (Huang et al. 2010). The most exhaustive review of biotechnological applications of halophilic microorganisms was undertaken by Oren (2002). Within this work, no mention was given to any halophilic cellulolytic organisms thus highlighting the embryonic stage of cellulase research and development within this particular group of extremophiles. However, based on the many instances of other enzymatic types such as DNAases, lipases, amylases, gelatinases, and proteases already employed commercially from halophilic organisms, there is great anticipation that cellulase derived from halophiles will offer up disruptive technologies in the production of cellulosic based biofuels and chemicals (van den Burg 2003; Setati 2010).

Screening for unique cellulases from extremophiles in general and halophiles specifically has been hampered by the difficulty in culturing the organisms within the laboratory in sufficient quality and quantity to perform adequate analysis. An extreme example of this was documented in the 2 years it took for investigators to culture samples of sufficient quantity for analysis (Vreeland et al. 1998). Furthermore, standard chromatographic techniques are rendered ineffective as the halophile enzymes generally exhibit high solubility in high concentrations of NaCl (Eisenberg et al. 1992). Table 18.1 illustrates the specifically identified species of

Table 18.1 Listing of halotolerant and halophilic cellulases detected from environmental DNA libraries

Name	Metagenome-derived[a,b]	*Salinivibrio* sp.[c]	*Halomonas* sp.[d]	*Vibrio* sp.[e]
Species I.D.	n/a	NTU-5	S66-4	G21
Enzyme name	Cel5A		Cel8H	Cel5A
Enzyme size (kDa)	42.1	29	36	47.8
Enzyme stability				
Temperature range (°C)	30–50	10–40	20–50	
pH range	5–9.7	6.5–8.5	6.4–8.4	5.5–10.5
Salinity range (M NaCl)	3[f]	0–4.3	0.9–3.4	0–4
Enzyme activity				
Optimum temperature (°C)	45	35	45	50[g] (55[h])
Optimum pH	6.5	7.5	5	6.5–7.5
Optimum salinity (M NaCl)	n/a	0.9	1	1

[a]Voget et al. (2006)
[b]The cellulase was metagenome-derived; has 86% similarity and 77% to *Cellvibrio mixtus*
[c]Wang et al. (2009)
[d]Huang et al. (2010)
[e]Gao et al. (2010)
[f]Halotolerance was tested by assaying residual activity towards carboxymethylcellulose after incubation in 3 M NaCl. No optimum enzymatic activity in a range of salinities was conducted was determined
[g]Maximum activity at 0 M NaCl
[h]Maximum activity at 0.4 M NaCl

halophilic bacteria strains that have undergone complete experimental assays for cellulase activity.

18.2.3 Specific Halophilic and Halotolerant Cellulases

Halocella cellulolytica is recognized as the first true halophilic cellulolytic bacterium identified within the literature (Bolobova et al. 1992). However, the advent of genomic DNA libraries of cellulases from extremophilic environments has accelerated the discovery of halophilic cellulases. Since the first metagenome-derived halotolerant cellulase was described, there have been three other halophilic cellulases defined (Rees et al. 2003). These techniques rely on recombinant expression of isolated clones to more easily produce the cellulase enzymes rather than growing up whole cultures and isolating the enzymes from them (Rees et al. 2003). Endoglucanases are often selectively screened (Voget et al. 2006). It should be noted that the use of carboxymethylcellulose (CMC) is not ideal for enzymatic studies of cellulase production as many organisms can hydrolyze CMC but cannot degrade insoluble cellulose (Lynd et al. 2002). Such false positives are thought to be an important issue since extremophiles tend to utilize of a very diverse range of carbohydrates due to the small amounts of cellulose and few competing species in

Table 18.2 Surveys of halophilic cellulase activities in extreme environments

Location	Number of species	Cellulase assay method	References
The Waste Isolation Pilot Plant (WIPP), USA	121	Coughlan and Mayer (1992)	Vreeland et al. (1998)
Şereflikoçhisar Salt Lake, Turkey	14	n/a[a]	Birbir and Sesal (2003)
Tuzkoy Salt Mine, Turkey	10	Prescott et al. (1993)	Birbir et al. (2004)
Lonar Lake, India	6	Incubation with CMC	Kanekar et al. (2008)
Sua pan evaporator ponds, Botswana	4[b]	Bailey et al. (1992)	Govender et al. (2009)
Howz Soltan Lake, Iran	68	Zhou et al. (2004)	Rohban et al. (2009)

[a]Method employed reported only as "standard" in reference (Birbir and Sesal (2003))
[b]Estimated from xylanase assay

their native habitat (Lynd et al. 2002). Table 18.1 presents the known halophilic endoglucanases so far presented in the literature.

Reports from surveys of hydrolytic enzymes from extreme habitats like soda lakes, salterns, and mines have revealed that there may be many more species of halophilic organisms awaiting formal review of their cellulolytic enzymes. Table 18.2 provides reports of cellulase activities discovered from surveys of hypersaline environments. Researchers using basic assays (also listed in Table 18.2) as a proxy for cellulase activity have reported impressive numbers of halophiles that express cellulase activity. It is apparent that cellulase activities occur in extreme natural environments and halophilic organisms possess cellulolytic systems. However, the reports are fairly shallow in terms of useful biotechnical information. As an example, the report of halophilic bacterial communities in a Turkish salt lake reported that there were 14 strains collected that demonstrated cellulase activity (Birbir and Sesal 2003). However, the report lacked any specific information of the biochemical methods used to deduce the cellulase activities. While such surveys could be construed to identify high incidence of halophilic cellulase, problems with over generalized assays and techniques may be inflating the reported numbers. Conversely, the identification of so many environmental isolates producing cellulase indicates that there are opportunities to discover biotechnologies that have not been demonstrated by culture collection strains (Sanchez-Porro et al. 2003).

18.3 Degradation of Hemicellulose by Halophilic Bacteria

The diversity and mechanisms of hemicellulases has been studied extensively in mesophilic bacteria and fungi as well as certain categories of extremophilic bacteria (Gilbert and Hazlewood 1993; Shallom and Shoham 2003; Collins et al. 2005). Recent interest in halophilic bacteria has resulted in the description of novel hemicellulose degrading halophilic bacteria and archaea as well as the characterization of purified halophilic and halotolerant enzymes.

18.3.1 Xylan Degradation

Xylan is the most abundant hemicellulose component in land plants (Joseleau et al. 1992). Xylan is a heterogeneous polysaccharide composed primarily of D-xylose linked by β-1,4 glycosidic bonds. These xylopyranoside units are often substituted by a variety of acetyl, arabinosyl, and glucuronosyl groups. The composition of xylan varies greatly between plant species with regard to both backbone composition and side group residues. The degradation of xylan requires endo-β-xylanases (EC 3.2.1.8) to fragment the chain, β-xylosidases (EC 3.2.1.37) to convert short xyloogliomers into monomeric units, and auxiliary enzymes to remove side group residues (Kulkarni et al. 1999).

Within the last decade, several species of halophilic and halotolerant microorganisms have been identified with the ability to degrade xylan and other polysaccharides. These include bacteria of the genera *Gracilibacillus*, *Chromohalobacter*, *Bacillus*, *Halomonas*, and *Marinimicrobium*, as well as the archaeon *Halorhabdus* (Wainø and Ingvorsen 2003; Wejse et al. 2003; Prakash et al. 2009; Giridhar and Chandra 2010; Møller et al. 2010; Wang et al. 2010a; Yang et al. 2010). Consistent characteristics of these organisms include the ability to grow in broad ranges of NaCl as well as moderate to high alkalinity.

In addition to identifying new halophilic species capable of xylan degradation, several xylanase and xylosidase enzymes have been characterized. The size of xylanases isolated from halophilic bacteria ranged from 15 kDa in the case of *Chromohalobacter* sp. TPSV101 to 62 kDa in the case of strain CL8 (Wejse et al. 2003; Prakash et al. 2009). Two xylanases of 45 and 67 kDa were identified in the archaeon *Halorhabdus utahensis* (Wainø and Ingvorsen 2003). As observed in the archaeon, several xylanases were identified in many of the bacterial isolates. The presence of several xylanases is consistent with findings in mesophilic and as well as extremophilic bacteria (Wong et al. 1988; Collins et al. 2005). These xylanases and xylosidases show activity under a wide range of physiological conditions including temperature, NaCl concentration, and pH. Activity over a wide range of NaCl concentrations in exemplified in the xylanase isolated from the moderately halophilic bacterium *Gracilibacillus* sp. TSCPVG, that demonstrated sustainable activity over 0–30% NaCl (Giridhar and Chandra 2010). These enzymatic properties would be essential for extracellular activity in the extreme environments from which these organisms were isolated.

In addition to xylanases, xylosidase enzymes were either characterized or their presence inferred from hydrolysate composition. β-Xylosidase was identified in both the bacterium *Gracilibacillus* sp. TSCPVG and archaeon *Halorhabdus utahensis* (Wainø and Ingvorsen 2003; Giridhar and Chandra 2010). Due to the complete degradation of xylan to D-xylose in other cultures, one can infer the presence of a xylosidase-like enzyme in other species. While none have been specifically identified, other enzymes may exist in these halophilic microorganisms that are involved in breaking down side groups and modifications. Other potential enzymes may include α-glucuronidase (EC 3.2.1.139), acetyl xylan esterase (EC 3.1.1.72), and ferulic and *p*-cumaric acid esterases (EC 3.1.1.73).

18.3.2 Mannan Degradation

Mannan is a polysaccharide composed of D-mannose monomers linked by β-1,4 glycosidic bonds. The mannan homopolymer is a major component of cell walls in certain species of algae and found in some seeds (Frei and Preston 1968; Whitney et al. 1998). Frequent substitutions in mannan by D-glucose results in the heteropolymer glucomannan, a component of hemicellulose, particularly in soft trees (Whitney et al. 1998). Mannan and glucomannan can be degraded by the enzymes β-1,4 mannanase (EC 3.2.1.78), β-1,4 mannosidase (EC 3.2.1.25), and β-glucosidase (EC 3.2.1.21). As with xylanases, mannanases are usually extracellular enzymes that first break the polymer into smaller ogliomers, followed by hydrolysis to monomeric sugars (Dhawan and Kaur 2007).

While not as thoroughly investigated as xylan degraders, several species of halophilic bacteria have been isolated with the ability to metabolize mannan heteropolymers. These include organisms of the genus *Marinimicrobium*, *Pantoea*, as well an organism designated as strain NN (DSM 11805) (Wainø and Ingvorsen 1999; Møller et al. 2010; Wang et al. 2010b). *Pantoea agglomerans* was shown to degrade both glucomannan and galactomannan substrates, while *Marinimicrobium haloxylanilyticum* and strain NN were shown to only metabolize glucomannan and galactomannan, respectively. As with the xylanases previously described, these enzymes were shown to function over a broad range of NaCl, pH and temperature ranges. As an insight into the physiology of these metabolisms, it was demonstrated that the majority of mannanase and especially mannosidase activity was localized to the cell wall and membrane of strain NN (Wainø and Ingvorsen 1999). As more mannan degrading enzymes are characterized, their ecological role will become more apparent.

18.4 Lignin Degradation by Halophilic and Halotolerant Microorganisms

The abundance of plants and therefore lignin in aquatic systems such as in coastal regions and salt marshes has indicated that the majority of lignin in these environments is degraded by microbial activity (Benner et al. 1986). Since very little is known about the lignin biogeochemical cycling in such ecosystems, the identification and isolation of microorganisms from high salt environments would be of ecological importance (González et al. 1997a). Fungi and bacteria are the two main groups of microorganisms involved in ligninolytic activity and both can act either individually or together to cause complete degradation of lignin. From an industrial standpoint, lignin and lignin-based derivatives are major components in the list of waste materials generated by paper and pulp industries, tanneries, molasses-based distilleries and textile mills, and can be a serious contamination problem due to their low biodegradability and high color (Raghukumar et al. 2008). Ligninolytic enzymes have been shown to be among the best decolorizing agents for treating these wastes. The waste streams in many of these industries generally have a high salinity and hence using halotolerant or halophilic microorganisms and/or enzymes would be a favorable choice for treatment. Also, in paper industries that recycle water containing high concentration of salts like NaCl such as the pulp and paper industry, it would be practical to use enzymes that are more salt tolerant so as to have effective biopulping (converting wood chips into pulp) and biobleaching (decolorizing by using enzymes) processes (Li et al. 2002).

18.4.1 Lignin Degradation by Fungi

Of the two microbial groups capable of ligninolytic activity, fungi are known to possess powerful and extensive mechanisms of lignin degradation. By far, the most commonly studied and successful lignin degradation activity has been demonstrated in many species of white rot basidiomycetous fungi under aerobic conditions. Many marine fungal species such as *Digitatispora marina*, *Halocyphina villosa*, and *Nia vibrissa* grow on decaying lignocellulosic substrates. A particular marine, hypersaline-tolerant isolate, *Phlebia* sp. strain MG-60, isolated from mangrove stands in Okinawa, Japan, has been shown to produce many enzymes for lignin degradation. One of its enzymes, manganese peroxidase, was specifically screened for its expression under varying conditions of salt and nitrogen. Its expression is regulated by the presence of Mn^{2+} and a considerable increase in the level of two gene transcripts was detected when more salt was added (Kamei et al. 2008). Any excess addition of Mn^{2+} or NH_4^+ inhibited the enzyme production but the addition of salt reversed the inhibition partially or completely (Li et al. 2003). Other activity included laccase and manganese peroxidase that were capable of effectively decolorizing the dye Poly R-478, an indicator of delignification by fungi. However, an increase in salt concentration decreased this decolorization property. Furthermore MG-60 was able to whiten unbleached pulp more efficiently than the commonly studied white rot fungus for lignin degradation, *Phanerochaete chrysosporium*, under increased hypersaline conditions (Bucher et al. 2004). The activity of laccases, in a similar study on marine fungi, seemed to be enhanced by the addition of copper or a combination of copper and guaiacol in a nitrogen rich medium.

The lignin degrading mechanism can be applied towards bioremediation efforts as shown by the halotolerant fungus, *Pestalotiopsis* sp. SN-3. An efficient lignin degrader, SN-3 produces laccase and possessed a significant potential for the degradation of toxic substances in marine environments. This laccase was extremely tolerant to high salt conditions and also had high enzymatic activity. Two novel thermo-halotolerant laccase isoforms were produced by *Pycnoporus sanguineus*. The purified laccases showed high thermal stability and hence can have a longer shelf-life (González et al. 2008). An advantage of using laccase over other enzymes is that laccase does not require components such as manganese or hydrogen peroxide and hence preferred in environmental applications. Several researchers are actively exploring fungal species that can grow under hypersaline conditions while simultaneously breaking down lignin and dye products. For example, a fungal species of mitosporic taxa, *Ulomyces chlamydosporum* and two from the ascomycetes taxa, *Emericella nidulans* and *Aspergillus phoenicis*, isolated from the Dead Sea, showed very high potential as decolorizing agents (Molitoris et al. 2000). Certain marine species such as *Ascocratera manglicola, Astrosphaeriella striatispora, Cryptovalsa halosarceicola, Linocarpon bipolaris* and *Rhizophila marina*, have also shown significant amounts of lignin solubilization.

18.4.2 Bacterial Lignin Degradation

Bacterial ligninolytic activity has not received much focus until recent times. In environments such as anaerobic sediments, waterlogged wood, coastal seawater and sediments and salt marshes, bacteria dominate fungi in terms of biodegrading lignin and its polymers (González et al. 1997b). A marine, aerobic species, *Sagittula stellata* was identified and shown to be capable of breaking down lignin (González et al. 1997a). In addition to looking for bacterial species that possess ligninolytic activity, much effort has been undertaken to purify and characterize specific enzymes as in the case of a halophilic archaeon, *Haloferax volcanii* (Uthandi et al. 2009). Laccase (LccA), purified from this archaea, was able to oxidize a variety of organic substrates such as bilirubin, syringaldazine, etc. and was tolerant of high salt (0.1–1.4 M NaCl), mixed organosolvents and high temperature (55°C). Laccase has also been isolated and purified from halotolerant bacterial species such as *Streptomyces psammoticus* (Niladevi et al. 2008).

Cloning and expression of ligninolytic genes are the next step in the further study of lignin degradation by halotolerant/halophilic microorganisms. For example, a marine *Bacillus* sp., was successfully cloned and expressed with genes encoding for protocatechuate 3,4-dioxygenase. The enzymes isolated from such a bacterium were very efficient in cleaving the bonds within the lignin. The search for novel microorganisms and enzymes that can degrade lignin, along with the potential applications that this naturally occurring process holds, has lead to an improved understanding of the properties and functioning of lignin metabolism and the biogeochemical flow of complex organic molecules in hypersaline ecosystems.

18.5 Biodiesel Production by Halophilic/Halotolerant Algae

As enumerated previously, many halophilic prokaryotes have developed enzymatic moieties for utilizing polysaccharides central to biofuel development. Conversely, there are several primary producers of such polymers within highly saline environments that are of equal if not more importance in potential to supply feedstocks for biomass derived chemicals and fuels. Considerable development pressures on arable land around the globe have resulted from unsustainable development for first generation biofuel crops such as palm, sugarcane, soybeans, and corn. Combined with finite arable land, the problems with unsustainable crop derived feedstocks for biofuels has led to the coining of the term "peak soil" (Schenk et al. 2008). Microalgal culture aims to reduce the arable land and freshwater requirements of biomass for fuel and chemicals and thus thwart the "peak soil" issue. As a result, more emphasis has been placed on prospecting for phototrophic and heterotrophic halophilic primary producers that not only create unique feedstocks but also to provide an avenue for more sustainable cultivation (Mutanda et al. 2011). Central to the sustainability of industrial algal production for biofuels is utilization of marginal or even polluted sources. Halophilic cyanobacteria and algae are compatible with the effluent from pickling and tannery industries (Oren 2010), saline aquifers, produced water from oil and gas extraction, and the utilization of abandoned mine pools as water sources (Alleman 2009).

18.5.1 Microalgae

As true eukaryotes, the principal microalgae present in moderately high salinities (1–3.5 M NaCl) are from the genus *Dunaliella* (*D. salina*, *D. parva*, and *D. viridis*) as well as *Asteromona gracilis* and they form the basis of the food chain for higher animals such as brine shrimp and brine flies (DasSarma and Arora 2002). It is also likely that these algae are one of the primary sources of polysaccharides for non-marine, hypersaline environments because they commonly contain cellulose, and to a lesser extent mannans and xylans as fibrillar constituents of their cell walls (Barsanti and Gualtieri 2006; Lee 2008). While *Dunaliella* have traditionally been a commercial source for β-carotene (Ben-Amotz and Avron 1983), this halophilic green alga has also been the subject of much research in relation to biofuels. Under normal growth conditions, *Dunaliella* has shown promise of producing a high quality hydrocarbon from the pyroloysis of the protein fraction of the biomass (Ginzburg 1993). Under stressed conditions, such as osmotic and nitrogen deficiency, *Dunaliella* can produce significant quantities of carbohydrates that are ideal for methane and ethanol production (Feinberg 1984).

Diatoms from the *Amphora*, *Nitzschia* and *Entomoneis* genera have also been widely found in environments up to 3 M NaCl, but not in the abundance as *Dunaliella* (Clavero et al. 2008). Furthermore, the research on the specific polysaccharides found in halophilic diatoms is limited. However, some species have been reported to produce proline and oligosaccharides under different osmotic and nutrient stress (DasSarma and Arora 2002).

18.5.2 Cyanobacteria

These are the dominant organism in most all moderate to extreme saline impoundments and most often are comprised of *Aphanothece halophytica*, *Dactylococcopsis salina*, *Phormidium ambiguum* and many members of the order *Oscillatoriales* (*O. neglecta*, *O. limnetica*, and *O. salina*) (DasSarma and Arora 2002). Cellulose biosynthesis has been confirmed in many cyanobacteria including members of Oscillatoriales, but not in any specific halophilic species (Nobles et al. 2001). Furthermore, with genetic modification, it has been demonstrated that the production of osmolites may be manipulated in such a way to favor production of useful feedstocks like sucrose. Engineered mutants of organisms that typically produce glucosyl glycerol have been made to produce a simple carbohydrate like sucrose (Miao et al. 2003). Further development into the enhancement of quality and quantity of yields of both carbohydrates would create a distinct advantage for current fermentative biofuel technologies like bioethanol and biobutanol.

18.6 Conclusions

Though little work has been done to demonstrate the use of halophilic organisms for biofuel production, these organisms can be applied in this manner in two distinct ways. Halophilic algae can possibly be used to produce biodiesel and other biomass feedstocks while lowering the amount of water required for this process. Halophilic/halotolerant bacteria can possibly be used to breakdown non-food biomass that has been exposed to alkaline pretreatment for subsequent fermentation to biofuel products such as alcohols and hydrogen. Furthermore, the cellulolytic enzymes these organisms possess possibly can be used as a less harsh pretreatment of lignocellulosic biomass. Significant amounts of information have been discovered about extremophilic lignocellulolytic systems. However, investigations of halophilic lignocellulolytic degradation are nascent by comparison. For example, there have been very few studies on strict halophiles that possess cellulase complexes (Bolobova et al. 1992). Additionally, modern recombinant screening techniques and data mining of genomes has the potential to yield many halophilic lignocellulytic enzymes. This is an exciting period as there is still much to learn about these enzymes in halophilic microorganisms. Furthermore, there is great potential for developing biotechnologies with halophilic lignocellulytic enzymes in regards to treating biomass, leading to innovations in fields such as bio-fuel production.

References

Alleman D (2009) Non-traditional sources of water for coal-fired power plants. Ground Water Protection Council Annual Form, Salt Lake City, UT

Averhoff B, Müller V (2010) Exploring research frontiers in microbiology: recent advances in halophilic and thermophilic extremophiles. Res Microbiol 161:506–514

Bailey MJ, Biely P, Poutanen K (1992) Interlaboratory testing of methods for assay of xylanase activity. J Biotechnol 23:257–270

Barsanti L, Gualtieri P (2006) Algae: anatomy, biochemistry, and biotechnology. CRC/Taylor & Francis, Boca Raton, FL

Beguin P, Aubert J-P (1994) The biological degradation of cellulose. FEMS Microbiol Rev 13:25–58

Ben-Amotz A, Avron M (1983) Accumulation of metabolytes by halotolerant algae and its industrial potential. In: Ornston LN, Balows A, Baumann P (eds) Annual review of microbiology, vol 37. Annual Review Inc, Palo Alto, CA, pp 95–119

Benner R, Moran MA, Hodson RE (1986) Biogeochemical cycling of lignocellulosic carbon in marine and freshwater ecosystems: contributions of relative procaryotes and eucaryotes. Limnol Oceanogr 31:89–100

Bhat MK, Bhat S (1997) Cellulose degrading enzymes and their potential industrial applications. Biotechnol Adv 15:583–620

Biely P, Puls J, Schneider H (1985) Acetyl xylan esterases in fungal cellulolytic systems. FEBS Lett 186:80–84

Birbir M, Sesal C (2003) Extremely halophilic bacterial communities in Şereflikoçhisar Salt Lake in Turkey. Turk J Biol 27:7–22

Birbir M, Ogan A, Calli B, Mertoglu B (2004) Enzyme characteristics of extremely halophilic archaeal community in Tuzkoy Salt Mine, Turkey. World J Microbiol Biotechnol 20: 613–621

Bolobova AV, Simankova MC, Markovich NA (1992) Cellulase complex of a new halophilic bacterium *Halocella celluloytica*. Microbiology (Russia) 61:557–562

Bronnenmeier K, Kern A, Liebl W, Staudenbauer WL (1995) Purification of *Thermotoga maritima* enzymes for the degradation of cellulosic materials. Appl Environ Microbiol 61:1399–1407

Bucher VVC, Hyde KD, Pointing SB, Reddy CA (2004) Production of wood decay enzymes, mass loss and lignin solubilization in wood by marine ascomycetes and their anamorphs. Fungal Divers 15:1–14

Clavero E, Hernandez-Marine M, Grimalt JO, Garcia-Pichel F (2008) Salinity tolerance of diatoms from thalassic hypersaline environments. J Phycol 36:1021–1034

Collins T, Gerday C, Feller G (2005) Xylanases, xylanase families and extremophilic xylanases. FEMS Microbiol Rev 29:3–23

Coughlan MP, Mayer F (1992) The cellulose-decomposing bacteria and their enzyme systems. In: Balows A, Trüper HG, Dworkin M, Harder W, Schleifer K-H (eds) The prokaryotes, vol 1, 2nd edn. Springer, New York, pp 461–467

Cox PM, Betts RA, Jones CD, Spall SA, Totterdell JJ (2000) Acceleration of global warming due to carbon-cycle feedbacks in a coupled climate model. Nature 408:184–187

Danson MJ, Hough DW (1997) The structural basis of protein halophilicity. Comp Biochem Physiol A Physiol 117:307–312

DasSarma S, Arora P (2002) Halophiles. In: Encyclopedia of life sciences, vol 8. Nature Publishing Group, London, pp 485–466

Dhawan S, Kaur J (2007) Microbial mannanases: an overview of production and applications. Crit Rev Biotechnol 27:197–216

Eisenberg H, Mevarech M, Zaccai G (1992) Biochemical, structural, and molecular genetic aspects of halophilism. Adv Protein Chem 43:1–62

Feinberg DA (1984) Fuel options from microalgae with representative chemical compositions. Solar Energy Research Institute, U.S. DOE, Golden, CO

Frei E, Preston RD (1968) Non-cellulosic structural polysaccharides in algal cell walls. III. Mannan in siphoneous green algae. Proc R Soc Lond B 169:127–145

Fujinami S, Fujisawa M (2010) Industrial applications of alkaliphiles and their enzymes – past, present and future. Environ Technol 31:845–856

Fukushima T, Mizuki T, Echigo A, Inoue A, Usami R (2005) Organic solvent tolerance of halophilic α-amylase from a haloarchaeon, *Haloarcula* sp. strain S-1. Extremophiles 9:85–89

Gao Z, Ruan L, Chen X, Zhang Y, Xu X (2010) A novel salt-tolerant endo-β-1,4-glucanase Cel5A in *Vibrio* sp. G21 isolated from mangrove soil. Appl Microbiol Biotechnol 87:1373–1382
Gilbert HJ, Hazlewood GP (1993) Bacterial cellulases and xylanases. J Gen Microbiol 139: 187–194
Ginzburg B-Z (1993) Liquid fuel (oil) from halophilic algae: a renewable source of non-polluting energy. Renew Energy 3:249–252
Giridhar PV, Chandra TS (2010) Production of novel halo-alkali-thermo-stable xylanase by a newly isolated moderately halophilic and alkali-tolerant *Gracilibacillus* sp. TSCPVG. Process Biochem 45:1730–1737
Gomes J, Steiner W (2004) The biocatalytic potential of extremophiles and extremozymes. Food Technol Biotechnol 42:223–235
González JM, Mayer F, Moran MA, Hodson RE, Whitman WB (1997a) *Sagittula stellata* gen. nov., sp. nov., a lignin-transforming bacterium from a coastal environment. Int J Syst Bacteriol 47:773–780
González J, Mayer F, Moran MA, Hodson RE, Whitman WB (1997b) *Microbulbifer hydrolyticus* gen. nov., sp. nov., and *Marinobacterium georgiense* gen., nov., sp. nov., two marine bacteria from a lignin-rich pulp mill waste enrichment community. Int J Syst Bacteriol 47:369–376
González ED, Vallejo OD, Anaya CM, Sánchez MM, González MC (2008) Production of two novel laccase isoforms by a thermotolerant strain of *Pycnoporus sanguineus* isolated from an oil-polluted tropical habitat. Int Microbiol 11:163–169
Govender L, Naidoo L, Setati ME (2009) Isolation of hydrolase producing bacteria from Sua pan solar salterns and the production of endo-1,4-β-xylanase from a newly isolated haloalkaliphilic *Nesterenkonia* sp. Afr J Biotechnol 8:5458–5466
Huang X, Shao Z, Hong Y, Lin L, Li C, Huang F, Wang H, Liu Z (2010) Cel8H, a novel endoglucanase from the halophilic bacterium *Halomonas sp.* S66-4: molecular cloning, heterogonous expression, and biochemical characterization. J Microbiol 48:318–324
Joseleau J, Comtat J, Ruel K (1992) Chemical structure of xylans and their interactions in the plant cell walls. In: Kusters van Someren MA, Beldman G, Visser J, Voragen AGJ (eds) Xylans and xylanases, progress in biotechnology. Elsevier, Amsterdam, pp 1–15
Kamei I, Daikoku C, Tsutsumi Y, Kondo R (2008) Saline-dependent regulation of manganese peroxidase genes in the hypersaline-tolerant white rot fungus *Phlebia* sp. strain MG-60. Appl Environ Microbiol 74:2709–2716
Kanekar PP, Joshi AA, Kelkar AS, Borgave SB, Sarnaik SS (2008) Alkaline Lonar Lake, India – a treasure of alkaliphilic and halophilic bacteria. In: Sengupta M, Dalwani R (eds) Proceedings of Taal 2007: 12 World Lake Conference. The Ministry of Environment and Forests of the Government of India, Jaipur, pp 1765–1774
Karbalaei-Heidari HR, Ziaee A-A, Amoozegar AM (2007) Purification and biochemical characterization of a protease secreted by the *Salinivibrio* sp. strain AF-2004 and its behavior in organic solvents. Extremophiles 11:237–243
Kulkarni N, Shendye A, Rao M (1999) Molecular and biotechnological aspects of xylanases. FEMS Microbiol Rev 23:411–456
Lee RE (2008) Phycology, 3rd edn. Cambridge University Press, New York
Li LH, Flora RM, King KW (1965) Individual roles of cellulase components derived from *Trichoderma viride*. Arch Biochem Biophys 111:439–447
Li X, Kondo R, Sakai K (2002) Studies on hypersaline-tolerant white-rot fungi I: screening of lignin degrading fungi in hypersaline conditions. J Wood Sci 48:147–152
Li X, Kondo R, Sakai K (2003) Studies on hypersaline-tolerant white-rot fungi IV: effects of Mn^{2+} and NH_4^+ on manganese peroxidase production and Poly R-478 decolorization by the marine isolate *Phlebia* sp. MG-60 under saline conditions. J Wood Sci 49:355–360
Li J, Yuan H, Yang J (2009) Bacteria and lignin degradation. Front Biol 4:29–38
Lynd LR, Weimer PJ, van Zyl WH, Pretorius IS (2002) Microbial cellulose utilization: fundamentals and biotechnology. Microbiol Mol Biol Rev 66:506–577

Mevarech M, Frolow F, Gloss LM (2000) Halophilic enzymes: proteins with a grain of salt. Biophys Chem 86:155–164

Miao X, Wu Q, Wu G, Zhao N (2003) Sucrose accumulation in salt-stressed cells of agp gene deletion-mutant in cyanobacterium *Synechocystis* sp. PP 6803. FEMS Microbiol Lett 218: 71–77

Molitoris HP, Buchalo AS, Kurchenko I, Nevo E, Rawal BS, Wasser SP, Oren A (2000) Physiological diversity of the first filamentous fungi isolated from the hypersaline Dead Sea. Fungal Divers 5:55–70

Møller M, Kjeldsen K, Ingvorsen K (2010) *Marinimicrobium haloxylanilyticum* sp. nov., a new moderately halophilic, polysaccharide-degrading bacterium isolated from Great Salt Lake, Utah. Antonie Leeuwenhoek 98:553–565

Mosier N, Wyman C, Dale B, Elander R, Lee YY, Holtzapple M, Ladisch M (2005) Features of promising technologies for pretreatment of lignocellulosic biomass. Bioresour Technol 96: 673–686

Mutanda T, Ramesh D, Karthikeyan S, Kumari S, Anandraj A, Bux F (2011) Bioprospecting for hyper-lipid producing microalgal strains for sustainable biofuel production. Bioresour Technol 102:57–70

Niladevi KN, Jacob N, Prema P (2008) Evidence for a halotolerant-alkaline laccase in *Streptomyces psammoticus*: purification and characterization. Process Biochem 43:654–660

Nobles DR, Romanovicz DK, Brown RM Jr (2001) Cellulose in cyanobacteria. Origin of vascular plant cellulose synthase? Plant Physiol 127:529–542

Oren A (2002) Halophilic microorganisms and their environments. Kluwer Academic Publishers, Dordrecht

Oren A (2010) Industrial and environmental applications of halophilic microorganisms. Environ Technol 31:825–834

Pikuta EV, Hoover RB, Tang J (2007) Microbial extremophiles at the limits of life. Crit Rev Microbiol 33:183–209

Pimentel D, Marklein A, Toth MA, Karpoff MN, Paul GS, McCormack R, Kyriazis J, Krueger T (2009) Food versus biofuels: environmental and economic costs. Hum Ecol 37:1–12

Prakash S, Veeranagouda Y, Kyoung L, Sreeramulu K (2009) Xylanase production using inexpensive agricultural wastes and its partial characterization from a halophilic *Chromohalobacter* sp. TPSV 101. World J Microbiol Biotechnol 25:197–204

Prescott LM, Harley JP, Klein DA (1993) Microbial growth and metabolism and microorganisms and the environment. In: Prescott LM, Harley JP, Klein DA (eds) Microbiology, 2nd edn. Wm. C. Brown Publishers, Dubuque, pp 176–179

Puls J, Schuseil J (1993) Chemistry of hemicelluloses: relationship between hemicellulose structure and enzymes required for hydrolysis. In: Coughlan M, Hazlewood G (eds) Hemicellulose and hemicellulases. Portland Press, London, pp 1–27

Raghukumar C, Ticlo DD, Verma AK (2008) Treatment of colored effluents with lignin-degrading enzymes: an emerging role of marine-derived fungi. Crit Rev Microbiol 34:189–206

Rees H, Grant S, Jones B, Grant WD, Heaphy S (2003) Detecting cellulase and esterase enzyme activities encoded. Extremophiles 7:415–421

Reese ET (1976) History of the cellulase program at the U.S. Army Natick Development Center. Biotechnol Bioeng Symp 6:9–19

Rohban R, Amoozegar MA, Ventosa A (2009) Screening and isolation of halophilic bacteria producing extracellular hydrolyses from Howz Soltan Lake, Iran. J Ind Microbiol Biotechnol 36:333–340

Saha B (2003) Hemicellulose bioconversion. J Ind Microbiol Biotechnol 30:279–291

Sanchez-Porro C, Martin S, Mellado E, Ventosa A (2003) Diversity of moderately halophilic bacteria producing extracellular hydrolytic enzymes. J Appl Microbiol 94:295–300

Schenk PM, Thomas-Hall SR, Stephens E, Marx UC, Mussgung JH, Posten C, Kruse O, Hankamer B (2008) Second generation biofuels: high-efficiency microalgae for biofuel production. Bioenerg Res 1:20–43

Schwarz WH (2001) The cellulosome and cellulose degradation by anaerobic bacteria. Appl Microbiol Biotechnol 56:634–649

Setati ME (2010) Diversity and industrial potential of hydrolase-producing halophilic/halotolerant eubacteria. Afr J Biotechnol 9:1555–1560

Shallom D, Shoham Y (2003) Microbial hemicellulases. Curr Opin Microbiol 6:219–228

Uffen RL (1997) Xylan degradation: a glimpse at microbial diversity. J Ind Microbiol Biotechnol 19:1–6

Uthandi S, Saad B, Humbard MA, Maupin-Furlow JA (2009) LccA, an archaeal laccase secreted as a highly-stable glycoprotein into the extracellular medium of *Haloferax volcanii*. Appl Environ Microbiol 76:733–743

van den Burg B (2003) Extremophiles as a source for novel enzymes. Curr Opin Microbiol 6:213–218

Voget S, Steele HL, Streit WR (2006) Characterization of a metagenome-derived halotolerant cellulase. J Biotechnol 126:26–36

Vreeland RH, Piselli AF Jr, McDonnough S, Meyers SS (1998) Distribution and diversity of halophilic bacteria in a subsurface salt formation. Extremophiles 2:321–331

Wainø M, Ingvorsen K (1999) Production of halostable β-mannanase and β-mannosidase by strain NN, a new extremely halotolerant bacterium. Appl Microbiol Biotechnol 52:675–680

Wainø M, Ingvorsen K (2003) Production of β-xylanase and β-xylosidase by the extremely halophilic archaeon *Halorhabdus utahensis*. Extremophiles 7:87–93

Wang C-Y, Hsieh Y-R, Ng C-C, Chan H, Lin H-T, Tzeng W-S, Shyu Y-T (2009) Purification and characterization of a novel halostable cellulase from *Salinivibrio* sp. strain NTU-05. Enzyme Microb Technol 44:373–379

Wang C-Y, Chan H, Lin H-T, Shyu Y-T (2010a) Production, purification and characterisation of a novel halostable xylanase from *Bacillus* sp. NTU-06. Ann Appl Biol 156:187–197

Wang J, Shao Z, Hong Y, Li C, Fu X, Liu Z (2010b) A novel β-mannanase from *Pantoea agglomerans* A021: gene cloning, expression, purification and characterization. World J Microbiol Biotechnol 26:1777–1784

Wejse P, Ingvorsen K, Mortensen K (2003) Purification and characterisation of two extremely halotolerant xylanases from a novel halophilic bacterium. Extremophiles 7:423–431

Whitney SEC, Brigham JE, Darke AH, Grant Reid JS, Gidley MJ (1998) Structural aspects of the interaction of mannan-based polysaccharides with bacterial cellulose. Carbohydr Res 307: 299–309

Wong KK, Tan LU, Saddler JN (1988) Multiplicity of β-1,4-xylanase in microorganisms: functions and applications. Microbiol Mol Biol Rev 52:305–317

Wood TM (1975) Properties and mode of action of cellulases. Biotechnol Bioeng Symp 5:111–137

Woolard CR, Irvine RL (1995) Treatment of hypersaline wastewater in the sequencing batch reactor. Water Res 29:1159–1168

Yang C, Wang Z, Li Y, Niu Y, Du M, He X, Ma C, Tang H, Xu P (2010) Metabolic versatility of halotolerant and alkaliphilic strains of *Halomonas* isolated from alkaline black liquor. Bioresour Technol 101:6778–6784

Zhou X, Chen H, Li Z (2004) CMCase activity assay as a method for cellulase adsorption analysis. Enzyme Microb Technol 35:455–459

Chapter 19
Enzymes from Halophilic Archaea: Open Questions

María José Bonete and Rosa María Martínez-Espinosa

19.1 Introduction

The prokaryotic extremophiles have become a focus of scientific interest owing to their unique properties in terms of physiology, ecology, biochemistry and genetics. Most of the extremophiles that have been identified up to now belong to the Archaea domain (Woese et al. 1990). Halophilic Archaea constitute the family *Halobacteriaceae*, the only family in the order *Halobacteriales*. The haloarchaea have been an early topic of research due to their role in salted food deterioration (Rodríguez-Valera 1992). These microorganisms form thriving and dominant populations in many hypersaline environments (Oren 2002). Studies on the physiology and enzymology of haloarchaea are still scarce in comparison with studies on Bacteria; however, research on the Archaea has greatly increased, in part initiated by genomic science as well as by a continuing interest in their proteomics (Joo and Kim 2005; Karadzic and Maupin-Furlow 2005; Kirkland et al. 2008), biochemistry and metabolism (Verhees et al. 2003; Cabello et al. 2004; Martínez-Espinosa et al. 2006; Bonete et al. 2007; Bonete et al. 2008; Falb et al. 2008; Johnsen et al. 2009; Martínez-Espinosa et al. 2009). On the other hand, protein engineering and direct evolution provide new approaches to better understand enzyme stability and to allow researchers to modify enzymatic specificity in ways that may not exist in the natural world (Hough and Danson 1999; Esclapez et al. 2007; Pire et al. 2009).

19.1.1 Extremozymes

The use of enzymes as biotransformation catalysts is very well defined and established so far, but the use of these biocatalysts in chemical reactions is generally

M.J. Bonete (✉) • R.M. Martínez-Espinosa
División de Bioquímica y Biología Molecular, Facultad de Ciencias, Universidad de Alicante, Ap. 99, E-03080 Alicante, Spain
e-mail: mjbonete@ua.es

A. Ventosa et al. (eds.), *Halophiles and Hypersaline Environments*,
DOI 10.1007/978-3-642-20198-1_19, © Springer-Verlag Berlin Heidelberg 2011

restricted to mild conditions of pH, temperature, pressure, ionic strength, etc. Despite the fact that thousands of enzymes have been identified and many of them have found their role into biotechnological and industrial applications, the present enzyme toolbox is still not sufficient to meet all the industrial demands. The major reason for this fact is due to enzyme inhibition or instability in the industrial reaction conditions (van den Burg 2003).

Enzymes isolated from different living organisms inhabiting extreme environments are not constrained to these mild conditions needed for the majority of enzymes. These enzymes, termed extremozymes, provide information to improve stability, activity and specificity of many enzymes used thus far for industrial biocatalysts. On the other hand, the extremozymes, as isolated, constitute a new source of enzymes with novel activities and applications. On the bases of their characteristics, the extremozymes may be classified as summarized in Table 19.1.

Table 19.1 Classification of the extremozymes from the Archaea domain

Extremozyme group	Characteristics	Typical genera	Enzymes	Applications
Psychrophilic	Active at temperatures approaching the freezing point of water	To be defined	Proteases	Detergents
			Dehydrogenases	Food
			Glycosyl hydrolases	Cosmetics
				Biosensors
			Lipases	Textile
Halophilic	High activity and stability in salt solutions	*Haloarcula*, *Halobacterium*, *Haloferax*, *Halorubrum*	Proteases	Peptide synthesis
			Dehydrogenases	Biocatalysis in organic media
			Oxido-reductases	Bioremediation
			Glycosyl hydrolases	Starch processing
Thermophilic	High activity and stability at high temperatures	*Thermoplasma*	Proteases	Detergents
			Glycosyl hydrolases	Hydrolysis in food and feed, brewing, baking
			Chitinases	Textiles
			Xylanases	Paper bleaching
			Lipases, esterases	Molecular biology (PCR)
			DNA polymerases	
			Dehydrogenases	
Piezophilic	Active at high pressure	To be defined	Proteases	Food processing
Adicidophilic	Active and stable at pH lower than 4	*Adicianus*, *Sulfolobus*	Glycosyl hydrolases	Antibiotic production
				Desulfurization of coal
			Proteases	Starch processing
			Oxidases	Feed component
Alkaliphilic	Active and stable at pH higher than 9	*Natronobacterium*, *Natronococcus*	Proteases	Detergents
			Cellulases	Food and feed

Several genes encoding extremozymes have been sequenced, isolated, cloned and overexpressed (heterologous and homologous overexpression) (De Castro et al. 2008; Zafrilla et al. 2010). These studies and the structures solved of some extremozymes have provided opportunities to design more stable versions of enzymes that can function in extreme salinity, pH, temperatures and non-aqueous media. Selectivity of extremozymes (enantioselectivity, regioselectivity and chemoselectivity) can also be adjusted through judicious use of particular enzymes in organic solvents or water/solvent mixtures. Finally, the detailed analysis of the complete genome sequences from Archaea genera, involving comparison with gene sequences of proteins of known functions from other organisms, could result in the identification of interesting enzymes. Nevertheless, although the putative function of a gene product can be identified by this methodology, the analysis of the protein product is needed as well. Also, studies on the expression product are essential to obtain additional information such as substrate specificity, stability, etc. Regarding to this last aspect, it is interesting to note that although the number of haloarchaeal genomes sequenced is increasing, the analysis of the expression products is still scarce. Consequently, the potential uses of the haloarchaeal enzymes are poorly known at the time this review was written.

19.2 Enzymes from Haloarchaea

The haloarchaea accumulate high KCl concentrations within the cells to be isotonic with the growth medium. This strategy requires extensive adaptation of the intracellular enzymatic machinery to the presence of salt, as the proteins should maintain their proper conformation and activity at near-saturating salt concentrations (Oren 2008). Haloarchaeal enzymes, while performing identical enzymatic functions as their non-halophilic counterparts, have been shown to exhibit substantially different properties: requirement for high salt concentrations (in the 1–4 M range) for activity and stability, and a high excess of acidic over basic amino residues (Mevarech et al. 2000; Britton et al. 2006). The high negative surface charge of halophilic proteins makes them more soluble and renders them more flexible at high salt concentrations, conditions under which non-halophilic proteins tend to aggregate and become rigid. This high surface charge is neutralized mainly by tightly bound water dipoles. The requirement of high salt concentration for the stabilization of halophilic enzymes, on the other hand, is due to a low affinity binding of the salt to specific sites on the surface of the folded polypeptide, thus stabilizing the active conformation of the protein (Mevarech et al. 2000; Britton et al. 2006). Using mutational studies on the protein surfaces, it has been shown that it is possible to decrease the salt dependence of a typical halophilic protein to the level of a mesophilic form and engineer a protein from a mesophilic organism into an obligate halophilic form (Esclapez et al. 2007; Tadeo et al. 2009).

Although demand for salt-tolerant enzymes in manufacturing or related processes has been limited up to now, several industrial sectors such as waste-waters treatment or chemistry in nonaqueous media are developing strategies to better optimize their processes using halophilic enzymes. One of the most important problems of maintaining the enzymatic activity in high organic solvents concentrations is the low water activity, but haloarchaea enzymes are able to work properly under conditions of very low water activity. In general terms, enzyme catalysis in organic solvents offers benefits such as enhanced stability, alterations in substrate and enantiomeric specificities, and increased product yield (Klibanov 1989). So, enzymes from haloarchaea could be the most suitable option for application in nonaqueous media (Marhuenda-Egea and Bonete 2002; Pire et al. 2004).

Because of the extreme properties of the enzymes from halophilic Archaea and similar catalytic abilities to their bacterial counterparts, is interesting to focus attention on their potential biotechnological applications (Ding and Lai 2010). The halophilic Archaea display a considerable extent of enzyme diversity, but the presence or absence of certain enzymatic activities is closely linked with the taxonomic status of the strains investigated (Oren 1994). Thus, different species will be the best candidates for different specific applications. For example, haloarchaea denitrifiers could be excellent for nitrate/nitrite removal from salty wastewaters (Martínez-Espinosa et al. 2007b). Related to that aspect, it has been shown that some haloarchaea such as *Haloarcula marismortui* and *Haloferax mediterranei* are able to reduce nitrate under anaerobic conditions (Yoshimatsu et al. 2002; Lledó et al. 2004) thanks to the formally called NarGH (respiratory nitrate reductases). It is interesting to note that haloarchaea Nar-type nitrate reductases have the active site on the outside of the cytoplasmic membrane ($\Delta\psi^{+}$). On the bases of these results, it has been proposed to adopt a location-based classification of nNar for a system in which the NarG subunit is located on the membrane potential-negative ($\Delta\psi^{-}$) side and pNarG for a system in which the NarG subunit is located on the membrane potential-positive ($\Delta\psi^{+}$) side (Martínez-Espinosa et al. 2007a).

The nitrite produce by NarGH is further reduce to nitric oxide by nitrite reductases which are poorly known in Archaea at the time of writing this work. General biochemical and physiological characterizations carried out with *Hfx. mediterranei* as halophilic denitrifier model have revealed that this specie is a complete denitrifier, i.e., the nitrite produce by NarGH is further reduce to nitric oxide thanks to the respiratory nitrite reductase, which is a Cu-type Nir in the mentioned species (data not published). Nitric oxide is then reduced to nitrous oxide by nitric oxide reductase and finally, nitrous oxide reductase produces dinitrogen from nitrous oxide. As a consequence of this metabolic pathway, *Hfx. mediterranei* can be use to remove nitrate and nitrite from brines, obtaining dinitrogen and nitrate/nitrite free brines as products. The isolated enzymes could be immobilized and used in biosensors. Although several examples of this technology were documented (Dinçkaya et al. 2010; Silveira et al. 2010), no studies on biosensors using enzymes from halophilic denitrifiers have been reported thus far. Another nice example of haloarchaea applications is the use of several *Halobacterium*, *Haloarcula* and *Haloferax* species to break down halogenated hydrocarbons (Patzelt et al. 2001). Although this capability has been demonstrated so far, the metabolic machinery involved in this process is still unknown in haloarchaea.

19.3 Biotechnological Applications of Haloarchaeal Enzymes

Halophilic Archaea have been successfully tested for biotechnological applications during the last 10 years. Protein engineering, gene expression libraries, complete genome sequences and more recently, proteome data bases from haloarchaea, could help us to discover new extremozymes to be applied in biocatalytic processes that are faster, more accurate, specific and environmentally friendly (van den Burg 2003). In addition, the bioinformatics approach will facilitate to predict the function of novel proteins of haloarchaea (Joo and Kim 2005). However, in comparison with the thermophilic and the alkaliphilic extremophiles, haloarchaea have as yet found relatively few biotechnological applications (Oren 2010), mainly due to the scarce knowledge we have of their enzymology and protein structures. Besides, analysis of the stability of the halophilic proteins in a matrix could shed light on the real applications as reusable biocatalysts in biotechnological processes involving extreme environmental conditions (D'Souza et al. 1997). Some of the most interesting biotechnological applications of the haloarchaeal enzymes are summarized below.

19.3.1 Enzymes for Bioremediation

Haloarchaea have been proposed as good candidates for bioremediation of saline and hypersaline wastewaters, thanks to their high tolerance to salt, metals and organic pollutants such as 1,2-dichloroethane, naphthalene/anthracene or benzoate degradation via CoA ligation (Schiraldi et al. 2002; Ding and Lai 2010).

On the other hand, haloarchaea denitrifiers such as *Haloferax mediterranei* or *Haloarcula marismortui* have become a good source for biotechnological applications in wastewater treatment due to their capacity to reduce nitrate and nitrite to NOx. Cells grown in salty wastewaters containing nitrate, nitrite or even ammonia, are able to use these nitrogen sources for growth and denitrification, causing removal of these nitrogen compounds from the water (Martínez-Espinosa et al. 2006, 2007a, b). The oxy-anions nitrate and nitrite may cause eutrophication in lakes or rivers (Howarth 2004) and, if present in drinking water, increase the risk of methaemoglobinemia in infants and of intestinal carcinogenic nitrosamine in adults (Greer and Shannon 2005; van Grinsven et al. 2010). Enzymes such as nitrate reductases or nitrite reductases isolated from haloarchaea could be good examples of enzymes to construct a potentiometric ion selective electrode to detect nitrate or nitrite. On the other hand, these enzymes, when immobilized, could be use to remove nitrate or nitrite from saline wastewaters.

Recently, it has been described that few species of Archaea are able to detoxify inorganic forms of arsenic by volatilization involving methylation to volatile arsines or by converting them to less-toxic non-volatile species (Bini 2010). Regarding this detoxification, ArsM is a bacterial homolog of the rat methyltransferase and catalyzes the formation of trimethylarsine from arsenite. A knockout of the gene encoding ArsM in *Halobacterium salinarum* resulted in sensitivity to arsenite, confirming its role as a detoxifying enzyme (Wang et al. 2004). This enzyme could be also used for bioremediation purposes.

19.3.2 Enzymes for Industry

Mesophilic and thermophilic amylases, proteases, lipases and esterases have been used in textile, cosmetic, pharmaceutics and food industries since the beginning of the twentieth century. Whole haloarchaeal cells have been used for the degradation of organic pollutants and in the treatment of concentrated textile waste waters, particularly the first dye bath liquor from the dying process, which contains a high salt load (Margesin and Schinner 2001), but only few studies have been reported on their enzymes. Thus, few biochemical studies have been carried out with amylases (Good and Hartman 1970; Pérez-Pomares et al. 2003, 2009) or proteases (Vidyasagar and Prakash 2006) from haloarchaea. These enzymes could keep high activity and stability in high salt environments and could have potential application value. α-Amylases are used to obtain fructose and glucose syrup from starch, to improve flour in the baking industry, to produce modified starches for the paper industry or to replace the malt in the brewing industry. For the proteases for example, it has been proposed that an enzyme isolated from haloarchaea can be a suitable source for preparation of fish sauce (Vidyasagar and Prakash 2006).

Regarding to lipases and esterases, these are enzymes in demand for industrial processes developed in low water activity media; however, studies of these enzymes from haloarchaea are scarce (Boutaiba et al. 2006). Although the analysis of the genome sequences from *Halobacterium* sp. NRC-1 and *Har. marismortui* (http://www.genome.ad.jp) reveals genes coding putatives lipases and esterases, biochemical characterization on those proteins is only beginning to emerge. Haloarchaeal enzymes are not yet employed in food industries, although some dehydrogenases involved in alcohol, carbohydrates or ammonium metabolism could have relevant roles in such processes (Ferrer et al. 1996; Cao et al. 2008a, 2008b).

19.3.3 Enzymes Involved in Polyesters and Exopolysaccharides Production

Biopolymers, such as biosurfactants and exopolysaccharides, are of interest for biotechnology. Biosurfactant-producing haloarchaea can play a significant role in the accelerated remediation of oil-polluted saline environments, but only few studies have been performed on that topic (Banat et al. 2000). Members of the genus *Haloferax* produce exopolysaccharides but the enzymes catalyzing the reactions have not been extensively described (Parolis et al. 1996).

Some haloarchaea are able to produce polyesters (PHA: polyhydroxyalkanoates and PHB: polyhydroxybutyrates). Polyhydroxyalkanoates are accumulated as carbon and energy sources in many Archaea growing under nutrient-limiting conditions with excess carbon source (Legat et al. 2010). These polyesters are biodegradable thermoplastics. Although the production and the structure of these alkanoates has been analyzed mainly from members of the genus *Haloferax*, the enzymes involved in their synthesis are poorly known and no large-scale applications have yet been reported (Oren 2010). *Hfx. mediterranei* accumulates large quantities of poly(beta-hydroxybutyrate) (PHB) as intracellular granules when the cells grow under phosphate limitation. The PHB production in continuous cultures is stable over a 3-month period (Lillo and Rodriguez-Valera 1990). Due to its high growth rate, metabolic versatility, and genetic stability, *Hfx. mediterranei* has become an interesting microorganism to produce this bioplastic. The PHA accumulated by *Hfx. mediterranei* was first reported to be poly(3-hydroxybutyrate) (PHB) (Fernandez-Castillo et al. 1986), but it was re-evaluated as poly(3-hydroxybutyrate-co-3-hydroxyvalerate) (PHBV) (Don et al. 2006). Although the knowledge on PHA and PHB synthesis in haloarchaea is increasing (Lu et al. 2008; Han et al. 2009, 2010), the regulation of their biosynthetic pathway remains poorly described.

19.4 Further Questions on Haloarchaeal Enzymes

Several haloarchaea genera produce other biopolymers of high interest in biotechnology such as carotenoids, antibiotics, etc. However, the enzymes involved in the synthesis or degradation of such compounds have been poorly described. Below are some of the most interesting fields to be explored.

19.4.1 Haloarchaeal Antigens

The similarities between archaeal and eukaryotic cells at the level of cell division justify the use of haloarchaea in the pre-screening for anti-cancer drugs. Some of the haloarchaeal antigens could be used for cancer diagnosis (Ben-Mahrez et al. 1988); nevertheless, applications in medical fields are still scarce.

19.4.2 Enzymes Involved in Carotene Synthesis

The production of β-carotene by eukaryotic halophiles has been extensively reported (Oren 2005) and the production of this pigment at industrial scale is very successful. Haloarchaea species are also able to produce pigments such as β-carotene, astaxantin, or bacterioruberin (Calo et al. 1995). Nevertheless, carotenoid metabolism in halophilic Archaea has not been completely described. Recently, it has been reported that *Hfx. mediterranei* produces three red C_{50} carotenoid pigments: bisanhydrobacterioruberin, monoanhydrobacterioruberin and bacterio-ruberin, as well as a C_{45} carotenoid: 2-isopentenyl-3,4-dehydrorhodopin (Fang et al. 2010). These are studies in which haloarchaeal carotenoid production in response to nutritive factors is described; however, the enzymes catalyzing the synthesis of these carotenoids have not been characterized.

19.4.3 Enzymes Involved in Archaeocins Synthesis

Peptide or protein antibiotics have been discovered in all three domains of life, and their production is nearly universal. Bacteriocin and eucaryocin research is well established, while research on archaeocins is still in its infancy. To date, only few archaeocins (halocins and sulfolobicins) have been partially or fully characterized, but hundreds of archaeocins are believed to exist, especially within the haloarchaea. No information has been reported up to now on the archaeocin biosynthesis pathways. Archaeocin research will provide excellent opportunities for discovery of novel antibiotics that may have clinical applications (O'Connor and Shand 2002).

19.4.4 Haloarchaeal Enzymes and Astrobiology

A more recent application for the haloarchaea group is related to its position in the tree of the life. The haloarchaea are among some of the earliest life-forms and are able to tolerate high doses of UV and radiation. These properties make them of interest in the field of Astrobiology, including discussions or research on the origins of life and survival on Mars (Litchfield 1998; Peeters et al. 2010). Several research lines have been focused on *Halorubrum chaoviator* as a model to analyze life under the exposure of the cells to conditions of outer space in the Biopan facility, finding that the cells of this specie survived exposure to conditions of outer space for 2 weeks (Mancinelli et al. 2009; Olsson-Francis and Cockella 2010).

19.5 Feasibility of Large-Scale Application of Haloarchaeal Enzymes

In general, the feasibility of industrial processes based on the use of extremozymes depends on productivity of the fermentation processes (usually fermentation processes are limited by low growth rates and low biomass yields). To overcome these limitations, attention has been focused on the study of the physiology of haloarchaea of biotechnological interest and on designing bioreactors and bioprocesses that increase the productivity of the fermentation process. Because of the unconventional environmental conditions needed for the cultivation of halophiles, contamination problems are minimal (Schiraldi et al. 2002); however, the conditions required for growth (containing up to 5 M NaCl) often are incompatible with standard industrial fermentation and the downstream processing plant. Because of these problems and the high cost of enzyme purification from the haloarchaea, most applications rely on the expression of the corresponding gene in a mesophilic host. This strategy gives nice results when working with thermophilic enzymes. However, expression of halophilic enzymes in a mesophilic host is a much more complex problem. Haloarchaea cells accumulate KCl at concentrations up to 5 M, and this salt concentration is required for the enzymes to be active and stable as mentioned before. The expression of the genes coding for haloenzymes in a mesophilic host such as *Escherichia coli* takes place in a low ionic strength environment and as a consequence, the protein produced is either insoluble or soluble but inactive (Esclapez et al. 2006).

Fortunately, unfolding or solubilization in the presence of a denaturant (urea, guanidine hydrochloride, etc.), followed by renaturation in the presence of NaCl, results in an active protein with properties similar to the native enzyme. On the other hand, homologous expression of genes from halophiles in haloarchaea such as *Hfx. volcanii* is also a successful way to over-express proteins of interest in biotechnology.

19.6 Conclusions and Perspectives

In theory, enzymes of haloarchaea have a huge potential for industrial, agricultural, chemical and even pharmaceutical or medical applications. The more new species will be described and our knowledge on the genetics and biochemistry of the halophilic Archaea will be expanded, the exploration and application of haloarchaeal enzymes will increase. Thus, it is clear that a large number of haloenzymes are as yet undiscovered and our understanding of the structural basis of stability and activity of halophilic enzymes is far from complete. Studies focused on biochemical characterizations as well as descriptions of metabolic pathways will increase the number of applications of haloenzymes in biotechnology.

References

Banat IM, Makkar RS, Cameotra SS (2000) Potential commercial applications of microbial surfactants. Appl Microbiol Biotechnol 53:495–508

Ben-Mahrez K, Thierry D, Sorokine I, Danna-Muller A, Kohiyama M (1988) Detection of circulating antibodies against c-myc protein in cancer patient sera. Br J Cancer 57:529–534

Bini E (2010) Archaeal transformation of metals in the environment. FEMS Microbiol Ecol 73:1–16

Bonete MJ, Camacho M, Martínez-Espinosa RM, Esclapez J, Bautista V, Pire C, Zafrilla B, Díaz S, Pérez-Pomares F, Llorca F (2007) In the light of the haloarchaea metabolism. In: Méndez-Vilas A (ed) Communicating current research and educational topics and trends in applied microbiology. Formatex, Badajoz, pp 170–183

Bonete MJ, Martínez-Espinosa RM, Pire C, Zafrilla B, Richardson DJ (2008) Nitrogen metabolism in haloarchaea. Saline Systems 4:9

Boutaiba S, Bhatnagar T, Hacene H, Mitchell DA, Baratti JC (2006) Preliminary characterization of a lipolytic activity from an extremely halophilic archaeon, *Natronococcus* sp. J Mol Catal B Enzym 41:21–26

Britton KL, Baker PJ, Ficher M, Ruzheinnikov S, Gilmour DJ, Bonete MJ, Ferrer J, Pire C, Esclapez J, Rice DW (2006) Analysis of protein solvent interactions in glucose dehydrogenase from the extreme halophile *Haloferax mediterranei*. Proc Natl Acad Sci USA 103: 4846–4851

Cabello P, Roldán MD, Moreno-Vivián C (2004) Nitrate reduction and the nitrogen cycle in archaea. Microbiology 150:3527–3546

Calo P, de Miguel T, Sieiro C, Velazquez JB, Villa TG (1995) Ketocarotenoids in halobacteria: 3-hydroxy-echinenone and trans-astaxanthin. J Appl Microbiol 79:282–285

Cao Y, Liao L, Xu XW, Oren A, Wu M (2008a) Aldehyde dehydrogenase of the aloalkaliphilic archaeon *Natronomonas pharaonis* and its function in ethanol metabolism. Extremophiles 12:849–854

Cao Y, Liao L, Xu XW, Oren A, Wang C, Zhu XF, Wu M (2008b) Characterization of alcohol dehydrogenase from the haloalkaliphilic archaeon *Natronomonas pharaonis*. Extremophiles 12:471–476

D'Souza SE, Altekar W, D'Souza SF (1997) Immobilization of *Haloferax mediterranei* aldolase by cross-linking in a proteinic matrix: stability and halophilic characteristics. World J Microbiol Biotechnol 13:561–564

De Castro RE, Ruiz DM, Giménez MI, Silveyra MX, Paggi RA, Maupin-Furlow JA (2008) Gene cloning and heterologous synthesis of a haloalkaliphilic extracellular protease of *Natrialba magadi* (Nep). Extremophiles 12:677–687

Dinçkaya E, Akyilmaz E, Sezgintürk MK, Ertaş FN (2010) Sensitive nitrate determination in water and meat samples by amperometric biosensor. Prep Biochem Biotechnol 40:119–128

Ding JY, Lai MC (2010) The biotechnological potential of the extreme halophilic archaea Haloterrigena sp. H13 in xenobiotic metabolism using a comparative genomics approach. Environ Technol 31:905–914

Don TM, Chen CW, Chan TH (2006) Preparation and characterization of poly(hydroxyalkanoate) from the fermentation of *Haloferax mediterranei*. J Biomater Sci Polym 17:1425–1438

Esclapez J, Bonete MJ, Camacho M, Ferrer J, Pire C, Bautista V, Martínez-Espinosa RM, Zafrilla B, Pérez-Pomares F, Díaz S (2006) An optimized method to produce halophilic proteins in *Escherichia coli*. Microb Cell Fact 2006, 5(Suppl 1):S22

Esclapez J, Pire C, Bautista V, Martínez-Espinosa RM, Ferrer J, Bonete MJ (2007) Analysis of acidic surface of *Haloferax mediterranei* glucose dehydrogenase by site-directed mutagenesis. FEBS Lett 581:837–842

Falb M, Müller K, Königsmaier L, Oberwinkler T, Horn P, von Gronau S, Gonzalez O, Pfeiffer F, Bornberg-Bauer E, Oesterhelt D (2008) Metabolism of halophilic archaea. Extremophiles 12:177–196

Fang CJ, Ku KL, Lee MH, Su NW (2010) Influence of nutritive factors on C50 carotenoids production by *Haloferax mediterranei* ATCC 33500 with two-stage cultivation. Bioresour Technol 101:6487–6493

Fernandez-Castillo R, Rodríguez-Valera F, Gonzalez-Ramos J, Ruiz-Berraquero F (1986) Accumulation of poly(β-hydroxybutyrate) by halobacteria. Appl Environ Microbiol 51:214–216

Ferrer J, Pérez-Pomares F, Bonete MJ (1996) NADP-glutamate dehydrogenase from the halophilic archaeon *Haloferax mediterranei*: enzyme purification. N-terminal sequence and stability. FEMS Microbiol Lett 141:59–63

Good WA, Hartman PA (1970) Properties of the amylase from *Halobacterium halobium*. J Bacteriol 104:601–603
Greer FR, Shannon M (2005) Infant methemoglobinemia: the role of dietary nitrate in food and water. Pediatrics 116:784–786
Han J, Lu Q, Zhou L, Liu H, Xiang H (2009) Identification of the polyhydroxyalkanoate (PHA)-specific acetoacetyl coenzyme A reductase among multiple FabG paralogs in *Haloarcula hispanica* and reconstruction of the PHA biosynthetic pathway in *Haloferax volcanii*. Appl Environ Microbiol 75:6168–6175
Han J, Li M, Hou J, Wu L, Zhou J, Xiang H (2010) Comparison of four *phaC* genes from *Haloferax mediterranei* and their function in different PHBV copolymer biosyntheses in *Haloarcula hispanica*. Saline Systems 6:9
Hough DW, Danson MJ (1999) Extremozymes. Curr Opin Chem Biol 3:39–46
Howarth RW (2004) Human acceleration of the nitrogen cycle: drivers, consequences, and steps toward solutions. Water Sci Technol 49:7–13
Johnsen U, Dambeck M, Zaiss H, Fuhrer T, Soppa J, Sauer U, Schönheit P (2009) D-Xylose degradation pathway in the halophilic archaeon *Haloferax volcanii*. J Biol Chem 284: 27290–27303
Joo WA, Kim CW (2005) Proteomics of halophilic archaea. J Chromatogr B Analyt Technol Biomed Life Sci 815:237–250
Karadzic IM, Maupin-Furlow JA (2005) Improvement of two-dimensional gel electrophoresis proteome maps of the haloarchaeon *Haloferax volcanii*. Proteomics 5:354–359
Kirkland PA, Humbard MA, Daniels CJ, Maupin-Furlow JA (2008) Shotgun proteomics of the haloarchaeon *Haloferax volcanii*. J Proteome Res 7:5033–5039
Klibanov AM (1989) Enzymatic catalysis in anhydrous organic solvents. Trends Biochem Sci 14:141–144
Legat A, Gruber C, Zangger K, Wanner G, Stan-Lotter H (2010) Identification of polyhydroxyalkanoates in *Halococcus* and other haloarchaeal species. Appl Microbiol Biotechnol 87: 1119–1127
Lillo JG, Rodríguez-Valera F (1990) Effects of culture conditions on poly(beta-hydroxybutyric acid) production by *Haloferax mediterranei*. Appl Environ Microbiol 56:2517–2521
Litchfield CD (1998) Survival strategies for microorganisms in hypersaline environments and their relevance to life on early Mars. Meteorit Planet Sci 33:813–819
Lledó B, Martínez-Espinosa RM, Marhuenda-Egea FC, Bonete MJ (2004) Respiratory nitrate reductase from haloarchaeon *Haloferax mediterranei*: biochemical and genetic analysis. Biochim Biophys Acta 1674:50–59
Lu Q, Han J, Zhou L, Zhou J, Xiang H (2008) Genetic and biochemical characterization of the poly (3-hydroxybutyrate-co-3-hydroxyvalerate) synthase in *Haloferax mediterranei*. J Bacteriol 190:4173–4180
Mancinelli RL, Landheim R, Sánchez-Porro C, Dornmayr-Pfaffenhuemer M, Gruber C, Legat A, Ventosa A, Radax C, Ihara K, White MR, Stan-Lotter H (2009) *Halorubrum chaoviator* sp. nov., a haloarchaeon isolated from sea salt in Baja California, Mexico, Western Australia and Naxos, Greece. Int J Syst Evol Microbiol 59:1908–1913
Margesin R, Schinner F (2001) Potential of halotolerant and halophilic microorganisms for biotechnology. Extremophiles 5:73–83
Marhuenda-Egea FC, Bonete MJ (2002) Extreme halophilic enzymes in organic solvents. Curr Opin Biotechnol 13:385–389
Martínez-Espinosa RM, Richardson DJ, Butt JN, Bonete MJ (2006) Respiratory nitrate and nitrite pathway in the denitrifier haloarchaeon *Haloferax mediterranei*. Biochem Soc Trans 34: 115–117
Martínez-Espinosa RM, Dridge EJ, Bonete MJ, Butt JN, Butler CS, Sargent F, Richardson DJ (2007a) Look on the positive side! The orientation, identification and bioenergetics of 'Archaeal' membrane-bound nitrate reductases. FEMS Microbiol Lett 276:129–139

Martínez-Espinosa RM, Zafrilla B, Camacho M, Bonete MJ (2007b) Nitrate and nitrite removing from salted water by *Haloferax mediterranei*. Biocatal Biotransform 25:295–300

Martínez-Espinosa RM, Lledó B, Marhuenda-Egea FC, Díaz S, Bonete MJ (2009) NO_3^-/NO_2^- assimilation in halophilic archaea: physiological analysis, *nasA* and *nasD* expressions. Extremophiles 13:785–792

Mevarech M, Frolow F, Gloss LM (2000) Halophilic enzymes: proteins with a grain of salt. Biophys Chem 86:155–164

O'Connor EM, Shand RF (2002) Halocins and sulfolobicins: the emerging story of archaeal protein and peptide antibiotics. J Ind Microbiol Biotechnol 28:23–31

Olsson-Francis K, Cockella CS (2010) Experimental methods for studying microbial survival in extraterrestrial environments. J Microbiol Methods 80:1–13

Oren A (1994) Enzyme diversity in halophilic archaea. Microbiologia 10:217–228

Oren A (2002) Diversity of halophilic microorganisms: environments, phylogeny, physiology, and applications. J Ind Microbiol Biotechnol 28:56–63

Oren A (2005) A hundred years of *Dunaliella* research: 1905–2005. Saline Systems 1:2

Oren A (2008) Microbial life at high salt concentrations: phylogenetic and metabolic diversity. Saline Systems 4:2

Oren A (2010) Industrial and environmental applications of halophilic microorganisms. Environ Technol 31:825–834

Parolis LAS, Boan IF, Rodríguez-Valera F, Widmalm G, Manca MC, Jansson PE, Sutherland IW (1996) The structure of the exopolysaccharide produced by the halophilic archaeon *Haloferax mediterranei* strain R1 (ATCC 33500). Carbohydr Res 295:147–156

Patzelt H, Kessler B, Oesterhelt D (2001) Descomposición de hidrocarburos halogenados por bacterias Archaea halófilas. Patent 16.09.2001-2159119. BOPI: 16.09.2001. Spain

Peeters Z, Vos D, ten Kate IL, Selch F, van Sluis CA, Sorokin DY, Muijzer G, Stan-Lotter H, van Loosdrecht MC, Ehrenfreund P (2010) Survival and death of the haloarchaeon *Natronorubrum* strain HG-1 in a simulated martian environment. Adv Space Res 46: 1149–1155

Pérez-Pomares F, Bautista V, Ferrer J, Pire C, Marhuenda-Egea FC, Bonete MJ (2003) α-amylase activity from the halophilic archaeon *Haloferax mediterranei*. Extremophiles 7:299–306

Pérez-Pomares F, Díaz S, Bautista V, Pire C, Bravo G, Esclapez J, Zafrilla B, Bonete MJ (2009) Identification of several intracellular carbohydrate-degrading activities from the halophilic archaeon *Haloferax mediterranei*. Extremophiles 13:633–641

Pire C, Marhuenda-Egea FC, Esclapez J, Alcaraz L, Ferrer J, Bonete MJ (2004) Stability and enzymatic studies of glucose dehydrogenase from the archaeon *Haloferax mediterranei* in reverse micelles. Biocatal Biotransform 22:17–23

Pire C, Esclapez J, Diaz S, Pérez-Pomares F, Ferrer J, Bonete MJ (2009) Alteration of coenzyme specificity in halophilic $NAD(P)^+$ glucose dehydrogenase by site-directed mutagenesis. J Mol Catal B Enzym 59:261–265

Rodríguez-Valera F (1992) Biotechnological potential of halobacteria. Biochem Soc Symp 58:135–147

Schiraldi C, Giuliano M, De Rosa M (2002) Perspectives on biotechnological applications of archaea. Archaea 1:75–86

Silveira CM, Gomes SP, Araújo AN, Montenegro MC, Todorovic S, Viana AS, Silva RJ, Moura JJ, Almeida MG (2010) An efficient non-mediated amperometric biosensor for nitrite determination. Biosens Bioelectron 25:2026–2032

Tadeo X, López-Méndez B, Trigueros T, Laín A, Castaño D, Millet O (2009) Structural Basis for the amino acid composition of proteins from halophilic archaea. PLoS Biol 7:e1000257

van den Burg B (2003) Extremophiles as a source for novel enzymes. Curr Opin Microbiol 6:213–218

van Grinsven HJ, Rabl A, de Kok TM (2010) Estimation of incidence and social cost of colon cancer due to nitrate in drinking water in the EU: a tentative cost-benefit assessment. Environ Health 9:58

Verhees CH, Kengen SWM, Tuininga JE, Schut GJ, Adams MWW, de Vos WM, van der Oost J (2003) The unique features of glycolytic pathways in Archaea. Biochem J 375:231–246

Vidyasagar M, Prakash SB (2006) Optimization of culture conditions for the production of haloalkaliphilic thermostable protease from an extremely halophilic archaeon *Halogeometricum* sp. TSS101. Lett Appl Microbiol 43:385–391

Wang G, Kennedy SP, Fasiludeen S, Rensing C, DasSarma S (2004) Arsenic resistance in *Halobacterium* sp. strain NRC-1 examined by using an improved gene knockout system. J Bacteriol 186:3187–3194

Woese CR, Kandler O, Wheelis ML (1990) Towards a natural system of organisms-proposal for the domains Archaea, Bacteria and Eukarya. Proc Natl Acad Sci USA 87:4576–4579

Yoshimatsu K, Iwasaki T, Fujiwara T (2002) Sequence and electron paramagnetic resonance analyses of nitrate reductase NarGH from a denitrifying halophilic euryarchaeote *Haloarcula marismortui*. FEBS Lett 516:145–150

Zafrilla B, Martínez-Espinosa RM, Esclapez J, Pérez-Pomares F, Bonete MJ (2010) SufS protein from *Haloferax volcanii* involved in Fe-S cluster assembly in haloarchaea. Biochim Biophys Acta 1084:1476–1482

Chapter 20
A Short History of the Symposia on Halophilic Microorganisms: From Rehovot 1978 to Beijing 2010

Aharon Oren

> "Everyone knows how God destroyed Sodom and Gommorrah, and all the Cities of the Plain; ... But no one knows, until now, of a conversation that took place in Heaven on that same day," a conversation between the Archangel Basil "who had been put in charge of biological classification and who had, thus far, found it a heavy job" and Little Vishnik, "one of Vishnu's many manifestations, a visiting god from a neighbouring heaven." The need was felt to create a new organism:
> "'It should ask questions we can't answer,' said Little Vishnik.
> 'Absolutely!' The Archangel keyed this into his computer. 'And it should like salt.'
> 'Very much,' Little Vishnik agreed. 'But it should be very small.'
> ...
> 'It should have a coat, with a honeycomb pattern and purple patches.'
> 'Why not?' said the Archangel, pushing another key. 'Should it move?'
> 'It should certainly swim.'
> 'Right,' the Archangel said. 'Anything else?'
> 'It should like salt.'
> 'I've already said that.' But the Archangel pushed another key. 'Now it will like salt *very much*. Should it sing?'
> 'One of these days, perhaps.' Little Vishnik winked.
> The Archangel pushed more keys. 'Now it's all arranged.'
> The two heavenly beings beamed at each other, well pleased with the new creature they had designed. They were especially pleased that it had been programmed into the celestial computer, whose circuits cannot be altered. Now they waited with pleasant anticipation for the Lord to awake and turn His bright, burning eyes on the doomed Cities of the Plain."

Thus ends the tale "The first halophile" about the creation of *Halobacterium*, presented by the late Donn Kushner at the symposium on "General & Applied Aspects of Halophilic Microorganisms" that took place in Alicante, Spain, in 1989 (Kushner 1991).

A. Oren (✉)
Department of Plant and Environmental Science, The Institute of Life Sciences, and the Moshe Shilo Minerva Center for Marine Biogeochemistry, The Hebrew University of Jerusalem, Jerusalem 91904, Israel
e-mail: orena@cc.huji.ac.il

A. Ventosa et al. (eds.), *Halophiles and Hypersaline Environments*,
DOI 10.1007/978-3-642-20198-1_20, © Springer-Verlag Berlin Heidelberg 2011

Fig. 20.1 The participants of the EMBO Workshop on Halophilism, held in May 1978 in Rehovot, Israel

The "Halophiles 2010" symposium, held in Beijing in June–July 2010, marks 32 years since the first scientific meeting entirely devoted to halophilic microorganisms was held, as far as I have been able to ascertain. The meeting on "Energetics and Structure of Halophilic Microorganisms" in Rehovot, Israel, in May 1978 (Fig. 20.1) was the first of a series of meetings and workshops (Table 20.1) aiming at solving the "questions we can't answer," asked by those tiny organisms – quoting Donn Kushner's tale again. The halophiles keep challenging scientists with unresolved questions, and although considerable progress has been made in the past 30 years, we probably will never be able to find all the solutions and obtain a full understanding of the secrets of those diverse microorganisms that inhabit salt lakes and other hypersaline environments with salt concentrations up to saturation.

The Rehovot 1978 meeting was surely very timely. Just a few years had passed since Walther Stoeckenius and Dieter Oesterhelt had discovered the structure and function of bacteriorhodopsin (Oesterhelt and Stoeckenius 1971). This discovery had brought *Halobacterium* and other extreme halophiles to the forefront of scientific interest, after a long period in which little attention had been devoted to the halophiles. It is, therefore, not surprising that most presentations dealt with bacteriorhodopsin and other halophile retinal pigments. Out of the 58 papers published in the proceedings volume (Caplan and Ginzburg 1978), no less than 28 centered on retinal proteins.

In their foreword to the proceedings book, the organizers of the meeting Roy Caplan and Margaret Ginzburg wrote: "The purple pigment now tends to overshadow its producers and their study: in this sense it has become a classical case of 'the tail wagging the dog.' Although workers in the field of halophilism must be grateful to be allowed to bask in the giant shadow of their recent offspring, yet it should be emphasized that studies of halophilic microorganisms and of their modes of adaptation to life in a highly saline milieu are of fundamental intrinsic interest to biology." Very true indeed! Bacteriorhodopsin is highly unusual even among the proteins of *Halobacterium*, as it is fully functional in the absence of salt. It cannot

Table 20.1 Symposia and workshops on halophilic microorganisms, 1978–2010

Dates	Title	Organizers	Venue	Approximate number of attendees	Publication of proceedings
May 14–19, 1978	Energetics and structure of halophilic microorganisms	S.R. Caplan, M. Ginzburg	Rehovot, Israel	79	Book
1981	EMBO Workshop on halophilic microorganisms	J.K. Lanyi	Ischia, Italy	unknown	–
September 1–6, 1985	The molecular basis of haloadaptation in microorganisms	W.D. Grant, M. Kogut, K. Wegmann	Obermarchtal, Germany	52	Special issue of FEMS Microbiology Reviews
March 23–28, 1986	Aspects of halophilism	H. Eisenberg	Jerusalem, Israel	46	–
March 26–April 5, 1989	Modern aspects of halophilism – the twelfth Edmond de Rothschild School in Molecular Biophysics	H. Eisenberg	Neve Ilan and Rehovot, Israel	58	–
September 17–22, 1989	General & applied aspects of halophilic microorganisms	F. Rodríguez-Valera	Alicante, Spain	97	Book
November 15–22, 1992	Halophilic bacteria: Research priorities and biotechnological potential for the 1990s	R.H. Vreeland	Williamsburg, VA, USA	75	Contributed papers published in Experientia
June 22–26, 1997	Microbiology and biogeochemistry of hypersaline environments	A. Oren	Jerusalem, Israel	69	Book
September 23–27, 2001	Halophiles 2001	A. Ventosa	Sevilla, Spain	141	Book
September 4–9, 2004	Halophiles 2004	N. Gunde-Cimerman, A. Plemenitaš	Ljubljana, Slovenia	104	Book
September 2–6, 2007	Halophiles 2007	T.J. McGenity	Colchester, UK	125	Contributed papers published in Saline Systems
June 29–July 3, 2010	Halophiles 2010	Y. Ma	Beijing, P.R. China	166	Book; A meeting report was published in Applied and Environmental Microbiology (Ma et al. 2010)

be denied, however, that the discovery of bacteriorhodopsin, and later of the chloride pump halorhodopsin and the sensory rhodopsins, has contributed much to the popularity of the extreme halophiles as research objects.

The year 1978 was an important landmark in the study of halophiles for another reason as well: that year saw the publication of the paper by Magrum, Luehrsen and Woese entitled: "Are extreme halophiles actually 'bacteria'?" (Magrum et al. 1978). Shortly before, Carl Woese had realized that the methanogens belong to a phylogenetically distinct lineage, and based on this observation he created the concept of the third domain of life, the Archaebacteria or Archaea. Now it appeared that also *Halobacterium* and *Halococcus* belonged to this new group. Thanks to the ease with which the extreme halophiles can be handled in the laboratory, they rapidly became popular model organisms for studies on the archaeal domain.

One of the attendees of the Rehovot meeting was present at the Beijing 2010 symposium as well: it was Janos Lanyi who keeps surprising us with novel discoveries on the retinal pigments present in different groups of halophiles. At the time of the Rehovot 1978 meeting I was a Ph.D. student in Jerusalem, and I was finalizing my thesis on photosynthesis and sulfide metabolism in cyanobacteria. I had no special interest in salt adaptation then, and therefore I did not attend the symposium. In retrospect, I surely should have attended the event: my model organism was *Oscillatoria limnetica* (*Phormidium hypolimneticum*), a filamentous cyanobacterium isolated from Solar Lake on the Sinai Peninsula, a hypersaline lake with salt concentrations up to 18%, and I grew the organism at salt concentrations twice as high as seawater. My true interest in the halophiles only started in 1980 when I initiated my studies on the Dead Sea.

Having attended all meetings listed in Table 20.1 from 1985 onwards, I hopefully have obtained a more or less complete picture of the scientific activities of the ever growing community of halophile microbiologists, enabling me to write this short overview on over 30 years of symposia on the adaptation of microorganisms to life at high salt concentrations. The views expressed and the features highlighted in this short essay are of course to a large extent based on the personal view of the author, and undoubtedly not all colleagues will share all of my views. I apologize for any omissions or factual errors.

I personally consider the meeting on "The Molecular Basis of Haloadaptation in Microorganisms," held in 1985 in the beautiful baroque monastery of Obermarchtal, a tiny village in the south of Germany (Fig. 20.2) as the first truly interdisciplinary symposium on all aspects of microbial life at high salt concentrations; it was also the place where I met most of those colleagues with whom I have been interacting now for over a quarter of a century, as apparent from the group pictures of later meetings (Figs. 20.3–20.9). When the Rehovot meeting in 1978 was biased in favor of presentations on bacteriorhodopsin and other retinal proteins, the Obermarchtal symposium did not have a single lecture on this topic, and this was intentionally so. The meeting organizers and guest editors of the meeting proceedings wrote: "The organizers felt that certain features of archaebacterial halophiles represented a special case of halophily and specifically excluded consideration of halobacterial photoactive pigments, although certain other aspects were

Fig. 20.2 Group picture of the participants in the September 1985 Obermarchtal, Germany meeting, taken in front of the late baroque church of Birnau on the shore of the Lake of Constance

Fig. 20.3 The participants in the workshop on "Aspects of Halophilism," Jerusalem, Israel, 1986, photographed in front of the building of the Israel Academy of Sciences and Humanities

included, since these unusual prokaryotes may provide us with particular insights into general aspects of halophilic life" (Grant and Kogut 1986). In retrospect this statement was a prophetic one: since proteorhodopsin was discovered in certain marine bacteria (Béjà et al. 2000), it has become clear that retinal proteins that function analogously to bacteriorhodopsin are not restricted to the world of the extreme halophiles.

Fig. 20.4 The 12th Edmond de Rothschild School on Modern Aspects of Halophilism, Neve Ilan and Rehovot, Israel, March–April 1989

Fig. 20.5 The group picture of the symposium on General & Applied Aspects of Halophilic Microorganisms, Alicante, September 1989. Photograph, courtesy of Prof. Francisco Rodríguez-Valera, Alicante

I am pleased to note that the organizers of all later meetings were happy to restore the prominent role that the retinal pigments play in halophile research when scheduling the lecture programs. This still does not imply that all meetings did indeed cover all aspects of halophilic life. The title of the Williamsburg 1992 meeting ("Halophilic Bacteria: Research Priorities and Biotechnological Potential for the 1990s") explicitly excluded coverage of algae and other eukaryotes that inhabit hypersaline environments. Although eukaryotic algae were mentioned in two out of the 50 abstracts submitted, none of the oral and poster presentations offered centered on the eukaryote world.

Fig. 20.6 The attendees of the symposium on "Microbiology and Biogeochemistry of Hypersaline Environments," Jerusalem, Israel, June 1997

Fig. 20.7 The participants in the "Halophiles 2004" symposium, Ljubljana, Slovenia, September 2004

Fig. 20.8 The group picture of "Halophiles 2007" symposium, Colchester, UK, September 2007

Fig. 20.9 The group picture of "Halophiles 2010" symposium, Beijing, China, June–July 2010

Halophile meetings were often the place where major breakthroughs were communicated prior to their publication in the scientific literature. The Sevilla 2001 meeting will be remembered as the event in which the properties of *Salinibacter*, the extremely red halophilic representative of the *Bacteroidetes*, were disclosed to the scientific community. It was also the place where Nina Gunde-Cimerman, Ana Plemenitaš, and their colleagues from Ljubljana convinced us that halophilic fungi and yeasts, long neglected, also deserve our full attention. Being a newcomer to the community of halophile scientists, Gunde-Cimerman became so enthusiastic during the Sevilla meeting that, with her characteristic spontaneity, she invited all to come to Ljubljana for the next meeting in 2004. This course of events also illustrates the fact that the series of halophile meetings is not linked to an established society, but rather is the result of the activity of highly motivated individuals who have taken the initiative to organize such events at more or less regular intervals. The Ljubljana meeting will remain famous as there the isolation and cultivation of Walsby's gas-vacuolate square archaeon, now *Haloquadratum walsbyi*, was proudly announced, simultaneously and independently by the group of Henk Bolhuis and Francisco Rodríguez-Valera and by the group of Mike Dyall-Smith. One of the novel aspects featured in Colchester was the intriguing physiology of the recently discovered anaerobic "polyextremophilic" haloalkalithermophiles, presented by Juergen Wiegel and Noha Mesbah. Different presentations in Colchester symposium dealt with the biology of halophilic viruses, a topic on which much progress has been made recently as shown by several speakers at the Beijing meeting (Ma et al. 2010).

The symposia also gave us the opportunity to get to know our colleagues not only on account of their scientific achievements, but also for their other highly diverse talents. It became apparent that many "halophilologists" like to spend part of their precious time with activities unrelated to science. Those old-timers who attended the 1985 Obermarchtal meeting will undoubtedly remember the magic show by master magician Edwin (Eddy) Dawes, who also has authored a number of books about magic in addition to his many publications about bacterial storage polymers. At that same meeting it was discovered that several colleagues shared a love for chamber music. The result: after the performance of Beethoven's spring sonata in Jerusalem in 1986 by Donn Kushner and me, we had a concert by the "OKKK Piano Quartet" (Aharon Oren – Morris Kates – Donn Kushner – Masamichi Kohiyama) in Alicante in 1989. In Williamsburg 1992 and in Jerusalem 1997 Larry Hochstein, who had been a professional clarinet player in Hollywood movie studio orchestras before he started a career in science, joined the chamber music club, with his performances in Mozart's Kegelstatt trio and in Brahms's first clarinet sonata. His dream to find a good soprano/halophile microbiologist for a joint performance of Schubert's "The shepherd on the rock" was unfortunately never realized. Donn Kushner, not only a scientist and a musician but also an author of a number of best-selling children's books, entertained us in Alicante in 1989 with the little tale, of which a fragment was cited at the beginning of this chapter, about the Archangel Basil and Little Vishnik who jointly created the first extreme halophile. Many will also remember the performance at the Jerusalem 1997 meeting of the "Why a duck" scene from the Marx Brothers 1929 movie "The Cocoanuts" by Ron Oremland as Groucho, complete with black mustache and big cigar, partnered by my daughter Dina as Chico. Those who had been present in Ljubljana in 2004 will not forget Nina Gunde-Cimerman's flawless tango dancing. And last but not least, Graham Underwood surprised us all with the performance of his morris dancing troupe at the 2007 meeting in Colchester.

Most meetings included one or more excursions in the often attractive areas around the congress venues. Some of these excursions were held to sites related to salt and salt production. Thus, the participants in the Jerusalem 1986 workshop on "Aspects of Halophilism" visited the Dead Sea and the Dead Sea Works potash plant. The tour to the salterns of Sečovlje, on the Slovenian Adriatic coast, during the 2001 Ljubljana congress was a wonderful opportunity to explore a solar salt production facility that had been operative since the Middle Ages using traditional technology. Many participants used the occasion to collect samples, hoping to isolate novel interesting halophiles from them.

There has been a strong tradition of publication of the proceedings of the halophile meetings. The presentations in Rehovot 1978, Alicante 1989, Jerusalem 1997, Sevilla 2001, Ljubljana 2004, and Beijing 2010 have all been published in books (Caplan and Ginzburg 1978; Rodriguez-Valera 1991; Oren 1999; Ventosa 2004; Gunde-Cimerman et al. 2005; Ventosa et al. 2011). The proceedings of the Obermarchtal 1986 symposium appeared in a special issue of FEMS Microbiology Reviews (Grant and Kogut 1986), selected presentations of the Williamsburg 1992 meeting were published in two issues of Experientia, and the open-access journal Saline Systems was used as the framework for publication of the proceedings of the Colchester 2007 symposium. At least one of the Williamsburg 1992 papers (Galinski 1993) has become a bestseller, based on the number of citations it has received since its publication.

The organization of such meetings of course requires considerable funds. Over the years a large number of agencies have contributed to the financial basis that made the events possible. These included FEMS, EMBO, NATO, IUBMB, the Foundation Edmond de Rothschild, the Israel Science Foundation, the National Science Foundation, the United States Office of Naval Research, and the Chinese

Academy of Sciences, to name just a few. Sometimes major sponsoring came from commercial firms as well, and not only from companies involved in biological sciences: the generous contribution from the Slovenian mobile operator Mobitel toward the organization of the Ljubljana 2004 event was highly appreciated.

When I look back to "my" first halophilologists meeting – Obermarchtal 1985, I fondly remember how excited I was to meet all the renowned halophile experts in person. From that time onwards they ceased to be just names below the titles of journal articles, but they had become colleagues and friends with whom I could interact. For me it was a supreme thrill to meet old and famous scientists such as Helge Larsen (Obermarchtal 1985) and Ben Volcani (Alicante 1989), colleagues who in past generations made major contributions to the development of the field. I am sure that our young colleagues who attended their first halophile meeting in Beijing in 2010 have felt some of the same excitement.

References

Béjà O, Aravind L, Koonin EV, Suzuki MT, Hadd A, Nguyen LP, Jovanovich S, Gates CM, Feldman RA, Spudich JL, Spudich EN, DeLong EF (2000) Bacterial rhodopsin: evidence for a new type of phototrophy in the sea. Science 289:1902–1906

Caplan SR, Ginzburg M (eds) (1978) Energetics and structure of halophilic microorganisms. Elsevier/North-Holland Biomedical Press, Amsterdam

Galinski EA (1993) Compatible solutes of halophilic eubacteria: molecular principles, water-solute interaction, stress protection. Experientia 49:497–498

Grant WD, Kogut M (1986) The molecular basis of haloadaptation in microorganisms. FEMS Microbiol Rev 39:1

Gunde-Cimerman N, Oren A, Plemenitaš A (eds) (2005) Adaptation to life at high salt concentrations in Archaea, Bacteria, and Eukarya. Springer, Dordrecht

Kushner DJ (1991) Halophiles of all kinds: what are they up to now and where do they come from? In: Rodriguez-Valera F (ed) General and applied aspects of halophilic microorganisms. Plenum, New York, NY, pp 63–71

Ma Y, Galinski EA, Grant WD, Oren A, Ventosa A (2010) Meeting review. Halophiles 2010: life in saline environments. Appl Environ Microbiol 76:6971–6981

Magrum LJ, Luehrsen KR, Woese CR (1978) Are extreme halophiles actually "bacteria"? J Mol Evol 11:1–8

Oesterhelt D, Stoeckenius W (1971) Rhodopsin-like protein from the purple membrane of *Halobacterium halobium*. Nature 233:149–152

Oren A (ed) (1999) Microbiology and biogeochemistry of hypersaline environments. CRC, Boca Raton, FL

Rodriguez-Valera F (ed) (1991) General and applied aspects of halophilic microorganisms. Plenum, New York, NY

Ventosa A (ed) (2004) Halophilic microorganisms. Springer, Berlin

Ventosa A, Oren A, Ma Y (2011) Halophiles and hypersaline environments: current research and future trends. Springer, Berlin

Index

A. Ventosa et al. (eds.), *Halophiles and Hypersaline Environments*,
DOI 10.1007/978-3-642-20198-1, © Springer-Verlag Berlin Heidelberg 2011

O

P

R

S

T

U

V